HITES 2012: 'Horizons of Innovative Theories, Experiments, and Supercomputing in Nuclear Physics'

Journal of Physics: Conference Series Volume 403

New Orleans, Louisiana, USA
4-7 June 2012

ISBN: 978-1-62276-869-1
ISSN: 1742-6588

Printed from e-media with permission by:

Curran Associates, Inc.
57 Morehouse Lane
Red Hook, NY 12571

Some format issues inherent in the e-media version may also appear in this print version.

Printed by Curran Associates, Inc. (2013)

For permission requests, please contact the Institute of Physics
at the address below.

Institute of Physics
Dirac House, Temple Back
Bristol BS1 6BE UK

Phone: 44 1 17 929 7481
Fax: 44 1 17 920 0979

techtracking@iop.org

Additional copies of this publication are available from:

Curran Associates, Inc.
57 Morehouse Lane
Red Hook, NY 12571 USA
Phone: 845-758-0400
Fax: 845-758-2634
Email: curran@proceedings.com
Web: www.proceedings.com

Public Lecture

Emerging Symmetries in Nuclear Physics

Nuclear Structure Approaches for the 21st Century and Breakthroughs via Innovative Computational Approaches

Weakly-Bound Systems, Continuum Effects, and Reactions

Nuclear Astrophysics: Structure and Reactions for Understanding the Universe

Special Lectures

Summary Talk

Preface

This volume contains the contributions of the speakers of an international conference in honor of Jerry Draayer's 70th birthday, entitled "Horizons of Innovative Theories, Experiments and Supercomputing in Nuclear Physics". The list of contributors includes not only international experts in these fields, but also many former collaborators, former graduate students, and former postdoctoral fellows of Jerry Draayer, stressing innovative theories such as special symmetries and supercomputing, both of particular interest to Jerry. The organizers of the conference intended to honor Jerry Draayer not only for his seminal contributions in these fields, but also for his administrative skills at departmental, university, national and international level.

Signed:
Ted Hecht
University of Michigan

Scientific Advisory Committee

Ani Aprahamian	University of Notre Dame
Baha Balantekin	University of Wisconsin
Bruce Barrett	University of Arizona
Umit Catalyurek	Ohio State Unversity
David Dean	Oak Ridge National Laboratory
Jutta Escher (Chair)	Lawrence Livermore National Laboratory
Jorge Hirsch	UNAM, Mexico
David Rowe	University of Toronto
Brad Sherill	Michigan State University
Joel Tohline	Louisiana State University
Edward Zganjar	Lousiana State University

Organizing Committee

Jeff Blackmon	Louisiana State University
Mark Caprio	University of Notre Dame
Tomas Dytrych	Louisiana State University
Ana Georgieva	INRNE, Bulgaria
Kristina Launey (Co-chair)	Louisiana State University
Gabriella Popa	Ohio University Zanesville
James Vary (Co-chair)	Iowa State University

Local Organizing Committee

Laura Linhardt	Louisiana State University
Charlie Rasco	Louisiana State University
Karen Richard (Coordinator)	Louisiana State University

List of Participants

Yoram Alhassid	Yale University
Akito Arima	Chairman, Japan Science Foundation
Naftali Auerbach	Tel Aviv University
Chairul Bahri	Universityof Notre Dame
Bruce Barrett	Universityof Arizona
Roelof Bijker	UNAM
Peter Bjonerud	SURA
Jeff Blackmon	Louisiana State University
Andrey Blokhin	SAP
Richard Boyd	Lawrence Livermore National Laboratory
Mark Caprio	Universityof Notre Dame
Octavio Castanos	UNAM
Umit Catalyurek	Ohio State University
Michael Cherry	Louisiana State University
David Dean	Oak Ridge National Laboratory
Catherine Deibel	Louisiana State University
Jerry Draayer	Louisiana State University
Alison Dreyfuss	Keene State University
Tomas Dytrych	Louisiana State University
Charlotte Elster	Ohio University
David Ernst	Vanderbilt University
Jutta Escher	Lawrence Livermore National Laboratory
Mia Ferriss	Louisiana State University
Amy Flatten	Director of International Affairs, APS
Martin Freer	University of Birmingham
Carla Frohlich	North Carolina State University
Stefano Gandolfi	Los Alamos National Laboratory
Doron Gazit	Hebrew University of Jerusalem
Ana Georgieva	INRNE
Joe Ginocchio	Los Alamos National Laboratory
Walter Greiner	Frankfurt Institute for Advanced Studies
Xin Guan	Liaoning Normal University
Vesselin Gueorguiev	UC Merced
Gaute Hagen	Oak Ridge National Laboratory
Joseph Hamilton	Vanderbilt University
Wick Haxton	UC Berkeley

Anna Hayes	Los Alamos National Lab
Ted Hecht	University of Michigan
Kristina Launey	Louisiana State University
Randall Laviolette	Advanced Scientific Computing Research, DOE
Elizabeth Lawson	SURA
Yorck Leschber	SAP
Laura Linhardt	Louisiana State University
Fengqiao Luo	Universityof Notre Dame
Tom Luu	Lawrence Livermore National Laboratory
Kevin Macon	Louisiana State University
Pieter Maris	Iowa State University
Gabriel Martinez-Pinedo	GSI
Hrayr Matevosyan	University of Adelaide, Australia
Robert McKeown	Jefferson Lab
Hai Ah Nam	Oak Ridge National Laboratory
Witold Nazarewicz	University of Tennessee and ORNL
Thomas Neff	GSI
Nikola Nikolov	Louisiana State University
Filomena Nunes	Michigan State University
Kostas Orginos	College of William and Mary and Jefferson Lab
Erich Ormand	Lawrence Livermore National Laboratory
Feng Pan	Liaoning Normal University
Thomas Papenbrock	University of Tennessee
Stuart Pittel	University of Delaware
Marek Ploszajczak	GANIL
Gabriela Popa	Ohio University Zanesville
Ravi P Rau	Louisiana State University
Rene Reifarth	Goethe University Frankfurt
David Richards	Jefferson Lab
Lee Riedinger	University of Tennessee
Julie Roche	Ohio University
George Rosensteel	Tulane University
Robert Roth	TU Darmstadt
David Rowe	University of Toronto
Bradley Sherrill	Michigan State University
Andrey Shirokov	Moscow State University
Ionel Stetcu	Los Alamos National Laboratory
Yasuyuki Suzuki	Niigata University

Gregory Tobin	Louisiana State University
Joel Tohline	Louisiana State University
Dirk Troltenier	SAP
James Vary	Iowa State University
Robert Wiringa	Argonne National Laboratory
John Wood	Georgia Tech
Edward Zganjar	Louisiana State University

Introduction

More than hundred people, including some of the most prominent scientists from the U.S. and abroad, paid tribute to Jerry P. Draayer's world-wide recognized research and academic achievements, honored during "Horizons of Innovative Theories, Experiments, and Supercomputing in Nuclear Physics" (HITES 2012), an international conference held from June 4 to June 7, 2012, in the historic New Orleans French Quarter. The conference focused on a forefront challenge in physics, namely to achieve better comprehension of the origin, structure and evolution of baryonic matter in the universe, including descriptions of unstable nuclei and complex nuclear substructures, starting from considerations based on emerging symmetries and tied to a fundamental quark/gluon level.

The conference was an enormous success and promoted strong inter-disciplinary dialogs. In particular, it brought together not only experts in nuclear theory, but also in experimental nuclear physics, high-energy physics, astrophysics, and, above all, in computer science, as well as leaders of various organizations that deal with present-day challenges on education, science policy, and international outreach. The conference welcomed many young emerging scientists, as well as undergraduate students from universities in Baton Rouge, New Orleans, and New Hampshire, together with graduate students in theoretical and experimental nuclear physics.

Jerry Draayer offered a public lecture "What a Wonderful World – Simplicity within Complexity".

During the conference, Jerry Draayer received certificates of appreciation and commemorative plaques in recognition of his exceptional contribution to the Thomas Jefferson National Accelerator Facility, Southeastern Universities Research Association (SURA), American Physical Society, Louisiana Board of Regents, and the Department of Physics & Astronomy at Louisiana State University. Indeed, throughout his career, Jerry has made important contributions to nuclear structure physics and computational physics, mentored a number of students and postdocs, participated in many international collaborations, and served the science community in a number of leadership positions. Prior to his current leadership as the President of SURA, Jerry was on the Louisiana Board of Regents as the Associate Commissioner for Sponsored Programs Research & Development; this followed his directorship of the Office of Federal Programs for the Regents. In his 30+ years as a faculty member in the Department of Physics and Astronomy at LSU, Jerry has served as Chair of the Department, Vice-president of the Faculty Senate, and Chair of the Council for the College of Basic Sciences. He has co-authored four books, published over 250 refereed journal articles, and is a regular invited lecturer at national and international scientific conferences which has resulted in additional 200 publications. Jerry is the recipient of numerous awards and distinctions, including LSU Distinguished Research Master, Francis G. Slack Fellow (APS), and Roy P. Daniels Professor of Physics.

The conference topics – reflecting Jerry's outstanding research – encompass a broad area of theoretical and experimental nuclear physics, fundamental physics, and high-performance computing. The topics include: (1) Emerging Symmetries in Nuclear Physics; (2) Nuclear Structure Approaches for the 21st Century; (3) Weakly-Bound Systems, Continuum Effects, and Reactions; (4) Nuclear Astrophysics: Structure and Reactions for Understanding the Universe; (5) Breakthroughs via Innovative Computational Approaches; (6) Fundamental Principles of Nature: Hadronic Physics and Physics at Jefferson Lab; and (7) Beyond Basic-Science Research (Education, Science Policy, and the like). For each of these themes, an overview talk was presented by keynote speakers, David Rowe, Witold Nazarewicz, Marek Ploszajczak, Wick Haxton, Randall Laviolette, and Robert McKeown. There were also two special lectures by Akito Arima and Walter Greiner who presented reviews on exciting developments in nuclear physics in the last decades. In addition, Amy Flatten (Director of International Affairs, APS) addressed issues regarding challenges for science in the emerging global economy and Lee Riedinger (Director of Center for Interdisciplinary Research and Graduate Education) talked on the role of university consortia in the national science agenda. Stuart Pittel gave a presentation, "The atomic nucleus at 100 and Jerry Draayer at 70: Two histories interwoven", that paid tribute to Jerry's research and his role in the development of nuclear physics. The conference ended with a summary talk offered by David Dean.

During the conference, Jerry presented a public lecture "What a Wonderful World – Simplicity within Complexity" that opened with the tunes of "What a Wonderful World" by New Orleans' favorite Louis Armstrong and focused on discovering emerging symmetries, or simplicities, within complex nuclear systems, which in turn, can pinpoint the fundamentals of nature. This lecture is included as part of the present proceedings volume. Also, the volume includes a Preface by Ted Hecht, one of Jerry's most valued collaborators and mentors.

We would like to express our appreciation to all the contributors to the HITES 2012 Proceedings for making this volume an invaluable collection of excellent manuscripts on forefront research.

Kristina Launey
Mark Caprio
Jutta Escher
Jorge Hirsch
Calvin Johnson

The conference was sponsored by Louisiana State University; Center for Computation and Technology (CCT), LSU; Department of Physics and Astronomy, LSU; Lawrence Livermore National Laboratory (LLNL); International HPC Initiative for Advanced Research; and J. Bennett Johnston, Sr., Center for Advanced Microstructures & Devices (CAMD), LSU.

What a wonderful world – simplicity within complexity

Jerry P Draayer

Department of Physics & Astronomy, Louisiana State University, Baton Rouge, LA 70803, USA

E-mail: draayer@lsu.edu

Abstract. In this lecture,[1] Louis Armstrong's transformational Jazz,[2] as exemplified by his signature recital of "What a Wonderful World", will take us to the ever evolving World of Nuclear Physics. In particular, I will focus on the discovery of simplicities, or symmetry patterns, in complex nuclear systems. And as Jazz is to music, so too Nuclear Physics is a restless endeavor – the dissonance of forlorn flattened notes tracking with the intrigue of symmetry, and its breaking in nuclei. I will also discuss some recent approaches to nuclear structure at this dawn of the 21st century, linking the intrigue of quarks and gluons with the fundamental science of the strong and weak interactions to the emergence of simplicity within complexity unveiled in atomic nuclei.

1. Emerging symmetries within complex nuclear systems

A forefront challenge in nuclear physics is to advance our understanding of atomic nuclei based on the nature of the strong force and how it impacts complex nuclear dynamics, keying on features that are central to nuclear reactions, astrophysics, neutrino and cosmology research, as well as to applied studies of nuclear energy and radioactive isotopes for medicine. Indeed, while the non-perturbative nature of Quantum Chromodynamics (QCD) in the regime of low energies together with the computationally intensive many-body problem place tremendous difficulties on achieving a description of nuclei starting from QCD itself ("*ab initio*"), establishing a connection from the quark/gluon foundation to nuclear properties with their diverse characteristics is essential for understanding nuclear structure at a fundamental level.

Notwithstanding, complex nuclear systems often display striking simplicities. Furthermore, experimental evidence supports the fact that low-lying nuclear states in light and medium-mass nuclei favor a dominance of low spin and high deformation, which has been also recently demonstrated through our first-principle studies [1, 2]. We have also showed the power of using symmetry-dictated subspaces to reach new domains of nuclear structure currently inaccessible by *ab initio* calculations. This, in turn, points to a remarkable new insight into the symmetry

[1] Public lecture presented at the "Horizons of Innovative Theories, Experiments, and Supercomputing in Nuclear Physics", HITES, in New Orleans, June 4-7, 2012. The able assistance of Kristina D. Launey in the planning, presentation and publication of the lecture is gratefully acknowledged.

[2] Louis Armstrong, whose bigger than life statue with trumpet in hand greets visitors to the New Orleans International Airport, is a favorite son of Louisiana, known for his "... foundational influence in jazz, shifting the music's focus from collective improvisation to solo performance ... "(Wikipedia).

patterns observed in atomic nuclei, namely, understanding the mechanism on how such simple structures emerge from the fundamental level of the underlying quark/gluon physics.

The symmetry-guided approach we have developed utilizes symmetry to reduce the dimensionality of the model space through a very structured winnowing of the basis states to physically relevant subspaces. We achieve significantly enhanced reach beyond that of the current no-core shell model (NCSM) by taming the scale explosion problem through the development of a practical, *ab initio* symmetry-adapted no-core shell model (SA-NCSM) framework. This framework exploits our knowledge of dominant symmetries, first of the interaction itself and then of those that emerge as a result of the many-body dynamics, and utilizes the power of the current and unfolding next-generation of high performance computing (HPC) facilities. This combination opens up new regions of the periodic chart, the '*sd*-shell', to investigation with *ab initio* methods. Our approach enables the inclusion of higher-order correlations as required, for example, for understanding excited 0^+ states – e.g., the elusive Hoyle state – in the lighter nuclei, and sets the stage for advancing the applicability of the current *ab initio* methods to systems beyond (e.g., 'fp-shell' nuclei, and even beyond to 'upper fp-shell' and 'lower gds-shell' as well as rare-earth and actinide species where deformation plays an even more dominant role) that are necessary, for example, for a better understanding of astrophysical processes among higher-mass nuclei.

Furthermore, we find that the results of various symmetry-adapted, no-core shell-model calculations for light and medium-mass nuclei are independent of whether the system Hamiltonian is phenomenological in nature or derived from realistic nucleon-nucleon (NN) interactions. They imply that only a small fraction of the full model space is needed for a description of low-lying states in these nuclei, and this in turn points to the importance of using a symmetry-adapted, no-core shell-model framework for describing such nuclei, one based on an LS coupling scheme with spatial configurations organized according to deformation.

The outcome confirms the relevance of the pioneering work of early developers of the field, J.P. Elliott and his SU(3) model [3] and M. Moshinsky with his U(3) many-body oscillator work [4], extended to open, multi-shell environments. Specifically, the results demonstrate the relevance of these algebraic methods in this region, and in particular, how they can be used to quell the combinatorial growth in the dimensionality of the problem with the addition of higher oscillator shells to the model space. Indeed, the results show the utility of a symmetry-adapted, no-core shell-model approach for light nuclei that takes maximum advantage of group-theoretical methods as well as advanced computational techniques.

The theory is applied to a study of the structure of various light nuclei, e.g. ^{6}Li, ^{6}He, ^{8}Be, ^{16}O and presented here for the case of ^{12}C, as well as to first studies of *sd*-shell nuclei such as ^{24}Si and ^{22}Mg, now rendered feasible due to symmetry-based considerations. Comparisons with results from no-core shell model analyses illustrate the efficacy of the symmetry-adapted framework. In addition, a simple algebraic interaction, which reduces to the Elliott model in its single-shell limit, is shown to reproduce all features of the elusive Hoyle state in ^{12}C. While this requires that the no-core underpinning of the model be extended beyond the current reach of the standard no-core shell model to excitations as high as eighteen major oscillator shells, only the most deformed configurations in the extended region are required. This further affirms the importance of exploiting algebraic methods in exploring special correlations in many-particle and multi-shell environments. The latter also suggests a possible path forward for realizing macroscopic theories that target collective degrees of freedom in terms of more microscopic ones that build forward from the NN interaction itself.

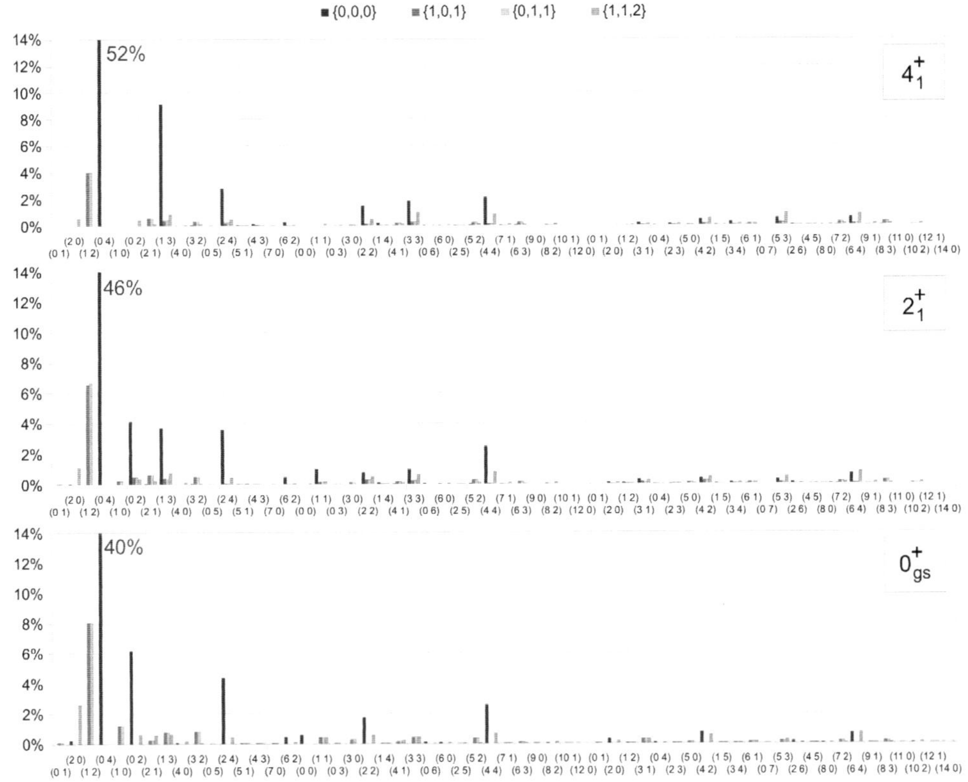

Figure 1. Probability distribution of the ground-state rotational band (0^+, 2^+, and 4^+ states) for ^{12}C over deformed subspaces labeled by $(\lambda\,\mu)$ for four of the most important spin components $\{S_\pi, S_\nu, S\} = \{0,0,0\}, \{1,0,1\}, \{0,1,1\}$, and $\{1,1,2\}$. The wavefunction was obtained using the $N_{\max} = 6$ SA-NCSM with the JISP16 bare interaction and $\hbar\Omega = 15$ MeV.

2. Symmetry-guided framework

2.1. Symmetry patterns revealed in results of ab initio studies and the SA-NCSM model

The *ab initio* symmetry-adapted no-core shell model (SA-NCSM) [2] adopts the first-principle concept and utilizes a many-particle basis that is reduced with respect to the physically relevant SU(3)$\supset$SO(3) subgroup chain. This allows the full model space to be down-selected to the physically relevant space [1, 5]. The significance of the SU(3) group for a microscopic description of the nuclear collective dynamics can be seen from the fact that it is the symmetry group of the Elliott model [6, 7, 8], and a subgroup of the Sp(3, $\mathbb{R}$) symplectic model [9].

The conventional NCSM basis spaces are constructed using harmonic oscillator (HO) single-particle states and are characterized by the $\hbar\Omega$ oscillator strength as well as by the cutoff in total oscillator quanta, $N_{\max}$, above the lowest energy configuration for a given nucleus. The basis states of the SA-NCSM for a given $N_{\max}$ are constructed in the proton-neutron formalism using also HO single-particle states and are labeled by the SU(3)$\supset$SO(3) subgroup chain quantum numbers $(\lambda\,\mu)\kappa L$, together with proton, neutron, and total intrinsic spins S_π, S_ν, and S. The orbital angular momentum L is coupled with S to the total orbital momentum J and its projection M_J. Each basis state in this scheme is labeled schematically as

$|\vec{\gamma}\,(\lambda\,\mu)\kappa L;(S_\pi S_\nu)S;JM_J\rangle$. The label κ distinguishes multiple occurrences of the same L value in the parent irrep $(\lambda\,\mu)$, and $\vec{\gamma}$ distinguishes among configurations carrying the same $(\lambda\,\mu)$ and $(S_\pi S_\nu)S$ labels.

The *ab initio* SA-NCSM results for *p*-shell nuclei reveal a dominance of shapes of large deformation (typically large $|\lambda\ -\mu|$) in the $0\hbar\Omega$ subspace. For example, the *ab initio* $N_{\rm max}=6$ SA-NCSM results with the bare JISP16 realistic interaction [10] for the 0^+ ground state (*g.st.*), first 2^+ and first 4^+ states of ^{12}C reveal the dominance of the $0\hbar\Omega$ component with the foremost contribution coming from the leading $(0\,4)$ $S=0$ irrep (Fig. 1). Furthermore, we find that important SU(3) configurations are then organized into structures with $\mathrm{Sp}(3,\mathbb{R})$ symplectic symmetry, that is, the $(0\,4)$ symplectic irrep gives rise to $(0\,2)$ and $(2\,4)$ configurations in the $2\hbar\Omega$ subspace, and those configurations indeed realize the major components of the wavefunction in this subspace. This further confirms the significance of the symplectic symmetry to nuclear dynamics. Similar results are observed for other *p*-shell nuclei. The outcome points to the fact that the relevant model space can be systematically determined by down-selecting to important spin configurations in lower subspaces while expanded to include a limited set of strongly deformed configurations in the higher $N_{\rm max}$ regime.

In short, the SA-NCSM advances an extensible microscopic framework for studying nuclear structure and reactions that capitalizes on advances being made in *ab initio* methods while exploiting symmetries – exact and partial, known to dominate the dynamics.

2.2. Symmetry patterns of realistic nucleon-nucleon interactions

The nucleon-nucleon interaction itself possesses a clear structure when its SU(3) content is studied. This is observed in the decomposition of the NN interaction into $\mathrm{SU}(3)\times\mathrm{SU}(2)_S\times\mathrm{SU}(2)_T$ tensors (isoscalar interactions will be henceforth considered). This is analogous to the unitary transformation of a V_{2b} two-body interaction represented in a m-scheme harmonic oscillator (HO) basis to a JT-coupled basis, which renders V_{2b} as only one $\mathrm{SU}(2)_J\times\mathrm{SU}(2)_T$ tensor of rank $J_0=0$ and $T_0=0$ (a scalar with respect to rotations in coordinate and isospin space). For example, the scalar interaction part of $(\lambda_0\,\mu_0)=(0\,0)$ does not mix nuclear shape deformations in analogy to the isoscalar part of an interaction that does not mix isospin values. In addition, the $(\lambda_0\,\mu_0)$ interaction parts with $\lambda_0=\mu_0$ are almost diagonal, that is, connect configurations within a few shells, while interaction parts with a large difference $|\lambda_0-\mu_0|$ typically couple low-lying and higher-lying shell-model configurations.

This decomposition organizes the interaction into only a small number of pieces of information that bring forward important physics. In particular, as a measure of the strength or "size" of each interaction tensor, we use its Hilbert-Schmidt norm, which is directly related to the square of the $(\lambda_0\,\mu_0)S_0$ reduced matrix elements. For example, we find a dominance of the $(0\,0)$ scalar part followed by the symplectic-like modes of $(0\,2)$, and equally, the conjugate $(2\,0)$, and then tensors as $(1\,1)$, $(2\,2)$, $(3\,3)$, and etc., which typically dominate for the pairing interaction or contact term (see Fig. 2 for the bare JISP16). These results, we find, repeat for various realistic bare and renormalized interactions.

2.3. Symmetry-guided shell-model approach and alpha clustering phenomena: NCSpM model

The no-core symplectic shell model (NCSpM) is a fully microscopic no-core shell model that uses a symplectic $\mathrm{Sp}(3,\mathbb{R})$ basis and $\mathrm{Sp}(3,\mathbb{R})$-preserving interactions. The NCSpM employed within a full model space up through a given $N_{\rm max}$ coincides with the NCSM for the same $N_{\rm max}$ cutoff. However, in the case of the NCSpM, the symplectic irreps divide the space into 'vertical slices' that are comprised of basis states of a definite deformation $(\lambda\,\mu)$. Hence, the model space can be reduced to only a few important configurations that are chosen among all possible $\mathrm{Sp}(3,\mathbb{R})$ irreps within the $N_{\rm max}$ model space. The NCSpM, while selecting the most relevant symplectic configurations, is employed to provide shell model calculations beyond current NCSM limits,

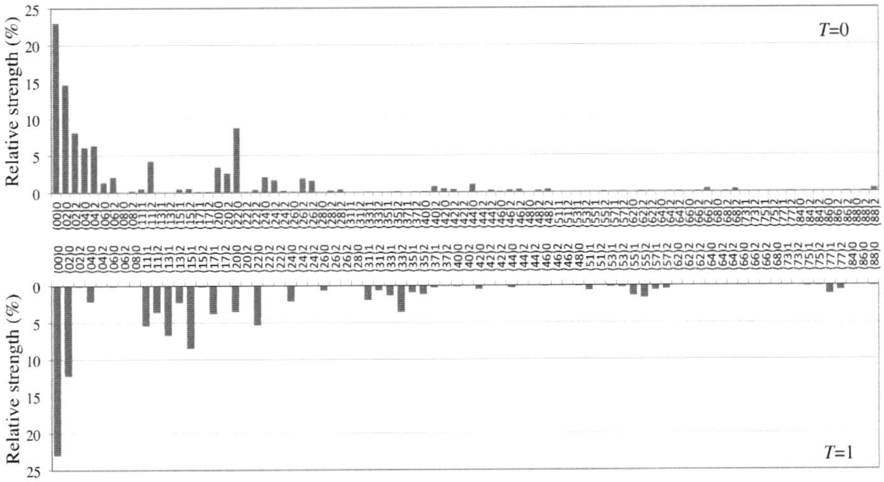

Figure 2. Relative strengths of the $T = 0$ and $T = 1$ bare JISP16 interaction tensors labeled by $(\lambda_0\,\mu_0)$ and S_0 for the case shown in Fig. 1, namely, $\hbar\Omega = 15$ MeV and $N_{\max} = 6$ for p-shell nuclei.

namely, up through $N_{\max} = 20$ for ^{12}C, the model spaces we found sufficient for the convergence of results.

We employ a very simple Hamiltonian with an effective interaction derived from the long-range expansion of the two-body central nuclear force,

$$H_{\text{eff}} = H_0 - \frac{\chi}{2}\frac{1}{\gamma}\left(e^{\gamma(Q.Q - \langle Q.Q\rangle_n)} - 1\right), \tag{1}$$

which includes the spherical HO potential (which together with the kinetic energy yields the HO Hamiltonian, H_0) and the $Q.Q$ quadrupole-quadrupole interaction not restricted to a single shell. For the latter term, the average contribution, $\langle Q.Q\rangle_n$, of $Q.Q$ within a subspace of n HO excitations is removed [11], that is, the trace of $Q.Q$ divided by the space dimension for a fixed n. Hence, the large monopole contribution of the $Q.Q$ interaction is removed, which, in turn, helps eliminate the considerable renormalization of the zero-point energy, while retaining the $Q.Q$-driven behavior of the wavefunctions.

This Hamiltonian in its zeroth-order approximation (for parameter $\gamma \to 0$) and for a valence shell goes back to the established Elliott model [6, 8]. We take the coupling constant χ to be proportional to $\hbar\Omega$ and, to leading order, to decrease with the total number of HO excitations, as shown by Rowe [12] based on self-consistent arguments. It is important to note that such a model Hamiltonian (1) fixes the χ strength of the $Q.Q$-term by the value of $\hbar\Omega$, thereby rendering the eigenstates $\hbar\Omega$-independent.

As the interaction and the model space are carefully selected to reflect the most relevant physics, the outcome reveals a quite remarkable agreement with the experiment. The low-lying energy spectrum and eigenstates for ^{12}C were calculated using the NCSpM with H of Eq. (1) for $\hbar\Omega = 18$ MeV given by the empirical estimate $\approx 41/A^{1/3} = 17.9$ MeV. The results are shown for $N_{\max} = 20$, which we found sufficient to yield convergence. This $N_{\max}$ model space is further reduced by selecting the most relevant symplectic irreps, namely, the spin-zero ($S = 0$) $0\hbar\Omega$ 0p-0h $(0\,4)$, $2\hbar\Omega$ 2p-2h $(6\,2)$, and $4\hbar\Omega$ 4p-4h $(12\,0)$ symplectic bandheads together with all

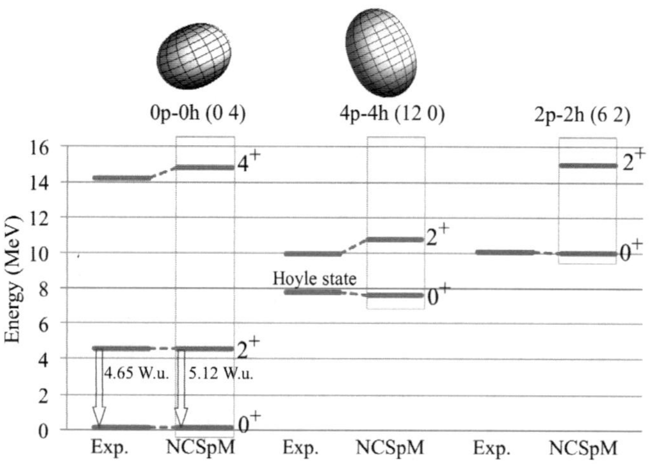

Figure 3. ^{12}C energy spectrum calculated by the $N_{\max} = 20$ NCSpM and compared to experiment. The NCSpM calculations are performed using Eq. (1) with $\gamma = -1.71 \times 10^{-4}$ and three Sp(3, $\mathbb{R}$) irreps, $0\hbar\Omega$ 0p-0h (0 4) and $4\hbar\Omega$ 4p-4h (12 0) (both shown with the bandhead shape deformation according to the mapping to the collective model), as well as $2\hbar\Omega$ 2p-2h (6 2). Experimental energies for all states, except the latest results for the 10-MeV 2^+ and 0_3^+ [13], are from [14].

multiples thereof up through $N_{\max} = 20$ of total dimensionality of 4.5×10^3. In comparison to the experimental energy spectrum (Fig. 3), the outcome reveals that the lowest 0^+, 2^+, and 4^+ states of the $0\hbar\Omega$ 0p-0h (0 4) symplectic irrep calculated for $\gamma = -1.71 \times 10^{-4}$ closely reproduce the *g.st.* rotational band, while the calculated lowest 0^+ states of the $4\hbar\Omega$ 4p-4h (12 0) and the $2\hbar\Omega$ 2p-2h (6 2) slices are found to lie close to the Hoyle state and the 10-MeV 0^+ resonance (third 0^+ state), respectively. We note that the NCSpM energies given in Fig. 3 are rescaled by an overall factor of ~ 2. This factor is determined by fixing the lowest 2^+ by its experimental value. This, however, has no implications on the underlying physics, as an overall factor for H does not change its properties and eigenstates, together with associated observables. Indeed, the model successfully reproduces other observables for ^{12}C that are informative of the state structure, such as mass rms radii, electric quadrupole moments and $B(E2)$ transition strengths (Fig. 3).

In accordance with the mapping [15] between the shell-model $(\lambda\,\mu)$ SU(3) labels and the shape variables of the Bohr-Mottelson collective model [16], the symplectic basis states bring forward important information about the nuclear shapes and deformation. For example, as shown in Fig. 3, the (0 4) symplectic bandhead of the ground-state rotational band and the (12 0) bandhead of the Hoyle-state rotational band describe oblate and prolate shapes, respectively. While a preponderance of the (0 4) shape (of $\sim 65\%$ probability) is observed for the former band (Fig. 4, red/left bars), the latter band includes shapes of even larger deformations (more prolate) with the largest contribution ($\sim 30\%$) of (16 0).

In addition, a close similarity is observed when the probability distributions for the *g.st.* rotational band are compared to *ab initio* results when only configurations of zero proton and neutron spins ($S_{\pi,\nu} = 0$) are selected (Fig. 4). In particular, NCSpM eigenstates, which are $\hbar\Omega$-independent, are compared to SA-NCSM calculations with the bare JISP16 realistic interaction for $\hbar\Omega = 20$ MeV (around the minimum of the calculated binding energy for ^{12}C) and a $N_{\max} = 8$ model space. This space appears to be sufficient to yield convergent results for the *g.st.* rotational band for both models. The close agreement points to the fact that the schematic interaction used in NCSpM has effectively captured most of the underlying physics of the realistic interaction important to the low-energy nuclear dynamics.

While the model includes an adjustable parameter, γ, this parameter only controls the

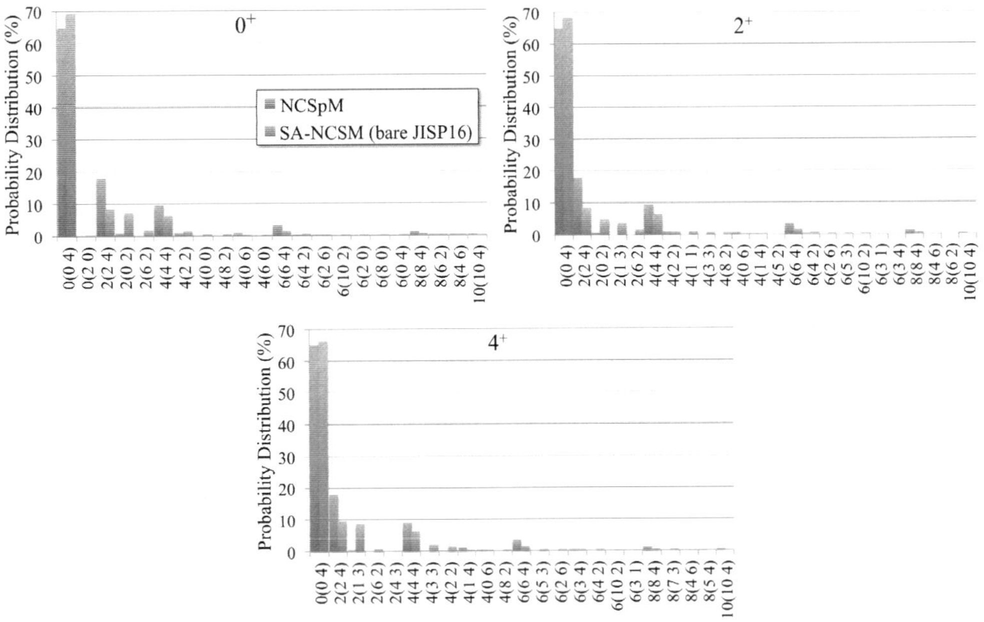

Figure 4. Probability distribution for ^{12}C as a function of the n total excitations and SU(3) labels, that is, $n(\lambda\,\mu)$, of the lowest 0^+, 2^+, and 4^+ states as calculated by the NCSpM (left) and the $N_{\max} = 8$ SA-NCSM using the bare JISP16 for $\hbar\Omega = 20$ MeV (right).

decrease rate of the $Q.Q$ interaction with increasing n. The entire many-body apparatus is fully microscopic and no adjustments are possible. Hence, as γ varies, there is only a small window of possible γ values that, for large enough $N_{\max}$, closely reproduces the relative positions of the three lowest 0^+ states.

In short, the present microscopic study with a schematic many-nucleon interaction shows how both collective and cluster-like structures emerge out of a no-core shell-model framework, which can extend to and take into account essential high-lying shell-model configurations.

3. Conclusion

Symmetry-adapted, no-core shell-model calculations with SU(3) the underpinning symmetry were presented. We showed that employing symmetry considerations is effective in providing an efficient description of low-lying eigenstates of ^{12}C, which also holds for various other p-shell nuclei. A symmetry-guided framework was suggested based on our recent findings of low-spin and high-deformation dominance in realistic NCSM results. This holds promise to significantly enhance the reach of *ab initio* shell models toward heavier nuclear systems as well as to achieve descriptions of collective and cluster phenomena from first principles.

Acknowledgments

This work was supported by the U.S. NSF (0904874), the U.S. DOE (DE-SC0005248 & FG02-95ER-40934), and SURA. We also acknowledge DOE/NERSC and LSU/LONI for providing HPC resources.

References

[1] Dytrych T, Sviratcheva K D, Bahri C, Draayer J P and Vary J P 2007 *Phys. Rev. Lett.* **98** 162503
[2] Dytrych T *et al.* 2012 *to be submitted to Phys. Rev. Lett.*
[3] 1992 *Proceedings of the international symposium in honor of K.T. Hecht, 19-21 September 1991, Ann Arbor, MI* ed Draayer J P and Janecke J (World Scientific)
[4] 2004 *Proceedings of the Symposium in Honor of Jerry P. Draayer's 60th Birthday, 18-21 February 2003, Playa del Carmen, Mexico* ed Escher J, Castaños O, Hirsch J G, Pittel S and Stoitcheva G (World Scientific)
[5] Dytrych T, Sviratcheva K D, Draayer J P, Bahri C and Vary J P 2008 *J. Phys. G: Nucl. Part. Phys.* **35** 123101
[6] Elliott J P 1958 *Proc. Roy. Sot. A* **245** 128
[7] Elliott J P 1958 *Proc. Roy. Sot. A* **245** 562
[8] Elliott J P and Harvey M 1962 *Proc. Roy. Sot. A* **272** 557
[9] Rosensteel G and Rowe D J 1977 *Phys. Rev. Lett.* **38** 10
[10] Shirokov A M, Vary J P, Mazur A I and Weber T A 2007 *Phys. Lett.* **B 644** 33
[11] Rosensteel G and Draayer J P 1985 *Nucl. Phys. A* **436** 445
[12] Rowe D J 1967 *Phys. Rev.* **162** 866
[13] Itoh M *et al.* 2011 *Phys. Rev. C* **84** 054308
[14] Ajzenberg-Selove F 1990 *Nucl. Phys. A* **506** 1
[15] Castaños O, Draayer J P and Leschber Y 1988 *Z. Phys. A* **329** 33
[16] Bohr A and Mottelson B R 1969 vol 1 (Benjamin, New York)

The many roles of symmetry in nuclear physics

David J Rowe

Department of Physics, University of Toronto, Toronto, ON M5S 1A7, Canada

E-mail: `rowe@physics.utoronto.ca`

Abstract. Much of my talk is not new. But only recently has its substance become recognized as significant due to the powerful methods, developed by J.P. Draayer and his LSU group, and the ubiquitous occurrence of deformation in essentially all nuclei.

1. Introduction

Everyone has a sense of what symmetry means in terms of patterns that are repeated under certain transformations of a system. However, the extraordinary power of group theory when allowed to include transformations that take place continuously and in time as well as in space is less widely recognised. Symmetries need not be static. For example, one finds symmetries in repeated structures in architecture and in evolving motifs in music. We are then speaking of dynamical symmetries which prove to be important in nuclear physics.

The symmetries of a system are what give it structure and form whether it be a physical theory or a work of art. Without symmetry of any kind, a system would be amorphous and chaotic. Music that is lacking in symmetry, or whose structure is not evident to the listener, is heard as noise. However, the ability to appreciate the structure of a theory or work of art depends very much on the beholder.

The focus of this presentation will be on a symmetry-based paradigm for nuclear physics, Basically, the paradigm follows the sequence: (i) Collect data. Examine it in many ways, until patterns appear. (ii) Construct models to explain the patterns. Predict consequences and identify observations that would invalidate the model or require its modification. (iii) Repeat (i) and (ii) to develop a model and correct its failings. Determine the model's limitations and domain of validity. (iv) Show how a model, with suitable adjustment if necessary, can be embedded in a fundamental theory of the system which may also need adjustment.

This review illustrates the path by which the simplest model of nuclear collective phenomena, namely the Bohr model, evolves by use of symmetry methods into a microscopic theory.

2. Lie groups, symmetry groups, dynamical groups and spectrum generating algebras

Definition: A Lie group is a group and an analytic manifold.

The content of this definition highlights the fact that the concept of a Lie group embodies the three most important fields of mathematics that are used extensively in physics: namely algebra, analysis and geometry.

For example, the group SO(3) can be realised as the geometrical space of all orientations of a rigid body. The geometry and connectivity of this space is hard to visualize. However, once it is identified with the group SO(3), it is handled with ease. It is a smooth differentiable manifold on which rotational wave functions can be expanded by harmonic analysis (akin to Fourier expansions). As a Lie group, it has a Lie algebra of infinitesimal generators L_x, L_y, L_z with commutation relations $[L_x, L_y] = i\hbar L_z$.

Definition: A symmetry group $G_{\rm sym}$ is a group of transformations of a system that leave its Hamiltonian invariant.

Suppose, for example, that $\hat{H}$ is a Hamiltonian and $\{|\alpha\lambda\nu\rangle\}$ is an orthonormal basis for its Hilbert space, where λ labels an irrep $\hat{S}^\lambda$ of $G_{\rm sym}$, ν labels a basis for this irrep, and α indexes the multiplicity of equivalent λ irreps. Then

$$\hat{S}^\lambda(g)|\alpha\lambda\nu\rangle = \sum_\mu |\alpha\lambda\mu\rangle S^\lambda_{\mu\nu}(g), \quad \forall g \in G_{\rm sym}, \tag{1}$$

and

$$\hat{H}|\alpha\lambda\nu\rangle = \sum_\beta |\beta\lambda\nu\rangle H^\lambda_{\beta\alpha} \tag{2}$$

where

$$S^\lambda_{\mu\nu}(g) = \langle\alpha\lambda\mu|\hat{S}^\lambda(g)|\alpha\lambda\nu\rangle, \quad H^\lambda_{\beta\alpha} = \langle\beta\lambda\nu|\hat{H}|\alpha\lambda\nu\rangle. \tag{3}$$

It is seen that the Hamiltonian is block diagonal.

Definition: A dynamical group $G_{\rm dyn}$ for a Hamiltonian $\hat{H}$ is a Lie group of transformations for which the Hilbert space of $\hat{H}$ carries a sum of irreducible unitary representations with each irreducible subspace being spanned by eigenstates of $\hat{H}$.

In both cases the Hamiltonian is partially diagonalized. The difference is that the non-zero off-diagonal matrix elements are only between states of different but equivalent irreps of a symmetry group whereas they are only between states of the same irrep of a dynamical group. Moreover the Lie algebra of the dynamical group, which is usually a spectrum generating algebra, facilitates the task of completing the diagonalization. Thus, dynamical groups and symmetry groups serve complementary purposes

Definition: A spectrum generating algebra (SGA) for a class of Hamiltonians is a Lie algebra such that the Hamiltonians can be expressed (e.g., usually as polynomials) in its elements.

3. The Bohr model

The Bohr model was formulated [1] as a liquid-drop model with quadrupole shape coordinates $\{q_\nu\}$ defined by an expansion of the nuclear radius in terms of spherical harmonics

$$R(\theta,\varphi) = R_0\Big[1 + \sum_\nu q_\nu Y^*_{2\nu}(\theta,\varphi) + \dots\Big]. \tag{4}$$

Corresponding quantum mechanical momenta were then defined as operators on collective model wave functions by

$$\hat{\pi}^\nu = \mathrm{i}\hbar\frac{\partial}{\partial q_\nu}, \quad \nu = 0, \pm1, \pm2. \tag{5}$$

so that they would obey the commutation relations of a Heisenberg-Weyl HW(5) Lie algebra

$$[\hat{q}_\mu, \hat{\pi}^\nu] = \mathrm{i}\hbar\delta^\nu_\mu. \tag{6}$$

An algebraic expression of the Bohr model, known as the algebraic collective model or ACM, has recently been formulated [2] with an $\mathrm{SU}(1,1) \times \mathrm{SO}(5) \supset \mathrm{SO}(3)$ group structure in which $\mathrm{SU}(1,1) \times \mathrm{SO}(5)$ is a subgroup of a dynamical group that includes the Heisenberg-Weyl group HW(5), and SO(3) is a symmetry group. The remarkable fact is that the geometry and representations of SU(1,1) and SO(5) enable exact calculations of SO(5) spherical harmonics and Clebsch-Gordan coefficients and give analytical expressions for all reduced matrix elements of the SO(5) spherical harmonics and other matrix elements as needed [3, 4]. Thus, the ACM provides analytical expressions for the matrix elements of any rotationally-invariant Bohr model Hamiltonian which enables it to be diagonalized quickly and easily [5]. An example, given in figure 1, illustrates the basic concepts of axially symmetric rotors with β and γ vibrational bands.

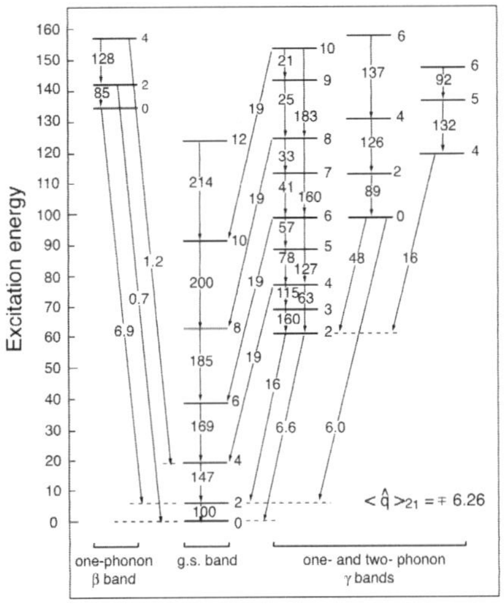

Figure 1. The Bohr model spectrum and E2 transition rates for the Hamiltonian $\hat{H}(B,\alpha,\chi,\kappa) = -\frac{\nabla^2}{2B} + \frac{1}{2}B\left[(1-2\alpha)\beta^2 + \alpha\beta^4\right] - \chi\beta\cos 3\gamma$ with $B = 20$, $\alpha = 1.5$ and $\chi = \pm 2$. Energies and E2 rates are given in units such that the excitation energy of the 2_1 state is 6 units and its E2 rate to the ground state is 100 units.

4. Progression to a microscopic collective model

The first step in the progression is a replacement of the surface shape coordinates $\{q_\nu\}$ by microscopic Cartesian monopole/quadrupole moments

$$Q_{ij} = \sum_{n=1}^{A} x_{ni}x_{nj}, \quad i,j = 1,2,3, \quad n = 1,\ldots,A. \tag{7}$$

Time derivatives and corresponding momentum observables are then given by

$$\dot{Q}_{ij} = \frac{dQ_{ij}}{dt} = \sum_n (\dot{x}_{ni}x_{nj} + x_{ni}\dot{x}_{nj}), \quad \hat{P}_{ij} = \sum_n (p_{ni}x_{nj} + x_{ni}p_{nj}). \tag{8}$$

These shape and momentum observables are now microscopic and have commutation relations,

$$[Q_{ij}, \hat{P}_{kl}] = \mathrm{i}\hbar\left[\delta_{il}\hat{Q}_{ik} + \delta_{ik}\hat{Q}_{il} + \delta_{jl}\hat{Q}_{jk} + \delta_{jk}\hat{Q}_{jl}\right]. \tag{9}$$

Together with the antisymmetric angular momenta $\hbar\hat{L}_k = \hbar\hat{L}_{ij} = \sum_n \left(\hat{x}_{ni}\hat{p}_{nj} - \hat{x}_{nj}\hat{p}_{ni}\right)$ they span a known CM(3) Lie algebra [6]. The dynamical content of the CM(3) model is characterized in figure 2. This model has not been used to describe data but it represents an important step towards a more fundamental microscopic model that provides a much deeper understanding of the structure of deformed states in nuclei.

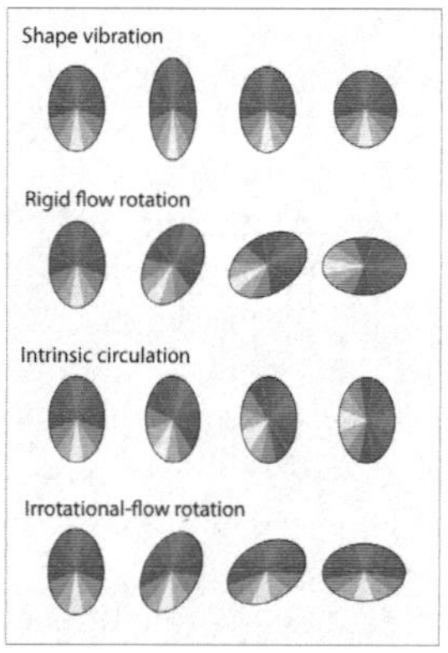

Figure 2. The CM(3) model is richer than the Bohr model. Its six $\{Q_{ij}\}$ observables characterize the shape and orientation of the model nucleus. The three angular momenta $\{L_{ij}\}$ are infinitesimal generators of rigid rotations. The three diagonal momenta $\{\hat{P}_{ii}\}$ are infinitesimal generators of shape vibrations and the three off-diagonal momenta $\{\hat{P}_{ij}; i \neq j\}$ are infinitesimal generators of irrotational-flow rotations. Intrinsic current circulations are generated by combinations of contrary rigid and irrotational-flow rotations.

5. The symplectic model

The symplectic model [7–9] is an algebraic model whose dynamical group Sp(3, $\mathbb{R}$) is the set of all linear canonical transformations of the single-particle phase-space observables

$$\hat{x}_{ni} \to \sum_j \big(a_{ij}\hat{x}_{nj} + b_{ij}\hat{p}_{nj}\big), \quad \hat{p}_{ni} \to \sum_j \big(c_{ij}\hat{x}_{nj} + d_{ij}\hat{p}_{nj}\big), \tag{10}$$

that preserve the commutation relations $[\hat{x}_{ni}, \hat{p}_{mj}] = \mathrm{i}\hbar\delta_{m,n}\delta_{i,j}$. Its Lie algebra comprises all elements of the CM(3) Lie algebra

$$Q_{ij} = \sum_{n=1}^{A} x_{ni}x_{nj}, \quad P_{ij} = \sum_n (p_{ni}x_{nj} + x_{ni}p_{nj}), \quad \hbar\hat{L}_{ij} = \sum_n \big(\hat{x}_{ni}\hat{p}_{nj} - \hat{x}_{nj}\hat{p}_{ni}\big), \tag{11}$$

plus the extra bilinear combinations of momentum operators

$$\hat{K}_{ij} = \sum_n \hat{p}_{ni}\hat{p}_{nj}. \tag{12}$$

Adding the extra $\hat{K}_{ij}$ operators is a major step forward for the following reasons.

(i) The group Sp(3, $\mathbb{R}$) is semi-simple (in fact, simple) whereas CM(3) is not.

(ii) The many-nucleon kinetic energy and harmonic-oscillator Hamiltonians

$$\sum_{n=1}^{A}\sum_{i=1}^{3} \frac{1}{2m}\hat{p}_{ni}^2, \quad \hat{H}^{(A)}_{\mathrm{DHO}} = \sum_{n=1}^{A}\left[\frac{1}{2m}\hat{\mathbf{p}}_n^2 + \frac{1}{2}m(\omega_1^2\hat{x}_{n1}^2 + \omega_2^2\hat{x}_{n2}^2 + \omega_3^2\hat{x}_{n3}^2)\right], \tag{13}$$

are elements of the Sp(3, $\mathbb{R}$) Lie algebra for all values of $\omega_1, \omega_2, \omega_3$. Thus, the symplectic model avoids the need for arbitrary inertia parameters and is compatible with both spherical and deformed harmonic-oscillator shell models

(iii) Sp(3,$\mathbb{R}$) is a dynamical group for any Hamiltonian of the form

$$\hat{H} = \frac{1}{2m}\sum_i \hat{K}_{ii} + V(Q). \tag{14}$$

(iv) It contains many useful subgroup chains, e.g.

$$\mathrm{Sp}(3,\mathbb{R}) \supset \mathrm{CM}(3) \supset \mathrm{ROT}(3) \supset \mathrm{SO}(3), \qquad \mathrm{Sp}(3,\mathbb{R}) \supset \mathrm{U}(3) \supset \mathrm{SU}(3) \supset \mathrm{SO}(3), \tag{15}$$

where ROT(3) is a dynamical group for a rigid-rotor model and SU(3) is the dynamical group of Elliott's SU(3) model. The first chain reflects its content in terms of the dynamical flows of a quantum fluid while the latter is more useful for the construction of a shell-model theory of deformed nuclei.

$\mathrm{Sp}(3,\mathbb{R})$ irreps in nuclei are labeled by the three quantum numbers $N_0(\lambda_0\mu_0)$ of its lowest weight U(3) sub-representations, with $(\lambda_0\mu_0)$ being standard SU(3) labels. There are three classes of irreps, as illustrated in figure 3: irreps with $(\lambda_0\mu_0) = (0\,0)$ correspond to spherical (closed-shell) nuclei; $(\lambda_0\,0)$ and $(0\,\mu_0)$ irreps correspond to axially symmetric rotational nuclei; and irreps with non-zero λ_0 and μ_0 correspond to triaxial rotational nuclei.

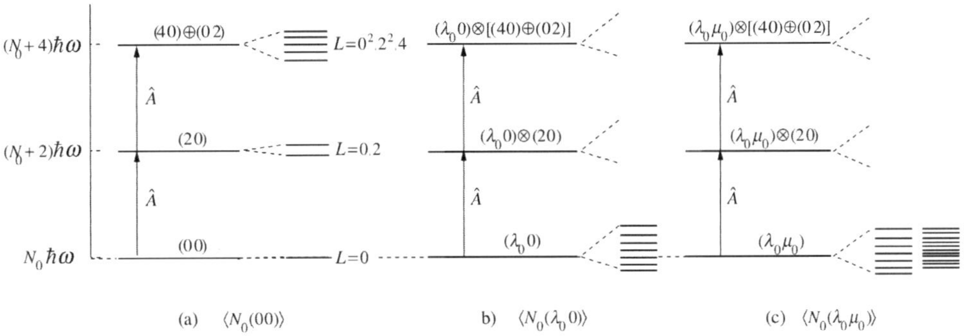

Figure 3. Basis states for the three classes of $\mathrm{Sp}(3,\mathbb{R})$ irreps

Much progress has been achieved [10, 11] at LSU by Jerry Draayer's group in the development of multi-shell calculations for light nuclei in SU(3) and $\mathrm{Sp}(3,\mathbb{R})$ bases. However, for heavy rotational nuclei, different techniques are needed. For one thing, it is a priori less evident what shell model spaces are required for realistic calculations. For example, a comparison of independent-particle energies in the spherical shell model with those of the Nilsson model, shown in figures 4 and 5, indicate that, due to the large amount of level crossing in the neighbourhood of the Fermi surface, the Nilsson-model state for ^{168}Er is already at energy $\gtrsim 12\hbar\omega$ above that of the spherical shell model ground state.

6. Geometrical and mean-field perspectives

It has been shown [12] that, if $|\sigma\rangle \equiv |N_0(\lambda_0\mu_0)\rangle$ is a lowest weight state of an $\mathrm{Sp}(3,\mathbb{R})$ irrep that is defined as a highest-weight state for its lowest-weight U(3) irrep, the Hilbert space for the $\mathrm{Sp}(3,\mathbb{R})$ irrep is spanned by the set of states

$$\mathcal{R} = \{\hat{R}(\Omega)\hat{U}(d)\hat{R}(\Omega')|\sigma\rangle;\ \Omega \in \mathrm{SO}(3),\ d = \mathrm{diag}(d_1, d_2, d_3),\ \Omega' \in \mathrm{SO}(3)\}, \tag{16}$$

where the product $\Omega d\Omega'$ is the factorization of a general linear matrix $g = \Omega d\Omega' \in \mathrm{GL}(3,\mathbb{R})$, $\hat{R}(\Omega)$ and $\hat{R}(\Omega')$ are rotation operators, and $\hat{U}(d)$ is a unitary scale transformation of the (x, y, z) coordinates. The sequence of transformations $\hat{R}(\Omega)\hat{U}(d)\hat{R}(\Omega')$ is illustrated in figure 6.

For a Hamiltonian $\hat{H}$, the minimum-energy

$$E(\Omega d\Omega') = \langle\sigma(\Omega d\Omega')|\hat{H}|\sigma(\Omega d\Omega')\rangle \tag{17}$$

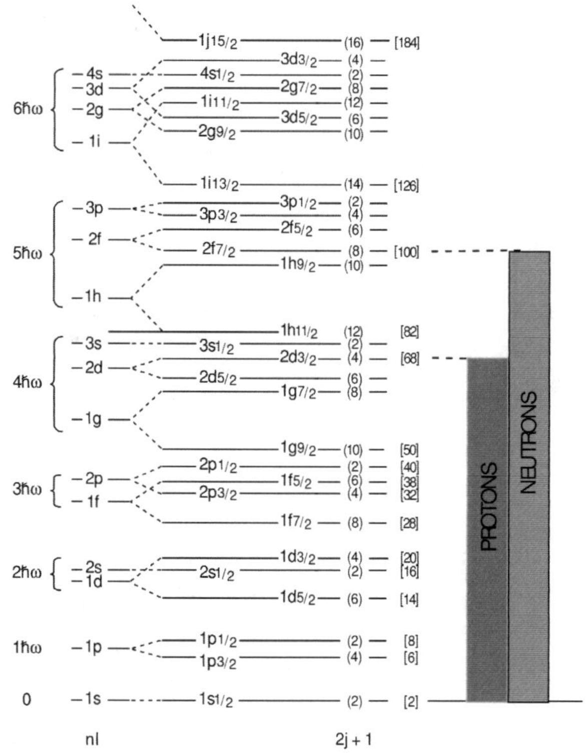

Figure 4. Single-particle energies in the spherical shell model. The red and green boxes indicate the levels that would be occupied by protons and neutrons, respectively, in this independent-particle model.

state $|\sigma(\Omega d\Omega')\rangle = \hat{R}(\Omega)\hat{U}(d)\hat{R}(\Omega')|\sigma\rangle$ on the manifold $\mathcal{R}$, is given by mean-field theory and should satisfy a self-consistency condition [13]. If $\hat{H}$ is rotationally invariant, $E(\Omega d\Omega')$ is independent of Ω. Consideration of figure 6(d) suggests that Ω' should be the identity matrix at the energy minimum. Thus, we seek a state $|\sigma(d)\rangle$ by self-consistency considerations.

By construction, the SU(3) highest-weight state $|\sigma\rangle$ is an eigenstate of a spherical harmonic oscillator. In contrast, the density distribution of the state $|\sigma\rangle$ is only spherical if $\lambda_0 = 0$ and $\mu_0 = 0$. Thus, a generic state $|\sigma\rangle$ does not satisfy shape-consistency with the Hamiltonian for which it is an eigenstate. However, following the scale transformation $\hat{U}(d)$, the state $|\sigma(d)\rangle$ becomes an eigenstate of a triaxial harmonic oscillator Hamiltonian with potential

$$V_d(x) = \frac{1}{2}\sum_n m(\omega_1 x_{n1}^2 + \omega_2 x_{n2}^2 + \omega_3 x_{n3}^2), \tag{18}$$

which can be chosen to satisfy shape consistency.

Self-consistency results were given to leading order in references [13, 14]. They are determined to all orders in N_1, N_2 and N_3 for which the state $|\sigma(d)\rangle$ becomes an eigenstate of the generally triaxial harmonic oscillator Hamiltonian with potential (18) and frequencies given by

$$N_1\omega_1 = N_2\omega_2 = N_3\omega_3 = k \tag{19}$$

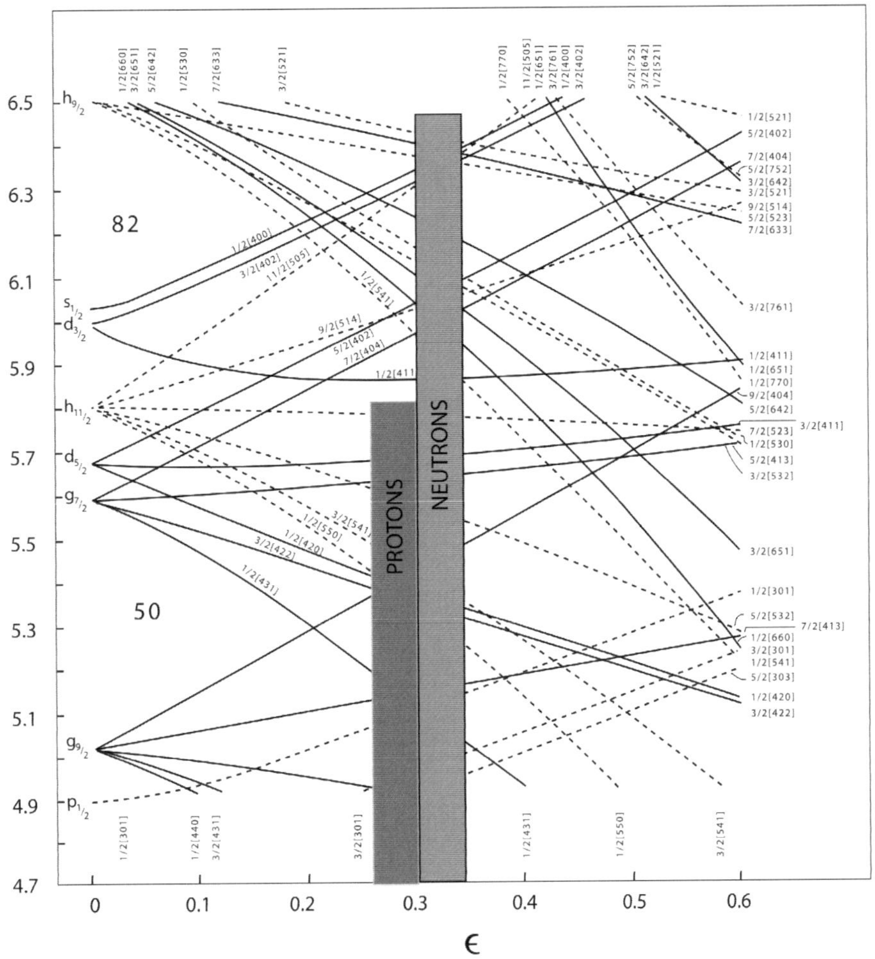

Figure 5. Single-particle energies in the in the Nilsson model. The red and green boxes indicate the levels that would be occupied by protons and neutrons, respectively, in this independent-particle model.

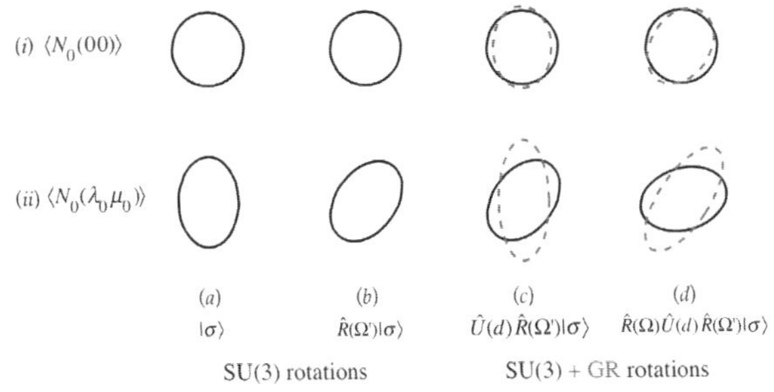

Figure 6. The dynamical degrees of freedom of the symplectic model: row (i) for an irrep $\langle N_0(00)\rangle$; row(ii) for a generic irrep $\langle N_0(\lambda_0\mu_0)\rangle$. Columns (a) and (b) illustrate the basic SU(3) shape orientations; (c) and (d) illustrate the extra vibrations and rotations associated with the giant-resonance (GR) degrees of freedom.

with k determined such that, for volume conservation, $\omega_1\omega_2\omega_3 = \omega^2$. It follows that the quadrupole moments of the SU(3) approximation become doubled and the harmonic oscillator energy of the lowest-weight state, which in the SU(3) approximation is given by $N_0\hbar\omega = (N_1 + N_2 + N_3)\hbar\omega$ becomes

$$E(N_1N_2N_3) = 3(N_1N_2N_3)^{1/3}\hbar\omega. \tag{20}$$

Sample results for the predicted lowest-energy Sp(3, $\mathbb{R}$) irreps are given in table 1. The irreps shown in blue prove to be remarkable consistent with observed low-energy states of the nuclei considered (cf. reference [15] for ^{168}Er). The irrep indicated in red for ^{12}C and ^{16}O are also consistent with the rotational bands based on the first excited states of these nucleus (cf. references [14, 16]).

Table 1. Spherical and self-consistent harmonic oscillator energies for the lowest-weight states of various Sp(3, $\mathbb{R}$) irreps in units of $\hbar\omega$.

^{12}C

N_0	λ	μ	$E(N_1N_2N_3)$
26	0	4	25.3
30	12	0	26.0
28	6	2	26.3
32	10	2	28.6

^{16}O

N_0	λ	μ	$E(N_1N_2N_3)$
36	0	0	36
40	8	4	37.3
38	4	2	37.3
40	7	3	38.1
48	24	0	38.1
44	16	2	38.3

^{168}Er

N_0	λ	μ	$E(N_1N_2N_3)$
814	30	8	812.6
826	96	20	812.9
822	70	28	813.0
818	52	20	813.1
816	42	12	813.1
830	114	16	813.2

7. A sample calculation

Figure 7 shows the results of an Sp(3, $\mathbb{R}$) calculation [17] for ^{166}Er. For comparison, results are also shown for SU(3) and rigid-rotor models. The remarkable fact is that they all agree with one another more precisely than they agree with the observed spectrum; in particular, none of the models exhibits the centrifugal stretching effects shown by the data. The absence of such stretching in the symplectic model is an artifact of a potential energy with an overly sharp minimum, which could be adjusted. What is important is that the microscopic symplectic

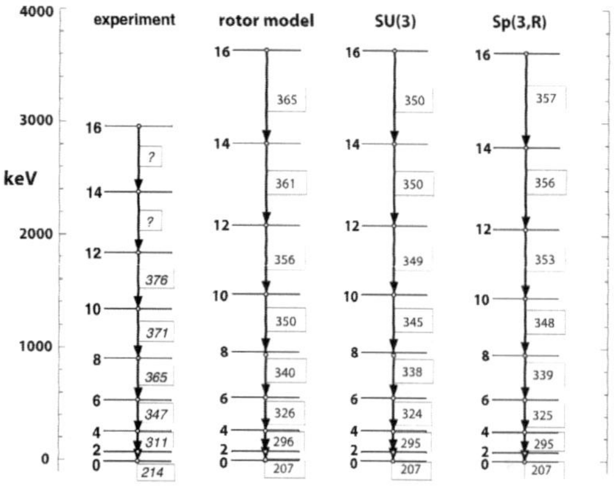

Figure 7. Sp(3, $\mathbb{R}$) model calculations for an irrep chosen by exptl. and Nilsson model considerations, for a Hamiltonian $\hat{H} = \hat{H}_{\rm ho} + \chi\,(Q\cdot Q + \epsilon/Q\cdot Q)$ where $\hat{H}_{\rm ho}$ is a spherical harmonic oscillator Hamiltonian, ϵ is such that the minimum of the potential is at the experimentally observed deformation, and χ is fixed by a self-consistency condition to ensure that the quadrupole moments of states that emerge from the calculation have the same deformation.

model reproduces the results of the phenomenological rotor model very precisely. Even more remarkable is the fact that it returns the observed moment of inertia with a kinetic energy defined purely by the nucleon mass and with a two parameter potential energy chosen to have a minimum at the observed deformation and a strength determined by self-consistency to get the observed E2 transition rates corresponding to such a potential. The success of the SU(3) model can be understood on the grounds that the lower angular momentum states of a large-dimensional representation of SU(3) contract to those of a rigid rotor [9, 18–20]. Nevertheless, the remarkable agreement between the three models is a strong implication that they are effectively describing the same rotational dynamics, albeit in the sense of a quasi-dynamical symmetry.

8. Quasi-dynamical symmetry

Figure 8 shows that the amplitudes in an SU(3) basis of the Sp(3, $\mathbb{R}$) states of figure 7 are essentially the same for all $L \leq 10$. This is in spite of the fact that the SU(3) spectrum of

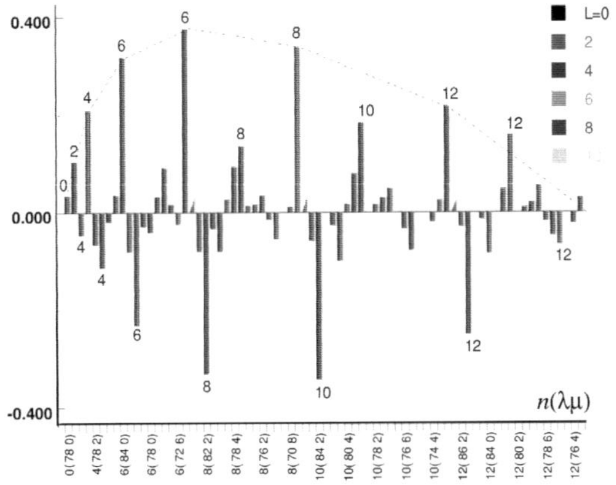

Figure 8. Amplitudes for the wave functions of the $L \leq 10$ Sp(3.R) model states of figure 7 in an SU(3) basis. The distribution of basis states span 6 major harmonic shell of positive parity. The remarkable fact is that the SU(3) amplitudes of all states are essentially the same for all L.

figure 7 is for the lowest-weight SU(3) irrep of the Sp(3, $\mathbb{R}$) calculation which, as figure 8 shows, contributes only a tiny component to any of the Sp(3, $\mathbb{R}$) states shown; the amplitude of this lowest-weight SU(3) irrep appears on the far left of figure 8. Thus, the SU(3) and Sp(3, $\mathbb{R}$) states of figure 7 have negligible overlaps with one another. Nevertheless, because of the near equality of the amplitudes of the many contributing SU(3) irreps, the Sp(3$\mathbb{R}$) results are essentially identical to that of the average SU(3) irrep.

The equality of amplitudes from different mixed irreps of a subgroup is a characteristic of a quasi-dynamical symmetry [21]. It is also observed in a pairing + quadrupole model [22] with Hamiltonian

$$\hat{H}(\alpha) = \hat{H}_0 + (1 - \alpha) V_{\rm su(2)} + \alpha V_{\rm su(3)} , \tag{21}$$

where $\hat{H}_0$ is a harmonic oscillator Hamiltonian, $V_{\rm su(2)} = -G\hat{S}_+\hat{S}_-$ is an SU(2) quasi-spin pairing interaction and $V_{\rm su(3)} = -\chi Q \cdot Q$ is an SU(3) interaction. The model has a USp(6) dynamical group. It also has two competing dynamical subgroups, SU(3) and SU(2), and an SO(3) symmetry group. The spectrum and wave functions for this model are shown for various values of α in figure 9.

Quasi-dynamical symmetry underlies the collective motions of a complex system with many independent modes which, in the absence of interactions between them, vibrate or rotate independently. However, unless it overwhelms their vibrational or rotational behavior, attractive interactions between the modes, tends to make them vibrate and/or rotate in unison in their lowest frequency states. This perspective and figure 9, suggest that the existence of dominant rotational dynamics tends to suppress the mixing of states of widely different deformation. Moreover, there is expected to be a suppression of the mixing of rotational states belonging

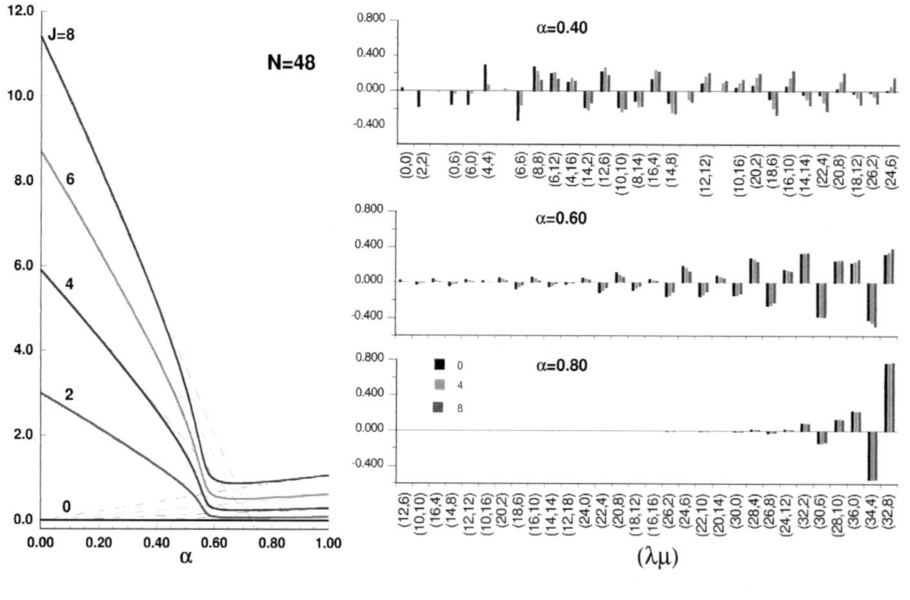

Figure 9. Energies (left) and wave functions in an SU(3) basis (right). The SU(2) dynamical symmetry dominates for $\alpha \lesssim 0.55$ and the SU(3) dominates for $\alpha \gtrsim 0.6$. However, the wave functions show that this dominance is characteristic of a quasi-dynamical symmetry.

to different $\mathrm{Sp}(3,\mathbb{R})$ irreps of large deformation when they appear among states of much lower deformation to which they are not directly connected by a two-body interaction.

9. Concluding remarks

The several avenues explored above, give hope that viable procedures for describing well-deformed nuclear states in terms of mixed symmetry shell-model calculations in an $\mathrm{Sp}(3,\mathbb{R}) \times \mathrm{SU}(2)$ basis might be achievable in the near future. In progressing towards this goal many model investigations of the optimal routes to be followed are needed. The following are a few.

It is of particular importance to discover the energy distribution of lowest-weight U(3) irreps in nuclei, in terms of an empirical formula or a simple model. Equation (20) is a start on this search. However, it needs to be checked out in mean-field calculations and extended, perhaps along the lines of the Nilsson model, to include strongly-coupled spin degrees of freedom.

Studies of the distributions of nuclear shapes in nuclei, already initiated as reviewed in reference [23, 24], need to be continued so that their systematics and patterns can be understood. Methods of approach have been been suggested by Jarrio *et al.* [15] and Carvalho [25], and extensions proposed by Hess *et al.* [26].

Two subjects that would naturally have been pursued in this contribution if there were time and space are: (i) a procedure for carrying out shell model calculations within the space of one or a few $\mathrm{Sp}(3,\mathbb{R})$ irreps; and (ii) a procedure for defining an effective interaction on a subspace spanned by lowest-weight U(3) irreps.

Because the states $|\sigma(\Omega d\Omega')\rangle = \hat{R}(\Omega)\hat{U}(d)\hat{R}(\Omega')|\sigma\rangle$ of the set (16) span an $\mathrm{Sp}(3,\mathbb{R})$ irrep, a good basis for diagonalizing a given microscopic Hamiltonian in such an irrep is given by a discrete subset of states on this manifold. This should give a good approximation if the selected states are centered about the minimum-energy state. Procedures for projecting out states of good angular momentum have been much investigated, e.g., by Cusson and Lee [27], and are known to simplify for eigenstates of a deformed harmonic oscillator [28]. Moreover, if the observed low-lying states imply states of relatively strong β rigidity, it is expected that a relative small number of states of different deformation will be sufficient. Such a method was proposed by Vassanji and Rowe [29] and pursued by Carvalho *et al.* [30] and others.

An effective interaction for a subspace spanned by lowest-weight U(3) irreps is, in principle, defined by the Suzuki-Lee projection method [31] and the fact that the projection of states in an $\mathrm{Sp}(3,\mathbb{R}) \supset \mathrm{U}(3)$ basis onto a lowest weight U(3) subspace would appear to be straightforward. However, subtle concerns will inevitably arise when mixed $\mathrm{Sp}(3,\mathbb{R})$ states are combined with spin wave function. Thus, again model investigations with schematic interactions need to be considered.

Finally, it is worthwhile to consider how it might be possible to take advantage of quasi-dynamical symmetry to define a shell-model basis in which calculations will be most rapidly convergent. For example, quasi-dynamical symmetry would seem to imply that if a calculation can be done in an $\mathrm{Sp}(3,\mathbb{R}) \supset \mathrm{U}(3)$ basis for $L = 0$, or more generally, for a minimum value of the total angular momentum J, then at least such a calculation will have defined the $\mathrm{U}(3) \times \mathrm{SU}(2)$ irreps that contribute most strongly for all low-lying states.

References

[1] Bohr A 1952 *Mat. Fys. Medd. Dan. Vid. Selsk.* **26** 1–40
[2] Rowe D J and Turner P S 2005 *Nucl. Phys.* **A753** 94–105
[3] Rowe D J, Turner P S and Repka J 2004 *J. Math. Phys.* **45** 2761–2784
[4] Caprio M A, Rowe D J and Welsh T A 2009 *Comp. Phys. Comm.* **180** 1150–1163
[5] Rowe D J, Welsh T A and Caprio M A 2009 *Phys. Rev. C* **79** 054304(1–16)
[6] Weaver L, Biedenharn L C and Cusson R Y 1973 *Ann. Phys. (N.Y.)* **77** 250–278
[7] Rosensteel G and Rowe D J 1977 *Phys. Rev. Lett.* **38**(1) 10–14
[8] Rosensteel G and Rowe D J 1980 *Ann. Phys. (NY)* **126** 343–370
[9] Rowe D J 1985 *Rep. Prog. Phys.* **48** 1418–1480
[10] Dytrych T, Sviratcheva K D, Bahri C, Draayer J P and Vary J P 2008 *J. of Phys.G: Nucl. Part. Phys.* **35** 095101(11)
[11] Dytrych T, Sviratcheva K D, Draayer J P, Bahri C and Vary J P 2008 *J. of Phys.G: Nucl. Part. Phys.* **35** 123101(47)
[12] Rowe D J and Rosensteel G 1980 *Ann. Phys. (N.Y.)* **126** 198–233
[13] Rowe D J 1967 *Phys. Rev.* **162** 866–871
[14] Rowe D J, Thiamova G and Wood J L 2006 *Phys. Rev. Lett.* **97** 202501(1–4)
[15] Jarrio M, Wood J L and Rowe D J 1991 *Nucl. Phys. A* **528** 409–435
[16] Dreyfuss A, Launey K, Bahri C, Dytrych T and Draayer J 2011 *APS Division of Nuclear Physics Meeting Abstracts*
[17] Bahri C and Rowe D J 2000 *Nucl. Phys. A* **662** 125–147
[18] Carvalho M J 1984 *Algebraic study of rotational features in nuclei* Ph.D. thesis University of Toronto Toronto
[19] Leschber Y, Draayer J P and Rosensteel G 1986 *J. of Phys.G: Nucl. Part. Phys.* **12** L179–L183
[20] Castaños O, Draayer J P and Leschber Y 1988 *Z. Phys. A Atomic Nuclei* **329** 33–43
[21] Rowe D J 2004 *Computational and Group-Theoretical Methods in Nuclear Physics* ed Escher J, Castaños O, Hirsch J G, Pittel S and Stoitcheva G (Singapore: World Scientific) pp 165–173 (arXiv.org/abs/1106.1607)
[22] Rowe D J, Bahri C and Wijesundera W 1998 *Phys. Rev. Lett.* **80** 4394–4397
[23] Heyde K K, Isacker P V, Waroquier M, Wood J L and Meyer R A 1983 *Phys. Rep* **102** 291–393
[24] Heyde K and Wood J L 2011 *Rev. Mod. Phys.* **83** 1467–1521
[25] Carvalho J and Rowe D J 1992 *Nucl. Phys. A* **548** 1–28
[26] Hess P O, Algora A, Hunyadi M and Cseh J 2002 *Eur. Phys. J.* **A 15** 449–454
[27] Cusson R Y and Lee H C 1973 *Nucl. Phys. A* **211** 429–462
[28] Rowe D J, Bartlett S and Bahri C 2000 *Phys. Lett. B* **472** 227–231
[29] Vassanji M G and Rowe D 1984 *Nucl. Phys. A* **426** 205–221
[30] Carvalho M J, Rowe D J, Karram S and Bahri C 2002 *Nucl. Phys. A* **703** 167–187
[31] Suzuki K and Lee S Y 1980 *Prog. Theor. Phys.* **64** 2091–2106

Quantum information approach to the description of quantum phase transitions

O Castaños, R López-Peña, E Nahmad-Achar and J G Hirsch

Instituto de Ciencias Nucleares, Universidad Nacional Autónoma de México, Apdo. Postal 70-543, 04510 México D F, México

E-mail: ocasta@nucleares.unam.mx

Abstract. The fidelity, linear entropy and entanglement entropy, concepts of information theory, are used to determine the localization of quantum phase transitions of the Lipkin-Meshkov-Glick (LMG) and the Dicke models for a finite number of particles. Another concept, the so-called fidelity susceptibility, gives us information of the scaling behavior of the quantum phase transitions of the models. In particular universal curves are presented for the specific susceptibility. It is shown that the control parameters of both models have a similar dependence in the number of particles when the systems approach to the thermodynamic limit.

1. Introduction

In the last thirty years the fields of quantum information and quantum computation have grown tremendously, and the community has been using quantum mechanics as a powerful framework to help the development of information technologies. Since the establishment of the Landauer principle in 1961, it is well understood that information has physical manifestations and, due to improvements in the storing of information, in ever smaller devices, this has led the science community to assume naturally that the physics that will rule such devices will be quantum mechanics [1, 2].

Recently, there have been remarkable discoveries that have called the attention of the physicist and mathematician communities, they are quantum teleportation [3], the use of quantum cryptography in commercial business [4] and, equally important, the appearance of Shor's factorization algorithm [5].

The purpose of this contribution is to show how basic concepts of quantum information can be used in many body physics. As examples we consider two Hamiltonian models used in different fields of physics, which are simple enough to avoid being lost in the calculations. A review of the LMG and Dicke Hamiltonian models [6, 7] is given, and the definitions of concepts such as the fidelity, and the fidelity susceptibility are introduced. We present briefly which are the values of the control parameters of the considered models that lead to quantum phase transitions. This is done by calculating the expectation value of the Hamiltonian with respect to appropriate coherent states which define a energy surface. This is a function depending on variables and control parameters, which is characterized by its singularities. The theory of singularities or catastrophes has been applied in many different fields [8]. For each of the two models considered, we have found their corresponding separatrix, that is the region of points in the control parameter space where there are sudden changes in the structure of the ground state of the systems [9, 10, 11]. Afterwards, the fidelity and fidelity susceptibility observables are calculated. We show that the minimum value of the fidelity or the maximum of the fidelity susceptibility as a function of the control parameter can be used to determine the localization of the quantum phase transitions for a finite number of particles. We are considering that there is a quantum phase transition when there is a sudden change in the ground state properties of a finite system. When the number of particles of the system is very large the size of the changes grows to infinity [12]. Additionally, the values of the parameter inducing the transition move to its thermodynamic limit. The fidelity susceptibility can be used to get the scaling behavior of the maximum susceptibility in terms of the number of particles N, and it can be used to determine how the coupling parameter γ_m of maximum susceptibility goes to its thermodynamic limit. Additionally, we use the definition of the specific susceptibility introduced in [13], which lets find universal curves in both models. Finally, we discuss the meaning of entanglement in a composite system, and its importance in quantum information. To measure the degree of entanglement of a bipartite system, we use the linear and von Neumann entropies. We calculate both measures of entanglement for the LMG and Dicke models and find that their maximum values can be used to localize finite quantum phase transitions.

This work is dedicated with appreciation to the seventieth birthday of Jerry P. Draayer for his great contributions in the fields of nuclear and computational physics.

2. Hamiltonian Models

The LMG model has been used in different contexts. It was introduced in nuclear physics as a toy model to test all kinds of many body theories because it is an exactly solvable model. The model assumes that the nucleus is a system of fermions which can occupy two levels with the same degeneracy Ω, separated by an energy $\hbar\omega$. One of the levels is just above the Fermi energy and the other also just below. There are residual interactions which scatter pairs of particles between the two levels without changing the total number of particles occupying the shells [6, 14]. Thus the LMG Hamiltonian can be written in terms of the angular momentum operators, using the quasispin formalism, as

$$H_{LMG} = \omega\, J_z + \bar{\gamma}_x\, J_x^2 + \bar{\gamma}_y\, J_y^2 \,, \tag{1}$$

with $\Omega = 2j + 1 \equiv N + 1$, where N denotes the number of particles.

In other fields like quantum optics and condensate matter it has been used to describe two level atoms in a cavity interacting with a classical electromagnetic field, It describes also a system of N particles with spin $s = 1/2$ mutually interacting with a magnetic field transverse to the interaction directions X, Y, or as a two mode Bose Einstein condensate by means of the Schwinger realization of the angular momentum operators [15, 16, 17, 18]. Notice that the unitary transformation $e^{-i\,\pi\,J_z}$ changes the operators $J_x \to -J_x$ and $J_y \to -J_y$ and so it leaves the Hamiltonian invariant. While the LMG model was initially used as a toy model, now it has a more realistic role because of the experimental and technological advances.

The Dicke Hamiltonian, which describes a cloud of cold two-level atoms interacting with a one mode electromagnetic field in an optical cavity in the dipolar approximation, is given by [7, 16]

$$H_D = \hat{a}^\dagger \hat{a} + \omega_A \hat{J}_z + \frac{\gamma}{\sqrt{N}}\left(\hat{a}^\dagger + \hat{a}\right)\left(\hat{J}_+ + \hat{J}_-\right) \,. \tag{2}$$

If we consider again the completely symmetric configuration then $N = 2\,j$. This Hamiltonian is invariant under the transformation $\exp{(i\,\pi\,\Lambda)}$, where the operator $\Lambda \equiv a^\dagger a + J_z + \sqrt{J^2 + 1/4} - 1/2$ denotes the total number of excitations of the system and is a conserved quantity within the Tavis-Cummings model [11]. For the Dicke model, the eigenvalue of Λ is denoted by λ and the eigenvectors must have definite parity [19, 20]. This Hamiltonian can be also interpreted as describing a one mode interacting boson-fermion model.

For these two models, we have followed a procedure established by Gilmore [8] which considers the algebraic structures of the models, constructed the associated coherent states as variational states, and by evaluating the energy surface of the model determine their separatrix, which is the place in the control parameter space where there is a quantum phase transition [9, 10].

In both cases, as it was mentioned previously, there is a residual symmetry which can be used to get a better agreement with the exact solution of the Hamiltonian models for a finite number of particles. This is done by taking linear symmetry adapted coherent states to the point symmetry of the cyclic group C_2 [18, 19, 20].

3. Fidelity and Fidelity Susceptibility

The fidelity is a basic concept in classical information theory, which measures the accuracy of a transmission, and which usually presents imperfection due mainly to the limitation of the resources and the random noise [21]. For the quantum case, it is known that for two pure states described by the density operators $\rho_1 = |\chi\rangle\langle\chi|$ and $\rho_2 = |\phi\rangle\langle\phi|$, the fidelity is defined by

$$F(|\chi\rangle\langle\chi|, |\phi\rangle\langle\phi|) = |\langle\chi|\phi\rangle|^2 .$$

It is a measure of the distance between the states. Then it has values in the range $0 \le F \le 1$, with $F = 1$ indicating that the states are the same up to a global phase and $F = 0$ when they are orthogonal. Thus, it means also the transition probability from one state to another and its geometric interpretation is the closeness of states [2].

For one pure state $|\chi\rangle$ and other mixed state ρ_2, one has

$$F(|\chi\rangle\langle\chi|, \rho_2) = \langle\chi|\rho_2|\chi\rangle ,$$

that denotes the probability to be a pure state. It is used for example to measure how reliable is the transmission of quantum information for a given communication protocol.

In general, for mixed states, it has been proved that [21, 22] the definition

$$F(\rho_1, \rho_2) = \left\{Tr\left(\sqrt{\sqrt{\rho_1}\,\rho_2\,\sqrt{\rho_1}}\,\right)\right\}^2 , \tag{3}$$

of the fidelity satisfies the properties:

$$\begin{aligned} &0 \le F(\rho_1, \rho_2) \le 1 , \\ &F(\rho_1, \rho_2) = F(\rho_2, \rho_1) , \\ &F(U\rho_1, U\rho_2) = F(\rho_1, \rho_2) . \end{aligned} \tag{4}$$

Therefore it gives a measure of how similar are two probability densities. It is currently used to establish the accuracy of quantum computations in the presence of errors induced by noise effects due to interaction with the environment or imperfections in the hardware. It has been used in the experimental generation of entanglement because the entangled state should be deterministic [23], that is, there must be a very large fidelity between the generated state and a Bell state. The fidelity has been also used as a measure of the accuracy with which the initial quantum state can be recovered by inverting the dynamics with a perturbed Hamiltonian, and in that case it is called Loschmidt echo, providing a criterion for stability of a quantum motion under changes in the initial conditions. A review of the fidelity approach to the characterization of quantum phase transitions is given in [24].

Here we are going to show that the fidelity [25] can be used to determine when the ground state of a quantum system presents a sudden change as function of a control parameter. If we denote that parameter by λ one has

$$F(\lambda, \lambda + \delta\lambda) = |\langle \psi(\lambda)|\psi(\lambda + \delta\lambda)\rangle|^2 ,$$

and its minimum value determines where the system is changing drastically, then it can be used to characterize the quantum phase transitions. For a first order quantum phase transition associated to a discontinuity on the first derivative of the energy of the system the states in the different sides of the transition are orthogonal and so the fidelity goes to zero.

From the expression of the fidelity, a Taylor series expansion can be done around the critical parameter λ_c,

$$F(\lambda_c, \lambda_c + \delta\lambda) = F(\lambda_c, \lambda_c) + \delta\lambda \left.\frac{dF}{d\lambda}\right|_{\lambda=\lambda_c} + \frac{1}{2!}\left.\frac{d^2F}{d\lambda^2}\right|_{\lambda=\lambda_c} (\delta\lambda)^2 + \cdots ,$$

and as we are looking for the minimum value, the first derivative of the fidelity is zero. Thus one can define the fidelity susceptibility by [26]

$$\chi_F = 2\frac{1 - F(\lambda_c, \lambda_c + \delta\lambda)}{(\delta\lambda)^2} . \tag{5}$$

It is dependent on the Hamiltonian term that causes the phase transition.

If a Hamiltonian can be written in the form $H = H_0 + \lambda H_I$, then the fidelity susceptibility can also be calculated by means of the first order perturbation expansion

$$\chi_F(\lambda) = \sum_{k\neq 0} \frac{|\langle \Psi_k(\lambda)|H_I|\Psi_0(\lambda)\rangle|^2}{(E_k(\lambda) - E_0(\lambda))^2} ,$$

where $|\Psi_0(\lambda)\rangle$ and $|\Psi_k(\lambda)\rangle$ denote the ground state and the k eigenstate of the system.

4. Separatrix

4.1. Lipkin Model

The expectation value of the Hamiltonian per particle with respect to the SU(2) coherent states is given by [9, 10]

$$\varepsilon(\theta, \phi) = -2\cos\theta + \gamma_x \sin^2\theta \cos^2\phi + \gamma_y \sin^2\theta \sin^2\phi , \tag{6}$$

where $((\theta, \phi)$ denote a point on the Bloch sphere. We use $\omega = 2$, $\bar{\gamma}_x = \frac{2}{N-1}\gamma_x$, $\bar{\gamma}_y = \frac{2}{N-1}\gamma_y$, and $\varepsilon(\theta, \phi) = \langle H_{LMG}/N\rangle$ except for constant terms. This energy surface allows to determine the critical points, bifurcation sets, and Maxwell sets. These sets organize all the critical points according to their stability within the control parameter space.

The minima of the energy surface can be denoted

$$\varepsilon(0, 0) = -2 , \quad \varepsilon(\theta_c, \phi_c) = \Gamma + \frac{1}{\Gamma} , \tag{7}$$

where we have defined

$$\begin{aligned} \Gamma &= \gamma_x , && \text{for} \ \ \gamma_x < -1 , \text{ and } \ \gamma_x < \gamma_y , \\ \Gamma &= \gamma_y , && \text{for} \ \ \gamma_y < -1 , \text{ and } \ \gamma_y < \gamma_x , \\ \Gamma &= \bar{\gamma} , && \text{for} \ \ \bar{\gamma} < -1 , \end{aligned}$$

where in the last expression we have $\gamma_x = \gamma_y = \bar{\gamma}$. The Maxwell sets are the loci in parameters space for which the classical energies at two or more critical points are equal. When a Maxwell set is crossed, the energy surface jumps from one critical branch to another, and phase transitions take place for $|\Gamma| \geq 1$. Phase transitions for the minimum values of the classical energy surface can only happen for $\Gamma \leq -1$. They correspond to the regions where $\gamma_x \leq -1$ and $\gamma_y \leq -1$.

Bifurcation sets are the loci in parameter space where the function ε changes because equilibria points are either created or destroyed. They are obtained from the vanishing of the determinant of the matrix of second derivatives of ε evaluated at the critical points, that is, $\det \varepsilon_{ij} = 0$, with $i, j = \theta, \phi$. The parameter values that satisfy these conditions are

$$\gamma_x = \pm 1 \; ; \quad \gamma_y = \pm 1 \; ; \quad \gamma_x = \gamma_y \text{ with } |\gamma_x| \geq 1 \, . \tag{8}$$

A second order phase transition of the Ginzburg-Landau type takes place when the straight lines $\gamma_x = -1$, $\gamma_y = -1$ or the point $(\gamma_x, \gamma_y) = (-1, -1)$ are crossed. The crossing of the straight line $\gamma_y = \gamma_x$ yields a first order transition. Special attention must be given to the crossing of the cusp point $(\gamma_x, \gamma_y) = (-1, -1)$ along the straight line $\gamma_y = -\gamma_x - 2$ because in that case there is a third order phase transition, related to a convergence of second order phase transitions [9, 10].

4.2. Dicke Model

The energy surface constructed by the tensorial product of Heisenberg-Weyl and SU(2) coherent states is given by [19, 20]

$$\mathcal{E}(q, p, \theta, \phi) = \frac{1}{4j}\left(p^2 + q^2\right) - \frac{\omega_A}{2}\cos\theta + \frac{1}{\sqrt{j}}\gamma\, q \sin\theta \cos\phi \, , \tag{9}$$

where we take the harmonic oscillator realization for the field part $\alpha = \frac{1}{\sqrt{2}}(q + i\,p)$ and the stereographic projection for the angular momentum part $\zeta = e^{-i\phi}\tan\frac{\theta}{2}$. The order parameters (q, p) correspond to the expectation values of the quadratures of the field, while (θ, ϕ) determine a point on the Bloch sphere.

The minima and degenerate critical points are obtained easily. By calculating the Hessian of the energy surface, we see that when $\gamma_c^2 = \omega_A/4$ the critical points degenerate and, for that value of the field-matter coupling, the phase transition from the normal to the superradiant behaviour of the atoms takes place. This value of the coupling parameter corresponds to the thermodynamic phase transition.

The critical points $(q_c, p_c, \theta_c, \phi_c)$ which minimize $\mathcal{H}$ are given by

$$\begin{aligned} &(0\,, 0\,, 0, \phi)\,, && \text{for } |\gamma| < \gamma_c\,, \\ &(-2\sqrt{j}\,\gamma\sqrt{1 - (\gamma_c/\gamma)^4}\cos\phi_c\,, 0\,, \arccos(\gamma_c/\gamma)^2, \phi_c)\,, && \text{for } |\gamma| > \gamma_c\,. \end{aligned} \tag{10}$$

The first row defines the minimum critical points for the normal phase where the variable ϕ is undetermined, while the second row describes those for the superradiant regime where one has two possibilities for the order parameter $\phi_c = 0$ or $\phi = \pi$. It is convenient to work with the variable $x = \gamma/\gamma_c$ in terms of which the energy values for the minima just described are

$$\mathcal{E}_{\text{normal}} = -2\,\gamma_c^2\,, \quad \mathcal{E}_{\text{superradiant}} = -\,\gamma_c^2\,x^2\left(1 + x^{-4}\right)\,. \tag{11}$$

Next, we are going to show how similar information about the localization of the quantum phase transitions can be obtained by means of the quantum information concepts.

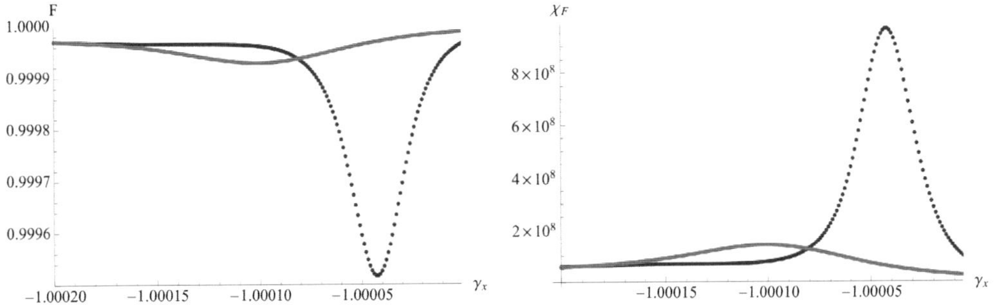

Figure 1. The fidelity and fidelity susceptibility as functions of γ_x for the even (dotted blue line) and odd (continuous red line) exact quantum solutions for $N = 2^{14}$. The quantum phase transition is localized at the minimum of the fidelity and the maximum of the fidelity susceptibility. For LMG model with $\gamma_y = -1$.

5. Results

As an example, we determine the fidelity and fidelity susceptibility for the LMG model. These are calculated for a number of atoms $N = 2^{14}$, keeping fixed the control parameter $\gamma_y = -1$. Then, we are going to study how sensitive is the Hamiltonian under the changes in the parameter γ_x.

The additional symmetry transformation of the LMG Hamiltonian, that is invariance under rotations $R(\pi) = e^{i\pi J_z}$, implies that the matrix Hamiltonian, in the angular momentum basis states $\{|jm\rangle\}$, can be broken into two parts: states with an even number of excited particles $n = j + m$ and states with an odd number.

The results for the fidelity and the fidelity susceptibility are shown in Fig. 1, for the ground and first excited states. The value of γ_x determined by the localization of the phase transition for the ground state is larger than the thermodynamic value $\gamma_{x\,c} = -1$, the obtained value for the first excited state is even larger. Then the values for the control parameter γ_x for the ground and first excited states are given by

$$\begin{aligned}
\gamma_x &= -1.000042\,,\; F = 0.999517\,,\; \chi_F = 9.66246 \times 10^8\,,\\
\gamma_x &= -1.000101\,,\; F = 0.999929\,,\; \chi_F = 1.41458 \times 10^8\,.
\end{aligned}$$

Thus, we have obtained the localization of the quantum phase transition for a large but finite number of particles. Now we want to determine how fast this value of $\gamma_{x\,m}$ approaches the thermodynamic limit when the number of particles changes.

Therefore, we are going to calculate the maximum fidelity susceptibility for different number of particles running from 2^{10} to 2^{16}. In Fig. 2 we show two log-log plots, the first one associated to the maximum value of the susceptibility as a function of the number of particles and the second one is related to $\log_2(\gamma_{x\,c} - \gamma_{x\,m})$ again versus $\log_2 N$. In both cases, one can fit straight lines which for the even and odd cases they take the form

$$\begin{aligned}
\chi_{max} &\approx 2^{-0.16}\, N^2\,, \qquad \gamma_{x\,c} - \gamma_{x\,m} \approx 2^{0.46}\, N^{-1}\,, \qquad \text{for the even case}\,,\\
\chi_{max} &\approx 2^{-2.95}\, N^2\,, \qquad \gamma_{x\,c} - \gamma_{x\,m} \approx 2^{1.71}\, N^{-1}\,, \qquad \text{for the odd case}\,,
\end{aligned}$$

where it is known that $\gamma_{x\,c} = -1$. The confidence limits of the fitted straight lines are related with the obtained variance of the order of 10^{-2}. They are given in the mentioned figure.

In conclusion, the maximum susceptibility of the interaction associated to the control parameter γ_x grows quadratically with the number of particles while the difference of the

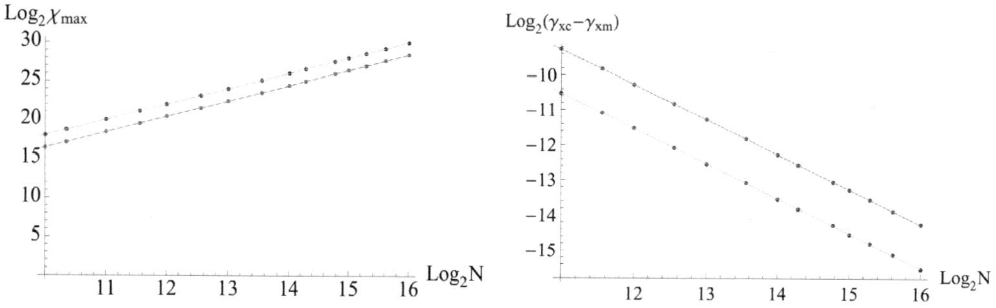

Figure 2. Log-log plots of the maximum fidelity susceptibility χ_{max} (left) and the relative coupling parameter $\gamma_c - \gamma_m$ as functions of the number of particles. We propose $\chi_{max} = 2^b N^a$ and $\gamma_c - \gamma_m = 2^d N^c$. For the LMG model with $\gamma_y = -1$. At the left, one has for the even case $b = -0.16 \pm 0.01$, and $a = 2.00 \pm 0.01$ while for the odd case $b = -2.95 \pm 0.01$ and $a = 2.00 \pm 0.01$. At the right, one has for the even case $c = -1.00 \pm 0.01$, and $d = 0.46 \pm 0.01$ while for the odd case $c = -1.00 \pm 0.01$ and $d = 1.71 \pm 0.01$.

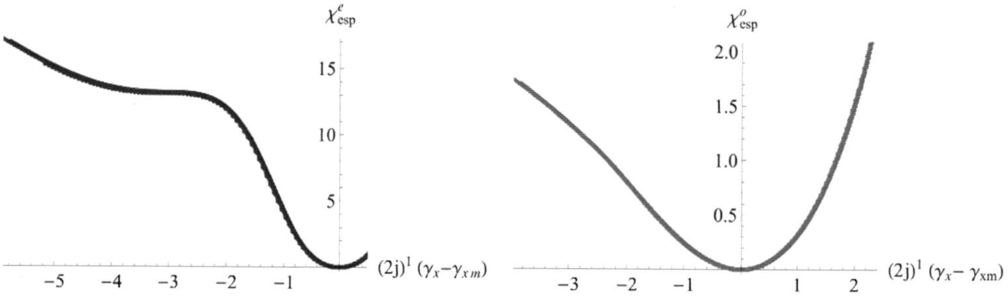

Figure 3. The specific susceptibility determines Universal curves for the even (blue line) and odd (red line) parity states. In the plots its value is shown for all the number of particles indicated in the text and one notices one curve only. For the LMG model with $\gamma_y = -1$.

thermodynamic value with γ_{xm}, the value which yields maximum fidelity susceptibility, is proportional to the inverse of the number of particles. This is true for the even and odd states of the system.

Following the work of Ho-Man Kwok et al [13], one can define a universal quantity called the specific susceptibility as follows

$$\chi_s^\delta = \frac{\chi_{max}(\gamma_{cm})}{\chi_F(\gamma_x)} - 1\,, \tag{12}$$

where $\delta = e$ denotes the specific susceptibility for the even case and $\delta = o$ denotes specific susceptibility for the odd case. If this function is plotted as a function of the variable $\gamma_x - \gamma_{xm}$ times the number of particles (see Fig. 3 and Fig. 5), one finds approximately the same values of the specific susceptibility for a large region of the abscissa. In the mentioned figure, there are plots of the specific susceptibility for the following number of particles

$$2048, 3000, 4096, 6000, 8192, 12000, 16584, 20000, 28000, 32768, 40000, 50000, 65536\,.$$

One can see that all the corresponding specific susceptibilities approximately fall along the same curve.

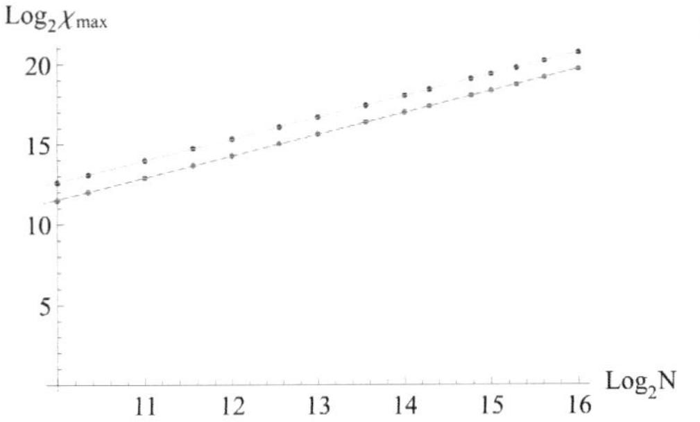

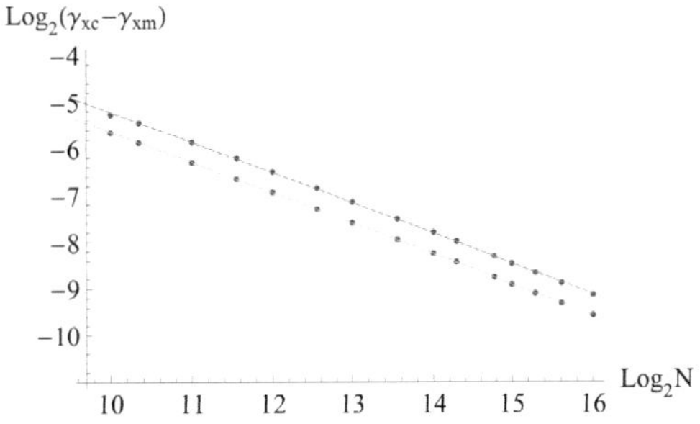

Figure 4. Log-log plots of the maximum fidelity susceptibility χ_{max} (left) and the relative coupling parameter $\gamma_c - \gamma_m$ as functions of the number of particles. We propose $\chi_{max} = 2^b N^a$ and $\gamma_c - \gamma_m = 2^d N^c$. For the LMG model with $\gamma_y = -0.5$. At the left, one has for the even case $b = -0.85 \pm 0.01$, and $a = 1.35 \pm 0.01$ while for the odd case $b = -2.04 \pm 0.01$ and $a = 1.36 \pm 0.01$. At the right, one has for the even case $c = -0.66 \pm 0.01$, and $d = 0.87 \pm 0.01$ while for the odd case $c = -0.65 \pm 0.01$ and $d = 1.26 \pm 0.01$.

When we consider other values of the parameter γ_y for the same set of number of particles, one finds a maximum susceptibility proportional to $N^{\frac{4}{3}}$ and that the relative critical coupling parameter $\gamma_{x\,c} - \gamma_{x\,m}$ is proportional to $N^{-\frac{2}{3}}$. As an example, we present in Fig. 4 the log-log plots for the maximum susceptibility as a function of the number of particles and for the $\gamma_{x\,c} - \gamma_{x\,m}$ also as a function of N, considering the value $\gamma_y = -0.5$. In both cases, one can fit straight lines which for the even and odd cases they take the form

$$\begin{aligned} \chi_{max} &\approx 2^{-0.85}\, N^{1.35}\,, \qquad \gamma_{x\,c} - \gamma_{x\,m} \approx 2^{0.87}\, N^{-0.65}\,, \qquad \text{for the even case}\,, \\ \chi_{max} &\approx 2^{-2.04}\, N^{1.36}\,, \qquad \gamma_{x\,c} - \gamma_{x\,m} \approx 2^{1.26}\, N^{-0.65}\,, \qquad \text{for the odd case}\,, \end{aligned}$$

where it is known that $\gamma_{x\,c} = -1$. The confidence limits of the fitted straight lines are indicated in the mentioned figure.

We expected a similar behavior because they all are second order quantum phase transitions. We conjecture that the result for $\gamma_y = -1$ is different because in that case we are moving towards the triple point of the model, that is $(\gamma_x = -1, \gamma_y = -1)$. The point is also special because it has a third order quantum phase transition when we are moving along the straight line $\gamma_y = -\gamma_x - 2$. This possible effect was corroborated by studying the fidelity susceptibility and the difference $-1 - \gamma_{ym}$ by fixing the value of $\gamma_x = -1$, and running along $-3 \leq \gamma_y \leq -1$. We get the same results than those indicated in Fig. 2 with $\gamma_y = -1$.

For this case, we construct the Universal curves associated to the calculation of the specific susceptibility plotted in Fig. 5 as function of $(2j)^{2/3}(\gamma_{x\,c} - \gamma_{x\,m})$, the plots are for the same number of particles than the previous case and, one sees approximately one curve only.

According to our results one concludes that, in general, the second order quantum phase transitions have the same behavior with the number of particles given by $\chi_{max} \approx N^{4/3}$ and $(\gamma_c - \gamma_{max}) \approx N^{-2/3}$. There are special cases, as for example moving towards a triple point, where one finds that $\chi_{max} \approx N^2$ and $(\gamma_c - \gamma_{max}) \approx N^{-1}$.

For the Dicke model, we have also an extra symmetry related with the parity of the total number of excitations. Thus we have to consider two cases: the even case related with the ground state of the model and the odd case with the first excited state. However in this contribution, we are going to restrict ourselves to the ground state of the system. We also fix the parameter $\omega_A = 1$, that is we consider the case of resonance. Then, we change the parameter γ in the

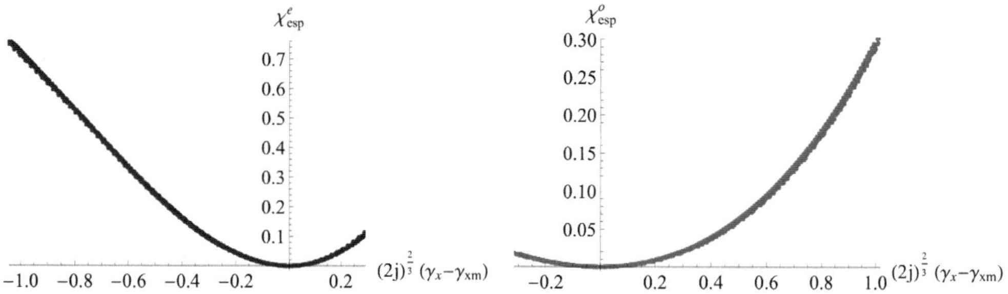

Figure 5. The specific susceptibility determines Universal curves for the even (blue line) and odd (red line) parity states. We use all the number of particles mentioned in the text and observe approximately one curve only. For the LMG model with $\gamma_y = -0.5$.

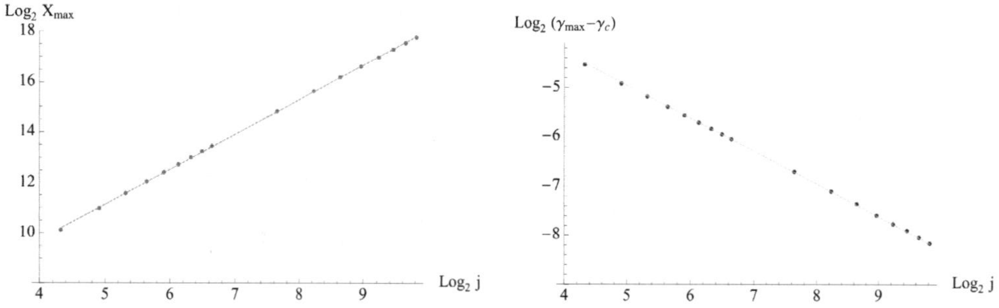

Figure 6. Log-log plots of the difference $\gamma_{max} - \gamma_c$ and χ_{max} as functions of j are displayed. A fit of a straight line of the form $b + ax$ was done and the results are: at left $a = 1.36 \pm 0.01$, and $b = 4.40 \pm 0.01$; at right $a \to -0.66 \pm 0.01$, and $b \to -1.67 \pm 0.01$. For the Dicke model.

interval $0.5 \leq \gamma \leq 0.6$, for a number of particles running from $N = 40$ up to $N = 2^{11}$. The log-log plots of the relative critical coupling matter-field strength $\gamma_{max} - \gamma_c$ and the maximum fidelity susceptibility as a function of $N/2$ are displayed in Fig. 6. Straight lines were fitted, at left with positive slope while at right with negative slope. They are approximately given by

$$\chi_{max} \approx N^{\frac{4}{3}}, \qquad (\gamma_{max} - \gamma_c) \approx N^{-\frac{2}{3}}. \tag{13}$$

In the Dicke model the thermodynamic quantum phase transitions is of second order and it is important to stress that the scaling behavior of the maximum fidelity susceptibility and the approaching to the coupling parameter strength satisfy the same relations that in the LMG model.

Finally, we want to discuss our results for the entanglement between the matter and field within the Dicke model. The entanglement concept was introduced by Schrödinger in the middle of the thirties and is closely related to the EPR paradox and Bell's theorem. It is considered as a resource that can be used in the same form that the energy to do, for example, quantum teleportation or send secure information[2]. For the Dicke model [27] and other algebraic models in molecular physics, as the vibron model [28] for example, the Renyi entropy has been calculated.

For a bipartite system, one can have complete information of the composite system but also

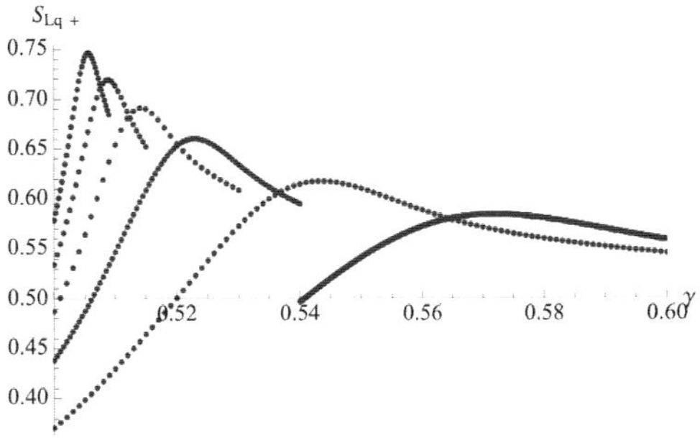

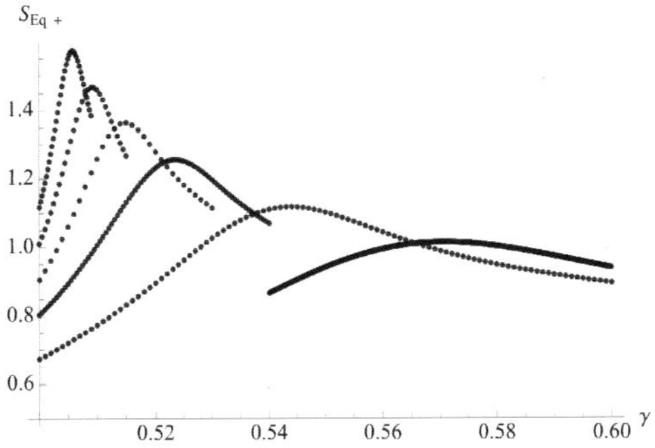

Figure 7. The linear and von Neumann entropies as functions of the coupling strength γ between the matter and the field are plotted.

complete ignorance about the parts of the system. Examples are the so called Bell states

$$
\begin{aligned}
|\phi_\pm\rangle &= \frac{1}{\sqrt{2}}\Big(|+,+\rangle \pm |-,-\rangle\Big), \\
|\psi_\pm\rangle &= \frac{1}{\sqrt{2}}\Big(|+,-\rangle \pm |-,+\rangle\Big),
\end{aligned}
$$

where the states can represent a system of 2 qubits, each of which has two possible eigenvalues, or it can represent a system of two particle with spin $s = 1/2$. For these states the linear and von Neumann entropies have maximum values. To illustrate how they can be calculated first we give their definitions: the linear entropy is defined by $S_L = 1 - Tr(\rho_2^2)$ and the entanglement entropy by $S_{VN} = -\sum_k \lambda_k \ln \lambda_k$.

For one of the Bell states of two qubits, the system is described by the density operator [2]

$$\rho = \frac{1}{2}\Big(|+,+\rangle\langle +,+| + |+,+\rangle\langle -,-| + |-,-\rangle\langle +,+| + |-,-\rangle\langle -,-|\Big).$$

Tracing over the first subsystem one gets the reduced density operator

$$\rho_2 = \frac{1}{2}\Big(|+\rangle\langle +| + |-\rangle\langle -|\Big),$$

and from it one can immediately calculate that $S_L = 1/4$. In the expression of the von Neumann entropy appear the eigenvalues of the reduced density matrix of the subsystem 2. Then, it is straightforward to evaluate that $S_{VN} = \ln 2 = 0.693$.

We calculate the entanglement of the matter with the one mode electromagnetic field for the model, and look numerically for its maximum values. The results are shown in Fig.7, where at left the positions of the maxima are given for the following set:

$$(N,\gamma) = \{(20, 0.572), (40, 0.543), (100, 0.523), (200, 0.514), (400, 0.509), (1000, 0.505),$$

while at the right one has

$$(N,\gamma) = \{(20, 0.571), (40, 0.544), (100, 0.524), (200, 0.515), (400, 0.509), (1000, 0.505)\}.$$

This coincides with the calculation of the minimum values of the fidelity to within the accuracy of the numerical calculation. The latter are given by

$$(N,\gamma) = \{(20, 0.568), (40, 0.543), (100, 0.524), (200, 0.515), (400, 0.509), (1000, 0.505)\}.$$

6. Conclusions

We have shown that the fidelity, the fidelity susceptibility, and the entanglement entropy concepts can be used to localize the position of finite quantum phase transitions in terms of the parameters of the different interaction terms of the Hamiltonian. This was done for the LMG and Dicke models. The agreement of the results with those previously obtained using the Ginzburg Landau formalism is complete.

For a finite number of atoms the localization of the quantum phase transitions is defferent from its value at the thermodynamic limit, and the N dependence of the maximum fidelity susceptibility was obtained. For a special crossing of the triple point of the LMG model we obtain

$$\chi_{max} \approx N^2 \, , \qquad (\gamma_{x\,c} - \gamma_{max}) \approx N^{-1} \, .$$

For other crossings of second order phase transitions one gets

$$\chi_{max} \approx N^{4/3} \, , \qquad (\gamma_{x\,c} - \gamma_{max}) \approx N^{-2/3} \, .$$

A similar result for the second order quantum phase transition of the Dicke model was obtained by means of the calculation of the von Neumann entropy. The calculated specific fidelity susceptibility, for both models, shows a universal character for all the number of particles.

Acknowledgments

This work was supported by CONACyT (under project 101541) and DGAPA-UNAM (under project 102811).

References

[1] Nielsen M A and Chuang I 2000 *Quantum Computation and Quantum Information*(Cambridge: Cambridge University Press)
[2] Benenti G and Casati G 2005 *Principles of quantum computation and information, Vol I and II* (Singapore: World Scientific)
[3] Bennett C H, Brassard G, Crepeau C, Jozsa R, Peres A and Wootters W 1993 *Phys. Rev. Lett.* **70** 1895
[4] Company: *ID Quantique SA*, Chemin de la Marbrerie 3, 1227 Carouge; Genve, Switzerland
[5] Shor P W 1994 *Algorithms for quantum computation: discrete logarithms and factoring* In Proceedings of the 35th Annual Symposium on Foundations of Computer Science
[6] Lipkin H J, Meshkov N and Glick A. J. 1965 *Nucl. Phys.* **62** 188
[7] Dicke R H 1954 *Phys. Rev.* **93** 99
[8] Gilmore R 1981 *Catastrophe Theory for Scientists and Engineers* (Wiley, New York)
[9] Castaños O, López-Peña R, Hirsch J G and López-Moreno E 2005 *Phys. Rev. B* **72** 012406
[10] Castaños O, López-Peña R, Hirsch J G and López-Moreno E 2006 *Phys. Rev. B* **74** 104118
[11] Castaños O, López-Peña R, Nahmad-Achar E and Hirsch J G 2009 *Phys. Scr.* **80** 055401
[12] Sadchev S 2001 *Quantum Phase Transitions* (Cambridge: Cambridge University Press)
[13] Kwok H M, Ning W Q, Gu S J and Lin H Q 2008 *Phys. Rev. E* **78** 032103
[14] Ring P and Schuck P 1980 *The nuclear many body problem* (Berlin: Springer-Verlag)
[15] Dodonov V V and Manko V I (eds) 2003 *Theory of Nonclassical States of Light* (London: Taylor and Francis)
[16] Vidal J, Dusuel S and Barthel T 2007 *J. Stat. Mech.* P01015
[17] Hines A P, McKenzie R H and Milburn G J 2011 *Phys. Rev. A* **83** 051601(R)
[18] Pérez-Campos C, González-Alonso J R, Castaños O and López-Peña R 2010 *Ann. Phys.* **325** 325
[19] Castaños O, Nahmad-Achar E, López-Peña R and Hirsch J G 2011 *Phys. Rev. A* **83** 051601(R)
[20] Castaños O, Nahmad-Achar E, López-Peña R and Hirsch J G 2011 *Phys. Rev. A* **84** 013819
[21] Jozsa R 1994 *J. Modern. Optics* **41** 2315
[22] Uhlmann A 1976 *Rep. Math. Phys.* **9** 273
[23] Turchete Q A et al 1998 *Phys. Rev. Lett.* **81** 3631
[24] Gu, S J 2010 *Int. J. of Mod. Phys. B* **24** 4371
[25] Zanardi P and Paunkovic N 2006 *Phys. Rev. E* **74** 031123
[26] You W L, Li Y W and Gu S J 2007 *Phys. Rev. E* **76** 022101
[27] Nagy Á and Romera E 2012 *Physica A* **391**, 3650
[28] Romera E and Nagy Á 2011 *Phys. Lett. A* **375** 3066

The Relativistic Origins of Pseudospin Symmetry in Nuclei

Joseph N Ginocchio

Theoretical Division, Los Alamos National Laboratory, Los Alamos, NM, 87544, USA

E-mail: gino@lanl.gov

Abstract. Jerry Draayer has worked in approximate pseudospin and pseudo- U(3) symmetry in nuclei in the non-relativistic shell model. We show that pseudospin symmetry is a SU(2) relativistic symmetry of the Dirac Hamiltonian for which the sum of the vector and scalar potentials is a constant. We show that the Dirac Hamiltonian for which the sum of the vector harmonic oscillator potential and a scalar harmonic oscillator is a constant has a pseudo-U(3) symmetry and we derive the generators for this symmetry.

1. Introduction

In 1968 Akito Arima and collaborators [1] and Ted Hecht and collaborators [2] discovered quasi-degeneracies in single particle energy levels which were not spin doublets. In fact the doublets differed in orbital angular momentum by units of two and radial nodes by one. These states have single nucleon quantum numbers $[n, \ell, \ell + \frac{1}{2}]$ and $[n-1, \ell+2, \ell+\frac{3}{2}]$ where n,ℓ are the radial and orbital quantum numbers, respectively, and the last quantum number in the brackets is the total angular momentum, j. For example, in the usual notation, $(1s_{\frac{1}{2}}, 0d_{\frac{3}{2}}), (1p_{\frac{3}{2}}, 0f_{\frac{5}{2}})$, etc, are pseudospin doublets [3]. These quasi-degeneracies persist in recent measurements in nuclei far from stability [4]. They realized that, if they define the average of the orbital angular momenta as a pseudo-orbital angular momentum ($\tilde{\ell}$) and then couple a pseudospin ($\tilde{s} = \frac{1}{2}$) to the pseudo-orbital angular momentum they will get the total angular momenta ($j = \tilde{\ell} \pm \frac{1}{2}$). For example, for the $(1s_{\frac{1}{2}}, 0d_{\frac{3}{2}})$ orbits, $\tilde{\ell} = 1$, which gives the total angular momenta $j = \frac{1}{2}, \frac{3}{2}$. Subsequently pseudospin doublets in deformed nuclei were discovered [5]. Jerry and collaborators examined pseudospin symmetry in triaxial nuclei by calculating the triaxial orbits in the triaxial Nilsson model [6].

2. Pseudospin symmetry in the non relativistic shell model

Jerry and collaborators showed that the non relativistic spin-orbit potential derived from relativistic mean field theory approximately conserves pseudospin symmetry[7].The transformation from spin variables to pseudospin variables was then an open question. Jerry and his collaborators used translational, parity, and time reversal symmetry to suggest the helicity transformation of the form $U_p = \vec{\sigma} \cdot \hat{p}$, where $\hat{p}$ is the unit momentum operator and the spin generators are $\vec{s} = \vec{\sigma}\ 2$ .With this transformation they were able to derive the conditions on the non-relativistic single-particle Hamiltonian for which pseudospin symmetry is a dynamical symmetry. If the strength of a constant single-particle spin orbit potential is C and the strength of constant single particle orbital angular momentum squared is D, then C = 4D produced pseudospin symmetry [8].

The pseudospin generators are then $\tilde{\vec{\sigma}} = U_p \, \vec{\sigma} \, U_p$. Since the helicity transformation is unitary this transformation preserves the spin commutation rules.The pseudospin doublets differ by two units of angular momentum and have different radial nodes.These generators intertwine momentum and spin.The momentum has the capability of changing the orbital angular momentum and the radial quantum number, and, when the details are worked out, these generators connect the pseudospin doublets in the correct way in the pseudospin limit.

3. Symmetries of the Dirac Hamiltonian

The Dirac Hamiltonian with a Lorentz scalar potential, $V_S(\vec{r})$, and a potential which is the fourth component of a Lorentz vector potential, $V_V(\vec{r})$, is

$$H = \vec{\alpha} \cdot \vec{p} + \beta(V_S(\vec{r}) + M) + V_V(\vec{r}), \tag{1}$$

where $\vec{\alpha}$, β are the Dirac matrices, $\vec{p}$ is the momentum, M is the mass, $\vec{r}$ is the radial coordinate, and the velocity of light is set equal to unity, c =1. The Dirac Hamiltonian has spin symmetry when the difference of the vector and scalar potentials in the Dirac Hamiltonian is a constant, $V_S(\vec{r})$ - $V_V(\vec{r}) = C_s$ [9]. Hadrons [10] and anti-nucleons in a nuclear environment have spin symmetry [11]. These are relativistic systems and normally, in such systems, we would expect large spin-orbit splittings, but, in this limit, spin doublets are degenerate.

3.1. Pseudospin Symmetry: A Symmetry of the Dirac Hamiltonian

In Ref [9], another SU(2) symmetry of the Dirac Hamiltonian was discovered to occur when the sum of the vector and scalar potentials in the Dirac Hamiltonian is a constant, $V_S(\vec{r}) + V_V(\vec{r}) = C_{ps}$. This symmetry was shown to be pseudospin symmetry [12]. The potentials can be spherical [1, 2], deformed [5], or triaxial, [6]. The generators for this SU(2) algebra $\tilde{\vec{S}}$, which commute with the Dirac Hamiltonian with any potentials that satisfy these conditions,$[H, \tilde{\vec{S}}] =$ 0, are given by [13]

$$\tilde{\vec{S}} = \begin{pmatrix} U_p \, \vec{s} \, U_p & 0 \\ 0 & \vec{s} \end{pmatrix}. \tag{2}$$

The eigenfunctions of the Dirac Hamiltonian in this limit will have degenerate doublets of states, one of which has pseudospin aligned and the other with pseudospin unaligned. The non-relativistic part of these relativistic pseudospin generators (upper quadrant of the matrix) are the generators proposed in [8].

These conditions hold for spherical, deformed, and triaxial potentials. If, in addition, the potentials are spherical, $V_S(\vec{r}) = V_S(r)$ and $V_V(\vec{r}) = V_V(r)$, then the pseudo-orbital angular momentum is conserved as well [13]. Its generators are given by

$$\tilde{\vec{L}} = \begin{pmatrix} U_p \, \vec{\ell} \, U_p & 0 \\ 0 & \vec{\ell} \end{pmatrix}, \tag{3}$$

where $\vec{\ell} = \frac{(\vec{r} \times \vec{p})}{\hbar}$.

The "upper" components of the degenerate doublets have the same quantum numbers of the spherical pseudospin doublets and the deformed pseudospin doublets. The "upper" matrix of the pseudospin generators in Eqs (2, 3), $U_p \, \vec{s} \, U_p$ and $U_p \, \vec{\ell} \, U_p$, have the spin intertwined with the momentum which enables the generators to connect the states in the doublet, which differ by two units of angular momentum and one unit of radial quantum number.

Hence pseudospin symmetry is a relativistic symmetry. The approximate equality in magnitude of the vector and scalar fields in nuclei and their opposite sign have been confirmed in relativistIc mean field theories [3] and in QCD sum rules [14, 3].

4. Consequences of Relativistic Pseudospin Symmetry

One immediate consequence of pseudospin symmetry as a relativistic symmetry is that the "lower" matrix of the pseudospin generators in Eqs (2, 3), $\vec{s}$ and $\vec{\ell}$ do not change the radial wavefunction of the "lower" component of the Dirac eigenfunctions. Hence this symmetry predicts that the radial wavefunctions of the "lower" component is the same for the two states in the doublet. Previous to this discovery many relativistic mean field calculations of nuclear properties have been made. Hence this prediction was tested with existing calculations and, indeed, these wavefunctions are very similar for both spherical [15, 16] and deformed nuclei [17, 18]. Because of the momentum dependence of the "upper" matrix of the generators the relationship between the "upper" components involves a differential equation but these have also been tested in spherical [19] and deformed nuclei [18] with success. There are also predictions about magnetic dipole transitions and Gamow-Teller transitions [20, 21], quadrupole transitions [3], and scattering matrices [22, 23]. These and other theoretical and experimental tests of pseudospin symmetry are not discussed in this paper because of limitation of space [3].

5. The Dirac Hamiltonian with Pseudo-U(3) Symmetry

It is well known that the Schrodinger equation with a non-relativistic spherical harmonic oscillator potential has a U(3) symmetry [24]. The energies depend linearly on the total number of harmonic oscillator quanta, $N = 2n + \ell$, and the generators of the U(3) symmetry are the angular momentum operator, a quadrupole operator, and a monopole operator which commute with the non-relativistic Hamiltonian. We know that the Dirac Hamiltonian for which the difference of the scalar and vector potentials is a constant has spin symmetry. But if, in addition, the potentials are spherical harmonic oscillators, the Dirac Hamiltonian can be solved analytically [25] and has an U(3) symmetry [26, 27]. The energies depend on the same harmonic quantum number N as the non-relativistic harmonic oscillator, and, therefore has the same degeneracies. However, the energy dependence on N depends on the ratio of the M to the strength of the potential. For M large, the dependence is linear just as in the non-relativistic oscillator. For M small, the energy goes like $N^{\frac{2}{3}}$, which means that, in the relativistic limit, the oscillator is not harmonic. Hence even a relativistic system can have U(3) symmetry.

If, on the other hand, the sum of the vector and scalar potentials is a constant and are spherical harmonic oscillators, the Dirac Hamiltonian, $\tilde{H}$, has a pseudospin symmetry and a pseudo-U(3) symmetry. The energy depends on the total number of pseudo-quanta, $\tilde{N} = 2\tilde{n} + \tilde{\ell}$. The quantum numbers, $\tilde{N}, \tilde{n}, \tilde{\ell}$, refer to the "lower" component of the Dirac eigenfunction. In terms of the non-relativistic quantum numbers, $\tilde{n} = n$ of the spin unaligned partner. The eigenvalue equation in the pseudospin limit is [3, 27]

$$\sqrt{E_{\tilde{N}} - M}\ (E_{\tilde{N}} + M) = \sqrt{2\ \hbar^2\omega^2 M}\ (\tilde{N} + \frac{3}{2}). \tag{4}$$

This eigenvalue equation is solved on Mathematica,

$$E_{\tilde{N}} = \frac{M}{3}\left[3B(A_{\tilde{N}}) - 1 + \frac{4}{3\ B(A_{\tilde{N}})}\right], \tag{5}$$

where $B(A_{\tilde{N}}) = \left[\frac{A_{\tilde{N}} + \sqrt{A_{\tilde{N}}^2 + \frac{32}{27}}}{2}\right]^{\frac{2}{3}}$, and $A_{\tilde{N}} = \frac{\sqrt{2}h\omega}{M}(\tilde{N} + \frac{3}{2})$. The spectrum goes like $\tilde{N}$ for M small and $\tilde{N}^{\frac{2}{3}}$ for M large; i.e., the relativistic harmonic oscillator is not harmonic.

The pseudo-U(3) generators which commute with the Dirac Hamiltonian, are the pseudo-orbital angular momentum given in Eq (3), and the

$$\tilde{Q}_m = \sqrt{\frac{3}{M\omega^2\hbar^2(\tilde{H}-M)}} \begin{pmatrix} M\omega^2\ ([pp]^{(2)}_m & \vec{\sigma}\cdot\vec{p}\ \not{M}\ ^2\ [rr]^{(2)}_m \\ M\omega^2\ [rr]^{(2)}_m\ \vec{\sigma}\cdot\vec{p} & \not{M}\ ^2\ (M\omega^2\ r^2 - 2M)[rr]^{(2)}_m\ + [pp]^{(2)}_m \end{pmatrix}, \quad (6)$$

along with the pseudo-quantum number operator

$$\tilde{\mathcal{N}} = \frac{\sqrt{\tilde{H}-M}(\tilde{H}+M)}{\hbar\sqrt{2M\omega^2}} - \frac{3}{2}. \quad (7)$$

These generators satisfy the commutation relations of a U(3) algebra,

$$[\tilde{\vec{L}}, \tilde{\vec{L}}]^{(t)} = -\sqrt{2}\ \tilde{\vec{L}}\ \delta_{t,1},\ [\tilde{\vec{L}}, \tilde{Q}]^{(t)} = -\sqrt{6}\ \tilde{Q}\ \delta_{t,2},\ [\tilde{Q}, \tilde{Q}]^{(t)} = 3\sqrt{10}\ \tilde{\vec{L}}\ \delta_{t,1}, \quad (8)$$

$$[\tilde{\mathcal{N}}, \tilde{\vec{L}}] = [\tilde{\mathcal{N}}, \tilde{Q}_m] = 0. \quad (9)$$

In the above we use the coupled commutation relation between two tenors, $T_1^{(t_1)}, T_2^{(t_2)}$ of rank t_1, t_2 which is $[T_1^{(t_1)}, T_2^{(t_2)}]^{(t)} = [T_1^{(t_1)} T_2^{(t_2)}]^{(t)} - (-1)^{t_1+t_2-t}[T_2^{(t_2)} T_1^{(t_1)}]^{(t)}$ [28]. Theses generators, along with the peudospin, commute with the Hamiltonian

$$[\tilde{H}, \tilde{\vec{S}}] = [\tilde{H}, \tilde{\vec{L}}] = [\tilde{H}, \tilde{Q}_m] = [\tilde{H}, \tilde{\mathcal{N}}] = 0, \quad (10)$$

The pseudospin generators commute with the pseudo-U(3) generators as well as the Dirac Hamiltonian, and so the invariance group is pseudo-U(3)×pseudo-SU(2), where the pseudo-SU(2) is generated by the pseudospin generators. Hence pseudo-U(3) is a relativistic symmetry.

However, the eigenfunctions of $\tilde{H}$ are not valence Dirac eigenfunctions, but hole Dirac eigenfunctions. Or in common parlance, "negative" energy states. The energies are not negative, though, since the harmonic oscillator is not finite. But the nodal structure of the eigenfunctions is that of hole states [3]. There are no bound valence states of $\tilde{H}$ and hence these solutions do not correspond to realistic nuclei. Of course, for real nuclei, the scalar and vector potentials are not exactly equal in magnitude, and so pseudospin is approximately conserved since the ratio of the sum of the potentials to their difference is approximately .03. For this reason we are attempting to solve analytically the Dirac Hamiltonian for general scalar and vector harmonic oscillator potentials.

Because, in general, the pseudospin symmetry limit has no bound valence Dirac eigenfunctions, it is not possible to do perturbation theory for approximate pseudospin symmetry around the exact pseudospin symmetry limit. Since this feature is unique to pseudospin symmetry, pseudospin symmetry is an interesting topic to understand better.

6. Summary

We have shown the pseudospin symmetry and pseudo-U(3) for which Jerry has made many contributions are relativistic symmetries. This is surprising since nuclei have been described well for most properties by non relativistic quantum mechanics. Perhaps there is a more fundamental rationale for pseudospin symmetry at the QCD level.

Acknowledgments

This research was supported by the US Department of Energy, Contract No. W-7405-ENG-36.

References

[1] Arima A Harvey M Shimizu K 1969 *Phys. Lett. B* **30** 517
[2] Hecht K Adler A 1969 *Nucl. Phys. A* **137** 129
[3] Ginocchio J N 2005 *Phys.Rep.* **414** 165
[4] Sorlin O Porquet M 2008 *Prog. Part. Nucl. Phys.* **61** 602
[5] Bohr A Hamamoto I Mottelson B 1982 *Phys. Scr.* **26** 267
[6] Beuschel T Blokhin A L Draayer J P 1997 *Nucl. Phys.* **619** 119
[7] Bahri C Draayer J P Moszkowski S A 1992 *Phys. Rev. Lett.* **68** 2133
[8] Blokhin A L Bahri C Draayer J P 1995 *Phys. Rev. Lett.* **74** 4149
[9] Bell J S Ruegg H 1997 *Nucl.Phys. B* **98** 151
[10] Page P R Goldman T Ginocchio J N 2001 *Phys.Rev. Lett.* **86** 204 (2001)
[11] Ginocchio J N 1999 *Phys.Rep.* **315** 231
[12] Ginocchio J N 1997 *Phys.Rev. Lett.* **78** 436
[13] Ginocchio J N Leviatan A 1998 *Phys. Lett. B* **425** 1
[14] Cohen T D *et al*, 1995 *Prog. Part. Nucl. Phys.* **35** 221
[15] Ginocchio J N Madland D 1998 *Phys. Rev. C* **57** 1167
[16] Meng J Sugawara-Tanabe K Yamaji S Ring P Arima A 1998 *Phys. Rev. C* **58** R628
[17] Sugawara-Tanabe K Yamaji S and A. Arima, 2002 *Phys. Rev. C* **65** 054313
[18] Ginocchio J N Leviatan A Meng J Zhou S G 2004 *Phys. Rev. C* **69** 03430
[19] Ginocchio J N 2002 *Phys. Rev. C* **66** 064312
[20] Ginocchio J N 1999 *Phys. Rev. C* **59** 2487
[21] von Neumann-Cosel P Ginocchio J N 2000 *Phys. Rev. C* **62** 014308
[22] Ginocchio J N 1999 *Phys. Rev. Lett.* **82** 4599
[23] Leeb H Sofianos S A 2004 *Phys. Rev. C* **69** 054608
[24] Elliott J P 1958 *Proc. R. Soc. London, Ser. A* **245** 128
[25] Ginocchio J N 2004 *Phys. Rev. C* **69** 034318
[26] Ginocchio J N 2005 *Phys. Rev. Lett.* **95** 252501
[27] Ginocchio J N 2011 *Journal of Physics* **267** 012037
[28] French J B 1966 *Many-Body Description of Nuclear Structure and Reactions*, edited by C. Bloch (New York: Academic Press)

The Oblique Basis Method from an Engineering Point of View

V G Gueorguiev
California State University Stanislaus, Turlock, CA 95382
E-mail: vesselin@mailaps.org

"Take the world from another point of view." - R. Feynman

Abstract. The oblique basis method is reviewed from engineering point of view related to vibration and control theory. Examples are used to demonstrate and relate the oblique basis in nuclear physics to the equivalent mathematical problems in vibration theory. The mathematical techniques, such as principal coordinates and root locus, used by vibration and control theory engineers are shown to be relevant to the Richardson - Gaudin pairing-like problems in nuclear physics.

1. Introduction

It is well known and commonly experienced observation that "... mathematical concepts turn up in entirely unexpected connections ..." [1]. The surprise of finding a mathematical structure applicable to seemingly unrelated fields of study is somehow proportional to our ignorance of one or the other subject, even though we may have studied the underlining mathematical field previously. The following contribution attests to this situation by drawing the analogies and pointing examples from my research topic in nuclear physics and a more recent engineering course in vibration and control theory which I had the pleasure to teach. This reinforces my view that one should keep an open mind and must often try to "Take the world from another point of view" [2]. In this respect, this article is devoted to the discussion of two major applications of a linear algebra: matrix diagonalization and location of zeros of complex algebraic systems. The first application is in the subject of complex many-body systems like nuclei [3] while the second is their application in the filed of vibration and control theory [4]. Although presenting the classical mechanics problem followed by its quantum mechanics framework within which the mathematical tools reside may be the better pedagogical point of view, we will use the opposite order of the exposition, due to the fact that the audience to which the material was presented is more familiar with my nuclear physics research framework.

In what follows, the article justifies the oblique basis approach to complex problems and introduces the two important collective models commonly used in nuclear physics. Then the oblique basis method is stated as it is pertained to nuclear physics. The engineering problem is stated next and the relevant mathematical techniques, such as Evans root-locus method [5] for studying the stability of a system, are outlined. Finally, we go back to the nuclear physics subject and show the relevance of the Evans root-locus method for the study of exactly solvable nuclear pairing interaction.

Before continuing the article with its main subject, I would like to acknowledge that, even though the oblique basis idea has been developed mostly as a collaborative effort [3, 6, 7], in my view it has been stimulated and supported strongly by my Ph.D. adviser Professor Jerry Draayer, in whose honor this conference has been held.

2. Important Exactly Solvable 2-body Nuclear Interactions

Exactly solvable problems such as harmonic oscillator and other group-symmetry based models in physics have been essential in the qualitative and quantitative understanding of physics phenomena. The need of better, more accurate, and easily replicable quantitative results has naturally introduced the use of powerful computers to the physics research. Never the less, exactly solvable problems are still essential to our understanding of nuclear structure [8] and reactions [7], especially in the limit of weak perturbation to the exact limit [6]. In particular, the nuclear shell-model in the limit of independent single particle motion is exactly solvable; however, when a two-body perturbation interaction is added then it becomes the cornerstone of the configuration mixing approach to the corresponding many-body problem [8]. When the interaction between the nucleons becomes significant the perturbation method is no-longer applicable and cannot reproduce the correct behavior of the system. Examples of such cases are well-deformed nuclei where the rotational structure of the spectra becomes more prominent or near-closed-shell nuclei where pairing phenomenon take over the low-energy spectra of the nuclei.

By using Fock space techniques within the independent particle model with traditional single-particle level labels, one can consider two important collective interactions in nuclei. These interactions represent two well known exactly solvable systems:

- the Elliott's SU(3) model with degenerate single-particle harmonic oscillator energies ($\varepsilon_i = \varepsilon_j = \varepsilon$) and quadrupole-quadrupole interaction $Q \cdot Q$ [8]:

$$H = \varepsilon N - \frac{\chi}{2} Q \cdot Q,$$

- the nuclear pairing (P^+P) interaction with non-degenerate single-particle energies (ε_i):

$$H = \varepsilon_i N_i + G P^+ P.$$

3. T=1 Proton-Neutron Pairing - SO(5) RG-model

While the exact solvability of the Elliott's SU(3) model is more or less well understood from group-theoretical point of view, the exact solvability of the particle number conserving pairing interaction has much less known implementations, even though the solution has been published since 1962 by Richardson [9]. The exploration and implementation of the Richardson-Gaudin particle number conserving pairing interaction may have been hindered by its more complicated algebraic structure as seen in the work of Gaudin [10] and Ushveridize [11]. However, the advancement of the computer power and recent understanding of the relevant mathematical and computational structure has brought a renewed interest in the topic [12, 13] along with the understanding of the integrability of the pairing interaction [14, 15]. The pairing has even been extended to an A-body generalized pairing interaction [16]. Finally, the full T=1 proton-neutron pairing has been formulated within the frame work of the Richardson-Gaudin algebras [17].

Here we briefly summarize the main results on the Richardson-Gaudin pairing with the purpose of pointing out some of the possible practical implementation issues. We recall the 10 generators of the SO(5) algebra in a representation suited for nuclear physics problems [17]. The three $T = 1$ pair-creation operators are defined as: $\hat{b}^\dagger_{-1,i} = \hat{n}^\dagger_i \hat{n}^\dagger_i$, $\hat{b}^\dagger_{0,i} = (\hat{n}^\dagger_i \hat{p}^\dagger_i + \hat{p}^\dagger_i \hat{n}^\dagger_i)/\sqrt{2}$, and $\hat{b}^\dagger_{+1,i} = \hat{p}^\dagger_i \hat{p}^\dagger_i$, where n and p refer to neutrons and protons, respectively, and i labels a single-particle basis (with $\bar{i}$ its time-reversed state) that may be associated with the spherical shell-model basis $i \equiv jm$ or with an axially-symmetric deformed basis $i \equiv \alpha m$. The three pair-annihilation operators are $\hat{b}_{-1,i}$, $\hat{b}_{0,i}$, and $\hat{b}_{+1,i}$. The three components of the isospin operator $[\hat{T}_{+,i} = (\hat{p}_i^\dagger \hat{n}_i + \hat{p}_{\bar{i}}^\dagger \hat{n}_{\bar{i}})/\sqrt{2}$, $\hat{T}_{0,i} = (\hat{p}_i^\dagger \hat{p}_i + \hat{p}_{\bar{i}}^\dagger \hat{p}_{\bar{i}})/2 - (\hat{n}_i^\dagger \hat{n}_i + \hat{n}_{\bar{i}}^\dagger \hat{n}_{\bar{i}})/2$, and $\hat{T}_{-,i} = (\hat{n}_i^\dagger \hat{p}_i + \hat{n}_{\bar{i}}^\dagger \hat{p}_{\bar{i}})/\sqrt{2}]$ close the $\mathrm{SU}_T(2)$ subalgebra of SO(5). These 9 operators together with the number operator $\hat{N}_i = \hat{p}_i^\dagger \hat{p}_i + \hat{p}_{\bar{i}}^\dagger \hat{p}_{\bar{i}} + \hat{n}_i^\dagger \hat{n}_i + \hat{n}_{\bar{i}}^\dagger \hat{n}_{\bar{i}}$ define the SO(5) algebra. The pn-pairing Hamiltonian with a pairing strength g and isospin breaking Δ is given by the expression [17]:

$$\hat{H} = \sum_j \varepsilon_j \left(\hat{N}_j + \Delta \hat{N}_{p,j}\right) + \frac{g}{2}\hat{T} \cdot \hat{T} + g \sum_{\mu j m j' m'} \hat{b}^\dagger_{\mu,jm} \hat{b}_{\mu,j'm'}, \tag{1}$$

with the eigenvalues of (1) given by the expression:

$$E = \sum_{\alpha=1}^{M} e_\alpha + \frac{\Delta}{2} \sum_{\beta=1}^{M+T_0+t} \omega_\beta + \sum_j \varepsilon_j \left[\frac{v_j}{2}(2+\Delta) - \Delta t_j\right] + \frac{g}{2} T_0(T_0 - 1), \tag{2}$$

where v_i is the seniority of each i level, *i.e.* the number of fermions not paired in time-reversed states with isospin $T = 1$, t_i is the isospin of the unpaired fermions, $t = \sum_i t_i$, M is the number of $T = 1$ time-reversed pairs, and T_0 is the z-axes component of the total isospin, *i.e.* the eigenvalue of the operator $\hat{T}_0 = \sum_i \hat{T}_{0,i}$. The total number of nucleons is $N = N_p + N_n = 2M + \sum_i v_i$ whereas their difference is $N_p - N_n = 2T_0$. The quantum numbers M, T_0, v_i, and t_i are conserved; T is also conserved if $\Delta = 0$.

The variables e_α and ω_β are solutions of the corresponding Richardson-Gaudin equations [17]:

$$\frac{1}{g} = \frac{1}{2}\sum_{i=1}^{L} \frac{v_i + 2t_i - 2}{z_i - e_\alpha} + \sum_{\alpha'(\neq\alpha)=1}^{M} \frac{2}{e_{\alpha'} - e_\alpha} - \sum_{\beta=1}^{M+T_0+t} \frac{1}{\omega_\beta - e_\alpha}, \tag{3}$$

$$\frac{\Delta}{g} = \sum_{\beta'(\neq\beta)=1}^{M+T_0+t} \frac{2}{\omega_{\beta'} - \omega_\beta} - \sum_{\alpha=1}^{M} \frac{2}{e_\alpha - \omega_\beta} - \sum_{i=1}^{L} \frac{2t_i}{z_i - \omega_\beta}. \tag{4}$$

where variables z_j are related to the single particle energies ε_j ($z_j = 2\varepsilon_j$) and the indexes are specialized to a spherical basis $i \equiv jm$.

Thus, the problem of solving the pairing hamiltonian is reduced to solving the above non-linear equations for e_α and ω_β. In practical terms, without computers in the old days, this would mean hard work by hand. Nowadays, there are sufficient computer power to solve these equations by starting at a small values of coupling constants g and Δ and gradually increasing these values to the desired magnitudes. However, there is an apparent singularity problem when some of the pair energies become close to the a single particle energy level, that is, when some of the denominators get close to zero. This problem can be resolved by trial and error search on the values pass the singularity point - this is similar to the problem of finding initial configurations for the values of e_α and ω_β to start the process with. Alternatively, one can use small complex component of the single particle energies or the Evans root-locus method to be touched upon in the last section.

4. Oblique Basis in Nuclear Physics

In second quantized form the effective Hamiltonian for a nuclear system is:

$$H = \sum_i \varepsilon_i a_i^+ a_i + \frac{1}{4} \sum_{i,j,k,l} V_{ij,kl} a_i^+ a_j^+ a_k a_l. \tag{5}$$

Here, ε_i are single-particle energies derived from excitation spectra of one valence particle system and $V_{kl,ij}$ are two-body matrix elements viewed as a residual interaction beyond the single-particle mean-field confining potential.

When the residual interaction can be treated as perturbation then one can apply perturbation theory using the basis of the exactly solvable limit. If there are two or more exactly solvable limits then there are naturally two or more possible perturbation treatments that are based on their corresponding basis sets. However, if the strengths of the interactions are such that perturbation theory is not applicable, that is, there is a strong mixing, then a combination of relevant basis sets can be used to approach the problem. Thus, the oblique basis consists of two or more basis sets. In this sense the oblique basis idea has been applied to the case when the Elliott's SU(3) is broken due to the non-degenerate single-particle energies [3] and also illustrated on simple toy model of a particle in a box [6].

In our simple example [3] the two basis sets were: spherical shell-model states (ssm-states) expressed in spherical single-particle coordinates (nlj) that are eigenvectors of the single particle non-interaction hamiltonian, while the second set has a good SU(3) structure (su3-states) that track nuclear deformation. The SU(3) basis set is given in cylindrical single-particle coordinates. Such oblique calculations result in a nonorthogonal oblique basis and require an evaluation of the matrix elements of physical operators plus a knowledge of the scalar product $(e_\alpha \cdot E_i)$ related to the overlap matrix. Schematically, these basis vectors are labeled by $\alpha = 1,\ldots,$ dim(ssm-basis) and $i = 1,\ldots,$ dim(su3-basis) and the corresponding matrixes can be represented in the following way:

$$\text{basis vectors} \quad : \quad \begin{pmatrix} e_\alpha : \text{ssm} & - & \text{basis} \\ E_i : \text{su3} & - & \text{basis} \end{pmatrix}, \tag{6}$$

$$\text{overlap matrix} \quad : \quad \Theta = \begin{pmatrix} \mathbf{1} & \Omega \\ \Omega^+ & \mathbf{1} \end{pmatrix}, \qquad \Omega_{\alpha i} = e_\alpha \cdot E_i, \tag{7}$$

$$\text{Hamiltonian matrix} \quad : \quad H = \begin{pmatrix} H_{ssm\times ssm} & H_{ssm\times su3} \\ H_{su3\times ssm} & H_{su3\times su3} \end{pmatrix} = \begin{pmatrix} H_{\alpha\beta} & H_{\alpha j} \\ H_{i\beta} & H_{ij} \end{pmatrix}. \tag{8}$$

In general the oblique basis results in a non-orthogonal basis that may even become linearly dependent and over-complete. Thus, the matrix equation includes an overlap matrix $(\Theta_{ij} = \langle i|j\rangle)$ and has the form:

$$\sum_j (H_{ij}v_j - \lambda\Theta_{ij}v_j) = 0. \tag{9}$$

In analogy to the SU(3) symmetry breaking in the pf-shell, which provides symmetry-adapted guidance to the relevant collective basis, one can expect that in a multi-shell space one would naturally use the symplectic symmetry group [18, 19].

Furthermore, the oblique basis method can be extended and applied to the study of particle transfer reactions in strongly deformed nuclei where harmonic oscillator bound states techniques can be employed to compute the single particle binding energies in mean-field Woods-Saxon potential and then the proper tail of the wave function can be constructed using Sturmian basis that results in an equivalent oblique basis [7].

5. The Engineering Problems in Vibrations and Control Theory

A system in mechanical engineering is often represented by a set of masses, springs, and dumpers [4]. The state of the system corresponds to a vector $\vec{x}$ that describes the coordinates of the masses. For our simple example we can neglect the dumper effects which usually corresponds to velocity dependent force $[c]\dot{\vec{x}}$. In this case the Newton's second law of motion is written as a matrix equation using mass $[m]$ and spring $[k]$ matrices:

$$[k]\vec{x} + [m]\ddot{\vec{x}} = 0$$

5.1. Normal Modes for a Simple System
By using the method of separation of variables and the reasonable assumption that the system will exhibit some periodic motion near its equilibrium position, one sets $\vec{x} = \vec{X}T(t)$ and $\ddot{T} = -\omega^2 T$. The resulting equation for the frequency of oscillations ω becomes effectively a generalized eigenvalue equation when the mass matrix is not diagonal. Simple example of such non-diagonal $[m]$ is a wagon with a compound pendulum inside [4].

Solutions ω_j and vectors $\vec{X}_i$ to the generalized eigenvalue equation:

$$[k]\vec{X}_i - \omega_i^2[m]\vec{X}_i = 0, \tag{10}$$

form a $[k]$ or $[m]$ "orthonormal basis" when $\omega_j \neq \omega_i$ ($\vec{X}_i^{\,T}[m]\vec{X}_j = 0$ and $\vec{X}_i^{\,T}[k]\vec{X}_j = 0$). Thus, in terms of the normal mood matrix $\hat{X} = \{\vec{X}_1, \vec{X}_2, \ldots, \vec{X}_n\}$ the super-matrices: $M = \hat{X}^T[m]\hat{X}$ and $K = \hat{X}^T[k]\hat{X}$ become diagonal. Since the mass matrix is positive definite due to its relation to the kinetic energy of the system, then one can consider $[m]$ normalized normal modes $\vec{X}_i^{\,T}[m]\vec{X}_j = \delta_{ij}$ which results in a diagonal super-matrices $M_{ij} = \delta_{ij}\omega_i^2$. In terms of the normal modes the system of n-coupled masses becomes a system of n-uncoupled masses which is equivalent to the study of n one-dimensional oscillators. Identifying the normal modes of a mechanical system is very useful once one introduces back general damping and other possible forces to the problem. Such additional forces can be treated in a perturbative like regime using the corresponding normal modes of the system.

5.2. Evans Root-Locus Method
How to influence the behavior of particular system to achieve the desired response is the subject of the Control Theory. In particular, by going back to our simple system but keeping only one degree of freedom we have a simple second-order differential equation. From the general theory of these equations, we know that if the system is displaced from its equilibrium there would be some oscillations and some damping that will determine the transition behavior. The response of the system to a Dirac delta input is the transfer function. Since any function can be viewed as a collection of Dirac-delta impulses the output to a general input is essentially related to the structure of the transfer function which is in essence the ratio of the output to the input signal but in the Laplace transform variable s. Usually systems have a proportional controller K and a response of the controlled plant G(s) that is a simple polynomial function of s which result in an overall transfer function $\mathcal{T}(s) = KG(s)$.

A system is stable if all the poles of the transfer function $\mathcal{T}(s)$ are in the left half complex plane. A pole that has a real negative value corresponds to a pure damping without oscillations while a pole with non zero imaginary part corresponds to an oscillatory behavior. If the system has a pole with a positive real value then the system become unstable and the amplitude of a small input signal can grow to infinity. To prevent such behavior many systems include sensors $H(s)$ and negative feedback loops that reduce such output grow. Including a negative feedback modifies the transfer function of the system to become $\mathcal{T}(s) = KG(s)/(1 + KG(s)H(s))$.

Finding how the poles of the transfer function move in the complex plane is essential for controlling the response of a system. Often the combination of transfer functions of various control components results in a transfer function that is a ratios of polynomials. Were the controller has a simple proportional response K to the input signal. The poles of the transfer function $\mathcal{T}(s) = KG(s)/(1+KG(s)H(s))$ are clearly related to the zeros of $1+KG(s)H(s) = 0$. Understanding the zeros as K changes is thus essential. If one sets $G(s)H(s) = L(s) = \frac{b(s)}{a(s)}$ then Evans root-locus method gives rules for the location of the roots of the equation:

$$L(s) = -\frac{1}{K}$$

as K grows from 1 to infinity. Since $L(s)$ is considered to be a proper fraction the polynomials $b(s) = \prod_{i=1}^{m}(s - z_i)$ and $a(s) = \prod_{i=1}^{n}(s - p_i)$, that is, the degree n of $a(s)$ is bigger or equal to the degree m of $b(s)$ then $L(s)$ has the form:

$$L(s) = \sum_{i=1}^{n} \frac{A_i}{s - p_i} = -\frac{1}{K}. \tag{11}$$

This equation the main equation of the stability analysis in control theory based on Evans root-locus method [5].

The above mathematical expression and its corresponding mathematical problem is practically the same as our Richardson-Gaudin pairing equations discussed earlier (3). In particular, if we restrict the pairing to the problem of identical particles. Then we get the Richardson original equations [9] that only contains the pair energies e_α and the single particle energies $z_i = 2\epsilon_i$ as some of the poles of L:

$$\frac{1}{g} = \sum_{i=1}^{L} \frac{1}{e_\alpha - z_i} + \sum_{\alpha'(\neq\alpha)=1}^{M} \frac{2}{e_{\alpha'} - e_\alpha} \tag{12}$$

6. Conclusion

In the preceding text we have discussed the use of non-orthogonal basis sets as they emerge in the studies of complex systems like atomic nuclei and mechanical systems. In studying the nuclei, the motivation is to deal with strongly coupled competing modes where the system is far from any of the possible perturbation regimes. The well-known quadrupole-quadrupole and pairing interactions are examples of collective modes that can compete against each other as well as against the single particle spin-orbit interaction which is often considered as the main source of lifting the single particle energy degeneracy. In a sense, each of these three exactly solvable limits represents a "normal" mode of the system and can be used to address the strong coupling regime of the system. The similarity between the principle normal modes of a mechanical system and the oblique basis in nuclear physics is further enhanced by the analogy between the Richardson-Gaudin pairing equations and the Evans root-locus method from the control theory for studying the satiability of a mechanical system. The author hopes that the reader will find the analogies interesting and this work will stimulate future collaborative research on these topics and their implementations to nuclear physics and engineering problems.

Acknowledgments

The author, V. Gueorguiev, acknowledges the generous financial support for traveling and attending the *Horizons of Innovative Theories, Experiments, and Supercomputing in Nuclear Physics* (HITES 2012) conference in honor of his PhD adviser Jerry P. Draayer from Louisiana State University, Louisiana, USA; as well as financial support provided by the Professional Development Fund for lecturers from University of California, Merced.

References

[1] Wigner E 1960 *Comm. Pure Appl. Math.* **13** 1 (John Wiley & Sons, New York)
[2] Feynman R 1973 *Take the World from Another Point of View* (Television interview, Yorkshire Great Britain)
[3] Gueorguiev V G, Ormand W E, Johnson C W, and Draayer J P 2002 *Phys. Rev.* C **65** 024314 (*Preprint* nucl-th/0110047)
[4] Rao S S 2011 *Mechanical Vibrations* (Prentice Hall)
[5] Evans W R 1948 *Trans. AIEE* **67** 547
[6] Gueorguiev V G, Rau A R P, and Draayer J P 2006 *Am. J. of Phys.* **74** 394 (*Preprint* math-ph/0512019)
[7] Gueorguiev V G, Kunz P D, Escher J E, Dietrich F S 2007 (*Preprint* nucl-th/0706.2002)
[8] Gueorguiev V G 2002 Mixed-Symmetry Shell-Model Calculations in Nuclear Physics *Louisiana State University Ph.D. thesis* (*Preprint* nucl-th/0704.1108)
[9] Richardson R W 1963 *Phys. Lett.* **3** 277; 1966 *Phys. Rev.* **144** 874
[10] Gaudin M 1976 *Le Journal De Physique* **37** 1087
[11] Ushveridize AG 1994 *Quasi-Exactly Solvable Models in Quantum Mechanics* (Institute of Physics, Bristol)
[12] Pan F, Draayer J P, and Ormand W E 1998 *Phys. Lett.* B **422** 1 (*Preprint* nucl-th/9709036)
[13] Pan F and Draayer J P 2002 *Phys. Rev.* C**66** 044314
[14] Links J, Zhou H, Gould M D, and McKenzie R H 2002 *J. Phys.* A**35** 6459
[15] Asorey M, Falceto F, and Sierra G 2002 *Nucl. Phys.* B**622** 593
[16] Pan F, Gueorguiev V G, and Draayer J P 2004 *Phys. Rev. Lett.* **92** 112503 (*Preprint* nucl-th/0311075)
[17] Dukelsky J, Gueorguiev V G, Isacker P, Dimitrova S, Errea B, Lerma S 2006 *Phys. Rev. Lett.* **96** 072503 (*Preprint* nucl-th/0601082)
[18] Dytrych T, Sviratcheva K D, Bahri C, Draayer J P, Vary J P 2007 *Phys. Rev. Lett.* **98** 162503 (*Preprint* nucl-th/0704.1108)
[19] Draayer J P, Dytrych T, Launey K D, Langr D 2012 *Prog. Part. Nucl. Phys.* **67** 516

Wigner-Racah Algebra, Binary Correlations and Trace Propagation for Embedded Random Matrix Ensembles

V K B Kota

Physical Research Laboratory, Ahmedabad 380 009, India

E-mail: `vkbkota@prl.res.in`

Abstract.
For embedded unitary ensembles with $SU(\Omega) \times SU(r)$ embedding and generated by random two-body (in some situations k-body) interactions preserving $SU(r)$ symmetry, analytical formulas, in terms of $SU(\Omega)$ Racah coefficients, are derived for the lower order moments of the one and two-point functions in eigenvalues. This formulation unifies all known results and new results are derived for $r = 3$. Similarly, the method of binary correlation approximation developed for spinless identical fermion systems, has been extended to proton-neutron systems and it is used to show that the bivariate transition strength density appropriate for neutrinoless double beta decay will be close to a bivariate Gaussian. In addition, trace propagation methods of French are also used to derive analytical results for embedded ensembles and they have opened a new window to understand regular structures generated by random interactions.

1. Introduction

The GOE, now almost universally regarded as a model for a corresponding chaotic system [1] is an ensemble of multi-body, not two-body interactions. This difference, as pointed out first by J.B. French, shows up both in one-point (density of states) and two-point (fluctuations and smoothed transition strengths) functions generated by nuclear shell model. This led to the introduction of random matrix ensembles generated by random interactions. For random two-body interactions, we have GOE/GUE/GSE Hamiltonian matrix ensembles in two-particle spaces (similarly in k-particle spaces for k-body interactions) and these are then propagated to m particle spaces, using the geometry of m particle spaces, giving random matrix ensembles in many-particle spaces. As in these ensembles, in many-particle spaces, classical ensembles are embedded, these are generically called (introduced in 1975) embedded random matrix ensembles or in short, embedded ensembles (EE) [2]. With two-body interactions having shell model J (or JT) symmetry, these are called (introduced first in 1970), two-body random matrix ensembles (TBRE) [3, 4].

In the last 15 years strong motivation for understanding, analyzing and applying EE has come from: (i) statistical spectroscopy applications in nuclei and atoms [5]; (ii) experiments showing generic electron-electron interaction effects on transport properties of mesoscopic systems [6]; (iii) regular structures generated by random interactions in quantum many-body systems [7, 8]; (vi) possible applications to quantum gases and in quantum information science (thermalization, entanglement etc.) [9, 10]. Because of these, large variety of EE have been explored [11-22].

Most of the EE are based on symmetries as shown in Fig. 1. In 1972 Dyson has laid down criterion for the validity of a random matrix ensemble and they are [23]: (1) it predicts a global level-density distribution in agreement with observation; (2) it predicts a local statistical behavior of energy levels in agreement with Wigner distribution; (3) its definition is physically plausible; (4) its consequences are amenable to mathematical treatment. It is well established that EE satisfy (1), (2) and (3). A long standing question for EE is with regard to (4) and there is some progress using three different approaches as discussed in the next three sections. Most important result is that the Wigner-Racah algebra of the Lie algebra that generates the embedding in principle leads to analytical solubility of EE. Now we will give a preview.

Analytical approaches for EE are based on (i) Wigner-Racah algebra of the embedding algebra; (ii) extension of binary correlation approximation, first used by Wigner for GOE, giving asymptotic results; (iii) trace propagation over spaces defined by group symmetries. Results of (i) and (ii) with new examples are given in Sections 2 and 3 respectively. Section 4 gives an overview of (iii) and finally Section 5 gives conclusions. It should be emphasized that in (i) and (iii) employed are the results of two seminal papers by Draayer with Hecht [24] and Rosensteel [25] and in addition, it is conjectured that further progress in (ii) may come from another seminal paper by Draayer with French [26].

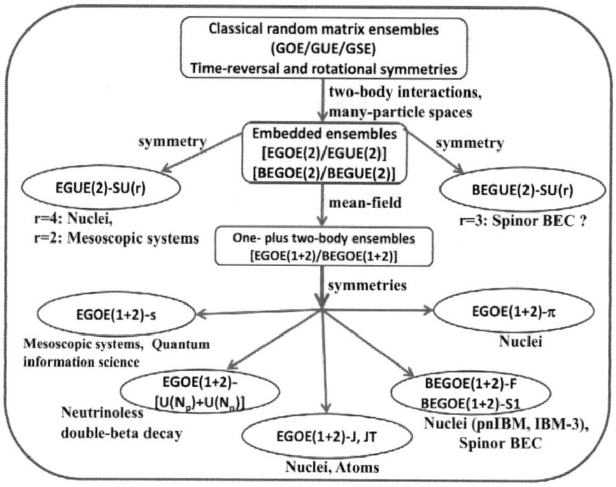

Figure 1. Embedded random matrix ensembles with symmetries. EGOE/EGUE correspond to fermionic systems and BEGOE/BEGUE correspond to bosonic systems. Physical systems where a given ensemble applies are mentioned below the ensemble. See text for other details.

2. Results from Wigner-Racah algebra of the embedding algebra: EGUE(2)-$SU(r)$

Consider a system of m fermions or bosons in Ω number of sp levels each r-fold degenerate. Then the SGA is $U(r\Omega)$ and a subalgebra of considerable interest is $U(r\Omega) \supset U(\Omega) \otimes SU(r)$ algebra. For random two-body Hamiltonians preserving $SU(r)$ symmetry, one can introduce embedded GUE with $U(\Omega) \otimes SU(r)$ embedding and this ensemble is called EGUE(2)-$SU(r)$. Ensembles with $r = 1$, 2 and 4 for fermions correspond to spinless fermions, fermions with spin ($\mathbf{s}$) as appropriate for mesoscopic systems and Wigner's spin-isospin $SU(4)$ symmetry used in nuclear structure respectively [18, 19, 20]. Similarly, for bosons $r = 1$, 2 and 3 correspond to spinless bosons, bosons with a fictitious spin (F) as appropriate for two spices boson systems and spin one bosons ($S = 1$) as appropriate for spinor BEC [18, 21, 15, 17]. It is important to note that the distinction between fermions and bosons is in the $U(\Omega)$ irreducible representations (irreps) that one has to consider [for m number of fermions the $U(r\Omega)$ irrep is $\{1^m\}$ and for m bosons it is $\{m\}$]. Now we will define EGUE(2)-$SU(r)$. Firstly, a general two-body Hamiltonian operator $\widehat{H}$ preserving $SU(r)$ symmetry can be written as, with $A^\dagger$ and A representing normalized two-particle creation and annihilation operators,

$$\widehat{H} = \widehat{H}_{\{2\}} + \widehat{H}_{\{1^2\}} = \sum_{f_2, v_2^i, v_2^f, \beta_2; f_2 = \{2\},\{1^2\}} H_{f_2 v_2^i v_2^f}(2)\; A^\dagger(f_2 v_2^f \beta_2)\, A(f_2 v_2^i \beta_2)\;. \tag{1}$$

Here f_2 are two-particle $SU(\Omega)$ irreps $\{2\}$ and $\{1^2\}$, in Young tableaux notation and they define uniquely the corresponding $SU(r)$ irreps $f_2^{(r)}$. Similarly, v_2's are f_2 sub-labels and β's belong to $f_2^{(r)}$. The two-body matrix elements (TBME) $H_{f_2 v_2^i v_2^f}(2) = \left\langle f_2 v_2^f \beta_2 \mid H \mid f_2 v_2^i \beta_2 \right\rangle$ are independent of the β_2's and the uniform sum over them ensures $SU(r)$ symmetry. Note that m particle states are by denoted by $\left| f_m v_m^f \beta_m^F \right\rangle$; $f_m^{(r)} = \widetilde{f}_m$ for fermions and $f_m^{(r)} = f_m$ for bosons. As $\widehat{H}$ is a $SU(r)$ scalar, the m particle H matrix will be a direct sum of matrices with each of them labeled by the f_m's with dimension $d_\Omega(f_m)$,

$$H(m) = \sum_{f_m} H_{f_m}(m) \oplus\;, \quad d_\Omega(f_m) = \prod_{i<j=1}^{\Omega} \frac{f_i - f_j + j - i}{j - i}\;. \tag{2}$$

Embedded random matrix ensemble EGUE(2)-$SU(r)$ for a $(m,\, f_m)$ system, i.e. $\{H_{f_m}(m)\}$, is generated by the ensemble of H operators given in Eq. (1) with $H_{\{2\}}(2)$ and $H_{\{1^2\}}(2)$ matrices replaced by independent GUE ensembles of random matrices so that the random variables defining the real and imaginary parts of the matrix elements of $H_{f_2}(2)$ are independent Gaussian variables with zero center and variance given by (with bar representing ensemble average) $\overline{H_{f_2 v_2^1 v_2^2}(2)\; H_{f_2' v_2^3 v_2^4}(2)} = \delta_{f_2 f_2'} \delta_{v_2^1 v_2^4} \delta_{v_2^2 v_2^3} (\lambda_{f_2})^2$. Typically, one is first interested in deriving the ensemble averaged forms of one (ρ) and two-point (S) functions in eigenvalues. These are derived via their moments M_p and Σ_{PQ} respectively,

$$\begin{aligned}
&\overline{\rho^{m,f_m}(E)} = \overline{\langle \delta(H - E) \rangle^{f_m}} \Longleftrightarrow M_p = \overline{\langle H^p \rangle^{f_m}}\;,\\
&S^{mf_m:m'f_{m'}}(E, E') = \overline{\rho^{m,f_m}(E)\rho^{m',f_{m'}}(E')} - \overline{\rho^{m,f_m}(E)}\;\; \overline{\rho^{m',f_{m'}}(E')} \Longleftrightarrow\\
&\widetilde{\Sigma}_{PQ}(m, f_m : m', f_{m'}) = \left[\left(\overline{\langle H^2 \rangle^{m,f_m}}\right)^{P/2} \left(\overline{\langle H^2 \rangle^{m',f_{m'}}}\right)^{Q/2} \right]^{-1} \overline{\langle H^P \rangle^{m,f_m}\;\langle H^Q \rangle^{m',f_{m'}}}\;.
\end{aligned} \tag{3}$$

Note that $\Sigma_{11} = \widetilde{\Sigma}_{11}$ and $\Sigma_{22} = \widetilde{\Sigma}_{22} - 1$. Under tensorial decomposition w.r.t $U(\Omega) \otimes SU(r)$, H will transform as the irrep $\{0\}$ w.r.t $SU(r)$ and w.r.t $SU(\Omega)$ each f_2 part transform as $\boldsymbol{F}_\nu$, $\nu = 0, 1, 2$ where $\boldsymbol{F}_0 = \{0\}$, $\boldsymbol{F}_1 = \{21^{\Omega-2}\}$ and $\boldsymbol{F}_2 = \{42^{\Omega-2}\}$ for $f_2 = \{2\}$. However, for $f_2 = \{1^2\}$ we have $\boldsymbol{F}_2 = \{2^2 1^{\Omega-4}\}$. Using tensorial decomposition and the Wigner-Eckart theorem, with the help of the Wigner-Racah algebra for $U(\Omega) \otimes SU(r)$ given in detail in [24], ensemble averages can be carried out. Final formulas for M_2, Σ_{11} and Σ_{22} involve functions P's, $\mathcal{Q}$'s and $\mathcal{R}$'s,

$$\begin{aligned}
M_2 &= \overline{\langle H^2 \rangle^{m,f_m}} = \sum_{f_2} \frac{(\lambda_{f_2})^2}{d_\Omega(f_2)} \sum_{\nu=0,1,2} \mathcal{Q}^\nu(f_2 : m, f_m)\;,\\
\overline{\langle H \rangle^{m,f_m}\;\langle H \rangle^{m',f_{m'}}} &= \sum_{f_2} \frac{(\lambda_{f_2})^2}{d_\Omega(f_2)} P^{f_2}(m, f_m)\; P^{f_2}(m', f_{m'})\;,\\
\Sigma_{22}(m, f_m; m', f_{m'}) &= \frac{X_{\{2\}} + X_{\{1^2\}} + 4X_{\{1^2\}\{2\}}}{\overline{\langle H^2 \rangle^{m,f_m}}\;\overline{\langle H^2 \rangle^{m',f_{m'}}}}\;;
\end{aligned}$$

$$
\begin{aligned}
X_{f_2} &= \frac{2(\lambda_{f_2})^4}{[d_\Omega(f_2)]^2} \sum_{\nu=0,1,2} [d_\Omega(\boldsymbol{F}_\nu)]^{-1}\, \mathcal{Q}^\nu(f_2 : m, f_m)\mathcal{Q}^\nu(f_2 : m', f_{m'})\;, \\
X_{\{1^2\}\{2\}} &= \frac{\lambda^2_{\{2\}}\lambda^2_{\{1^2\}}}{d_\Omega(\{2\})d_\Omega(\{1^2\})} \sum_{\nu=0,1} [d_\Omega(\boldsymbol{F}_\nu)]^{-1}\, \mathcal{R}^\nu(m, f_m)\, \mathcal{R}^\nu(m', f_{m'})\;.
\end{aligned} \tag{4}
$$

Here $d_\Omega(\boldsymbol{F}_\nu)$ is the dimension of the irrep $\boldsymbol{F}_\nu$. The functions P's, $\mathcal{Q}$'s and $\mathcal{R}$'s will contain only $SU(\Omega)$ Racah ($U-$) coefficients and dimension factors. For example,

$$
\begin{aligned}
&\mathcal{Q}^\nu(f_2 : m, f_m) = [F(m)]^2 \sum_{f_{m-2}, f'_{m-2}} \frac{\mathcal{N}_{f_{m-2}}}{\mathcal{N}_{f_m}} \frac{\mathcal{N}_{f'_{m-2}}}{\mathcal{N}_{f_m}} X_{UU}(f_2; f_{m-2}, f'_{m-2}; \boldsymbol{F}_\nu)\;; \\
&X_{UU}(f_2; f_{m-2}, f'_{m-2}; \boldsymbol{F}_\nu) = \\
&\sum_\rho \frac{U(f_m, \overline{f_2}, f_m, f_2; f_{m-2}, \boldsymbol{F}_\nu)_\rho U(f_m, \overline{f_2}, f_m, f_2; f'_{m-2}, \boldsymbol{F}_\nu)_\rho}{U(f_m, \overline{f_2}, f_m, f_2; f_{m-2}, \{0\}) U(f_m, \overline{f_2}, f_m, f_2; f'_{m-2}, \{0\})}\;.
\end{aligned} \tag{5}
$$

Here, $\mathcal{N}_f$ is the dimension of irrep f w.r.t the symmetric group $S(\Omega)$, $F(m) = -m(m-1)/2$ and ρ are multiplicity labels for the coupling $f_m \times \boldsymbol{F}_\nu \to f_m$. Using the tabulations in [27], Racah coefficients in Eq. (5) can be simplified. This formulation unifies all the results known for $r = 1, 2, 4$ for fermions [18-20] and $r = 1$ for bosons[18,21]. For example, applying Eq. (4) for $r = 1$ for bosons, the final formulas for Σ_{11} and Σ_{22} are,

$$
\begin{aligned}
&\Sigma_{11}(\{m\}, \{m'\}) = \frac{2\sqrt{m(m-1)(m')(m'-1)}}{\Omega(\Omega+1)\sqrt{(\Omega+m-1)(\Omega+m-2)(\Omega+m'-1)(\Omega+m'-2)}}\;, \\
&\Sigma_{22}(\{m\}, \{m'\}) = 2\left[36(\Omega+3)\binom{\Omega+2}{3}^2\binom{\Omega+m-1}{2}\binom{\Omega+m'-1}{2}\right]^{-1} \\
&\times\left[4\Omega^2(\Omega-1)\binom{\Omega+m+1}{2}\binom{\Omega+m'+1}{2} + 4(\Omega+2)^2(\Omega+3)\binom{m}{2}\binom{m'}{2}\right. \\
&\left.+4(\Omega^2-1)(\Omega+3)(m-1)(\Omega+m)(m'-1)(\Omega+m')\right]\;.
\end{aligned} \tag{6}
$$

More importantly, using this formulation results are derived for $r = 2$ and 3 for bosons. The results for $r = 2$ are verified using the corresponding results for fermion systems with $r = 2$ given in [19] and the well known $\Omega \to -\Omega$ symmetry. For $r = 3$, for example, for the three row irrep $\{n, n, n\}$ (then $m = 3n$) we have (with $\pi = 1$ for $\{2\}$ and -1 for $\{1^2\}$)

$$
\begin{aligned}
&P^{f_2}(m, \{n,n,n\}) = -\tfrac{6}{3-\pi} n(n-\pi)\;, \quad \mathcal{Q}^{\nu=0}(f_2 : m, \{n,n,n\}) = \left[P^{f_2}(m, \{n,n,n\})\right]^2, \\
&\mathcal{Q}^{\nu=1}(f_2 : m, \{n,n,n\}) = \frac{3(3+\pi)^2(\Omega+\pi)(\Omega-3)n(n-\pi)^2(\Omega+n)}{8(\Omega+2\pi)}\;, \\
&\mathcal{Q}^{\nu=2}(f_2 : m, \{n,n,n\}) = \frac{3(3+\pi)\Omega(\Omega-3+\pi)(\Omega-3)n(n-\pi)(\Omega+n)(\Omega+n+\pi)}{16(\Omega+2\pi)}\;, \\
&\mathcal{R}^{\nu=0}(m, \{n,n,n\}) = P^{\{2\}}(m, \{n,n,n\}) P^{\{1^2\}}(m, \{n,n,n\})\;, \\
&\mathcal{R}^{\nu=1}(m, \{n,n,n\}) = -\sqrt{\frac{\Omega^2-1}{\Omega^2-4}}\, 3(\Omega-3)n(n^2-1)(\Omega+n)\;.
\end{aligned} \tag{7}
$$

We have also worked out results for other irreps for $r = 3$; see for example [17].

Numerical calculations for $\Sigma_{PP}(m, f_m : m, f_m)$ for $P = 1, 2$ showed that there is increase in the magnitude of the fluctuations in energy centroids and spectral variances with increasing r value in direct correlation with the $H_{f_m}(m)$ matrices becoming more dense (implying stronger mixing) as we increase r in EGUE(2)-$SU(r)$ [20, 17]. Further investigation of this feature may establish that symmetries are responsible for chaos in finite quantum systems in general and nuclear shell model spaces in particular. Another important result is that $\Sigma_{PP}(m, f_m : m', f'_m)$

for $m \neq m'$ and/or $f_m \neq f'_m$ will be non-zero [see Eq. (6)] and thus there will be correlations between spectra with different quantum numbers. These give a unique signature for EE as these cross correlations vanish for classical (GOE/GUE/GSE) ensembles.

The formulation based on Wigner-Racah algebra has been used [18, 21, 17] for $r = 1$ to derive formulas for M_2, M_4, Σ_{11} and Σ_{22} for the more general EGUE(k) generated by k-body interactions. They have established Gaussian ($\mathcal{G}$) form for the one point function. However even for two-body interactions, for M_4 for $r > 1$ and Σ_{PQ} with $P, Q > 2$ for $r \geq 1$ could not be derived. Therefore the form of the one-point function is not be yet established, analytically, for $r > 1$ and the two-point function for any r. This is due to the fact that analytical formulas or numerical methods for the $SU(\Omega)$ Racah coefficients needed for these are not available [20].

3. Results from the binary correlation method: Application to neutrinoless double beta decay

Binary correlation approximation (BCA) was used first by Wigner to derive the moments of the eigenvalue density generated by GOE. This approach was extended to EGOE(k) [i.e. $r = 1$] for spinless (dilute) fermion systems by Mon and French [2]. In BCA, averages (moments) $\langle H^p \rangle^m$ are evaluated by retaining only terms that involve squares (but not any other power) of H matrix elements in the defining space. Given m fermions in N single particle states, for EGOE(2) BCA gives for the excess parameter γ_2 the result, $\gamma_2 \to \binom{m-2}{2}\binom{m}{2}^{-1} - 1 \sim -4/m$ showing that asymptotically (i.e. for dilute systems) the eigenvalue density approaches Gaussian. The BCA method was also applied to transition strength densities [28]. Given a transition operator $\mathcal{O}$, transition strength density $I_{biv:\mathcal{O}}(E_i, E_f) = I(E_f)| \langle E_f \mid \mathcal{O} \mid E_i \rangle |^2 I(E_i)$ where $I(E)$ are state (or eigenvalue) densities. Representing H and $\mathcal{O}$ by independent EGOEs, it was established that $I_{\mathcal{O}}$ takes bivariate Gaussian form. It should be recognized that for EGOE for spinless fermions, moments correspond to traces over the complete m-particle space, i.e. all particles are in a single unitary orbit. A problem of current interest is to derive the form of $I_{biv:\mathcal{O}}$ with $\mathcal{O}$ representing the transition operator appropriate for neutrinoless double beta decay (NDBD). This operator is a two-body operator changing two neutrons (n) into two protons (p) and the calculation of NDBD half-lives requires squares of matrix elements of $\mathcal{O}$ connecting the ground state of the parent nucleus to the ground state of the daughter nucleus [29]. Therefore we need to consider a system with protons and neutrons explicitly identified, i.e. two-orbit traces. Nuclear Hamiltonians are one plus two-body in nature, $H = h(1) + V(2)$. Then, $I_{biv:\mathcal{O}}$ can be written as a convolution of the corresponding strength density generated by the mean-field producing $h(1)$ part with a normalized spreading bivariate strength density ($\rho_{biv:\mathcal{O}:V}$) generated by the interaction $V(2)$ [28]. Two important questions are: (i) is $\rho_{biv:\mathcal{O}:V}$ close to a bivariate Gaussian (this can be established by calculating 4th and 6th order bivariate moments); (ii) is it possible to derive a formula for the bivariate correlation coefficient defining $\rho_{biv-\mathcal{G}:\mathcal{O}:V}$. From now on we put $H(2) = V(2)$. BCA is used to answer (i) and (ii).

With space #1 denoting protons and similarly space #2 neutrons, $(m_p, m_n) = (m_1, m_2)$ where m_p is number of valance protons and m_n valence neutrons for a given nucleus. We assume that protons are in N_1 number of single particle (sp) states and neutrons in N_2 number of sp states. Now, the general form of $H(2)$ and the transition operator $\mathcal{O}$ are,

$$H(2) = \sum_{i+j=2\,;\,\alpha,\beta,\gamma,\delta} \left[v_H^{\alpha\beta\gamma\delta}(i,j)\right] \alpha_1^\dagger(i)\beta_1(i)\gamma_2^\dagger(j)\delta_2(j)\ ; \quad \mathcal{O}(2) = \sum_{\gamma,\delta} v_{\mathcal{O}}^{\gamma\delta}(2)\, \gamma_1^\dagger(2)\delta_2(2)\,. \quad (8)$$

Here in general $\alpha_k^\dagger(l)$ is normalized l-particle creation operator in space $\#k$ and $\alpha_k(l)$ is the corresponding annihilation operator. Note that $v(2,0)$, $v(0,2)$ and $v(1,1)$ represent p-p, n-n and p-n parts of the Hamiltonian. Similarly $\mathcal{O}$ changes two neutrons into two protons. It is important to note that $H(2)$ given by Eq. (8) preserves (m_1, m_2). However, $\mathcal{O}$ and its hermitian conjugate $\mathcal{O}^\dagger$ do not preserve (m_1, m_2), $\mathcal{O}(2)\,|m_1, m_2\rangle = |m_1+2, m_2-2\rangle$ and $\mathcal{O}^\dagger(2)\,|m_1, m_2\rangle = |m_1-2, m_2+2\rangle$. Thus, given a (m_1, m_2) for an initial state, the $(m_1, m_2)_f$ for the final state generated by the action of $\mathcal{O}$ is uniquely defined. The bivariate moments of $\rho_{biv:\mathcal{O}:V}$ are defined by $\widetilde{M}_{PQ}((m_1, m_2)) = \overline{\langle \mathcal{O}^\dagger(2) H^Q(2) \mathcal{O}(2) H^P(2)\rangle^{(m_1,m_2)}}$. Asymptotic formulas for the moments are derived by representing $H(2)$ and $\mathcal{O}(2)$ operators by independent EGOEs and then applying BCA. The EGOE representation implies v_H^{---} are independent zero centered Gaussian variable with $\overline{[v_H^{\alpha\beta\gamma\delta}(i,j)]^2} = v_H^2(i,j)$, independent of the labels α, β, γ and δ. Similarly, $v_{\mathcal{O}}^{\gamma\delta}$ are zero centered Gaussian variables with $\overline{\{v_{\mathcal{O}}^{\gamma\delta}\}^2} = v_{\mathcal{O}}^2$. It is easy to see that (with $\widetilde{m}_i = N_i - m_i$),

$$\begin{array}{l}
\overline{\langle H^2(2)\rangle^{m_1,m_2}} = \displaystyle\sum_{i+j=2} v_H^2(i,j)\; T(m_1, N_1, i)\; T(m_2, N_2, j)\;, \\
\widetilde{M}_{00}(m_1, m_2) = \overline{\langle \mathcal{O}^\dagger(2)\mathcal{O}(2)\rangle^{m_1,m_2}} = v_{\mathcal{O}}^2 \binom{\widetilde{m}_1}{2}\binom{m_2}{2}\;, \\
\widetilde{M}_{20}(m_1, m_2) = \widetilde{M}_{00}(m_1, m_2)\; \overline{\langle H^2(2)\rangle^{m_1,m_2}}\;,\; \widetilde{M}_{02}(m_1, m_2) = \widetilde{M}_{00}(m_1, m_2)\; \overline{\langle H^2(2)\rangle^{m_1+2,m_2-2}}\;, \\
\widetilde{M}_{11}(m_1, m_2) = v_{\mathcal{O}}^2 \displaystyle\sum_{i+j=2} v_H^2(i,j) \binom{\widetilde{m}_1 - i}{2}\binom{m_2 - j}{2}\; T(m_1, N_1, i)\; T(m_2, N_2, j)\;.
\end{array} \tag{9}$$

The function T in Eq. (9) is $T(m, N, k) = \binom{\widetilde{m}+k}{k}\binom{m}{k}$. As $\widetilde{M}_{10}(m_1, m_2) = \widetilde{M}_{01}(m_1, m_2) = 0$, the correlation coefficient $\zeta_{biv}(m_1, m_2) = \{\widetilde{M}_{20}(m_1, m_2)\, \widetilde{M}_{02}(m_1, m_2)\}^{-1/2}\; \widetilde{M}_{11}(m_1, m_2)$ and this can be calculated using Eq. (9). With appropriate choice of sp orbits we have for example for ^{100}Mo, ^{150}Nd and ^{238}U, $(N_1, m_1, N_2, m_2) = (30, 2, 32, 8)$, $(32, 10, 44, 8)$ and $(44, 10, 58, 20)$ respectively. They will give, by using Eq. (9) with the assumption that $v_H^2(i,j)$ are independent of (i,j), $\zeta_{biv} = 0.57$, 0.72 and 0.83 respectively for the three nuclei. Thus in general $\zeta_{biv} \sim (0.6 - 0.8)$. More importantly, formulas for $\widetilde{M}_{PQ}((m_1, m_2))$ with $P+Q=4$ are derived using BCA but due to lack of space these are not presented here [30]. These will allow us to calculate the bivariate cumulants k_{rs} of order 4 and it should be noted that $|k_{rs}| \lesssim 0.3$ implies bivariate Gaussian form for $\rho_{biv:\mathcal{O}:V}$. For the four nuclei, the BCA formulas give $(k_{40}, k_{04}, k_{31}, k_{13}, k_{22}) = (-0.45, -0.42, -0.26, -0.24, -0.2)$, $(-0.27, -0.29, -0.2, -0.22, -0.19)$ and $(-0.18, -0.18, -0.15, -0.15, -0.13)$ respectively. Thus, $\rho_{biv:\mathcal{O}:V} \to \rho_{biv-\mathcal{G}:\mathcal{O}:V}$ and this result is being applied to calculate nuclear transition matrix elements for NDBD in ^{150}Nd.

It is important to recognize that the BCA method for EGOE/EGUE with symmetries such as $U(\Omega) \otimes SU(r)$ is not available (except for $r=1$ for fermions). Therefore, Gaussian form for eigenvalues densities could not be established even for EGOE with spin symmetry ($r=2$) for Fermi systems. Similarly, BCA method for boson systems is not available even for $r=1$. At present, Gaussian form even for spinless BEGOE(2) was established by using trace propagation formulas for fourth moments (see next section) and carrying out explicitly ensemble averaging [31]. These two are important open problems in EGOE theory and perhaps, the p-expansion method of French and Draayer [26] may prove to be useful in making advances in future.

4. Results from trace propagation

Propagation equations for energy centroids $\langle H\rangle^{m,f_m}$ and and spectral variances $\langle H^2\rangle^{m,f_m} - [\langle H\rangle^{m,f_m}]^2$ over irreps of group symmetries are successfully used to derive many important results for EGOEs. A simple counting theorem confirms if the propagators are Casimir invariants of the Lie algebras involved and otherwise the so-called integrity basis operators are needed [26, 25]. For EGOE(2), EGOE(2)-s, BEGOE(2) and BEGOE(2)-F, the propagation is simple [11, 13, 15, 32]. For EGOE(1+2)'s with random two-body interactions in the presence of a mean-field, in the strong coupling limit, as the eigenvalue densities approach Gaussian form, $\sigma^2(m,S)$ determines the ground state spin structure in fermion systems [$\sigma^2(m,F)$ for boson systems]. The variances also show that random interactions generate odd-even staggering in the ground state energies. More generally, propagation formulas for energy centroids and spectral variances will allow us understand a variety of regular structures generated by random interactions [33, 34, 13, 15]. Remarkably, EGOE(1+2)'s admit three transition markers, λ_c (for Poisson to GOE in fluctuations), λ_F (Breit-Wigner to Gaussian in strength functions) and λ_t (defining thermodynamic region) and their parametric (m, N, S or F etc.) dependence is determined by spectral variances [11, 13, 32, 15, 35].

5. Conclusions

Although advances are made in deriving analytical results for EE for certain quantities using (i) Wigner-Racah algebra, (ii) binary correlation approximation and (iii) trace propagation methods, mathematical tractability still remains a major issue in the study of EE. Further progress requires new developments in mathematical and computational methods and they are a challenge for future.

Acknowledgments

Thanks are due to Manan Vyas for many useful discussions and for collaboration.

References

[1] Haake F 2010 *Quantum Signatures of Chaos* (New York: Springer)
[2] Mon K K and French J B 1975 *Ann. Phys. (N.Y.)* **95** 90
[3] French J B and Wong S S M 1970 *Phys. Lett.* B **33** 449
[4] Bohigas O and Flores J 1971 *Phys. Lett.* B **34** 261
[5] Kota V K B and Haq R U 2010 *Spectral Distributions in Nuclei and Statistical Spectroscopy* (Singapore: World Scientific)
[6] Alhassid Y 2000 *Rev. Mod. Phys.* **72** 895
[7] Zelevinsky V and Volya A 2004 *Phys. Rep.* **391** 311
[8] Zhao Y M, Arima A and Yoshinaga N 2004 *Phys. Rep.* **400** 1
[9] Brown W G, Santos L F, Starling D J and Viola L 2008 *Phys. Rev.* E **77** 021106
[10] Santos L F, Borgonovi F and Izrailev F M 2012 *Phys. Rev.* E **85** 036209
[11] Kota V K B 2001 *Phys. Rep.* **347** 223
[12] Papenbrock T and Weidenmüller H A 2007 *Rev. Mod. Phys.* **79** 997
[13] Gómez J M G, Kar K, Kota V K B, Molina R A, Relaño A and Retamosa J 2011 *Phys. Rep.* **499** 103
[14] Vyas Manan, Kota V K B and Srivastava P C 2011 *Phys. Rev.* C **83** 064301
[15] Vyas Manan, Chavda N D, Kota V K B and Potbhare V 2012 *J. Phys. A: Math. Theor.* **45**, 265203
[16] Deota H N, Chavda N D, Kota V K B, Potbhare V and Vyas Manan 2012 arXiv:1207.7225 [nlin.CD]
[17] Vyas Manan and Kota V K B 2012 arXiv:1207.7032 [math-ph]
[18] Kota V K B 2005 *J. Math. Phys.* **46** 033514
[19] Kota V K B 2007 *J. Math. Phys.* **48** 053304
[20] Vyas Manan and Kota V K B 2010 *Ann. Phys. (N.Y.)* **325** 2451
[21] Agasa T, Benet L, Rupp T and Weidenmüller H A 2002 *Ann. Phys. (N.Y.)* **298** 229
[22] Benet L, Rupp T and Weidenmüller H A 2001 *Ann. Phys. (N.Y.)* **292** 67
[23] Dyson F J 1972 *J. Math. Phys.* **13** 90
[24] Hecht K T and Draayer J P 1974 *Nucl. Phys.* A **223** 285
[25] Draayer J P and Rosensteel G 1985 *Nucl. Phys.* A **439** 61
[26] French J B and Draayer J P 1979 *Symmetries and statistical behavior in Fermion systems* (Group Theoretical Methods in Physics) ed W Beiglbock, A Bohm and E Takasugi (Berlin: Springer), p 394
[27] Hecht K T 1974 *J. Math. Phys.* **15** 2148
[28] French J B, Kota V K B, Pandey A and Tomsovic S 1988 *Ann. Phys. (N.Y.)* **181** 235
[29] Barea J, Kotila J and Iachello F 2012 *Phys. Rev. Lett.* **109** 042501
[30] Vyas Manan and Kota V K B 2011 arXiv:1106.0395 [nucl-th]
[31] Patel K, Desai M S, Potbhare V and Kota V K B 2000 *Phys. Lett. A* **275** 329
[32] Vyas Manan, Kota V K B and Chavda N D 2010 *Phys. Rev.* E **81** 036212
[33] Kota V K B 2006 *Int. J. Mod. Phys.* E **15** 1869
[34] Kota V K B, Vyas Manan and Mayya K B K 2008 *Int. J. Mod. Phys.* E **17** 318
[35] Chavda N D, Kota V K B and Potbhare V 2012 arXiv:1205.6758 [nlin.CD]

The Heine-Stieltjes correspondence and the polynomial approach to the Gaudin-Richardson models

Feng Pan[1,2], **Kristina D Launey**[2], **and Xin Guan**[1]

[1]Department of Physics, Liaoning Normal University, Dalian 116029, China
[2]Department of Physics and Astronomy, Louisiana State University, Baton Rouge, LA 70803-4001, USA

E-mail: daipan@dlut.ecu.cn

Abstract. The Heine-Stieltjes correspondence is extended and applied to solve Bethe ansatz equations of the Gaudin-Richardson models, from which the extended Heine-Stieltjes polynomial approach to these models is proposed. As examples for the application of this approach, exact solutions of the standard two-site Bose-Hubbard model and the standard pairing model for nuclei are formulated from the corresponding polynomials.

1. Introduction

It is well known that there are only a few problems that can be solved analytically in quantum mechanics. The one-dimensional harmonic oscillator and the hydrogen atom are such typical systems, which become important examples to be illustrated in standard textbook of quantum mechanics. As noted by Wigner in 1960, "the miracle of the appropriateness of the language of mathematics for the formulation of the laws of physics is a wonderful gift which we neither understand nor deserve". But the application of mathematics to problems in physics can also, in turn, bring wider recognition to otherwise little-known mathematical concepts [1]. For example, the Bethe ansatz method for finding the exact eigenvalues and eigenvectors of the one-dimensional antiferromagnetic Heisenberg model Hamiltonian [2], the quantum inverse scattering method for solving integrable models in 1+1 dimensions introduced by Faddeev [3], and the Yang-Baxter equation for solvable problems in statistical mechanics and one-dimensional integrable systems [4, 5] finally led to formulation of quantum groups and non-commutative geometry in mathematics [6]. In the application of the Bethe ansatz method, Gaudin [7] and Richardson [8] independently found exact solutions for the spin or nuclear pairing many-body problems, which are referred to as the Gaudin-Richardson models. In this talk, we report our recent work on relation of solutions of the Bethe ansatz equations (BAE) for the Gaudin-Richardson models and zeros of the extended Heine-Stieltjes polynomials.

2. The Heine-Stieltjes correspondence

There is a large class of polynomials, called Heine-Stieltjes polynomials $y(x)$,which satisfy a second-order Fuchsian equation

$$A(x)\frac{d^2y(x)}{dx^2} + B(x)\frac{dy(x)}{dx} + V(x)y(x) = 0, \tag{1}$$

where $A(x)$ is a polynomial of degree m with $A(x) = \prod_{i=1}^{m}(x-a_i)$, $B(x)$ is a polynomial of degree $m-1$ such that for a set of real positive parameters $\{\gamma_i\}$ and another set of real parameters $\{a_i\}$ is given as

$$B(x)/A(x) = \sum_{i=1}^{m} \frac{\gamma_i}{x-a_i}, \tag{2}$$

and $V(x)$ is an unknown, but to be determined polynomial of degree $m-2$ that is allowed to depend on the solution $y(x)$. The latter are often called Van Vleck polynomials [9]. The case with $m = 2$ corresponds to the hypergeometric differential equation, while the case with $m = 3$ corresponds to the Heun equation [10]. Such Heine-Stieltjes polynomials and properties of their zeros have been studied extensively. If the polynomials $A(x)$ and $B(x)$ are algebraically independent, i.e. they do not satisfy any algebraic equation with integer coefficients, Heine proved that for every integer k there exist at most $\sigma(k) = (k+m-2)!/((m-2)!k!)$ different Van Vleck polynomials $V(x)$ such that $y(x)$ has a polynomial solution of degree k. As summarized in Szegö's work on orthogonal polynomials [11], for every set of nonnegative integers $(k_1, k_2, \cdots, k_m)$, there are uniquely determined real values of the parameters for the Van Vleck polynomials $V(x)$ such that $y(x)$ has a polynomial solution with k_j simple zeros in the open interval (a_{j-1}, a_j) for each $j = 1, 2, \cdots, m$. The polynomial $y(x)$ is uniquely determined up to a constant factor, and has the degree $k = k_1 + \cdots + k_m$. If $y(x)$ is a polynomial of degree k with simple zeros $\{x_1, x_2, \cdots, x_k\}$, one may write $y(x)$ as

$$y(x) = \prod_{j=1}^{k}(x-x_j). \tag{3}$$

Hence, at any zero x_i, $y(x)$ satisfies

$$\frac{y''(x_i)}{y'(x_i)} = \sum_{j\neq i} \frac{2}{x_i - x_j}. \tag{4}$$

Combining (1), (2), and (4), one obtains the following important relations among the zeros:

$$\sum_{j\neq i}^{k} \frac{2}{x_i - x_j} + \sum_{\mu=1}^{m} \frac{\gamma_\mu}{x_i - a_\mu} = 0 \tag{5}$$

for $i = 1, 2, \cdots, k$. It should be noted that the BAEs frequently appearing in search for exact solutions of quantum many-body problems, such as those associated with Gaudin type systems, are similar to the relations shown in (5). The link between Richardson's BCS pairing model for nuclei and the corresponding electrostatic problem was investigated in [12] based on an earlier unpublished preprint of Gaudin, which was then made clearer in [13]. A much more general approach to the pairing model was shown in [14] and [15]. Roots of the BAEs (5) simultaneously determine the eigenenergies and eigenstates of the corresponding quantum many-body problem. As an alternative, roots of the BAEs may also be calculated as zeros of the corresponding polynomial $y(x)$. In this way, a link between solutions of the Gaudin type for quantum many-body problems and the corresponding polynomials is established.

3. The standard two-site Bose-Hubbard model

As a simple extension of the Stieltjes correspondence, we revisit the Bethe ansatz solutions for the standard two-site Bose-Hubbard model studied in [16-18], for which the Hamiltonian is

$$\hat{H}_{\rm BH} = -t(c^\dagger d + d^\dagger c) + U(c^\dagger c c^\dagger c + d^\dagger d d^\dagger d), \tag{6}$$

where $c^\dagger$ (c) and $d^\dagger$ (d) are boson creation (annihilation) operators. The parameters t and U in the Bose-Hubbard model are related to the Josephson coupling and the charging energy, respectively. Following [16, 18], we use the unitary transformation for the boson operators with $c = (a - ib)/\sqrt{2}$ and $d = (a + ib)/\sqrt{2}$. Then, the Hamiltonian (6) can be rewritten in terms of a- and b-boson operators as

$$\hat{H}_{\rm BH} = t(b^\dagger b - a^\dagger a) - \frac{1}{2} U B^+(0) B^-(0) + U\hat{n}^2, \tag{8}$$

where $\hat{n} = a^\dagger a + b^\dagger b$ is the operator for the total number of bosons in the system, and

$$B^+(0) = b^{\dagger 2} + a^{\dagger 2}, \quad B^-(0) = b^2 + a^2 \tag{9}$$

are boson pairing operators. Let $|\nu_1, \nu_2\rangle = a^{\dagger \nu_1} b^{\dagger \nu_2}|0\rangle$ with $\nu_{1,2} = 0$ or 1 be the a- and b-boson pairing vacuum state satisfying

$$a^2|\nu_1, \nu_2\rangle = b^2|\nu_1, \nu_2\rangle = 0. \tag{10}$$

The algebraic Bethe ansatz implies that eigenvectors of (8) may be expressed as

$$|n, \ \zeta, \nu_1, \nu_2\rangle = B^+(x_1^{(\zeta)}) B^+(x_2^{(\zeta)}) \cdots B^+(x_k^{(\zeta)})|\nu_1, \nu_2\rangle \tag{11}$$

with $n = 2k + \nu_1 + \nu_2$, and

$$B^+(x_i^{(\zeta)}) = \frac{a^{\dagger 2}}{x_i^{(\zeta)} + 1} + \frac{b^{\dagger 2}}{x_i^{(\zeta)} - 1}, \tag{12}$$

in which $x_i^{(\zeta)}$ $(i = 1, 2, \cdots, k)$ are spectral parameters to be determined, and ζ is an additional quantum number for distinguishing different eigenvectors with the same quantum number k. It can then be verified by using the corresponding eigenequation that (11) is a solution when the spectral parameters $x_i^{(\zeta)}$ $(i = 1, 2, \cdots, k)$ satisfy the following set of BAEs:

$$U\left(\frac{2\nu_1 + 1}{x_i^{(\zeta)} + 1} + \frac{2\nu_2 + 1}{x_i^{(\zeta)} - 1}\right) + 2t + 4U \sum_{j(\neq i)} \frac{1}{x_i^{(\zeta)} - x_j^{(\zeta)}} = 0 \tag{13}$$

for $i = 1, 2, \cdots, k$, with the corresponding eigenenergy given by

$$E_n^{(\zeta)} = 2t \sum_{i=1}^{k} x_i^{(\zeta)} + t(\nu_2 - \nu_1) + Un^2. \tag{14}$$

Hence, once the ζ-th roots $\{x_i^{(\zeta)}\}$ are obtained from (13), the eigenenergy and the corresponding eigenstate are thus determined according to (14), (11), and (12).

In order to compare to the Heine-Stieltjes correspondence, we assume $t > 0$ and $U \neq 0$, and set

$$\alpha = \nu_2 + 1/2, \quad \beta = \nu_1 + 1/2, \quad \gamma = t/U. \tag{15}$$

Then, the BAEs (13) become

$$\left(\frac{\alpha}{x_i^{(\zeta)} - 1} + \frac{\beta}{x_i^{(\zeta)} + 1}\right) + \gamma + \sum_{j(\neq i)} \frac{2}{x_i^{(\zeta)} - x_j^{(\zeta)}} = 0. \tag{16}$$

When $U > 0$, all parameters α, β and γ are always positive. When $U < 0$, we interchange the boson operator $a^\dagger$ with $b^\dagger$ in (12). Then, one can easily verify that this yields the BAEs (13) with the interchange of ν_1 with ν_2 and $t \to -t$. Hence, in the $U < 0$ case, one obtains the same set of final BAEs as in (16) but with $\alpha \rightleftharpoons \beta$ and $\gamma = -t/U$, thereby keeping $\alpha > 0$, $\beta > 0$, and $\gamma > 0$. Hence, it is sufficient to consider the case for $U > 0$ only with BAEs given by (16). Although the $\gamma = 0$ result is trivial, according to the Stieltjes correspondence, the polynomial corresponding to (16) in this case is the Jacobi polynomial $P_k^{(\alpha-1,\beta-1)}(x)$ satisfying the well-known differential equation

$$\frac{d^2}{dx^2}P_k^{(\alpha-1,\beta-1)}(x) + \left(\frac{\alpha}{x-1} + \frac{\beta}{x+1}\right)\frac{d}{dx}P_k^{(\alpha-1,\beta-1)}(x) - \frac{k(k+\alpha+\beta-1)}{x^2-1}P_k^{(\alpha-1,\beta-1)}(x) = 0. \tag{17}$$

In this case, the Van Vleck polynomial $V(x)$ is trivially a k-dependent constant. Hence, there is only one set of zeros of $P_k^{(\alpha-1,\beta-1)}(x_i)$ with $i = 1, 2, \cdots, k$, satisfying the Bethe ansatz equation (16). The case with $\gamma \neq 0$ is non-trivial. According to (1)-(5), we write a differential equation corresponding to the Bethe ansatz equation (16) as

$$\frac{d^2y(x)}{dx^2} + \left(\frac{\alpha}{x-1} + \frac{\beta}{x+1} + \gamma\right)\frac{dy(x)}{dx} + \frac{V(x)}{x^2-1}y(x) = 0 \tag{18}$$

with the corresponding polynomials $A(x) = x^2 - 1$ and $B(x) = \gamma x^2 + (\alpha+\beta)x + \alpha - \beta - \gamma$ shown in (1). In contrast to the Heine-Stieltjes equation (1), however, in this case the polynomial $B(x)$ is of the same degree as that of $A(x)$. Therefore, the polynomials $y(x)$ determined by (18) should be similar to but different from those of Heine-Stieltjes type.

In the search for polynomial solutions of (18), we write [19]

$$y_k(x) = \sum_{j=0}^{k} b_j x^j. \tag{19}$$

Substitution of (19) into (18) determines the corresponding polynomial $V(x)$ as

$$V(x) = -\gamma k x + f, \tag{20}$$

where f is an undetermined constant depending on k, and together with the unknown expansion coefficients b_j satisfy the following four-term relations:

$$(j(\alpha+\beta+j-1)+f)\,b_j = (j+2)(j+1)b_{j+2} + (j+1)(\beta-\alpha+\gamma)b_{j+1} + \gamma(k-j+1)b_{j-1} \tag{21}$$

with $b_j = 0$ for $j \leq -1$ or $j \geq k+1$. It is clear that Eq. (21) can be represented as a matrix with $k+1$ eigenvalues labeled as $f^{(\zeta)}$ ($\zeta = 1, 2, \cdots, k+1$) and with corresponding eigenvectors $\{b_1^{(\zeta)}, \cdots, b_k^{(\zeta)}\}$, which determine the polynomial $y_k^{(\zeta)}(x)$ according to (19). The number of solutions for f is indeed the same as that of the eigenstates of the standard two-site Bose-Hubbard Hamiltonian (8) when $\gamma \neq 0$. The polynomials (19) with $\gamma \neq 0$ determined by (21) are called the extended Heine-Stieltjes polynomials.

4. The standard pairing model

The Hamiltonian of the standard pairing model is given by

$$\hat{H}_{\mathrm{SP}} = \sum_{j=j_1}^{j_n} \epsilon_j \hat{n}_j - G \sum_{jj'} S_j^+ S_{j'}^-, \tag{22}$$

where n is the total number of levels considered, $G > 0$ is the overall pairing strength, $\{\epsilon_j\}$ are unequal single-particle energies, $\hat{n}_j = \sum_m a_{jm}^\dagger a_{jm}$ is the number operator for the valence particles in the j-th level, and $S_j^+ = \sum_{m>0} (-)^{j-m} a_{jm}^\dagger a_{j\,-m}^\dagger$ ($S_j^- = (S_j^+)^\dagger$) are pair creation (annihilation) operators. Since the formalism for even-odd systems is similar, in the following, we only focus on the even-even seniority zero case. According to the Richardson-Gaudin method, k-pair eigenstates of (22) can be written as

$$|k; x\rangle = S^+(x_1) S^+(x_2) \cdots S^+(x_k)|0\rangle, \tag{23}$$

where $|0\rangle$ is the pairing vacuum state satisfying $S_j^-|0\rangle = 0$ for all j, and

$$S^+(x_i) = \sum_j \frac{1}{x_i - 2\epsilon_j} S_j^+, \tag{24}$$

in which x_i ($i = 1, 2, \cdots, k$) are spectral parameters to be determined. It can then be verified by using the corresponding eigenequation that (23) is the eigenstates of (22) only when the spectral parameters x_i ($i = 1, 2, \cdots, k$) satisfy the following set of BAEs:

$$1 - 2G \sum_j \frac{\rho_j}{x_i - 2\epsilon_j} - 2G \sum_{i'(\neq i)} \frac{1}{x_i - x_{i'}} = 0, \tag{25}$$

where $\rho_j = -(j + 1/2)/2$, with the corresponding eigenenergy given by $E_{n,\,k} = \sum_{i=1}^k x_i$.

According to Heine-Stieltjes correspondence, for nonzero pairing strength G, $A(x)$ and $B(x)$ of the Fuchsian equation (1), which corresponds to Eq. (25), are given by

$$A(x) = \prod_{j=j_1}^{j_n} (x - 2\epsilon_j) \tag{26}$$

and

$$B(x)/A(x) = \sum_{j=j_1}^{j_n} \frac{2\rho_j}{x - 2\epsilon_j} - \frac{1}{G}. \tag{27}$$

The corresponding Van Vleck polynomial $V(x)$ in this case is of degree $n-1$, which needs to be determined according to Eq. (1). In the original electrostatic analogue considered by Heine and Stieltjes, the parameters $\{\rho_j\}$ acting as fixed charges should all be positive with no external electrostatic field, $1/G \to 0$. Therefore, the polynomials $y(x)$ satisfying Eq. (1) with negative $\{\rho_j\}$ and $1/G \neq 0$ in this case are called the extended Heine-Stieltjes polynomials, which tend to be the original Heine-Stieltjes polynomials with negative $\{\rho_j\}$ in the $G \to \infty$ limit.

In the search for polynomial solutions of (1) for this case, we write [20]

$$y(x) = \sum_{j=0}^{k} a_j x^j, \quad V(x) = \sum_{j=0}^{n-1} b_j x^j, \tag{28}$$

where $\{a_j\}$ and $\{b_j\}$ are expansion coefficients to be determined. Substitution of Eqs. (26)-(28) into Eq. (1) yields two matrix equations. Namely, the condition that the coefficients in front of x^i $(i = 0, \cdots, k)$ must be zero generates a $(k+1) \times (k+1)$ matrix $\mathbf{F}$ with $\mathbf{F}\mathbf{v} = b_0\mathbf{v}$, where the eigenvector $\mathbf{v}$ of $\mathbf{F}$ is given by the expansion coefficients $\mathbf{v} = \{a_0, \cdots, a_k\}$, while the condition that the coefficients in front of x^i $(i = k+1, \cdots, n+k-1)$ must be zero generates another $(n-1) \times (k+1)$ upper-triangular matrix $\mathbf{P}$ with $\mathbf{P}\mathbf{v} = 0$, which provides a unique solution for $\{b_1, b_2, \cdots, b_{n-1}\}$ in terms of $\{a_j\}$. Entries of the two matrices are all linear in the coefficients $\{b_1, b_2, \cdots, b_{n-1}\}$. Matrices $\mathbf{F}$ and $\mathbf{P}$ can be easily constructed, for which a simple MATHEMATICA code is now available [21]. Applications of the method to Oxygen isotopes, ^{16}O to ^{28}O, reveal a good agreement of the calculated pairing gaps as compared to the corresponding experimental values (Fig. 1).

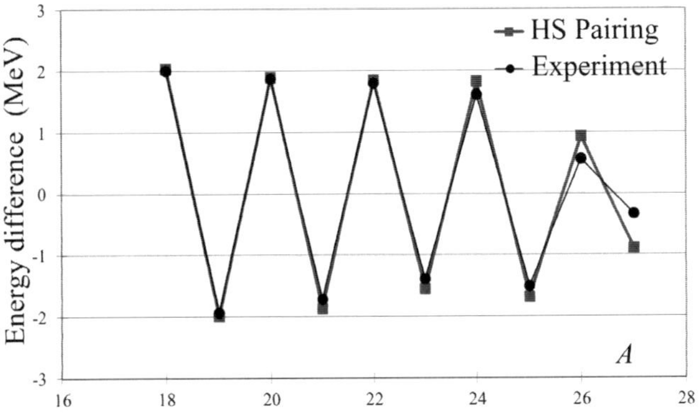

Figure 1. Energy difference, $(E(A-2){-}3E(A-1){+}3E(A){-}E(A+1))/4$ (with an absolute value yielding the pairing gap) of the E binding energies for Oxygen isotopes as a function of the mass number A. Calculations ("HS pairing") are performed for the *sd* shell using $G = 15/A$ MeV and single-particle energies derived from ^{16,17}O experimental data.

5. Summary

In this talk, the Heine-Stieltjes correspondence is extended and applied to solve Bethe ansatz equations of the Gaudin-Richardson models. From this, the extended Heine-Stieltjes polynomial approach to the Gaudin-Richardson models is proposed. Solutions of the standard two-site Bose-Hubbard model and the standard pairing model for nuclei are formulated from the corresponding extended Heine-Stieltjes polynomials. It follows that the eigenvalues and eigenstates of these models can be determined from the zeros of the corresponding polynomials. Furthermore, if these polynomials can be derived recursively, it also follows that it would be much easier to determine zeros from the polynomials of one variable than to solve a set of BAEs with a set of variables. The results by extension clearly show a new link between Bethe ansatz type solutions of a large class of Gaudin-Richardson quantum many-body problems and a class of extended Heine-Stieltjes polynomials satisfying second-order differential equations, which, in turn, opens a new path to find solutions of these BAEs from the zeros of the polynomials.

Acknowledgments
Support from the U.S. National Science Foundation (OCI-0904874), the U.S. Department of Energy (DE-SC0005248), the Southeastern Universities Research Association, the National Natural Science Foundation of China (11175078), the Doctoral Program Foundation of State Education Ministry of China (20102136110002), and the LSU–LNNU joint research program (9961) is acknowledged.

References
[1] Trabesinger A 2010 *Nature Phys.* **6** 560
[2] Bethe H 1931 *Z. Phys.* A **71** 206
[3] Faddeev L 1995 *Acta Applicandae Mathematicae* **39** 69
[4] Yang C N 1967 *Phys. Rev. Lett.* **19** 1312
[5] Baxter R J 1982 *Exactly solved models in statistical mechanics* (New York: Academic Press)
[6] Chari V and Pressley A 1994 *A Guide to Quantum Groups* (Cambridge: Cambridge University Press)
[7] Gaudin M 1976 *J. Phys. France* **37** 1087
[8] Richardson R W 1965 *J. Math. Phys.* **6** 1034
[9] Van Vleck E B 1898 *Bull. Amer. Math. Soc.* **4** 426
[10] Heine E 1878 *Handbuch der Kugelfunktionen Vol II second ed* (G Reimer, Berlin)
[11] Szegö G 1975 *Amer. Math. Soc. Colloq. Publ. Vol. 23* (Amer Math Soc, Providence, RI)
[12] Richardson R W 1977 *J. Math. Phys.* **18** 1811
[13] Gaudin M 1995 *Travaux de Michel Gaudin: Modeles Exactement Resolus* (Les Editions de Physique, France)
[14] Amico L, Di Lorenzo A, Mastellone A, Osterloh A, and Raimondi R 2002 *Ann. Phys.* **299** 228
[15] Dukelsky J, Esebbag C, and Pittel S 2002 *Phys. Rev. Lett.* **88** 062501
[16] Links J, Zhou H Q, McKenzie R H, and Gould M D 2003 *J. Phys.* A **36** R63
[17] Tonel A P, Links J, and Foerster A 2005 *J. Phys.* A **38** 1235
[18] Pan F and Draayer J P 2005 *Phys. Lett.* A **339** 403
[19] Pan F, Bao L, Zhai L, Cui X, and Draayer J P 2011 *J. Phys. A* **44** 395305
[20] Guan X, Launey K D, Xie M, Bao L, Pan F, and Draayer J P 2012 *Phys. Rev.* C **86** 024313
[21] Guan X and Launey K D 2012 *MATHEMATICA code for polynomial solutions to the exact pairing problem*, unpublished

Proton-neutron pairing correlations in the nuclear shell model

S Pittel[1], Y Lei[2,1], N Sandulescu[3], A Poves[4], B Thakur[1] and Y M Zhao[2]

[1] Bartol Research Institute and Department of Physics and Astronomy, University of Delaware, Newark, Delaware 19716, USA
[2] INPAC, Department of Physics and Shanghai Key Lab for Particle Physics and Cosmology, Shanghai Jiao Tong University, Shanghai, 200240, China
[3] Institute of Physics and Nuclear Engineering, 76900 Bucharest, Romania
[4] Departamento de Fisica Teorica and IFT UAM/CSIC, Universidad Autonoma de Madrid, 28049, Madrid Spain.

E-mail: pittel@bartol.udel.edu

Abstract. Systematic shell-model calculations in $2p1f$ shell nuclei are reported, to assess the relative importance of isoscalar and isovector pairing in the presence of nuclear deformation and spin-orbit effects. Results are presented for three $N = Z$ nuclei, ^{44}Ti, ^{46}V and ^{48}Cr.

1. Introduction

Pairing in nuclei is believed to derive from the short-range attraction between nucleons. To optimally exploit this force, nucleons form correlated Cooper pairs by rotating opposite to one another, thereby achieving net orbital angular momentum zero.

Since nucleons occur in two types (n and p), they can form several distinct types of Cooper pairs (nn, pp, or pn) with orbital angular momentum zero. In finite nuclei, however, the active neutrons and protons must be in the same major shell with $N \approx Z$ to exploit pn pairing. Elsewise, only nn and pp pairing will be important, as in those nuclei having a neutron excess.

Pairing correlations are traditionally treated using the Bardeen-Cooper-Schrieffer (BCS) approximation [1], in terms of a condensate of collective pairs. When necessary, mean-field correlations can be included on the same footing through the use of Hartree Fock Bogolyubov (HFB) theory [2]. In the presence of important pn pairing effects, BCS and HFB theory can be generalized to treat all pairing modes on an equal footing. Questions arise, however, as to whether these methods can adequately describe the physics of these competing modes without full restoration of symmetries [3].

The nuclear shell model, in contrast, can treat all pairing modes on the same footing without violation of symmetries. Furthermore, it can treat them in the presence of deformation, as is invariably important when $N \approx Z$. We thus carried out a systematic study of the various modes of pairing in $N \approx Z$ nuclei in the $2p1f$ shell using the nuclear shell model [4]. In this paper, we briefly describe our model and report selected results of this study.

2. Our model

Our model consists of neutrons and protons restricted to the orbitals of the $2p1f$ shell outside a doubly-magic ^{40}Ca core and subject to a schematic hamiltonian

$$H = \chi \left(Q \cdot Q + aP^{\dagger} \cdot P + bS^{\dagger} \cdot S + \alpha \sum_{i} \vec{l}_i \cdot \vec{s}_i \right) . \quad (1)$$

Here Q is the mass quadrupole operator, $P^{\dagger}$ creates a correlated $L = 0$, $S = 1$, $T = 0$ pair, $S^{\dagger}$ creates a correlated $L = 0$, $S = 0$, $T = 1$ pair, and the last term is the one-body part of a spin-orbit force.

We carry out calculations as a function of the various strength parameters and for various nuclei. We start with pure $SU(3)$ rotational motion [5] associated with the $Q \cdot Q$ interaction and then systematically add the various SU(3)-breaking terms to assess how they impact the rotational properties.

3. Calculations

3.1. An Optimal hamiltonian

It is useful to note that the hamiltonian (1) is capable of meaningfully describing the properties of nuclei in this region. With the set of parameters $\chi = -0.05$ MeV, $a = b = 12$, and $\alpha = 20$, we arrive at an *optimal hamiltonian* that acceptably reproduces the spectra of all nuclei we have studied, including the backbend that has been observed experimentally in ^{48}Cr. We refer to the parameter choice $a = b$ as $SU(4)$ pairing, from the dynamical symmetry that arises with this choice of parameters in the $SO(8)$ pairing model [3].

3.2. ^{44}Ti

The first nucleus we discuss is ^{44}Ti, with two neutrons and two protons outside the ^{40}Ca core. Figure 1 shows the energy splittings $E_I - E_{I-2}$ calculated for the ground (YRAST) band as a function of the parameters a and b of isoscalar and isovector pairing, respectively, *with no spin-orbit interaction present.* In these calculations we assume the optimal quadrupole strength of $\chi = -0.05$ MeV. In such a scenario, *i.e.* in the absence of a spin-orbit interaction, the isoscalar and isovector pairing interactions have precisely the same effect on the properties of the ground-state rotational band. Though already known from earlier works, this conclusion also emerges clearly in our study.

In figure 2 we show the corresponding results *with the optimal spin-orbit term included.* Now the symmetry between isoscalar and isovector pairing is broken, even though ^{44}Ti has $N = Z$, and isovector pairing dominates.

3.3. ^{46}V

Next we turn to ^{46}V, with one additional neutron and one additional proton. In figure 3, we show how the symmetry between isoscalar and isovector pairing in the absence of a spin-orbit force is reflected in this odd-odd $N = Z$ system. In the absence of isoscalar and isovector pairing, the $J = 1^+$ state and the $J = 0^+$ state form a degenerate doublet. With only isoscalar pairing (panel a), the $J = 1^+$ state is pushed down below the $J = 0^+$ state. With only isovector pairing (panel b) the reverse happens and the $J = 0^+$ becomes the ground state. In the $SU(4)$ limit (panel c) with equal isovector and isoscalar pairing strengths, the degeneracy remains for all values of the common pairing strength.

Figure 4 shows what happens *in the presence of the optimal spin-orbit interaction*, for equal isovector and isoscalar pairing strengths. Now the degeneracy is broken and the 0^+ state emerges as the ground state, as it does in experiment. The experimental splitting is 1.23 MeV, whereas our optimal hamiltonian produces a slightly smaller splitting of 1.05 MeV.

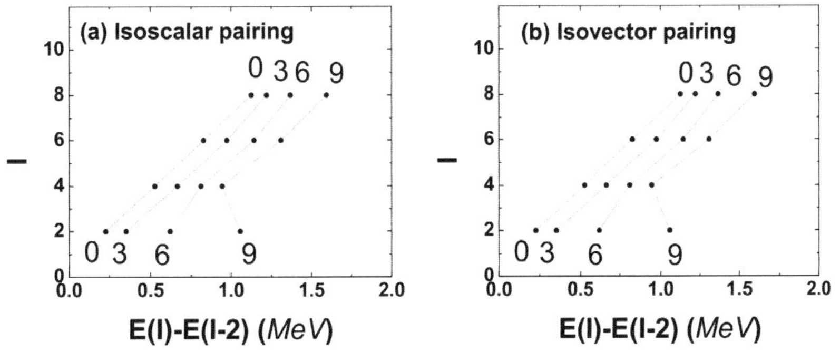

Figure 1. Calculated energy splittings $E(I) - E(I-2)$ (in MeV) in the ground band of ^{44}Ti as a function of (a) the isoscalar pairing strength and (b) the isovector pairing strength, with no spin-orbit splitting. The pairing strengths are shown at the ends of the lines, as elsewhere in the manuscript.

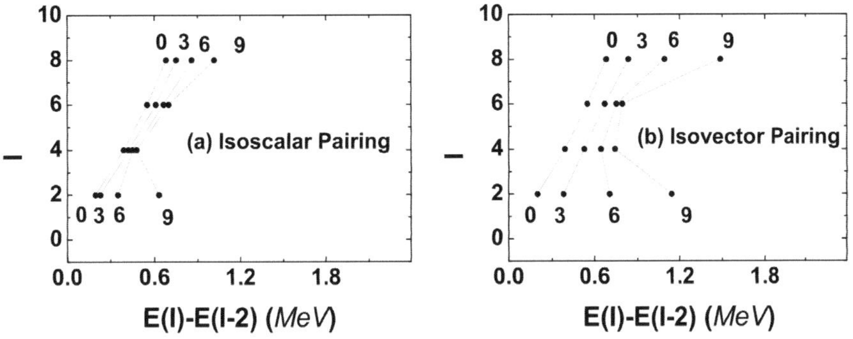

Figure 2. Spectra of the ground band of ^{44}Ti as a function of the strength of (a) isoscalar pairing interaction and (b) the isovector pairing interaction, with the optimal spin-orbit term present.

3.4. ^{48}Cr

Lastly, we consider ^{48}Cr, which also has $N = Z$ but now with two quartet-like structures present. Here we assume as our starting point both the optimal quadrupole-quadrupole force *and* the optimal one-body spin-orbit force and then ramp up the two pairing strengths from zero to their optimal values. The results are illustrated in figure 5, for scenarios in which we separately include isoscalar pairing (panel a) and isovector pairing (panel b) and then in which we include $SU(4)$ pairing with equal strengths (panel c). As noted earlier, the experimental spectrum for ^{48}Cr shows a backbend near $I = 12$. As is evident from panel c, this is nicely

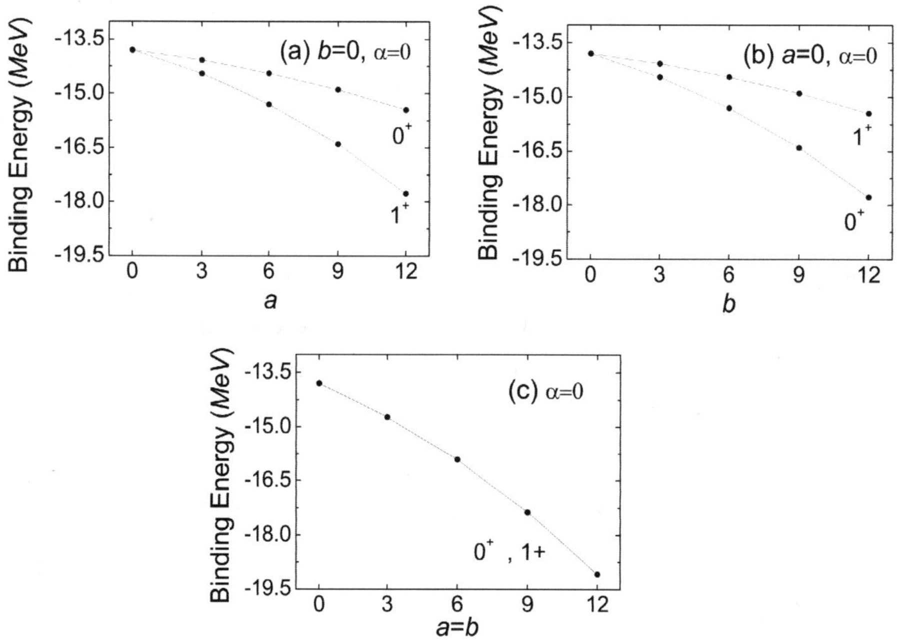

Figure 3. Calculated energies of the lowest $J^{\pi} = 0^{+}$ and 1^{+} states of ^{46}V with no spin-orbit term present, for (a) pure isoscalar pairing, (b) pure isovector pairing and (c) SU(4) pairing.

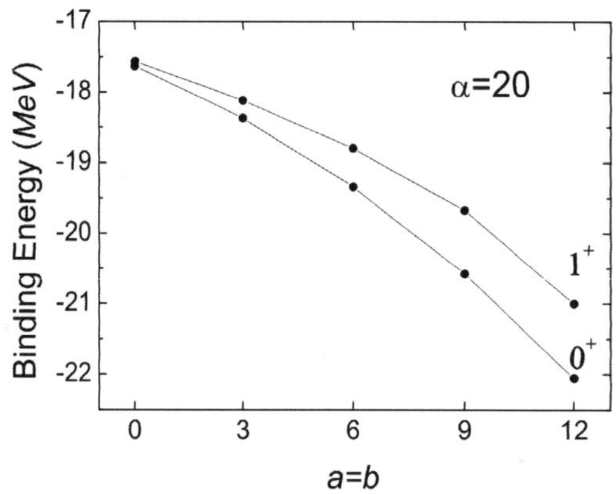

Figure 4. Calculated energies of the lowest $J^{\pi} = 0^{+}$ and $J^{\pi} = 1^{+}$ states of ^{46}V as a function of the equal strengths of isoscalar and isovector pairing, with the optimal spin-orbit term present.

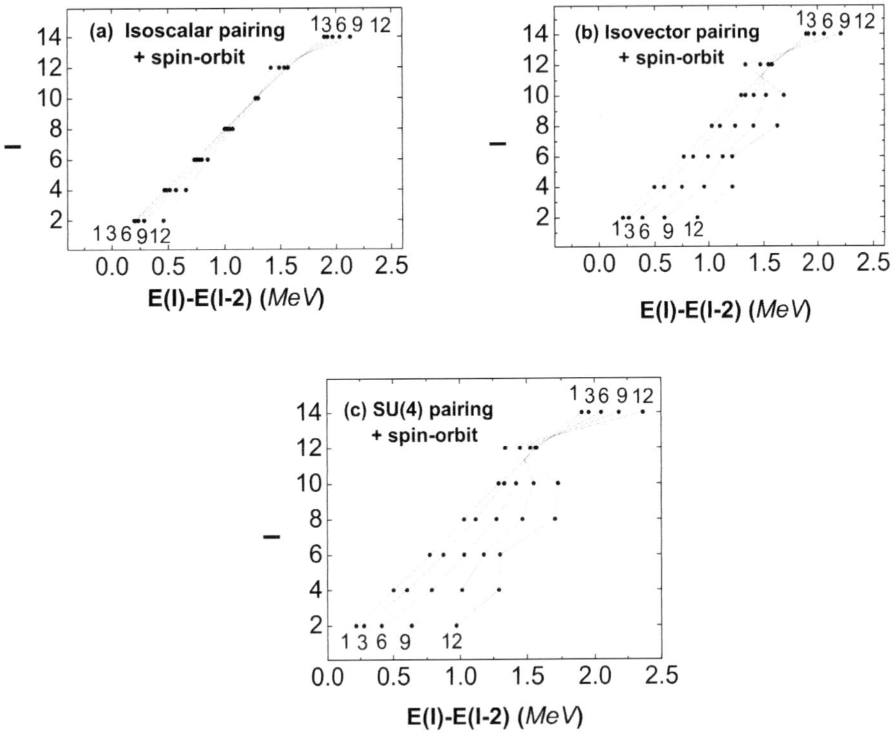

Figure 5. Calculated splittings in the ^{48}Cr ground band, for isovector, isoscalar, and SU(4) pairing, respectively, as described in the text.

reproduced by our *optimal* hamiltonian. From figure 5 we conclude that the backbend in ^{48}Cr cannot be reproduced with pure isoscalar pairing, but requires isovector pairing as well.

The backbend in ^{48}Cr was discussed in a fully realistic shell-model study in ref [6], where it was first shown that it derives from isovector pairing. Our results support that conclusion. To see more clearly how these different pairing modes are reflected in the ground band, we show in figure 6 the number of isovector $S^\dagger$ pairs and the number of isoscalar $P^\dagger$ pairs as a function of angular momentum. As in ref. [6], the contribution of isovector pairing in the $J = 0^+$ ground state is much larger than the contribution of isoscalar pairing. As the system cranks to higher angular momenta, however, the isovector pairing contribution falls off rapidly eventually arriving at a magnitude roughly comparable with the isoscalar pairing contribution at roughly $J^\pi = 10^+$. As the angular momentum increases even further we see a fairly substantial increase in the isovector pairing contribution at $J^\pi = 12^+$, which according to figure 5 is precisely where the backbend occurs. After the backbend, both isoscalar and isovector pairing contributions decrease to near zero as rotational alignment is achieved.

Since our calculations provide information not just on the the YRAST band but also on higher bands, we can readily address whether the origin of the backbend that occurs in ^{48}Cr in our calculations derives from band or level crossing. In figure 7, we compare the energies of

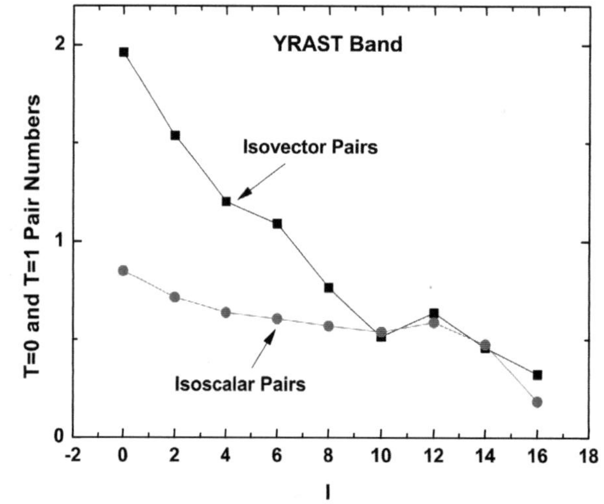

Figure 6. Calculated numbers of isovector $S^\dagger$ pairs and isoscalar $P^\dagger$ pairs in the ground (YRAST) band of ^{48}Cr for the optimal values of the hamiltonian parameters.

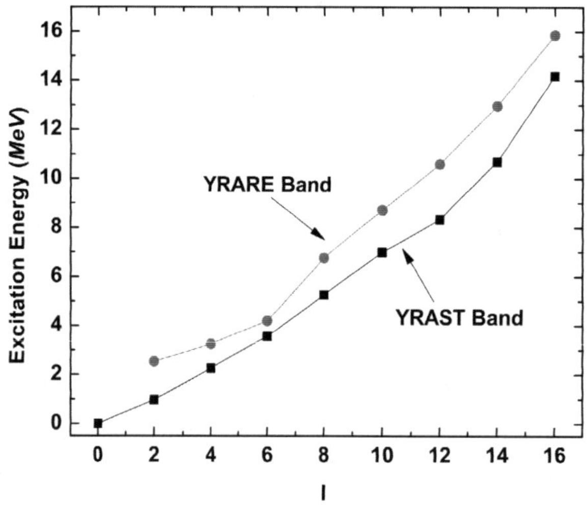

Figure 7. Calculated excitation energies of the ground (YRAST) band and the first excited (YRARE) band in ^{48}Cr for the optimal values of the hamiltonian parameters.

states in the YRAST band with those of the lowest excited (YRARE) band, a $K = 2^+$ band with similar intrinsic structure to the YRAST band. For the sake of comparison, we only show the even angular momentum states of the YRARE band. As can be seen from the figure, there is no evidence of the bands getting close together in energy in the vicinity of the backbend. On the basis of these results, we conclude that the backbend in the ground band of ^{48}Cr does not seem to derive from level crossing.

Indeed, the only behavior that is unusual in the vicinity of the backbend is the substantial increase in the number of isovector pairs (and to a lesser extent isoscalar pairs) in that region. This seems to be further confirmation of the critical role of pairing correlations in producing the backbend in ^{48}Cr and in our view merits further study.

4. Summary

We have reported a shell-model study of proton-neutron pairing in $2p1f$ shell nuclei using a hamiltonian that includes deformation, spin-orbit effects and isoscalar and isovector pairing. By working in a shell-model framework, we can assess the role of the various modes of pn pairing in the presence of deformation without violating symmetries.

Some of the key conclusions that emerged are: (a) the symmetry between isoscalar and isovector pairing effects disappears at $N = Z$ in the presence of a spin-orbit force and isovector pairing dominates, (b) that ^{46}V has a 0^+ ground state derives from the spin-orbit interaction and its relative effect on isoscalar and isovector pairing, (c) isovector pairing dominates in ^{48}Cr and produces its backbend, (d) the backbend in ^{48}Cr does not derive from the crossing of levels, and (e) isovector (and to a lesser extent isoscalar) pairing correlations exhibit a surprising behavior in the vicinity of the ^{48}Cr backbend.

Acknowledgments

This talk was presented by one of the authors (SP) at the conference *Horizons of Innovative Theories, Experiments, and Supercomputing in Nuclear Physics* honoring Jerry Draayer on his 70th birthday. The work of S.P., B.T. and Y.L. was supported in part by the National Science Foundation under grant # PHY-0854873, that of N.S. by the Romanian Ministry of Education and Research through CNCSIS grant Idei nr 57, that of A.P. by the projects FPA2009-13377 MICINN(Spain) and HEPHACOS S2009/ESP-1473 Comunidad de Madrid(Spain), and that of Y.M.Z. and Y.L. by the National Science Foundation of China under grant # 10975096 and by the Chinese Major State Basic Research Developing Program under grant # 2007CB815000.

[1] Bardeen J, Cooper L N and Schrieffer J R 1957 *Phys. Rev.* **108** 1175
[2] Goodman A L 1979 *Adv. Nucl. Phys.* **11** 263
[3] Dobeš J and Pittel S 1998 *Phys. Rev.* C **57** 688
[4] Lei Y, Pittel S, Sandulescu N, Poves A, Thakur B, Zhao Y M 2011 *Phys. Rev.* C **84** 044318
[5] Elliott J P 1958 *Proc. Roy. Soc. (London)* A **245**, 128, 562
[6] Poves A and Martinez-Pinedo G 1998 *Phys. Lett.* B **430** 203

Description of heavy deformed nuclei within the pseudo-SU(3) shell model

Gabriela Popa[1] **and Ana Georgieva**[2]
[1]Ohio University Zanesville, Zanesville, OH 44906, USA
[2]Bulgarian Academy of Science, Sofia, Bulgaria
E-mail: popag@ohio.edu

Abstract. We present a review of the pseudo-SU(3) shell model and its application to heavy deformed nuclei. The model have been applied to describe the low energy spectra, B(E2) and B(M1) values. A systematic study of each part of the interaction within the Hamiltonian was carried out. The study leads us to a consistent method of choosing the parameters in the model. A systematic application of the model for a sequence of rare earth nuclei demonstrates that an overarching symmetry can be used to predict the onset of deformation as manifested through low-lying collective bands.The scheme utilizes an overarching sp(4,R) algebraic framework.

1. Introduction

Over the past years many calculations were carried out by using the SU(3) shell model and the pseudo-SU(3) symmetry. Ongoing improvements and innovations in experimental techniques (4π detectors, radioactive beams, etc.) anticipate the identification of additional new phenomena in the near future, and more information about the existing ones. However, it is commonly accepted that the nuclear shell model should be able to address most of these issues and provide answers. The problem is that most shell-model theories are limited by the large dimensions of the required model spaces, which increases unmanageably with the number of nucleons A. Furthermore, more microscopic calculations are needed exactly in the region of heavy nuclei, in the mass range A $\geq$ 100.

Having this problem, it is essential to take full advantage of symmetries, those that are exact, as well as those which are only partially fulfilled.The selection rules that are associated with such symmetries generate, respectively, weakly coupled and disconnected subspaces of the full space and this allows for a significant reduction in the dimensionality of the model space. The symmetry used here is the pseudo-spin symmetry.

A shell model theory for heavy nuclei requires a severe truncation of the model space. To reproduce the essential physics found in the low-energy states of a large space in a smaller one, we have to select the basis states relative to those parts of the interaction that dominate the low-energy structure. Nuclear physics supports the view that the nuclear effective interaction appropriate to low-energy excitations must have a strong correlation with the pairing and quadrupole-quadrupole interactions. The $Q \cdot Q$ interaction, which dominates for near mid-shell nuclei, is known to introduce deformation and led to the introduction of the SU(3) shell model. This led also to a very natural way to truncate large model spaces, namely to consider few basis states in correspondence with the eigenvalue of $Q \cdot Q$ operator.

Based on the early contribution of Bohr and Mottelson[1], Elliott[2] developed an SU(3) symmetry truncation for light ds-shell nuclei. Raju, Draayer, and Hecht[3] developed a pseudo-SU(3) scheme for heavy nuclei. Ground and gamma band rotational structure of deformed nuclei were described in 1984 by Draayer and Weeks[4], in the framework of the pseudo-SU(3) model by using only one SU(3) irreducible representation (irrep), the one corresponding to the highest deformation, and considering the abnormal parity spaces to seniority zero configurations. The fits to the available data were very accurate ($|E_{exp} - E_{thy} \leq 25keV$). In this small space the ground and the gamma (K=2) bands were very well described, as well as the inter-band and intra-band B(E2) strengths. The model used the mathematical and computational advances made by Draayer and Williams[5], Akiyama and Draayer[6], Draayer, Weeks, and Hecht[7], and Draayer and Rosensteel[8].

Using mathematical and computational developments to the pseudo-SU(3) model, Castaños, Draayer, and Leschber improved it further to accommodate an increased model space, taking into consideration several protons and neutrons SU(3) irreps, as well as the total coupled representation in 1987[9, 10]. In this version, many bands can be calculated and as well the B(E2) and B(M1) transitions values within and between bands. Strong B(M1) transitions were predicted first in various Gadolinium, Erbium, Uranium, and Plutonium isotopes using the pseudo-SU(3) model in 1987[9]. Naqvi and Draayer introduced a shell-model operator for K-band splitting in [11]. The K^2 operator is used to resolve multiple occurrences of L-values in SU(3) representations. In 1994, Bahri and Draayer completed an improved code for faster calculations[12]. Later the effects of paring interaction on the K-band mixing were studied[13, 14] by Trotelnier, Bahri, Escher and Draayer within the framework of the pseudo-SU(3) model.

Considering such larger spaces with an SU(3) Hamiltonian plus proton and neutron single particle energies, and pairing interactions, an extended version of the model was applied[15, 16] to describe the low-lying energy structure and electromagnetic transitions probabilities of even-even nuclei in the rare-earth region .

The purpose of this work is to show that by using a realistic Hamiltonian based on proton and neutron single-particle energies, $Q \cdot Q$, and pairing interactions we can find systematically a set of parameters[16], in the chosen model space, to obtain a good agreement with the experiment in the even-even heavy nuclei. We are also studying the effect of each term in the Hamiltonian plus the rotor terms, namely J^2, K^2, and Q_3, on the energy spectra. The parameters of the rotor terms are fixed as the result of this study.

The role of these parameters is further revealed in a systematic study of the behavior of first excited $K^\pi = 0_2^+$, and $K^\pi = 2^+$ bands in deformed even-even nuclei of the rare-earth region. The nuclei are ordered in F-spin multiplets of a Sp(4,R) classification scheme [17]. The energy levels of the ground band (g.b.) $J = 2_1^+$, first excited $J^\pi = 0^+_{K^\pi=0_2^+}$, $J^\pi = 2^+_{K^\pi=2_1^+}$ states of nuclei that belong to the $F_0 = 0 = 1/2(N^\pi - N^\nu)$ multiplet (where N^π and N^ν are the number of valence proton and neutron pairs) are plotted in Fig. 3. The energies of the same set of levels for nuclei belonging to the $F_0 = 1$ multiplet are plotted in Fig. 4. Nuclei which belong to the $F_0 = 0$ multiplet have equal numbers of valence proton and neutron pairs, and those which belong to the $F_0 = 1$ multiplet vary by two pairs of protons. With such an investigation[18], we provide a microscopic interpretation of the complex and varying behavior of these collective excitations, by applying the algebraic shell model with pseudo SU(3) symmetry.

In the next section, the structure of the basis states is defined. An example is given for ^{156}Gd nucleus. The Hamiltonian is described in detail in section 3. The results of the analysis are explained in section 4. Also further applications of the model are presented.

2. Basis states

Building the many-body basis of the pseudo-SU(3) model relies on the proton-neutron degrees of freedom, pseudo-spin symmetry and group theoretical methods. The construction of the basis states starts with the deformed Nilsson single-particle levels for protons and neutrons, which are filled from the lowest energy upward, for a specific deformation parameter β for the corresponding nucleus. The completely filled shells are frozen and the remaining levels are taken into consideration. For heavy nuclei, the valence protons and neutrons fill different major shells. For a given nucleus there are two open shells (one for protons, one for neutron), each of them comprised of a set of normal and unique parity levels. The normal parity space is partitioned into pseudo SU(3) irreps, and the unique parity space is spanned by configurations of identical particles in a single j shell. It was shown that the seniority coupled scheme is appropriate for a description of the unique parity configuration[7]. Thus the low-energy structure of the normal parity part of the space will be dominated by a few irreps of pseudo SU(3). In the abnormal(unique) parity parts of the proton and neutron shells was shown that the configuration with low seniority are the most important ones, since those with high seniority are not favored (the pairing gap is large compared to the spacing of low-lying rotational bands.)

In this way, the occupation numbers of the normal (N) and unique (U)(also called intruder) parity states are determined for protons (π) and neutrons (ν), n_π^N, n_π^U, n_ν^N, n_ν^U. For example, for ^{156}Gd, $n_\pi^N = 8$ in $\tilde{\eta}_\pi = 3$, and $n_\nu^N = 6$ in $\tilde{\eta}_\pi = 4$ with degeneracies $\Omega_\pi = 20$ and $\Omega_\nu = 30$. Here, $\tilde{\eta}_\pi[\nu]$, represent the pseudo associated numbers. For 8 protons in $\tilde{\eta}_\pi = 3$ and 6 neutrons in $\tilde{\eta}_\nu = 4$ there are 32 possible SU(3) irreps in the proton space and 48 possible irreps in the neutron space, for proton and neutron pseudo-spin equal to zero. The corresponding irreps are ordered in descending C_2 value and the first 6 proton and 5 neutron irreps are chosen, that are given in Table 1. The direct products of these proton and neutron irreps give a large set of

Table 1. The SU(3) irreps for protons and neutrons corresponding to the largest C_2 in ^{156}Gd.

(λ_π, μ_π)	C_2	(λ_ν, μ_ν)	C_2
(10,4)	198	(18,0)	378
(12,0)	180	(15,3)	333
(0,12)	180	(12,6)	306
(8,5)	168	(13,4)	288
(5,8)	168	(14,2)	276
(9,3)	153		

possible strong coupled irreps from which we have selected the first 30 with the largest C_2 values. These irreps are given in Table 2. The many-particle states of n_α nucleons in a shell ($\alpha = \pi$ or ν) belong to the totally antisymmetric irreps $\{1^{n_\alpha}\}$ of the unitary group of $\Omega_\alpha^N = (\eta_\alpha + 1)(\eta_\alpha + 2)$ dimension. A complete classification of the basis states uses the quantum numbers following from subgroups of the unitary group.

$$\begin{array}{ccccccccc} \{1^{n_\alpha^N}\} & & \{\tilde{f}_\alpha\} & & \{f_\alpha\}\ \gamma_\alpha\ (\lambda_\alpha,\mu_\alpha)\ \tilde{S}_\alpha & & K_\alpha \tilde{L}_\alpha & & J_\alpha^N \\ U(\Omega_\alpha^N) & \supset & U(\Omega_\alpha^N/2)\times U(2) & \supset & SU(3)\times SU(2) & \supset & SO(3)\times SU(2) & \supset & SU_J(2), \end{array} \quad (1)$$

Above each group are given the quantum numbers that characterize its irreps with K_α and γ_α the multiplicity label of the indicated reduction. The eigenstates $|\psi_\alpha >$ of the Hamiltonian (2) are written as linear combination of the strong-coupled basis:

$$|\psi_\alpha >= \sum_i C_i^\alpha |\phi_i > \quad (2)$$

$$|\phi_i>=|\{n_\pi[f_\pi](\lambda_\pi,\mu_\pi),S_\pi;n_\nu[f_\nu](\lambda_\nu,\mu_\nu),S_\nu\}\rho(\lambda,\mu)KL\ S;JM>, \quad (3)$$

with two multiplicity labels: ρ that counts the number of times the irrep (λ,μ) occurs in the direct product $(\lambda_\pi,\mu_\pi)(\lambda_\nu,\mu_\nu)$ and K classifies the different occurrences of the orbital angular momentum L in (λ,μ).

Table 2. The first 30 pseudo SU(3) irreps used in the description of ^{156}Gd bands.

(λ_π,μ_π)	(λ_ν,μ_ν)	total (λ,μ)					
(10,4)	(18,0)	(28,4)	(26,5)	(27,3)	(24,6)		
(10,4)	(15,3)	(25,7)	(26,5)	(27,3)	(28,1)	(23,8)	(24,6)
(10,4)	(12,6)	(22,10)	(23,8)	(24,6)	(20,11)		
(10,4)	(13,4)	(23,8)	(24,6)				
(10,4)	(14,2)	(24,6)					
(12,0)	(18,0)	(30,0)	(28,1)				
(12,0)	(15,3)	(27,3)					
(12,0)	(12,6)	(24,6)					
(8,5)	(18,0)	(26,5)	(24,6)				
(8,5)	(15,3)	(23,8)	(24,6)				
(5,8)	(18,0)	(23,8)					

3. Model Interaction

We studied in detail the effect of each term of the Hamiltonian on the energy spectra. The relative strength of each term in the interaction affects not only the energy spectra, but also the strength of the B(E2) and B(M1) transitions among those states. The Hamiltonian consists of quadrupole-quadrupole $(Q\cdot Q)$, proton (π) and neutron (ν) single particle energies $(H^\pi_{sp}+H^\nu_{sp})$, proton and neutron pairing (H^π_P and H^ν_P), as well as the rotor part (H_{ROT}).

$$H=\chi Q\cdot Q+H^\pi_{sp}+H^\nu_{sp}-G_\pi H^\pi_P-G_\nu H^\nu_P+H_{ROT} \quad (4)$$

where

$$H_{ROT}=aJ^2+bK^2_J+c_3C_3+a_{sym}C_2 \quad (5)$$

C_2 and C_3 terms are the second and third order Casimir invariants of SU(3), which are related to the deformation of the nucleus. J^2 is the total angular momentum operator, and K^2_J is the K-band splitting operator(resolving the multiple occurrences of the same angular momentum within an SU(3) irrep). The single particle energies are calculated in the standard way:

$$H^\sigma_{sp}=H_0+c\hat{l}\hat{s}+d\hat{l}^2=H_{OSC}-\hbar\bar{\omega}_0\kappa(2\hat{l}.\hat{s}+\mu\hat{l}^2) \quad (6)$$

where $\sigma=\pi$ or ν, H_0 is the spherical oscillator for which SU(3) is an exact symmetry, and the constants values are taken from [19]. For ^{156}Gd the $\hbar\omega_0=41$ MeV $\times A^{-1/3}=7.616MeV$ is the oscillator constant for the equivalent spherical nucleus and the constants κ and μ have the values [19]: $\kappa_\pi=0.06370,\kappa_\nu=0.60,\mu_\pi=0.06370,\mu_\nu=0.42$

4. Results and conclusions

We studied the contribution of each term in the total Hamiltonian eq. (4) by varying the strength of each interaction and plotting the energy of the first few states of angular momentum $K=0,2,4,6,8$, and 1. In Figure 1 we present the energy values of few states of total angular momentum $J = 0, 2$, and 4 versus the strength of the proton and neutron paring. In the calculation, the strength of the proton and neutron paring is set the same and varied. The paring interaction is important to get the correct relative value of the states in the $K = 0^+_2$ band, versus the ground band. It is also important to get the relative ordering of the band head of the $K = 2^+$ band versus the ground and $K = 0^+_2$ bands.

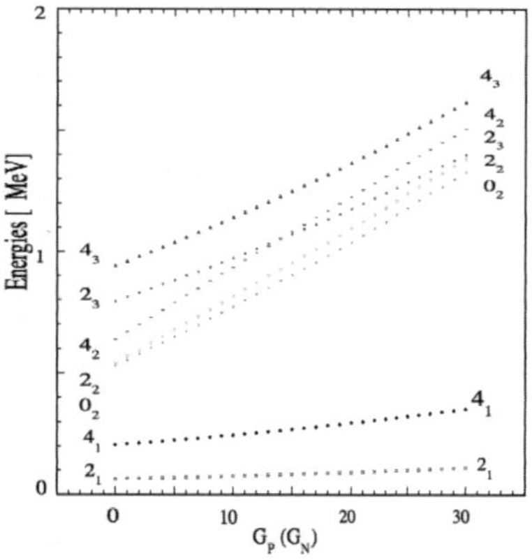

Figure 1. The energy values (in MeV) of the few three states of total angular momentum $J = 0, 2, 4$, versus the strength of the proton and neutron paring in ^{156}Gd.

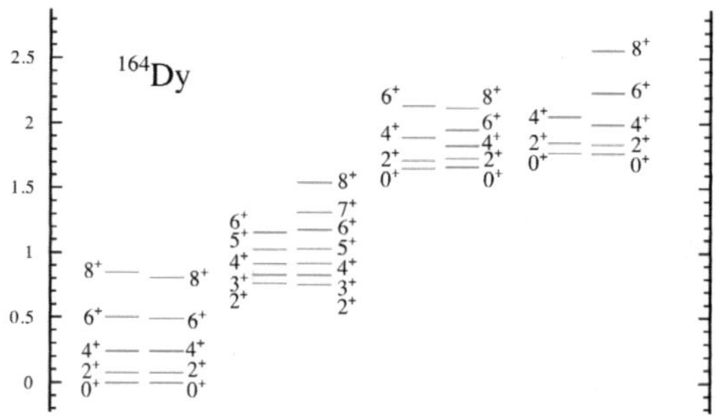

Figure 2. The energy values in MeV, in the first four bands in ^{164}Dy. The first one is the ground band, the second one is the $K = 2^+$ band, the third and fourth are the $K = 0^+_2$, and $K = 0^+_3$ bands. In the left hand side are given the experimental values[20], and in the right the calculated ones.

After varying one interaction strength at a time and studying the effect of each interaction on the energy levels and B(E2) transitions, we fixed the values for pairing ($G_\pi = 21/A$, $G_\nu = 17/A$), as well as for the quadrupole-quadrupole interaction strength ($\chi = 35A^{-5/3}$)[16]. The details of the study will be given elsewhere. The interaction strengths of the rotor terms were varied to give a best fit to the band heads of the first excited $K^\pi = 0^+$, $K^\pi = 2^+$ and $K^\pi = 1^+$ bands, as well as the moment of inertia of the g.s. band. Explicitly, the term proportional to K_J^2 breaks the $SU(3)$ degeneracy of the different K bands, the J^2 term represents a small correction to fine tune the moment of inertia, and the last term, C_2, is introduced to distinguish between $SU(3)$ irreps with λ and μ both even from the others with one or both odd, hence fine tuning the energy of the first excited $K^\pi = 1^+$ state.

With parameters fixed, the Hamiltonian is diagonalized. As an example, the energy values for the first four energy bands in ^{164}Dy are given in Figure 2. In the left hand side are given the experimental values [20], and in the right the calculated ones. Within this framework, the splitting and mixing of the pseudo-$SU(3)$ irreps are generated by the proton and neutron single particle terms ($H_{sp}^{\pi/\nu}$) and the pairing interactions. This mixing plays an important role in the reproduction of the behavior of the low-lying collective states in deformed nuclei [18].

This approach was employed for a systematic investigation of the experimental energies of the $J^\pi = 2^+_{g.s}$, $J^\pi = 0^+_{K^\pi=0^+_2}$, and $J^\pi = 2^+_{K^\pi=2^+_1}$ states in the deformed nuclei from the $F_0 = 0$ and $F_0 = 1$ multiplets of the rare earths, which are shown in Fig.3 and Fig. 4, where the rather irregular behavior of the band-heads of the first excited bands is quite obvious. The calculated energies in the framework of pseudo-SU(3) model, with above described fit of the respective parameter strengths, are compared to the experimental values in Tables 3 and 4. We calculated also all the states with $J \leq 8$ within these three bands. The obtained results are in

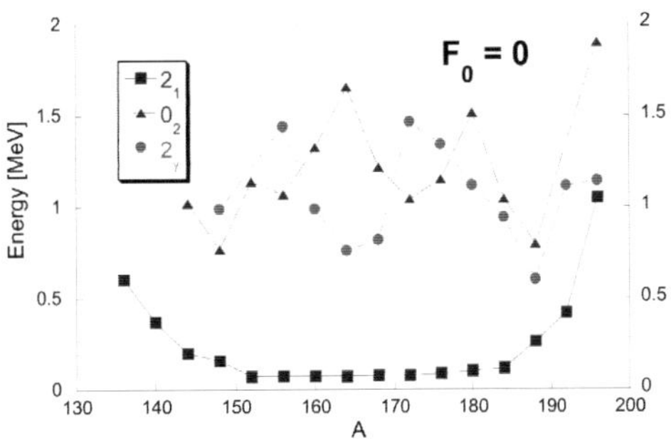

Figure 3. The experimental energy values [20] of the $J^{\pi} = 2_1^+$, $J^{\pi} = 0^+_{K^{\pi}=0_2^+}$, and $J^{\pi} = 2^+_{K=2_1^+}$ states in a sequence of nuclei for which the numbers of valence protons and neutrons are the same, ($F_0 = 0$).

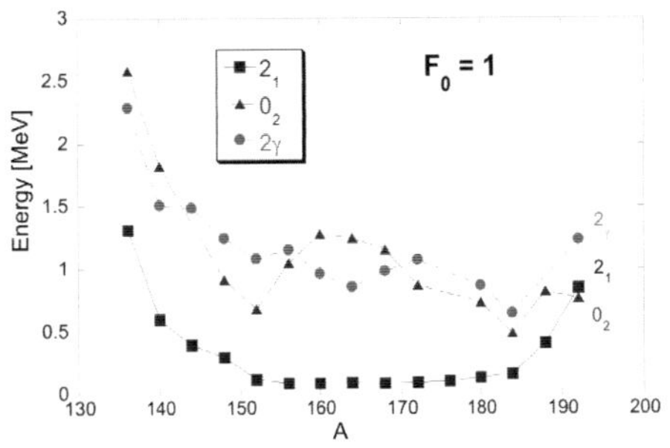

Figure 4. The experimental energy values [20] of the $J^{\pi} = 2_1^+$, $J^{\pi} = 0^+_{K^{\pi}=0_2^+}$, and $J^{\pi} = 2^+_{K=2_1^+}$ states in a sequence of nuclei for which the difference in the numbers of valence proton and neutron pairs is two, ($F_0 = 1$).

Table 3. Energy values for the $J^{\pi} = 2_1^+$, first excited $J^{\pi} = 0^+_{k^{\pi}=0_2^+}$ and $J^{\pi} = 2^+_{K^{\pi}=2_1^+}$ states for seven nuclei in the $F_0 = 0$ multiplet. The experimental values [20] are given in parenthesis. The numbers of valence proton and neutron pairs are given in the second and third columns. The resulted fitted parameters of the rotor part of the Hamiltonian are given in the fourth to seventh columns.

nucleus	N^{π}	N^{ν}	a	b	a_s	a_3 $\times$ 10^{-4}	$E(2_1)$ Th. (Exp.) [MeV]	$E(0^+_{K=0_2^+})$ Th. (Exp.) [MeV]	$E(2^+_{K^{\pi}=2_1^+})$ Th. (Exp.) [MeV]
^{152}Nd	5	5	0.000	0.00	0.0000	2.57	0.082(0.073)	1.14(1.14)	1.31(1.38)
^{156}Sm	6	6	0.000	0.55	0.0000	2.59	0.078(0.076)	1.07(1.07)	1.45(1.47)
^{160}Gd	7	7	0.001	0.15	0.0034	1.93	0.085(0.075)	1.33(1.33)	0.99(0.82)
^{164}Dy	8	8	-0.001	0.04	0.0008	0.65	0.073(0.073)	1.67(1.66)	0.76(0.76)
^{168}Er	9	9	-0.002	0.02	0.0008	0.75	0.089(0.080)	1.21(1.22)	0.81(0.82)
^{172}Yb	10	10	-0.001	0.12	0.0010	0.31	0.081(0.078)	1.04(1.04)	1.49(1.47)
^{176}Hf	11	11	-0.007	0.30	0.0060	0.43	0.178(0.088)	1.15(1.15)	1.26

very good agreement with experiment [18, 20]. The main reason for obtaining the position of each collective band with respect to each other, as well as of each level within the band is the specific content of the obtained SU(3) irreps into the collective states, which is related to their deformations. For nuclei from Table 3, in the middle of the shell, the ground and the γ band belong to the same (λ, μ) with $\lambda > \mu$. At the limits of the deformed region the ground band states have oblate deformation $(\lambda < \mu)$ and the $K^{\pi} = 0_2^+$ and the $K^{\pi} = 2_1^+$ bands are mixed in the same SU(3) irrep [18]. The analysis for nuclei from Table 4 is under investigation.

The correct description of collective properties of first excited $K^{\pi} = 0_2^+$, and $K^{\pi} = 2^+$ states is a result of representation mixing and state deformation induced by the model Hamiltonian. This study shows that pseudo-spin zero neutron and proton configurations with relatively few pseudo-SU(3) irreps with largest C2 values suffices to yield good agreement with known experimental

Table 4. Same as in Table 3 for four nuclei with $F_0 = 1$.

nucleus	N^π	N^ν	a	b	a_s	a_3 $\times$ 10^{-4}	$E(2_1)$ Th. (Exp.) [MeV]	$E(0^+_{K=0^+_2})$ Th. (Exp.) [MeV]	$E(2^+_{K^\pi=2^+_1})$ Th. (Exp.) [MeV]
^{156}Gd	7	5	0.001	0.10	0.0014	1.36	0.094 (0.089)	1.06 (1.05)	1.14 (1.15)
^{160}Dy	8	6	0.003	0.16	0.0004	0.94	0.092 (0.087)	1.30 (1.28)	0.99 (0.97)
^{168}Yb	10	8	0.00	0.01	0.0010	0.51	0.088 (0.088)	1.16 (1.15)	0.98 (0.98)
^{172}Hf	11	9	0.00	0.00	0.0000	0.51	0.119 (0.095)	0.87 (0.87)	1.08 (1.07)

energies.

In the presented applications of the pseudo-SU(3) model, we achieved a good description of the complex behavior of a series of heavy deformed nuclei using a small configuration space. We are working on further analyzing the effect of the different components of the interaction in the Hamiltonian, on the strength of the transition probabilities. Also, the size of the model space will be further investigated. Work in progress is on expanding the study to other groups of nuclei related to one another by an over-arching F-type symmetry.

Acknowledgments

This work is an appreciation of the numerous contributions of J. P. Draayer and his collaborators to the deeper understanding from symmetry principles of the nuclear structure, and the inspiration he provided in this way to the authors. We also acknowledge the support from the Bulgarian National Foundation (DID-02/16) and the Bulgarian-Rumanian agreement (DNTS-02/21), the U.S. National Science Foundation (PHY-0904874), Department of Energy (DE-SC0005248), and the Southeastern Universities Research Association.

References

[1] Bohr A and Mottelson B R 1953 Mat. Fys. Medd. **27** 16
[2] Elliott J P 1958 *Proc. Roy. Soc. (London)* A **245** 128, 562
[3] Raju R D, Draayer J P, and Hecht K T 1973 *Nucl. Phys.* A **202** 433
[4] Draayer J P and Weeks K J 1984 *Annals of Phys.* **156** 41
[5] Draayer J P and Williams S A 1968 *Nucl. Phys.* A **129** 647-665
[6] Akiyama Y and Draayer J P 1973 *Comp. Phys. Comm.* **5** 405-515
[7] Draayer J P, Weeks K J, and Hecht K T 1982 *Nucl. Phys.* A **381** 1
[8] Draayer J P and Rosensteel G 1982 *Nucl. Phys.* A **386** 189-199
[9] Castaños O, Draayer J P, and Leschber Y 1987 *Nucl. Phys.* A **473** 494-508
[10] Castaños O, Draayer J P, and Leschber Y 1987 *Ann. of Phys.* **180** 290
[11] Naqvi H A and Draayer J P 1990 *Nucl. Phys.* A **516** 351-364, **536** 297-308
[12] Bahri C and Draayer J P 1994 *Comp. Phys. Comm.* **83** 59-94
[13] Trotelnier D, Bahri C, and Draayer J P 1995 *Nucl. Phys.* A **589** 75
[14] Trotelnier D, Bahri C, Escher J, Draayer J P 1995 Z. Phys. A **354** 125-134
[15] Beuschel T, Hirsch J G, and Draayer J P 2000 *Phys. Rev.* C **61** 54307
[16] Popa G, Hirsch J G, and Draayer J P 2000 *Phys. Rev.* C **62** 064313
[17] Drenska S, Georgieva A, Gueorguiev V, Roussev R, and Raychev P 1995 *Phys. Rev.* C **52** 1853
[18] Popa G, Georgieva A, and Draayer J P 2004 *Phys. Rev.* C **69** 064307
[19] Ring P and Schuck P 1979 *The Nuclear Many-Body Problem* (Berlin: Springer)
[20] Firestone R B and Shirley V S 1996 *Table of Isotopes*, Vol. I, 8-th ed. (Wiley, New York) Vol. II

Gauge theory of collective modes

G Rosensteel

Physics Department, Tulane University, New Orleans, LA 70118 USA

E-mail: `george.rosensteel@tulane.edu`

Abstract. The classical theory of Riemann ellipsoids is formulated naturally as a gauge theory based on a principal G-bundle $\mathcal{P}$. The structure group $G = SO(3)$ is the vorticity group, and the bundle $\mathcal{P} = GL_+(3, \mathbb{R})$ is the connected component of the general linear group. The base manifold is the space of positive-definite real 3×3 symmetric matrices, identified geometrically with the space of inertia ellipsoids. Non-holonomic constraints determine connections on the bundle. In particular, the trivial connection corresponds to rigid body motion, the natural Riemannian connection to irrotational flow, and the invariant connection to the falling cat. The curvature form determines the fluid's field tensor which is an analogue of the familiar Faraday tensor. Associated G-bundles and the covariant derivative yield new quantum geometrical collective models that are a natural generalization of the Bohr model. These new geometric structures formulate the collective model as a Yang-Mills gauge theory.

1. Introduction

The general collective motion algebra $\mathfrak{gcm}(3)$ is a fifteen dimensional, noncompact subalgebra of the twenty-one dimensional symplectic Lie algebra. This subalgebra is a semi-direct sum of the nine-dimensional algebra of the general linear group $\mathrm{GL}_+(3,\mathrm{R})$ and the six dimensional abelian ideal $\mathbb{R}^6$ generated by the mass quadrupole and monopole tensors. The ideal may be identified with the symmetric 3×3 real matrices, considered as a vector space.

The irreducible unitary representations of $\mathfrak{gcm}(3)$ are indexed by a non-negative integer C [1, 2]. The Bohr-Mottelson quantum model of nuclear rotational and vibrational states is the particular irreducible representation corresponding to $C = 0$ [3]. The primary advantage to the $\mathfrak{gcm}(3)$ characterization of the Bohr collective model is that it provides, without any approximation, the model's microscopic foundation. This is achieved by viewing the algebra in three different ways.

First $\mathfrak{gcm}(3)$ is an abstract algebra of 6×6 matrices

$$\mathfrak{gcm}(3) = \left\{ S = \begin{pmatrix} X & -U \\ 0 & -X^t \end{pmatrix} \mid X, U \in M_3(\mathbb{R}), U^t = U \right\}, \tag{1}$$

where $M_n(\mathbb{R})$ denotes the set of real $n \times n$ matrices and a right superscript X^t indicates matrix transposition of X.

Secondly, a representation σ of the matrix algebra $\mathfrak{gcm}(3)$ provides the algebra's physical interpretation in quantum many-body physics. Let $(\hat{x}_{\alpha j}, \hat{p}_{\alpha j})$, $1 \leq j \leq 3$, denote the dimensionless Cartesian components of the position and momentum hermitian operators for particle α in a finite system of particles. They obey the canonical commutation relation,

$[\hat{x}_{\alpha j}, \hat{p}_{\beta k}] = i\,\delta_{\alpha\beta}\delta_{jk}$. The set of hermitian one-body operators,

$$\begin{aligned} \hat{Q}_{jk} &= \sum_{\alpha} \hat{x}_{\alpha j}\hat{x}_{\alpha k} \\ \hat{N}_{jk} &= \sum_{\alpha} \left(\hat{x}_{\alpha j}\hat{p}_{\alpha k} - \frac{i}{2}\delta_{jk} \right), \end{aligned} \tag{2}$$

close under commutation to form a Lie algebra. For A identical fermions these one-body operators act on the Hilbert space $\mathcal{H}$ that is the antisymmetrized tensor product of A-copies of the single-particle space. For each matrix S in $\mathfrak{gcm}(3)$, define the operator

$$\sigma(S) = i \sum_{jk} X_{jk}\hat{N}_{jk} + \frac{i}{2} \sum_{jk} U_{jk}\hat{Q}_{jk}. \tag{3}$$

When S is in $\mathfrak{gcm}(3)$, the operator $\sigma(S)$ is a skew-adjoint one-body operator on $\mathcal{H}$. This set of operators is a *reducible* representation of $\mathfrak{gcm}(3)$: $[\sigma(S_1), \sigma(S_2)] = \sigma([S_1, S_2])$.

The general collective motion group GCM(3) is a connected Lie matrix group given by exponentiation of $\mathfrak{gcm}(3)$,

$$\mathrm{GCM}(3) = \left\{ g = \begin{pmatrix} x & -x \cdot U \\ 0 & (x^t)^{-1} \end{pmatrix} \mid x \in GL_+(3,R), U \in M_3(R),\ U^t = U \right\}. \tag{4}$$

The connected general linear group $\mathrm{GL}_+(3,\mathrm{R})$ consists of the 3×3 real matrices with positive determinant; it is a subgroup of GCM(3) given by all $g = \mathrm{diag}(x, {}^T x^{-1}) \in \mathrm{GCM}(3)$ for $x \in \mathrm{GL}_+(3,\mathrm{R})$. The representation σ determines a unitary *reducible* representation of the Lie group GCM(3) on the quantum many-fermion Hilbert space $\mathcal{H}$.

Thirdly, the Mackey inducing construction determines all the irreducible unitary representations of GCM(3). For each nonnegative integer C, the Hilbert space of a GCM(3) irrep π is

$$\begin{aligned} H^C = & \Big\{ \Psi : GL_+(3,\mathbf{R}) \to \mathbf{C}^{2C+1} \mid \\ & (i)\Psi(gR) = \Psi(g)\mathcal{D}^C(R), \text{for } g \in GL_+(3,\mathbf{R}), R \in SO(3) \\ & (ii) \int_{GL_+(3,\mathbf{R})} \parallel \Psi(g) \parallel^2 d\nu(g) < \infty \Big\}, \end{aligned} \tag{5}$$

where $\mathcal{D}^C$ denotes the $2C+1$ dimensional unitary irreducible representation of SO(3), and $d\nu(g)$ is the invariant measure on $GL_+(3,\mathbf{R})$. The action of $GL_+(3,\mathbf{R})$ on H^C is given by

$$(\pi(x)\Psi)(g) = \Psi(x^{-1}g), \text{for } x, g \in GL_+(3,\mathbf{R}). \tag{6}$$

The Q tensor acts as a multiplication operator,

$$(\pi(Q_{ij})\Psi)(g) = (g\,{}^t g)_{ij}\Psi(g), \text{for } g \in GL_+(3,\mathbf{R}). \tag{7}$$

Every unitary irreducible representation of GCM(3) is equivalent to one of these representations on H^C for some non-negative integer C. Two irreducible representations defined by two different integers C are inequivalent.

Each GCM(3) irrep C determines a collective model. The $C = 0$ irrep is indistinguishable from the Bohr model, including both quadrupole and monopole collective motion.

2. What GCM(3) irreps are possible physical models?

The answer to this question is the solution to a well-defined group theoretical problem: What GCM(3) irreducible unitary representations C are part of the reduction of the GCM(3) reducible representation σ on many-fermion Hilbert space H?

In a study of the unitary $\mathfrak{sp}(3,\mathbb{R})$ discrete series representations and their $\mathfrak{gcm}(3)$ decomposition, Rowe and Repka [4] proved a reciprocity theorem that precisely answers the question. Consider a symplectic unitary irrep in the reduction of H. This irrep is labeled by the quantum numbers, $N_0(\lambda_0, \mu_0)$, of a "starting" U(3) irrep. The symplectic irrep is a reducible representation of the subgroup GCM(3) $\subset$ Sp(3,$\mathbb{R}$). The reciprocity theorem asserts that the number of times an infinite-dimensional GCM(3) unitary irrep, labeled by C, occurs in the reduction of one infinite-dimensional symplectic irrep equals the multiplicity of the angular momentum in the reduction of the finite-dimensional SU(3) irrep (λ_0, μ_0). For example, the leading symplectic irrep for Neon 20 has $(\lambda_0, \mu_0) = (8, 0)$; consequently, its GCM(3) decomposition is a direct sum of the $C = 0, 2, 4, 6, 8$ irreps.

Thus, except for systems of only a few fermions, every GCM(3) irrep is physically allowed. Whether or not a particular GCM(3) irrep C is discovered in its pure form for some isotope depends on the Hamiltonian. In general, a nuclear wave function will be a linear combination of many GCM(3) irreps with various values of C.

3. Riemann ellipsoids

In Fall 1987, I visited Jerry Draayer at LSU during a sabbatical year. We mainly collaborated on upgrading a symplectic model code to include symplectic tensor operators, especially the hexadecapole E4 operator, and I thoroughly enjoyed the intellectually stimulating environment of Jerry's group and department during my Baton Rouge visit. Norman Lebovitz, an astrophysicist at Chicago, gave a LSU departmental seminar about the Riemann ellipsoid model. Lebovitz had collaborated with Chandrasekhar in the 1960s about the foundations of this classical model and its applications to rotating galaxies and stars [5, 6, 7, 8]. During the seminar it became apparent to me that the Riemann ellipsoid model is the classical version of the GCM(3) quantum model [9]. The GCM(3) matrix group is the same, but it acts in the Riemann model as a group of canonical transformations of many-particle phase space.

Several key insights emerge from the group theoretic perspective on the Riemann model. In the GCM(3) set-up, the Riemann phase space is a co-adjoint orbit of GCM(3). The Riemann dynamical equations are a finite-dimensional Lax system, a considerable mathematical simplification of the Riemann equations [10]. By the way, the classical Lax system can only be derived and proven equivalent to the Riemann dynamical equations by using a simple expression from nuclear theory for the collective kinetic energy that Riemann and Chandrasekhar had overlooked.

An invariant of the Lax system is the GCM(3) Casimir which is a constant function on a co-adjoint orbit. This invariant is the Riemann fluid's Kelvin circulation and its value is C, a nonnegative real number. Only the integral values of C determine a phase space that satisfies the Bohr-Sommerfeld conditions for quantization. The quantized Hilbert space associated with the co-adjoint orbit is the same space H^C carrying a GCM(3) irrep. Thus the physical interpretation of the constant C as the Kelvin circulation comes from the classical Riemann model.

The Riemann model has another interesting aspect in which the circulation and vorticity are related to the angular momentum. The classical Riemann model introduces a real parameter f that determines this relationship in a so-called S-type ellipsoid for which the axis of rotation coincides with a principal axis. This raises a question: What is the fundamental significance of this relationship?

4. Differential geometry of the Riemann model

The relationship between the vortex velocity and the angular velocity determines a differential geometric connection on the group GCM(3). To achieve this understanding, the Lie group GCM(3) is viewed as a principal G bundle [11].

An ellipsoid is defined by a positive-definite real symmetric 3×3 matrix q; the points $X = (X_1, X_2, X_3)$ of its surface in Euclidean space are solutions to the quadratic equation

$$\sum_{ij} q_{ij}^{-1} X_i X_j = 1. \tag{8}$$

In particular, when $q = \mathrm{diag}(a_1^2, a_2^2, a_3^2)$, the surface is an ellipsoid with major axis half-lengths equal to a_1, a_2, a_3 and the ellipsoid's principal axes are aligned with the Cartesian axes. The space of all ellipsoidal surfaces is identified with the six dimensional manifold of such matrices q:

$$Q = \{q \in M_3(\mathbb{R}) \mid q^t = q, q > 0\}. \tag{9}$$

The configuration space for Riemann ellipsoids is the general linear group

$$\mathcal{P} = \{\xi \in M_3(\mathbb{R}) \mid \det \xi > 0\} = GL_+(3, \mathbb{R}), \tag{10}$$

because only linear motions are allowed.

The general linear group transforms an ellipsoid into another ellipsoid that is deformed and rotated. If $q \in Q$ and $\xi \in \mathcal{P}$, then this spatial transformation corresponds to $q \longmapsto \xi q \xi^t$. In particular, the unit sphere with $q = I$ is changed into the ellipsoid with $q = \xi\xi^t$. Thus, the natural mapping π from the group $\mathcal{P}$ onto the ellipsoidal space Q

$$\begin{array}{rcl} \pi : \mathcal{P} & \longrightarrow & Q \\ \xi & \longmapsto & q = \xi\xi^t \end{array} \tag{11}$$

is surjective (onto).

The group $\mathcal{P}$ also acts on itself. The group elements can be multiplied in two ways, on the left or on the right. Left multiplication of elements $\xi \in \mathcal{P}$ by $r \in SO(3)$

$$\xi \longmapsto L_r\xi = r\xi \tag{12}$$

corresponds to physical rotations r in three-dimensional Euclidean space. In contrast, right multiplication of $\xi \in \mathcal{P}$ by $g \in G = SO(3)$,

$$\xi \longmapsto R_g\xi = \xi g^{-1}, \tag{13}$$

corresponds to vortex motion. Although the groups G and $SO(3)$ are mathematically isomorphic, their distinct actions on $\mathcal{P}$ imply different physical interpretations.

With respect to left multiplication by the rotation group element $r \in SO(3)$, the ellipsoid with inertia tensor $q = \pi(\xi) = \xi\xi^t$ is transformed into the rotated ellipsoid with inertia tensor $\pi(L_r\xi) = r\xi\xi^t r^t = rqr^t$. But right multiplication by the vorticity group element $g \in G$ leaves the inertia ellipsoid invariant, $\pi(R_g\xi) = \xi g^{-1} g \xi^t = q$, since $g^{-1} = g^t$. This invariance is expressed more elegantly by the composition of mappings

$$\pi \circ R_g = \pi, \text{ for all } g \in G. \tag{14}$$

Hence G is a group of motions internal to the ellipsoidal surface.

These properties show that $\mathcal{P}$ is a principal G bundle where G is the structure or gauge group, Q is the base manifold, π is the projection of the bundle onto the base manifold, and π

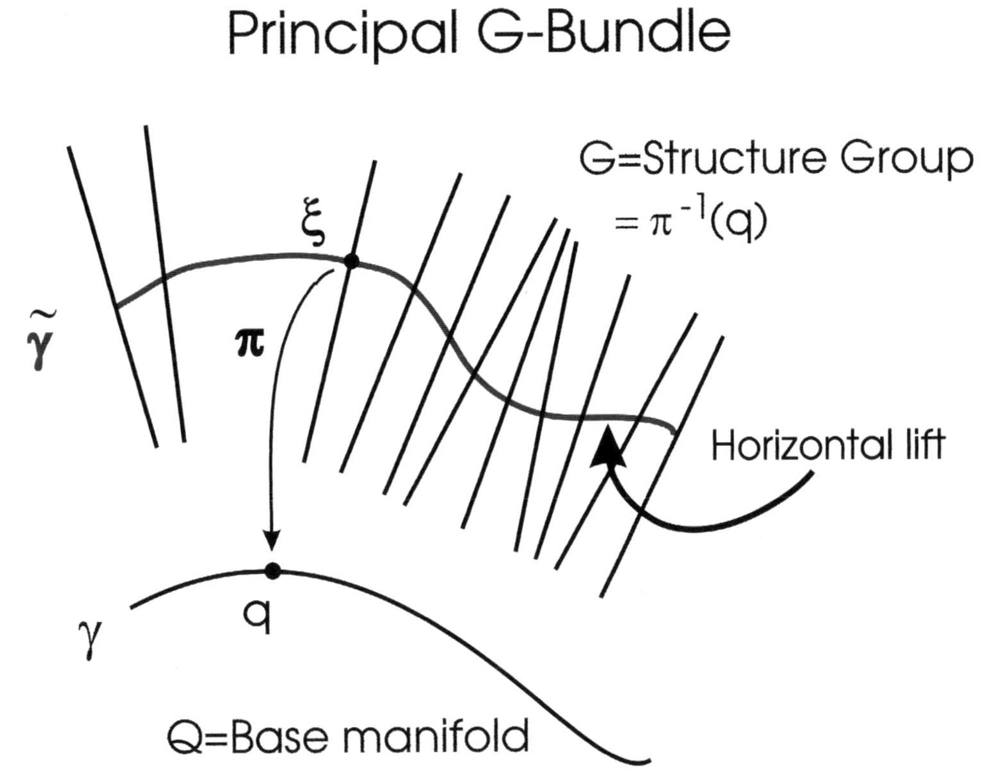

Figure 1. Horizontal lift

is right G-invariant. In Figure a time-dependent curve $\gamma(t)$ is drawn in the base manifold Q. As the time evolves, the curve describes the evolution of the orientation and deformation of a rotating ellipsoid. Above the base manifold are the fibres G of the bundle, indicated as straight lines in the figure. A point ξ of the bundle depends both on the deformation $q = \pi(\xi)$ and the internal vorticity measured by a structure group element.

As the ellipsoid is "cranked," in the usual nuclear physics sense, the system responds with internal vortex dynamics. This response determines a unique lift of the curve γ in the base manifold Q to a curve $\tilde{\gamma}$ in the bundle $\mathcal{P}$. In differential geometry, the curve $\tilde{\gamma}$ is called the *horizontal lift* of γ and it is mathematically equivalent to specifying a differential connection for the bundle [12].

5. Connections on the G principal bundle

A tangent vector to γ in the base manifold determines the angular velocity $\vec{\omega}$, and the horizontal lift determines the vortex velocity $\vec{\lambda}$. In concrete terms, the ratio of the components of these velocities define the Christoffel symbols for the connection:

$$\Gamma_i^a(q) = \delta_i^a \left(\frac{\lambda_i}{\omega_i}\right). \tag{15}$$

For an ellipsoid with principal axes half-lengths a_1, a_2, a_3, the special rotational modes for rigid, irrotational and the "falling cat" define the three connections

$$\Gamma_k^k = \begin{cases} 0 & \text{rigid} \\ 2a_i a_j/(a_i^2 + a_j^2) & \text{irrotational} \\ (a_i^2 + a_j^2)/(2a_i a_j) & \text{falling cat,} \end{cases} \tag{16}$$

where i, j, k are cyclic. A "falling cat" has internal vortex motion, yet no change in the angular momentum. The irrotational flow connection is associated with the Riemannian geometry on the bundle; the falling cat is the G-invariant connection. For an S-type ellipsoid the connection is a linear interpolation between the special rigid and irrotational flow connections.

What equations determine the connection? The answer must be independent of the coordinate system, and the natural geometrical object for the equations is the bundle curvature. One equation that the curvature satisfies is the Bianchi identity. A second equation relates the exterior derivative of the curvature tensor to the non-holonomic internal forces of the system.

The analogy with classical theory of electricity and magnetism is helpful. This is an Abelian gauge theory with space-time as the base manifold and $U(1)$ as the structure group. The curvature form is the Faraday tensor and the Bianchi identity is equivalent to two of Maxwell's equations, $\nabla \cdot \mathbf{B} = 0$ and $\nabla \times \mathbf{E} + \partial \mathbf{B}/\partial t = 0$. The other two of Maxwell's equations involve the current and are equivalent to a second independent equation for the Faraday tensor.

6. Covariant derivatives and the quantum model

The quantization of a gauge theory for a principal G bundle is achieved by constructing its associated bundles [11]. The domain of the wave functions is the base manifold and the wave functions are vector-valued. The vector values are in an irreducible representation of the structure group. For each irreducible representation of the structure group $G \simeq SO(3)$, an inequivalent associated bundle and a different quantization can be defined. An associated bundle for the Riemann model defines the space H^C of vector-valued wave functions.

A new ingredient of the bundle theory is that the classical connection determines the quantum covariant derivative. The Laplacian defined by the covariant derivative yields the collective kinetic energy. For irrotational flow, $C = 0$, and the Riemannian connection, the Laplacian is the familiar Bohr collective kinetic energy operator. But, when $C \neq 0$ and the connection is not Riemannian, a significant generalization of the Bohr collective model is attained.

Associated bundles and the covariant derivative are the essential structure of Yang-Mills theory. Thus the Bohr model is an example of a Yang-Mills theory. Indeed, the relationship is rather close because the Riemann bundle theory has the structure group SO(3) while the Yang-Mills electroweak theory uses the structure group U(2).

In a forthcoming article, the associated bundle theory for the Riemann model will be presented in some detail.

Acknowledgments

My thanks to my mentor and friend, David Rowe, who identified many of the key group theoretic ideas hidden in the original formulation of the Bohr collective model.

References

[1] Rosensteel G and Rowe D J 1976 *Ann. Phys., NY* **96** 1
[2] Weaver O L, Cusson R Y and Biedenharn L C 1976 *Ann. Phys., NY* **102** 493
[3] Bohr Å 1952 *Danski. Vid. Selsk. Mat. Fys. Medd.* **26**, No. 14; Bohr Å and Mottelson B 1954 *Rotational States in Atomic Nuclei* (Copenhagen: Ejnar Munksgaard)
[4] Rowe D J and Repka J 1998 *J. Math. Phys.* **39**, 6214
[5] Riemann B 1860 *Abh. K. Ges. Wis. Göttingen* **9** 3.

[6] Lebovitz N R 1967 *Ann. Rev. Astr. Astrophys.* **5** 465; 1984 *Ap. J.* **284** 364; 1990 *in* "Nonlinear astrophysical fluid dynamics" (New York: Ann. New York Acad. Sci.)
[7] Chandrasekhar S 1969 *Ellipsoidal Figures of Equilibrium* (New Haven: Yale Univ. Press)
[8] Dyson F J 1968 *J. Math. Mech.* **18** 91.
[9] Rosensteel G 1988 *Ann. Phys., NY* **186** 230
[10] Rosensteel G 1994 *Appl. Math. Lett.* **6** 55-58
[11] Kobayashi S and Nomizu K 1963 *Foundations of Differential Geometry* (New York: Interscience)
[12] Rosensteel G 2001 *J. Phys. A: Math. Gen.* **34** L169-L178

Do we understand excited 0^+ states in nuclei?

John L Wood[1]
School of Physics, Georgia Institute of Technology, Atlanta, Georgia, 30332-0430, USA

E-mail: john.wood@physics.gatech.edu

Abstract. Excited 0^+ states are the least understood of any low-energy degree of freedom in nuclei. Some examples of excited 0^+ states that are reasonably well understood are followed by details, from recent experimental work, of excited 0^+ states that are causing major reassessment of some models.

1. Introduction

Excited 0^+ states can arise in nuclei in association with the nucleon pairing degree of freedom and in model spaces with collective shape degrees of freedom. Models of pairing in nuclei are by now developed to the point that there is a wide consensus regarding the basic physics issues. Models of collective shape degrees of freedom are much further from a consensus on the basic physics issues. Collectivity in nuclei has been one of the major topics of nuclear structure research for sixty years. From Bohr's initial work [1], Bohr and Mottelson [2] working with their many collaborators at the Niels Bohr Institute in Copenhagen, extensively explored the case for low-energy collective motion in nuclei dominated by quadrupole shapes. This developed the Bohr model into a more unified perspective (with coupling of nucleon degrees of freedom) that has come to be known as the Bohr-Mottelson model. Finer details of the Bohr model were developed by Greiner and the Frankfurt School (see [3]).

Three further lines of development were pursued beyond the early work in Copenhagen and Frankfurt. The first attempted to arrive at the parameters of the Bohr-Mottelson-Frankfurt model by mapping from many-nucleon degrees of freedom: this approach was initiated by Kumar and Baranger [4]; and activity along this line continues intensively. The second adopted a boson approximation: the many boson model approaches are reviewed in Klein and Marshalek [5]. The most notable of the boson-based models is the interacting boson model (IBM) of Arima and Iachello [6]; and activity using this model remains intensive. The IBM has particularly introduced many practitioners of nuclear structure research, especially experimentalists, to the language of dynamical groups and spectrum generating algebras. The third approached the successes of the Bohr model from the fundamental view that it must be a submodel of the nuclear shell model. Building on work of Elliott [7] and Weaver, Biedenharn and Cusson [8], Rosensteel and Rowe [9] showed that the solution to this problem resided in the symplectic group, Sp(3,R).

This rich arsenal of collective models has provided the language with which theorists and experimentalists have expressed their ideas and discoveries for sixty years. But, do data point towards a preference between these models? The best-established mode of quadrupole collectivity in nuclei is

[1] To whom any correspondence should be addressed.

(near-rigid) quadrupole shapes and their rotational degrees of freedom. These can be fitted phenomenologically with high precision using the symmetric and asymmetric top limits of the Bohr model and their low-order corrections resulting from band mixing. (For recent critical assessments of "best case" axially symmetric and asymmetric rotors, see Kulp et al. [10] and Allmond et al. [11], respectively.) The symplectic model offers two dynamical insights into nuclear rotation that go beyond phenomenology: first it predicts that vorticity must be present in rotating nuclei, second it accommodates the Elliott [SU(3)] model. While the existence of vorticity in rotating nuclei remains an open question, SU(3) models of nuclear rotation have been thoroughly developed by Rowe and coworkers (see [12] and references therein) and Draayer and coworkers (see, e.g., [13]) and are strongly supported by data (see, e.g., [14]). Most notably, these descriptions show that it is not necessary to use effective E2 charges.

Beyond the basic rotational degrees of freedom exhibited by many nuclei, the most visible and widespread low-energy degrees of freedom in even-even nuclei are states with spin-parity 0^+. There appear to be nearly as many theories purporting to describe these states as there are experimentally well-characterized examples. Low-lying excited 0^+ states appear to arise in a variety of ways in even-even nuclei with little evident commonality in their structure. In the following, a brief summary of well-characterized excited 0^+ states is given. This is followed by details, based on very recent (and on-going) programs of experimental study, of two major redirections of interpretation that are necessitated: (1) It is questionable as to whether nuclei possess low-energy quadrupole vibrational degrees of freedom. (2) It is likely that a unified view of excited 0^+ states in nuclei based on pair excitations across shell and subshell gaps is supported.

2. Excited 0^+ states that are well understood

The 0^+ excited state with the most enduring interpretation is the first excited state in ^{16}O which was proposed by Morinaga [15] to be a proton-pair-neutron-pair excitation across the N, Z = 8 closed shells with associated deformation resulting from the interaction between protons and neutrons. This led to a similar interpretation of excited 0^+ states in ^{40}Ca and a plausible continuation in ^{56}Ni (see [16,17,18] for reviews of these developments). Another doubly closed-shell nucleus that exhibits an excited 0^+ state, which is reliably interpreted, occurs in ^{208}Pb [19]. The state results from a pair excitation across the N = 126 shell gap. However, this is not a deformed structure: this points to the interaction between active protons and neutrons being important for deformation.

In singly closed-shell nuclei the widespread occurrence of deformed bands built on low-lying excited 0^+ states has emerged to become a cornerstone of nuclear structure under the title of shape coexistence. These structures involve pair excitations across closed shells and occur lowest in energy near mid-shell for the other kind of nucleon with a "parabolic" energy trend as a function of open-shell nucleon number. Such structures persist in nuclei adjacent to closed shells. Shape coexistence has been reviewed, most recently, by Heyde and Wood [18].

Two types of low-lying excited 0^+ state that have been widely accepted among nuclear structure physicists are the two-phonon 0^+ state of the harmonic quadrupole vibrator model applied to nuclei believed to be spherical and the one-phonon beta vibrational state of the quadrupole deformed models applied to nuclei believed to be deformed. However, detailed investigations of some of the best candidate nuclei for exhibiting these modes show that vibrational behavior is untenable: this is taken up in the next section.

3. Excited 0^+ states that necessitate a reinterpretation

Extensive investigation of ^{110,112,114,116}Cd (see [20]) has revealed that these nuclei, far from being "text-book" cases of near harmonic spherical vibrators, exhibit serious disagreement with expected multi-phonon patterns of low-energy excitation. This is summarized in Fig. 1. The experimental work needed to explore collective structures at the level exhibited by Fig. 1 is highly demanding on experimental technique. Lifetimes of some of the levels are in the femtosecond range and need to be determined by Doppler shifted γ-ray energies following inelastic neutron scattering (see, e.g., [21]).

Important collective transitions are often low energy and compete with high-energy (non-collective) decay branches: this incurs the $(E_\gamma)^5$ factor for E2 transition rates which results in the γ-rays corresponding to the transitions of interest having decay branches < 10^{-3} of the strongest decay branches (see, e.g., [22]).

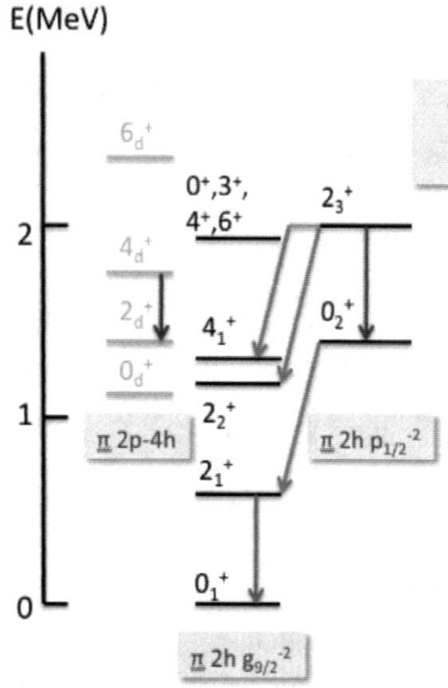

trans.	^{110}Cd	^{112}Cd	^{114}Cd	^{116}Cd	harm. vib.
$4_d 2_d$	115^{35}		119^{12}		
$2_3 0_2$	24^{2}	27^{8}	17^{5}	35^{10}	42
$2_3 4_1$	< 5	< 0.4	< 0.3	< 7	31
$2_3 2_2$	< 0.7^{6}	< 2	2.8^{4}	2.0^{6}	17
$0_2 2_1$	< 7.9	0.0099^{44}	0.0026^{4}	0.55^{4}	60
$2_1 0_1$	27	30	31	34	30 (norm)

Figure 1. Absolute B(E2) values in Weisskopf units (W.u.) for transitions between low-lying states in ^{110,112,114,116}Cd. The transitions indicated by diagonal (red) arrows disagree with a harmonic quadrupole vibrator interpretation of these nuclei. The band of states marked "π 2p-4h" is the deformed proton-pair excitation intruder band. The other 0^+ configurations are discussed in the text.

The state in ^{110,112,114,116}Cd which has popularly been considered to be a two-phonon 0^+ state is *systematically* observed to decay with a B(E2) less than 1% of the strength expected for a harmonic quadrupole vibrational two-phonon state. Attempts have been made to rationalize this with detailed IBM model calculations that take into account the mixing between the candidate vibrational states and the deformed intruder states (see [23] for the most recent investigation of this approach). However, such calculations require the fine-tuning of a number of model parameters that vary from nucleus to nucleus: this casts doubt on the validity of such a description.

The question arises as to the nature of the purported 0^+ two-phonon state. It is plausibly a proton-pair excitation involving an energy gap between the $p_{1/2}$ and $g_{9/2}$ proton subshells. However, this idea will need exploration via one- and two-proton transfer reaction spectroscopy. It should be noted that this 0^+ state has an associated 2^+ state with a B(E2) that is comparable to the ground-state collective structure (and a similar energy spacing). Thus, the structure would differ in pair structure, rather than deformation, relative to the ground state.

Multiple spectroscopic studies of ^{152}Sm reveal collectivity that supports two shape coexisting structures that strongly mix [24]. The interpretation of the first excited 0^+ state as a beta vibration is refuted by multi-step Coulomb excitation data: some details are shown in Fig. 2. If the 0^+ state at 685 keV were a beta vibration, the Coulomb excitation path from this state would connect to a two phonon

$K^\pi = 0^+$ band at an excitation energy of ~1370 keV (besides strong population of the rotational band built on the 685 keV state). Evidently, such an excitation is completely absent below an energy of ~2400 keV (deduced from the γ-ray line intensities and the energy range covered in the spectrum).

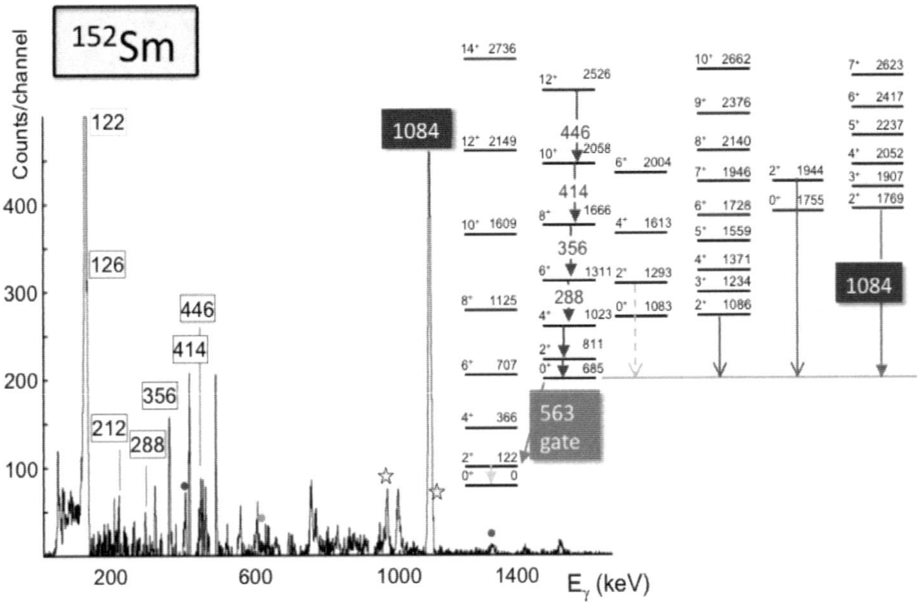

Figure 2. Response of the first excited 0^+ state in the nucleus ^{152}Sm to multi-step Coulomb excitation. The only state that is observed with strong population is the head of a $K^\pi = 2^+$ band at 1769 keV. (The population of the rotational band built on the first excited 0^+ state is masked by a decay branching ratio of 99.7% out of the band at 811 keV.)

A number of other features are revealed in Fig. 2. Two $K^\pi = 2^+$ bands occur with band-head energy separations that are nearly identical to that of the ground and first-excited $K^\pi = 0^+$ band-head energy separations. A similar pattern is exhibited by two $K^\pi = 0^-$ bands in ^{152}Sm [25]. The structure of ^{152}Sm appears rather as a nucleus possessing a "double-vacuum" [24] (see also [26]). There is an excited $K^\pi = 0^+$ band built on a 0^+ state at 1083 keV. However, this band has all of the properties of a "pairing isomer" [27, 28, 29]: such structures possess characteristic two-nucleon transfer strengths with an asymmetry between pick-up and stripping (see, e.g., [30, 31, 32]).

The evidence for shape coexistence comes from electric monopole (E0) transitions which systematically occur between the lowest $K^\pi = 0^+$ bands in N = 90 isotones [18]. The essential behavior is schematically depicted in Fig. 3. The underlying 0^+ excitation is plausibly a consequence of a subshell energy gap at Z = 64. There is accumulating evidence for E0 transition strength between pairs of $K^\pi = 2^+$ bands in the N = 90 isotones [33].

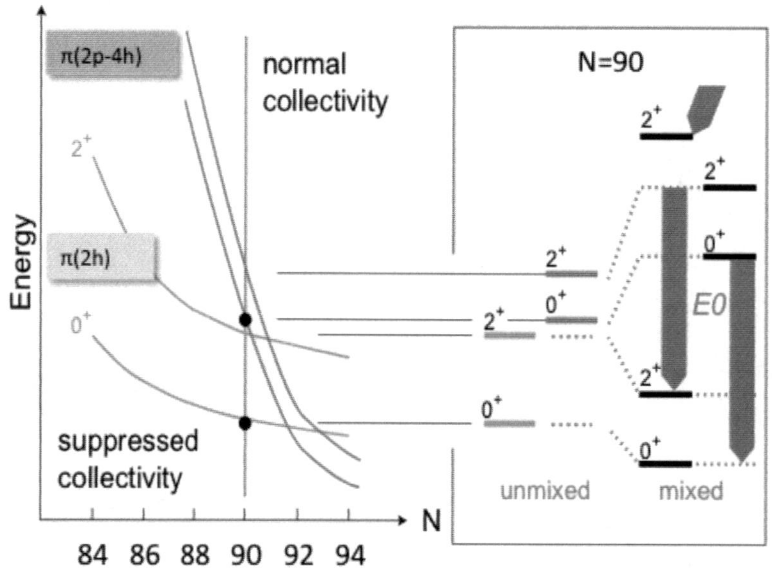

Figure 3. Schematic view of intruder states and shape coexistence at N = 90. The nuclei with N < 90 are weakly deformed, those with N > 90 are strongly deformed, and those at N = 90 exhibit near-degenerate coexisting structures; these strongly mix giving rise to strong electric monopole (E0) transition strength.

The E0 transition strengths between the lowest two $K^{\pi} = 0^+$ bands are shown in Fig. 4. The occurrence of E0 transition strength in nuclei is a model-independent measure of the coexistence of configurations with different mean-square charge radii that mix [34, 35].

4. Conclusions

The nature of excited 0^+ states in nuclei would appear to need further investigation. The historical interpretation of many of these states as resulting from vibrations of either spherical or deformed equilibrium shapes with a dominance of quadrupole shape degrees of freedom must be questioned. The above examples suggest that a unified view of excited 0^+ states in nuclei may be achievable by considering pair excitations across subshell energy gaps; similar to the occurrence of pair excitations across major shell energy gaps which give rise to shape coexistence.

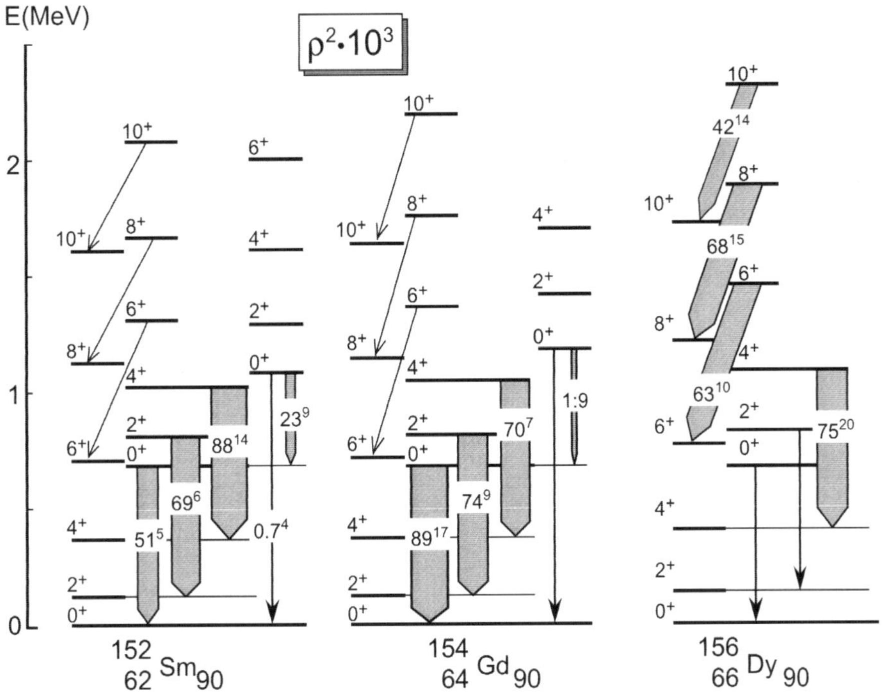

Figure 4. Electric monopole transition strengths between the lowest two $K^{\pi} = 0^{+}$ bands in the N = 90 isotones. (The figure is similar to one shown in ref. [18].)

Acknowledgements

The author wishes to acknowledge collaborations with Mitch Allmond, Paul Garrett, Kris Heyde, David Kulp, and Ed Zganjar who have significantly contributed to the ideas presented in this work.

5. References

[1] Bohr A 1952 *Mat. Fys. Medd. Dan. Vid. Selsk.* **26** No. 14
[2] Bohr A and Mottelson B 1953 *Mat. Fys. Medd. Dan. Vid. Selsk.* **27** No.16
[3] Eisenberg J M and Greiner W 1987 *Nuclear Models* (North Holland, Amsterdam), third edn.
[4] Kumar K and Baranger M 1967 *Nucl. Phys.* A **92** 608
[5] Klein A and Marshalek E 1991 *Rev. Mod. Phys.* **63** 375
[6] Arima A and Iachello F 1987 *The Interacting Boson Model* (Cambridge Univ. Press)
[7] Elliott P 1958 *Proc. Roy. Soc.* A **245** 128
[8] Weaver O L, Biedenharn L and Cusson R Y 1973 *Ann. Phys.* **77** 250
[9] Rosensteel G and Rowe D J 1977 *Phys. Rev. Lett.* **38** 10
[10] Kulp W D, Allmond J M, Hatcher P, Wood J L, Loats J, Schmelzenbach P, Stapels C J, Krane K S, Larimer R-M and Norman E B 2006 *Phys. Rev.* C **73** 014302
[11] Allmond J M, Zaballa R, Oros-Peusquens A, Kulp W D and Wood J L 2008 *Phys. Rev.* C **78** 014302
[12] Rowe D J 1996 *Prog. Nucl. Part. Phys.* **37** 225
[13] Leschber Y, Draayer J P and Rosensteel G 1996 *J. Phys.* G **75** L179

[14] Jarrio M, Wood J L and Rowe D J 1991 *Nucl. Phys.* A **528** 409
[15] Morinaga H 1956 *Phys. Rev.* **101** 254
[16] Heyde K, Van Isacker P, Waroquier M, Wood J L and Meyer R A 1983 *Phys. Repts.* **102** 291
[17] Wood J L, Heyde K, Nazarewicz W, Huyse M and Van Duppen P 1992 *Phys. Repts.* **215** 101
[18] Heyde K and Wood J L 2011 *Rev. Mod. Phys.* **83** 1467
[19] Igo G J, Barnes P D and Flynn E 1970 *Phys. Rev. Lett.* **24** 470
[20] Garrett P E and Wood J L 2010 *J. Phys.* G **37** 064028
[21] Garrett P E, Green K L, Lehmann H, Jolie J, McGrath C A, Yeh M and Yates S W 2007 *Phys. Rev.* C **75** 054310
[22] Green K L et al. 2009 *Phys. Rev.* C **80** 032502
[23] Garrett P E, Green K L and Wood J L 2008 *Phys. Rev.* C **78** 044307
[24] Kulp W D et al. 2008 *Phys. Rev.* C **77** 061301(R)
[25] Garrett P E et al. 1992 *Phys. Rev. Lett.* **103** 062501
[26] Sharpey-Schafer J F et al. 2011 *Eur. Phys. J.* A **47** 5
[27] Kulp W D, Wood J L, Krane K S, Loats J, Schmelzenbach P, Stapels C J, Larimer R-M and Norman E B 2006 *Phys. Rev. Lett.* **91** 102501
[28] Kulp W D et al. 2005 *Phys. Rev.* C **71** 041303(R)
[29] Ragnarsson I and Broglia R A 2009 *Nucl. Phys.* A **263** 315
[30] Friedman A M, Katori K, Albright D and Schiffer J P 1974 *Phys. Rev.* C **9** 760
[31] Casten R F, Flynn E R, Garrett J D, Hansen O, Mulligan T J, Bes D R, Broglia R A and Nilsson B 1972 *Phys. Lett.* B **40** 333
[32] Allmond J M et al. 2010 *Phys. Rev.* C **81** 064316
[33] Kulp W D, Wood J L, Krane K S, Loats J, Schmelzenbach P, Stapels C J, Larimer R-M and Norman E B 2006 *Phys. Rev.* C **69** 064309
[34] Kantele J, Julin R, Luontama M, Passoja A, Poikolainen T, Backlin A and Jonsson N-G 1979 *Z. Phys.* A **289** 157
[35] Wood J L, Zganjar E F, De Coster C and Heyde K 1999 *Nucl. Phys.* A **651** 323

Recent developments in the shell model Monte Carlo approach to nuclei

Y Alhassid[1], A Mukherjee[1], H Nakada[2] and C Özen[3]
[1]Center for Theoretical Physics, Sloane Physics Laboratory, Yale University, New Haven, Connecticut 06520, USA
[2]Department of Physics, Graduate School of Science, Chiba University, Inage, Chiba 263-8522, Japan
[3]Faculty of Engineering and Natural Sciences, Kadir Has University, Istanbul 34083, Turkey

E-mail: yoram.alhassid@yale.edu

Abstract. The shell model Monte Carlo (SMMC) approach provides a powerful method for the microscopic calculation of statistical and collective nuclear properties in model spaces that are many orders of magnitude larger than those that can be treated by conventional methods. We discuss recent applications of the method to describe the emergence of collectivity in the framework of the configuration-interaction shell model and the crossover from vibrational to rotational collectivity in families of rare-earth nuclei. We have calculated state densities of these rare-earth nuclei and find their collective enhancement factors to be correlated with the pairing and shape phase transitions. We also discuss an accurate method to calculate the ground-state energy of odd-even and odd-odd nuclei, circumventing the sign problem that originates in the projection on an odd number of particles. We have applied this method to calculate pairing gaps in families of isotopes in the iron region.

1. Introduction

A major challenge in nuclear theory is the derivation of nuclear properties from the underlying effective nuclear interactions. Advances in many-body methods and high-performance computing have enabled significant progress in the application of ab initio methods such as Green's function Monte Carlo [1], and the no-core shell model [2] and its symplectic extension [3]. However, these methods are limited to light nuclei. Density functional theory (DFT) has made much progress and is unique in providing a global theory of nuclei [4] but correlations beyond mean-field theory are often important and require extensions of DFT [5].

The configuration-interaction (CI) shell model approach provides an attractive alternative as it accounts for both shell effects and correlations [6]. With the help of increasingly more efficient matrix diagonalization algorithms, the CI approach has been successful in describing the low-energy spectroscopy of fp-shell nuclei. However, in heavier nuclei conventional diagonalization methods become limited by the combinatorial growth of the dimensionality of the many-particle space, hindering applications of the conventional CI approach in medium-mass and heavy nuclei.

The shell model Monte Carlo (SMMC) method [7, 8] has enabled calculations in model spaces that are many orders of magnitude larger than those that can be treated by conventional methods. Monte Carlo methods for fermions are often limited by the so-called sign problem.

However, the dominant components [9] of realistic effective nuclear interactions have a good sign and are often sufficient for realistic calculations of statistical and collective properties of nuclei. The smaller, bad sign components can be treated by the method of Ref. [8].

Here we discuss recent developments in SMMC, and in particular the application of SMMC in the study of collective and statistical properties of heavy nuclei [10, 11]. Another recent progress has been the introduction of a method to calculate the ground-state energy of odd-even and odd-odd nuclei, circumventing a sign problem that originates from the projection on an odd number of particles [12].

2. The shell model Monte Carlo approach

2.1. Hubbard-Stratonovich transformation

The SMMC is based on the Hubbard-Stratonovich (HS) transformation [13], in which the Gibbs operator $e^{-\beta H}$ of a system with an Hamiltonian H at an inverse temperature $\beta = 1/T$ is expressed as a functional integral

$$e^{-\beta H} = \int D[\sigma] G_\sigma U_\sigma \tag{1}$$

over external auxiliary fields $\sigma(\tau)$. Here G_σ is a Gaussian weight and U_σ is a one-body propagator describing non-interacting nucleons moving in external fields $\sigma(\tau)$.

Using the HS transformation, the thermal expectation value of an observable O can be written as

$$\langle O \rangle = \frac{\mathrm{Tr}\,(O e^{-\beta H})}{\mathrm{Tr}\,(e^{-\beta H})} = \frac{\int D[\sigma] W_\sigma \Phi_\sigma \langle O \rangle_\sigma}{\int D[\sigma] W_\sigma \Phi_\sigma} , \tag{2}$$

where $W_\sigma = G_\sigma |\,\mathrm{Tr}\; U_\sigma|$ is a positive-definite function and $\Phi_\sigma = \mathrm{Tr}\; U_\sigma / |\,\mathrm{Tr}\; U_\sigma|$ is known as the Monte Carlo sign function. The quantity $\langle O \rangle_\sigma = \mathrm{Tr}\,(O U_\sigma)/\mathrm{Tr}\, U_\sigma$ describes the thermal expectation value of the observable at a given configuration of the auxiliary fields σ.

In the SMMC approach H is a rotationally invariant shell-model Hamiltonian defined in a truncated many-particle space spanned by a specified set of N_s single-particle orbitals (this set can be different for protons and for neutrons). The quantities in the integrands of Eq. (2) can be expressed in terms of the $N_s \times N_s$ matrix $\mathbf{U}_\sigma$ representing U_σ in the single-particle space. For example, the grand-canonical trace of U_σ is given by $\mathrm{Tr}\; U_\sigma = \det(\mathbf{1} + \mathbf{U}_\sigma)$.

In the finite nucleus it is important to use the canonical ensemble with fixed number of protons and neutrons. This is done using particle-number projection, which for a finite number of orbitals can be expressed as a discrete Fourier sum. For example, the trace of U_σ at fixed particle number A is given by

$$\mathrm{Tr}_A\, U_\sigma = \frac{e^{-\beta\mu A}}{N_s} \sum_{m=1}^{N_s} e^{-i\phi_m A} \det\left(\mathbf{1} + e^{i\phi_m} e^{\beta\mu} \mathbf{U}_\sigma\right) , \tag{3}$$

where $\hat{A}$ is the particle-number operator, $\phi_m = 2\pi m/N_s$ $(m = 1, \ldots, N_s)$ are quadrature points and μ is a real chemical potential [required for the numerical stabilization of the sum in (3)].

2.2. Monte Carlo method

The imaginary-time interval $(0, \beta)$ is discretized in steps of $\Delta\beta$. The integration over the auxiliary fields is of very high dimension and is carried out by a Monte Carlo method. The integration over each σ field at a given time is approximated by a three-point quadrature formula and importance sampling is used to choose M uncorrelated configurations σ_k according to the positive-definite weight function W_σ. The expectation value in (2) is then estimated from

$$\langle O \rangle \approx \frac{\sum_k \langle O \rangle_{\sigma_k} \Phi_{\sigma_k}}{\sum_k \Phi_{\sigma_k}} . \tag{4}$$

Journal of Physics: Conference Series **403** (2012) 012012 doi:10.1088/1742-6596/403/1/012012

3. Collectivity in the configuration-interaction shell model approach

Heavy nuclei are known to exhibit collectivity in their low-energy spectra and electromagnetic transition rates. Phenomenological nuclear structure models describe well the various types of collectivity such as vibrational collectivity in spherical nuclei and rotational collectivity in deformed nuclei. However, much less is understood about the microscopic emergence of collectivity in the framework of a CI shell model approach. A shell model treatment of heavy deformed nuclei requires very large model spaces and that are accessible by the SMMC method.

We have extended the SMMC method to heavy nuclei using a proton-neutron formalism in which the protons and neutrons can occupy different shells [10]. In particular, we have applied the method to rare-earth nuclei using the $50-82$ major shell plus the $1f_{7/2}$ orbital for protons and the $80-126$ major shell plus the $0h_{11/2}$ and $1g_{9/2}$ orbitals for neutrons.

The SMMC method enables us to calculate thermal and ground-state observables in very large model spaces, but it does not provide detailed spectroscopic information. Since nuclear collectivity is often identified by its spectroscopic characteristics, it is necessary to find a thermal observable that is sensitive to the specific type of collectivity. We have identified such an observable in $\langle \mathbf{J}^2 \rangle_T$, where $\mathbf{J}$ is the total angular momentum of the nucleus [10, 11]. For an even-even nucleus and at sufficiently low temperatures, this observable is given by

$$\langle \mathbf{J}^2 \rangle_T \approx \begin{cases} 30 \dfrac{e^{-E_{2^+}/T}}{\left(1-e^{-E_{2^+}/T}\right)^2} & \text{vibrational band} \\ \dfrac{6}{E_{2^+}} T & \text{rotational band} \end{cases} \tag{5}$$

where E_{2^+} is the excitation energy of the first 2^+ level.

3.1. Rotational collectivity in heavy deformed nuclei

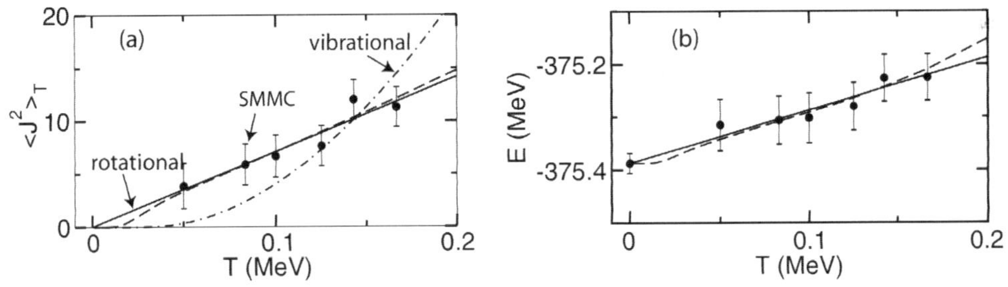

Figure 1. (a) $\langle \mathbf{J}^2 \rangle_T$ and (b) thermal energy $E(T)$ versus T for ^{162}Dy. The SMMC results (solid circles with error bars) are compared in (a) with fits to the rotational model (solid line) and vibrational model (dotted-dashed line) [see Eq. (5)], and in (b) with the rotational model (solid line). The dashed lines are obtained from the lowest five experimental rotational bands. Adapted from Ref. [10].

An important issue is whether we can describe the rotational collectivity of heavy deformed nuclei in the framework of a truncated spherical shell model space. In Ref. [10] we studied a strongly deformed rare-earth nucleus ^{162}Dy using the SMMC method and demonstrated its rotational character through the low-temperature behavior of $\langle \mathbf{J}^2 \rangle_T$ and the thermal energy $E(T)$. Figure 1 shows these observables as a function of temperature. For comparison we also show for $\langle \mathbf{J}^2 \rangle_T$ the fits to the rotational (solid line) and vibrational (dotted-dashed line) models in Eq. (5). The SMMC results clearly indicate the rotational character of ^{162}Dy in the truncated shell model space. Furthermore, the fit to the rotational model gives a ground-state moment of inertia of $I_{\rm g.s.} = 35.5 \pm 3.3$ MeV^{-1}, in agreement with the experimental value of $I_{g.s.} = 3/E_{2^+} = 37.2$ MeV^{-1}. The solid line in panel (b) is a fit to the rotational model prediction for the thermal energy $E(T) = E_0 + T$, where E_0 is the ground-state energy.

3.2. Crossover from vibrational to rotational collectivity

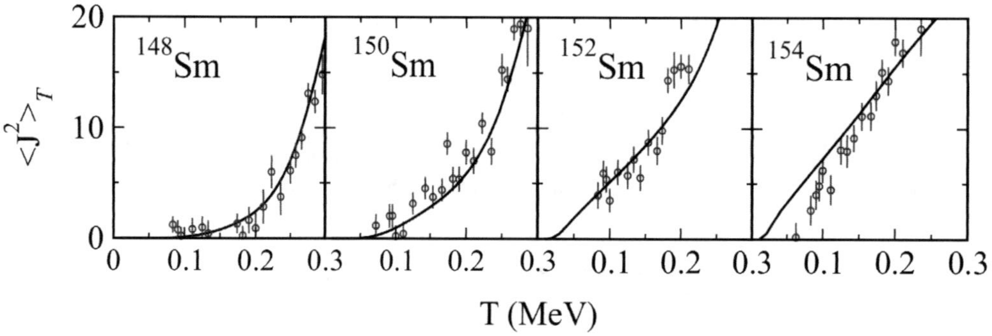

Figure 2. $\langle \mathbf{J}^2 \rangle_T$ as a function of temperature in the even $^{148-154}$Sm isotopes. The SMMC results (open circles with error bars) are compared with Eq. (6) using a complete set of experimentally known discrete nuclear levels and an experimentally determined BBF state density (except for ^{154}Sm, for which only discrete nuclear levels are used). Adapted from Ref. [11].

Even samarium and neodymium isotopes display a shape phase transition from spherical to deformed nuclei as the number of neutrons increases from shell closure towards the mid-shell region. In the SMMC approach, a signature of this transition can be observed in the low-temperature behavior of $\langle \mathbf{J}^2 \rangle_T$. In Fig. 2 we compare in the even $^{148-154}$Sm isotopes the SMMC results (circles) for $\langle \mathbf{J}^2 \rangle_T$ with results that are derived from the experimental data (solid lines) according to

$$\langle \mathbf{J}^2 \rangle_T = \frac{1}{Z(T)} \left(\sum_i^N J_i(J_i+1)(2J_i+1)e^{-E_i/T} + \int_{E_N}^{\infty} dE_x \, \rho(E_x) \, \langle \mathbf{J}^2 \rangle_{E_x} \, e^{-E_x/T} \right) . \quad (6)$$

Here $Z(T) = \sum_i^N (2J_i+1)e^{-E_i/T} + \int_{E_N}^{\infty} dE_x \rho(E_x) e^{-E_x/T}$ is the corresponding experimental partition function. The summations in (6) and in the expression for $Z(T)$ are carried over a complete set of experimentally known low-lying levels i with excitation energy E_i and spin J_i, and $\rho(E_x)$ is the back-shifted Bethe formula (BBF) with parameters that are determined from level counting and neutron resonance data. $\langle \mathbf{J}^2 \rangle_{E_x}$ (the average value of $\mathbf{J}^2$ at a given excitation energy E_x) is calculated from $\langle \mathbf{J}^2 \rangle_{E_x} = 3\sigma^2$, where σ^2 is the rigid-body spin-cutoff parameter. We find good agreement between the SMMC results and the experimental estimates. ^{148}Sm is spherical in its ground state and its $\langle \mathbf{J}^2 \rangle_T$ exhibits a "soft" response to temperature, typical of vibrational nuclei. In contrast, $\langle \mathbf{J}^2 \rangle_T$ in ^{154}Sm exhibits a linear response to temperature, indicating the rotational character of this nucleus. A similar study of even $^{144-152}$Nd isotopes [11] confirms that ^{144}Nd and ^{146}Nd are vibrational, while ^{150}Nd and ^{152}Nd are rotational.

4. State densities in heavy nuclei

4.1. State densities

The SMMC method has been successful in the calculation of state densities in medium-mass nuclei [14, 15]. In the SMMC method we calculate the average thermal energy $E(\beta)$ as a function

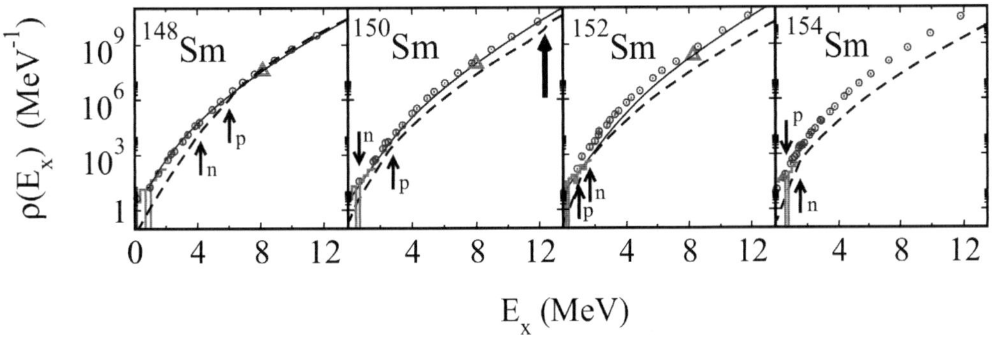

Figure 3. Total state densities of the even $^{148-154}$Sm isotopes. The SMMC results (open circles) are compared with the level counting data (histograms), the neutron resonance data (triangles), the experimentally parametrized BBF state densities (solid lines) and the HFB densities (dashed lines). Arrows denote proton and neutron pairing phase transitions and the thick arrow denotes a shape transition. Adapted from Ref. [11].

of β, from which we can compute the canonical partition function. The average state density is then determined by applying the saddle-point approximation to the integral that expresses the state density as the inverse Laplace transform of the canonical partition function.

In Fig. 3 we show the SMMC state densities (open circles) in the even $^{148-154}$Sm isotopes and compare them with level counting data at low energies (histograms) and neutron resonance data (triangles) when available. The solid lines are BBF state densities that are determined from fits to experimental data. The dashed lines are the densities obtained in the finite-temperature Hartree-Fock-Bogoliubov (HFB) approximation. The HFB results describe the intrinsic density of states, and the enhancement observed in the SMMC state density arises from collective states (vibrational and rotational) that are built on top of the intrinsic states. In the HFB approximation we observe proton and neutron pairing phase transitions (arrows) and shape phase transitions (thick arrows). ^{148}Sm is spherical in its ground state and exhibits pairing transitions only, while ^{150}Sm is deformed in its ground state and thus undergoes also a shape transition to a spherical shape at $E_x \approx 12.5$ MeV. Shape phase transitions also occur in ^{152}Sm and ^{154}Sm at excitation energies that increase with mass number (not shown in the figure).

4.2. Collective enhancement factors

Collective effects in the nuclear state density are difficult to model microscopically and are usually accounted for by phenomenological vibrational and rotational collective enhancement factors [16]. In Ref. [11] we have proposed to use the ratio of the SMMC and the HFB state densities, $K = \rho_{\rm SMMC}/\rho_{\rm HFB}$, as a microscopic measure of the collective enhancement factor. In Fig. 4 we show K as a function of the excitation energy E_x for the same samarium isotopes shown in Fig. 3. In ^{148}Sm, which is spherical in its ground state, any collectivity should be exclusively due to vibrational excitations, whereas the heavier deformed samarium nuclei are expected to have both vibrational and rotational collectivity. We observe that all collectivity disappears in ^{148}Sm (i.e., $K \approx 1$) at excitation energies above the pairing transitions, suggesting that the decay of the vibrational collectivity is associated with the pairing phase transitions. In the heavier samarium isotopes K reaches a local minimum above the pairing transition energies but increases at higher energies until it reaches a plateau and then decays to unity in the vicinity of the shape transition energy. This behavior can be explained by the contribution of rotational collectivity to K that decays to unity once the nucleus becomes spherical and can no longer

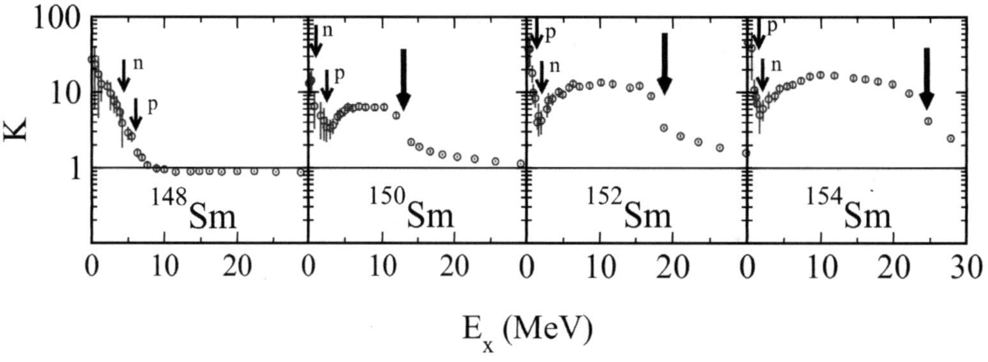

Figure 4. Total collective enhancement factor K (see text) in the even $^{148-154}$Sm isotopes as a function of excitation energy E_x. Arrows are as in Fig. 3. Adapted from Ref. [11].

support rotational bands.

5. Odd-even and odd-odd nuclei

The projection on an odd number of particles leads to a sign problem in SMMC (even for good sign interactions). This problem is particularly severe at low temperatures and makes it difficult to calculate the ground-state energy of odd-even and odd-odd nuclei. We have developed a method that circumvents this sign problem for the ground-state energy by using asymptotic properties of the single-particle Green's functions of the neighboring even-even nuclei [12].

5.1. Green's function method

For a rotationally invariant Hamiltonian, the scalar imaginary-time single-particle Green's functions are given by

$$G_\nu(\tau) = \frac{\mathrm{Tr}_{\mathcal{A}} \left[e^{-\beta H} \mathcal{T} \sum_m a_{\nu m}(\tau) a^\dagger_{\nu m}(0) \right]}{\mathrm{Tr}_{\mathcal{A}}\, e^{-\beta H}} , \tag{7}$$

where $\nu \equiv (nlj)$ labels the nucleon single-particle orbital with radial quantum number n, orbital angular momentum l and total spin j. Here $\mathcal{T}$ denotes time ordering and $a_{\nu m}(\tau) \equiv e^{\tau H} a_{\nu m} e^{-\tau H}$ is an annihilation operator of a nucleon at imaginary time τ ($-\beta \le \tau \le \beta$) in a single-particle state with orbital ν and magnetic quantum number m.

Consider an even-even nucleus $\mathcal{A} \equiv (Z, N)$, with Z protons and N neutrons. Assuming that the ground state of this nucleus has spin 0, the τ-dependence of the Green's function has the asymptotic form $G_\nu(\tau) \sim e^{-\Delta E_{J=j}(\mathcal{A}_\pm)|\tau|}$, where $\mathcal{A}_\pm$ denote the even-odd nuclei $(Z, N \pm 1)$ when ν is a neutron orbital and the odd-even nuclei $(Z \pm 1, N)$ when ν is a proton orbital. The $+$ and $-$ subscripts should be used for $\tau > 0$ and $\tau \le 0$, respectively. $\Delta E_{J=j}(\mathcal{A}_\pm)$ is the difference between the energies of the lowest spin J eigenstate of the $\mathcal{A}_\pm$-particle nucleus and the ground state of the $\mathcal{A}$-particle nucleus. In this asymptotic regime for τ, we can calculate $\Delta E_j(\mathcal{A}_\pm)$ from the slope of $\ln G_\nu(\tau)$. The minimum of $\Delta E_j(\mathcal{A}_\pm)$ over all possible values of j gives the difference between the ground-state energy of the $\mathcal{A}_\pm$ nuclei, $E_{\rm gs}(\mathcal{A}_\pm)$, and the ground-state energy of the $\mathcal{A}$ nucleus, $E_{\rm gs}(\mathcal{A})$. Both $E_{\rm gs}(\mathcal{A})$ and $G_\nu(\tau)$ are related to the even-even nucleus and can be calculated in SMMC without a sign problem (for a good sign interaction).

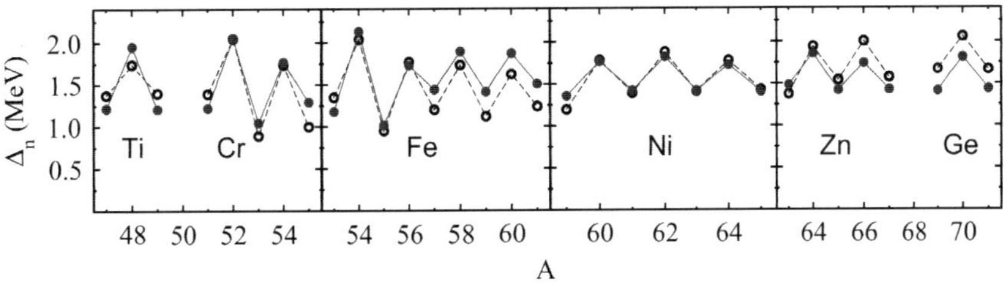

Figure 5. Neutron pairing gaps Δ_n as a function of mass number A in $fpg_{9/2}$-shell nuclei. The gaps calculated with the Green's function method (solid circles) are compared with the experimental gaps (open circles). Adapted from Ref. [12].

5.2. Pairing gaps in iron-region nuclei

The method of Sec. 5.1 is particularly useful for calculating pairing gaps (i.e., odd-even staggering of masses) because the odd-even ground state energy differences are calculated directly. In Fig. 5 we compare the SMMC pairing gaps of iron-region nuclei with their experimental values. We have used the complete $fpg_9/2$-shell model space with the Hamiltonian of Ref. [14]. The statistical errors for the SMMC pairing gaps ($\sim$ 0.01 MeV) are not visible in the figure. The theoretical gaps are in reasonable agreement with the experimental values.

6. Conclusion

We have discussed recent progress in the SMMC approach. We have shown that the crossover from vibrational to rotational collectivity in rare-earth nuclei can be described microscopically in the framework of the CI shell model approach. We have also described a method to calculate the ground-state energy of odd-even and odd-odd nuclei that circumvents the odd-particle sign problem and applied it to calculate pairing gaps of nuclei in the iron region.

Acknowledgments

It is a great pleasure for us to dedicate this article to Jerry Draayer on the occasion of his 70th birthday. This work was supported in part by the DOE grant DE-FG-0291-ER-40608, and by the JSPS Grant-in-Aid for Scientific Research (C) No. 22540266. Computational cycles were provided by the NERSC and Yale University High Performance Computing Center.

References

[1] Pieper S C and Wiringa R B 2001 *Ann. Rev. Nucl. Part. Sci.* **51** 53
[2] Navratil P, Quaglioni S, Stetcu I and Barrett B R 2009 *J. Phys.* G **36** 083101
[3] Dytrych T, Sviratcheva KD, Draayer JP, Bahri C and Vary JP 2008 *J. Phys. G* **35** 123101
[4] Bender M, Heenen P -H and Reinhard P -G 2003 *Rev. Mod. Phys.* **75** 121
[5] Bender M, Bertsch G F and Heenen P -H 2006 *Phys. Rev.* C **73** 034322
[6] Caurier E, Martinez-Pinedo G, Nowacki F, Poves A and Zuker A P 2005 *Rev. Mod. Phys.* **77** 427
[7] Lang G H, Johnson C W, Koonin S E and Ormand W E 1993 *Phys. Rev.* C **48** 1518
[8] Alhassid Y, Dean D J, Koonin S E, Lang G and Ormand W E 1994 *Phys. Rev. Lett.* **72** 613
[9] Dufour M and Zuker A P 1996 *Phys. Rev.* C **54** 1641
[10] Alhassid Y, Fang L and Nakada H 2008 *Phys. Rev. Lett.* **101** 082501
[11] Özen C, Alhassid Y and Nakada H 2012 arXiv:1206.6773.
[12] Mukherjee A and Alhassid Y 2012 *Phys. Rev. Lett.* **109** 032503
[13] Hubbard J 1959 *Phys. Rev. Lett.* **3** 77; Stratonovich R L 1957 *Dokl. Akad. Nauk. S.S.S.R.* **115** 1097
[14] Nakada H and Alhassid Y 1997 *Phys. Rev. Lett.* **79** 2939
[15] Alhassid Y, Liu S and Nakada H 1999 *Phys. Rev. Lett.* **83** 4265; Alhassid Y, Bertsch G F, Liu S and Nakada H 2000 *Phys. Rev. Lett.* **84** 4313; Alhassid Y, Liu S, and Nakada H 2007 *Phys. Rev. Lett.* **99** 162504
[16] Capote R et. al. 2009 *Nuclear Data Sheets* **110** 3107, and references therein

The no core shell model in an effective field theory framework

Bruce R Barrett

Department of Physics, PO Box 210081, University of Arizona, Tucson, AZ 85721-0081, USA

E-mail: bbarrett@physics.arizona.edu

Abstract. One of the outstanding problems in nuclear-structure theory is the construction of two-body (and higher-body) effective interactions in a model (or basis) space. In this presentation we discuss a recently developed approach to this problem, where one starts with an effective field theory (EFT), which contains only nucleonic fields and is formulated directly in a No-Core-Shell-Model (NCSM) space. Such an approach helps us to understand the gross features of nuclear systems from a QCD perspective. It also leads to a new method for the construction of effective interactions suitable for NCSM calculations. We then present applications to light nuclei within the pionless EFT and obtain reasonable results. Finally, we discuss future applications and extensions, such as testing the limits of the pionless EFT and extending the formalism to the pionfull EFT.

1. Introduction

A number of many-body techniques, such as the Green's Function Monte Carlo [1, 2, 3, 4], the Effective-Interaction Hyperspherical Harmonics [5] and the No Core Shell Model [6, 7, 8], now exist for calculating microscopically the structure of nuclei up to around mass $A = 16$. These approaches start with free realistic nucleon-nucleon (NN) potentials and, in many cases, also theoretically motivated three-nucleon (NNN) forces. These interactions are then employed in the different nuclear many-body formalisms to yield essentially parameter free results, which explain very well the properties of light nuclei, *i.e.*, $A \leq 16$. There is also the Coupled Cluster (CC) method [9, 10, 11], which can handle heavier nuclei, but generally for closed-shell nuclei or near closed-shell nuclei.

In general, these bare NN and NNN interactions need to be renormalized and/or truncated from the infinite Hilbert space to a finite model space, in which the many-nucleon calculations are performed. Thus, it is the effective NN and NNN interactions in these truncated model spaces that are necessary for carrying out most of the different many-body approaches for calculating microscopically nuclear properties. Because, in principle, all interactions are effective, their construction in free space, empirically and/or theoretically, is not necessary, especially if one can develop a formalism for directly constructing the effective interactions in the truncated model space. Not only would such a procedure be more efficient, but it would also have the possibility of yielding effective interactions that would converge more rapidly in the different many-body techniques, thereby allowing calculations to be performed for heavier-mass nuclei.

Published under licence by IOP Publishing Ltd

2. Motivation and goal

Why do we want to put the NCSM in an EFT framework? First, we want to have a formalism, which will provide an understanding of the gross features of nuclear systems from a QCD perspective. Second, as discussed above, we want to develop a new approach for the construction of effective interactions suitable for NCSM calculations. Consequently, our final goal is the formulation of an EFT with only nucleonic fields directly in the NCSM model space [12, 13, 14].

3. Formalism

3.1. Brief overview of Effective Field Theory (EFT)

EFT [15, 16, 17, 18] is based on a separation of scales, such as the nucleon mass or the pion mass, where the *details* of the physics at energies greater than the separation energy (representing the short-distance physics) are irrelevant for understanding the physics below the separation energy (representing the long-distance physics). In EFT the low-energy degrees of freedom are explicitly included, while the high-energy degrees of freedom are integrated out.

One then constructs a potential, in this case in the NCSM model space, which is consistent with the symmetries of the QCD Lagrangian and is expanded as a Taylor series in the momenta or as a contact gradient expansion in coordinate space. Each term in this potential expansion contains a constant. These constants are known as the low-energy constants (LECs). For now, the LECs are determined by fitting to experimental data.

In the case of the NCSM the calculations are performed in an harmonic-oscillator (HO) space for two-nucleons up to some maximum energy, $E_{max} = (N_{max} + \frac{3}{2})\hbar\omega$. Thus, the calculations are carried out in a truncated model space, so the interactions in this truncated space represent the *renormalized effective interactions.* In standard NCSM calculations one needs to repeat the calculations for increasing N_{max} to check that the results are converging in an appropriate manner. In EFT one introduces a cutoff Λ, known as the Ultraviolet (UV) cutoff, which depends on $N_{max}\hbar\omega$. To obtain a correct EFT result, one must achieve Renormalization Group Invariance (RGI). This means that the LECs, which are redetermined as Λ increases, become independent of Λ. In other words, for a high enough Λ, the LECs have absorbed the high-energy physics and the calculated physical observables are approximately independent of the cutoff Λ. The Taylor series expansion of the potential provides a power counting or hierarchy between the different contributions, which yield results that are improvable order by order, *i.e.*, leading order (LO) $>$ next-to-leading order (NLO) $>$ next-to-next-to-leading order (NNLO) $>$ *etc.*

4. Results

Our first investigation was for an EFT without pions, *i.e.*, a separation-scale energy of $\approx$ 100 MeV, which is known as the pionless EFT. In this case the Hamiltonian in the NCSM model space contains three potential terms consistent with the QCD symmetries, namely the NN potentials in the triple S and single S channels and a NNN potential in the $S = \frac{1}{2}$ channel. One then uses this Hamiltonian to compute the three LECs associated, respectively, with the three potential terms above. That is, the values of the LECs are adjusted until they reproduce, in our case, the binding energies of the deuteron, the triton and the α particle. Once these constants are determined, the Hamiltonian in the NCSM model space is completely known and can be used to compute other nuclear properties, such as the energy of the first-excited 0^+ state of the α particle or the binding energy of ^{6}Li. The determination of the Hamiltonian must be done for increasing values of Λ to achieve RGI. Because the calculations are performed using HO wave functions, *i.e.*, in an HO potential well, one must also take the limit as $\omega \rightarrow 0$ (the Infrared (IR) cutoff), since the nucleus is a self-bound system. Thus, the LECs are functions of both Λ and ω.

Figure 1 shows our results for the energy of the first-excited 0^+ state in the α particle, for increasing Λ and decreasing ω, which yields $E(J^\pi, T, {}^4He) = E(0_2^+, 0)_{theor} = 18.8$ MeV, in the

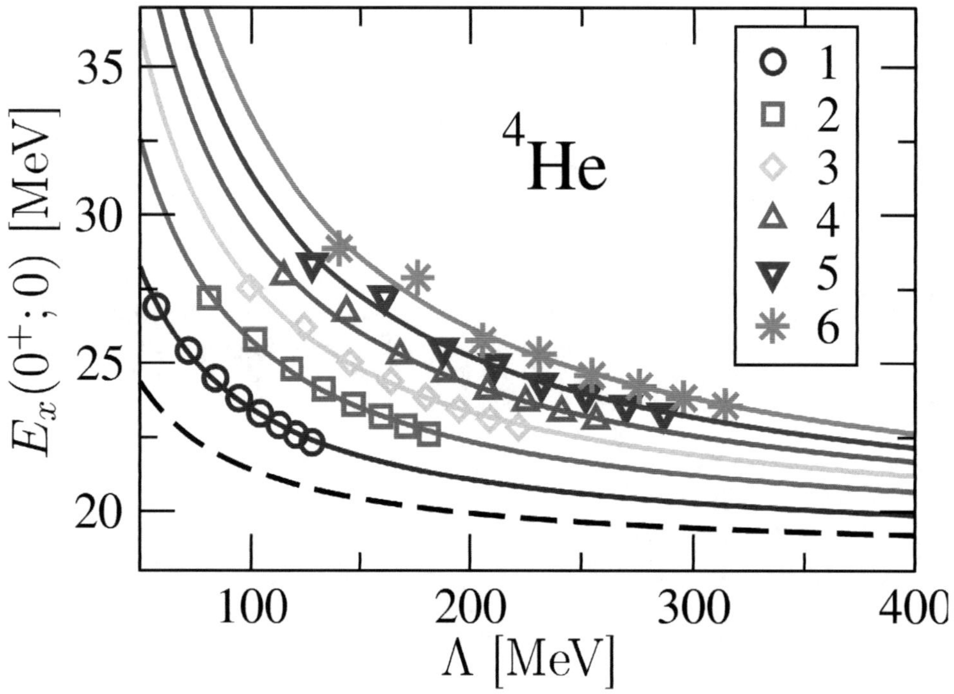

Figure 1. Energy of the first-excited $(0^+, 0)$ state in ^{4}He as a function of the UV cutoff Λ for different values of the frequency $\hbar\omega$ in MeV. The dashed curve marks the limit $\hbar\omega \to 0$. Taken from Ref. [12].

limits $\Lambda \to \infty$ and $\omega \to 0$, compared with $E(0_2^+, 0, ^4\text{He})_{expt} = 20.21$ MeV. These agree within 10%, which is remarkable agreement for a LO calculation and indicates that the description of this level is insensitive to details of the short-range physics.

Following the same procedure, we also evaluated for the first time in the pionless EFT the binding energy of ^{6}Li. For large values of the UV cutoff Λ and small values of the IR cutoff ω, we estimate the binding energy to be about 23 MeV, to be compared with the experimental result of 31.99 MeV. While not as precise as the first-excited-0^+-state energy in ^{4}He, this agreement between the theoretical and experimental values for the binding energy of ^{6}Li to within 30% is consistent with the expected errors in LO in the pionless EFT.

Obviously, we want to obtain better accuracy, which means that we need to include subleading contributions in the pionless EFT, *i.e.*, NLO, NNLO, *etc.* However, this becomes a very difficult task, because of lack of additional bound states in the NN system. The solution was to use the trick from lattice QCD of placing the system in a trap. In our case we use the Busch, *et al.* formula [19], which relates the energies of particles in an HO trap to their scattering phaseshifts. In the case of two trapped particles in an HO potential, their spectrum is entirely determined by the parameters of the effective range expansion (ERE). Thus, the scattering length yields the LO result; the effective range, the NLO result; *etc.* The energies obtained from the Busch formula are then used to determine the LECs.

This latter approach has been successfully applied [14] to the binding of the deuteron, NN phaseshifts and the $A = 3$ system for both $J^\pi = \frac{1}{2}^+, T = \frac{1}{2}$ and $J^\pi = \frac{3}{2}^+, T = \frac{1}{2}$ as well as to the neutron-deuteron scattering length. This conference contribution is too short to show all of

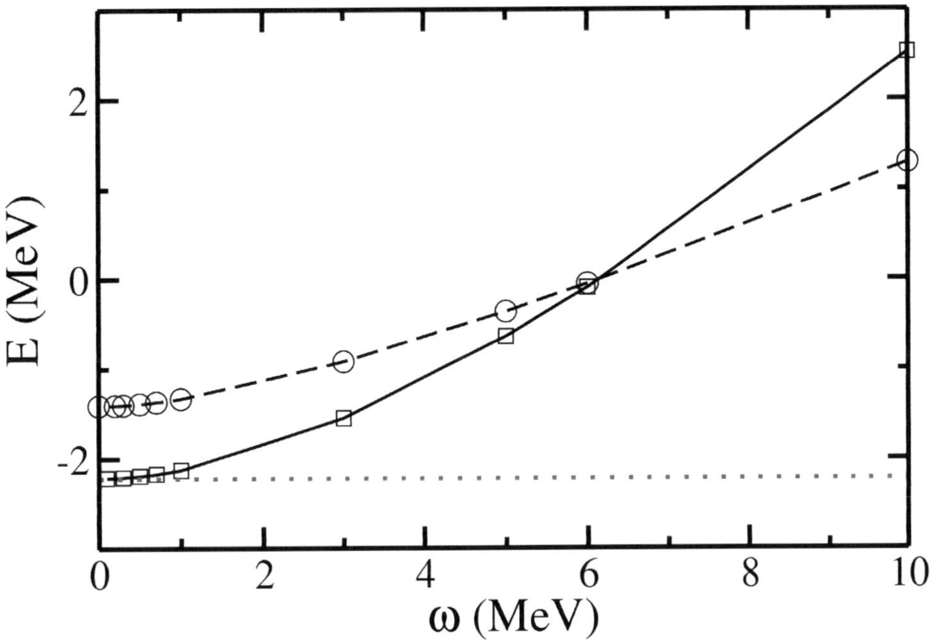

Figure 2. Ground-state energy of the deuteron in the HO trap as a function of the frequency ω. The energy of LO (NLO) is given by the ○ (□) connected by the - - - - (——) line. Taken from Ref. [14].

these results, so only a selected few will be given.

Figure 2 gives our result for the g.s. energy of the trapped NN system in the $^3S_1$ channel (*i.e.*, the deuteron) as a function of the frequency ω [14]. The energy at LO (NLO) is given by the dashed (solid) line. For small values of ω, the energy converges to the value in free space, which is, at NLO, indicated by the dotted line. This figure clearly shows how results in EFT systematically improve as one goes to higher orders in the expansion, *i.e.*, in the power counting.

Figure 3 illustrates our results for the g.s. energy in LO in the pionless EFT for three nucleons in an HO trap coupled to $J^\pi = \frac{3}{2}^+, T = \frac{1}{2}$, using only the NN force; there is no NNN force in this channel [14]. Each curve gives the result for the g.s. energy for a fixed value of the NN cutoff energy, N_2^{max}, as a function of increasing NNN cutoff energy, N_3^{max}, which starts at $N_3^{max} = N_2^{max}$. This figure shows a quite significant result, namely, that it is important in performing many-body calculations to have the many-body energy greater than the two-body energy. In Fig. 3 one observes that if N_3^{max} is the same as N_2^{max}, then all the degrees of freedom of the NNN system are not energetically available, as seen by the sudden and steep increase in the NNN binding energy as soon as N_3^{max} is increased beyond N_2^{max}, thereby significantly improving the convergence rate of the final answer. For example, looking at the results in Fig, 3, one sees that the result for $N_2^{max} = 8$ and $N_3^{max} = 10$ is essentially as good as that for $N_3^{max} = N_2^{max} = 16$, which is a much larger model space. Consequently, taking advantage of this effect allows for better converged results in much smaller model spaces.

5. Conclusions and outlook

We have presented our approach for formulating the NCSM in an EFT framework. After a brief description of EFT, we used the pionless EFT to calculate the first-excited-0^+-state of

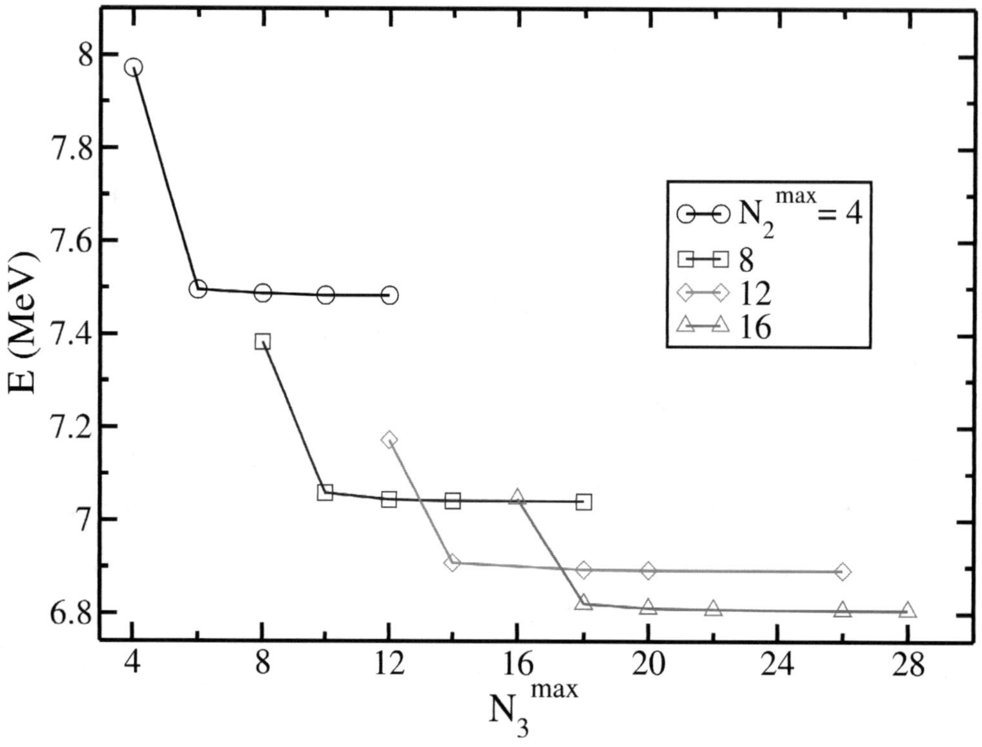

Figure 3. The g.s. energy in LO of the NNN system for $J^\pi = \frac{3}{2}^+, T = \frac{1}{2}$ as a function of the three-body model-space size N_3^{max}, for $\hbar\omega = 3$ MeV. Results are shown for different values of the two-body model-space size N_2^{max}. Adapted from Ref. [14].

^{4}He and the g.s. energy of ^{6}Li, obtaining quite reasonable results using only the leading order (LO) in the EFT. The extension to NLO in the pionless EFT was obtained by using the trick of binding the few-fermion systems in an HO trap, which allowed us to relate the energies of the fermions in the HO trap to the NN scattering phase shifts. Results for the $A = 3$ system in the $J^\pi = \frac{3}{2}^+, T = \frac{1}{2}$ channel showed the importance of having N_3^{max} for the three nucleons greater than N_2^{max}, in performing the many-body calculations, so as to take full advantage of all the degrees of freedom of the three nucleons. Doing this also leads to faster convergence in smaller model spaces, allowing converged calculations for heavier-mass nuclei.

Future plans include more calculations using the pionless EFT in NLO and for other light nuclei and in testing the limits of this approach. Work is also underway for extending our formalism to the EFT with pions, known as the pionfull EFT.

Acknowledgments
B.R.B. wishes to acknowledge his collaborators on this research project, M.C. Birse (U. Manchester, UK); J. Rotureau (Chalmers U. Technology, Sweden); I. Stetcu (Los Alamos National Laboratory); and U. van Kolck (U. Arizona and U. Paris-Sud 11, France) and also thanks J. Rotureau and I. Stetcu for their assistance with the figures and for helpful discussions. This work was supported in part by the NSF under Grant No. PHY-0854912.

References

[1] Carlson J 1987 *Phys. Rev.* C **36** 2026
[2] Carlson J and Schiavilla R 1998 *Rev. Mod. Phys.* **90** 743
[3] Schiavilla R, Wiringa R B and Carlson J 1993 *Phys. Rev. Lett.* **70** 3856
[4] Pieper S C and Wiringa R B 2001 *Annu. Rev. Nucl. Part. Sci.* **51** 53
[5] Barnea N, Leidemann W and Orlandini G 2000 *Phys. Rev.* C **61** 054001
[6] Navrátil P, Vary J P and Barrett B R 2000 *Phys. Rev. Lett.* **84** 5728
[7] Navrátil P, Vary J P and Barrett B R 2000 *Phys. Rev.* C **62** 054311
[8] Navrátil P, Quaglioni S, Stetcu I and Barrett B R 2009 *J. Phys. G: Nucl. Part. Phys.* **36** 083101
[9] Kümmel H, Lührmann K H and Zabolitzky J G 1978 *Phys. Rep.* **36** 1
[10] Kowalski K, Dean D J, Hjorth-Jensen M, Papenbrock T and Piecuch P 2004 *Phys. Rev. Lett.* **92** 132501
[11] Hagen G, Papenbrock T, Dean D J and Hjorth-Jensen M 2010 *Phys. Rev.* C **82** 034330
[12] Stetcu I, Barrett B R and van Kolck U 2007 *Phys. Lett* B **653** 358
[13] Stetcu I, Rotureau J, Barrett B R and van Kolck U 2010 *J. Phys. G: Nucl. Part. Phys.* **37** 064033
[14] Rotureau J, Stetcu I, Barrett B R and van Kolck U 2012 *Phys. Rev.* C **85** 034003
[15] Weinberg S 1990 *Phys. Lett* B **251** 459
[16] Weinberg S 1991 *Nucl. Phys.* B **363** 3
[17] Ordóñez C and van Kolck U 2002 *Phys. Lett* B **291** 459
[18] Bedaque P F and van Kolck U 2002 *Ann. Rev. Nucl. Part. Sci.* **52** 339
[19] Busch T, Englert B G, Rzazewski K and Wilkens M 1998 *Found. Phys.* **28** 549

HITES 2012 IOP Publishing
Journal of Physics: Conference Series **403** (2012) 012014 doi:10.1088/1742-6596/403/1/012014

The no-core shell model with general radial bases

M A Caprio[1], P Maris[2], J P Vary[2]

[1] Department of Physics, University of Notre Dame, Notre Dame, IN 46556-5670, USA
[2] Department of Physics and Astronomy, Iowa State University, Ames, IA 50011-3160, USA

Abstract. Calculations in the *ab initio* no-core shell model (NCSM) have conventionally been carried out using the harmonic-oscillator many-body basis. However, the rapid falloff (Gaussian asymptotics) of the oscillator functions at large radius makes them poorly suited for the description of the asymptotic properties of the nuclear wavefunction. We establish the foundations for carrying out no-core configuration interaction (NCCI) calculations using a basis built from general radial functions and discuss some of the considerations which enter into using such a basis. In particular, we consider the Coulomb-Sturmian basis, which provides a complete set of functions with a realistic (exponential) radial falloff.

1. Introduction

Significant advances presently are being made towards one of the basic goals of nuclear theory, namely, an *ab initio* understanding of the nucleus directly as a system of interacting protons and neutrons. Nuclear interactions motivated by quantum chromodynamics are being developed, via effective field theory methods [1, 2], to provide an underlying Hamiltonian. It is then necessary to solve the many-body problem for this Hamiltonian, obtaining nuclear eigenstates and predictions for observables. In no-core configuration interaction (NCCI) approaches such as the no-core shell model (NCSM) [3], the eigenproblem is formulated as a matrix diagonalization problem, in which the Hamiltonian matrix is represented with respect to a basis of antisymmetrized products of single-particle states, for the full A-body system of nucleons.

Actual NCCI calculations must be carried out in a finite, truncated space. The challenge is to reach a reasonable approximation of the converged results which would be achieved in the full, untruncated space for the many-body system. The success of the calculation is determined by the rate of convergence of calculated observables (*e.g.*, energies, radii, and electromagnetic moments and transition strengths) with increasing basis size and the ability to reliably extrapolate these results to the full, untruncated many-body space [4, 5]. Convergence rates are sensitive to the choice of single-particle states from which the many-body basis is constructed, as well as the truncation scheme used for the many-body basis.

In practice, such calculations have been based almost exclusively on a harmonic oscillator basis. It is worth recalling the special characteristics of this basis which make it particularly convenient for use in the nuclear many-body problem:

(1) An exact factorization of center-of-mass and intrinsic wavefunctions is obtained in many-body calculations when the oscillator basis is used in conjunction with the N_{max} truncation scheme, which is based on the total number of oscillator quanta. Thus, precise removal of, or correction for, spurious center-of-mass contributions to the dynamics is possible.

(2) Matrix elements of the nucleon-nucleon two-body interaction are naturally formulated in the relative oscillator basis. These matrix elements can easily be transformed to the two-body oscillator basis, of functions $\Psi_{n_1 l_1}(\mathbf{r}_1)\Psi_{n_2 l_2}(\mathbf{r}_2)$, by the Moshinsky transformation [6]. The simplicity of this transformation is lost with any other single-particle basis.

(3) The oscillator functions constitute a complete *discrete* basis for square-integrable functions. Many alternative bases, such as the eigenfunctions of the Schrödinger equation for a *finite-depth* potential such as the Woods-Saxon potential, do not provide this convenience, since for these bases a continuum of unbound single-particle states is needed for completeness.

However, there are also significant motivations for moving beyond the oscillator basis for the nuclear many-body problem [7]. The classic physical limitation of the oscillator basis, for application to the nuclear problem, lies in the Gaussian falloff ($\propto e^{-\alpha r^2}$) at large distance r. In contrast, for particles bound by a finite-range force, the actual asymptotics are exponential ($\propto e^{-\beta r}$). This mismatch in asymptotics, *i.e.*, the wavefunction tails, between the expansion basis and the physical system imposes a serious handicap on the convergence of calculations with increasing basis size. The problem is especially significant for observables, such as the root-mean-square radius or $E2$ strengths, which are sensitive to the large-r properties of the nuclear wavefunctions.

Here we consider an alternative basis for NCCI calculations, built from Coulomb-Sturmian functions [8, 9].[1] Although we focus on the Coulomb-Sturmian basis, many of the considerations addressed here are more broadly applicable to alternative single-particle bases for the NCCI problem. First, the procedures and results necessary for using the Coulomb-Sturmian basis for nuclear many-body calculations are outlined (section 2). Then, illustrative calculations for ^{6}Li are discussed (section 3). An expanded discussion may be found in [10].

2. Coulomb-Sturmian basis

The Coulomb-Sturmian functions [8, 9, 11], have previously been applied to few-body problems in atomic [8, 11–13] and hadronic [14–16] physics. They constitute a complete, discrete, orthogonal set of square-integrable functions, while also possessing realistic exponential asymptotics appropriate to the nuclear problem. The functions on $\mathbb{R}^3$ are given by $\Lambda_{nlm}(\mathbf{r}) = S_{nl}(r)Y_{lm}(\hat{\mathbf{r}})/r$, with radial wave function

$$S_{nl}(b;r) \propto (2r/b)^{l+1} L_n^{2l+2}(2r/b) e^{-r/b}, \tag{1}$$

where the L_n^α are generalized Laguerre polynomials, the Y_{lm} are spherical harmonics, n is the radial or node quantum number, l and m are the orbital angular momentum and its z-projection, and b is a radial scale parameter, analogous to the oscillator length parameter for the harmonic oscillator functions. The first several Coulomb-Sturmian radial functions, for $l = 0$, are plotted in figure 1(a). The radial functions for the harmonic oscillator are shown for comparison in figure 1(b).

Many-body calculations with the Coulomb-Sturmian basis can make use of the existing computational framework for NCCI calculations, but starting from the two-body matrix elements of the Hamiltonian in this new basis. The change of basis transformation has the form

$$\langle \bar{c}\bar{d}; J|V|\bar{a}\bar{b}; J\rangle = \sum_{abcd} \langle a|\bar{a}\rangle\langle b|\bar{b}\rangle\langle c|\bar{c}\rangle\langle d|\bar{d}\rangle\, \langle cd; J|V|ab; J\rangle, \tag{2}$$

[1] Since the Coulomb-Sturmian single-particle states arise [8] as solutions to a general Sturm-Liouville equation, rather than a Schrödinger equation or Hartree-Fock problem, it may be noted that they do not physically correspond to "shells" in the conventional sense. That is, they are not naturally interpreted as orbitals for independent-particle motion in some mean-field potential describing the zeroth-order dynamics of the system. Therefore, we will use the more inclusive term *configuration interaction*, rather than specifically *shell model*.

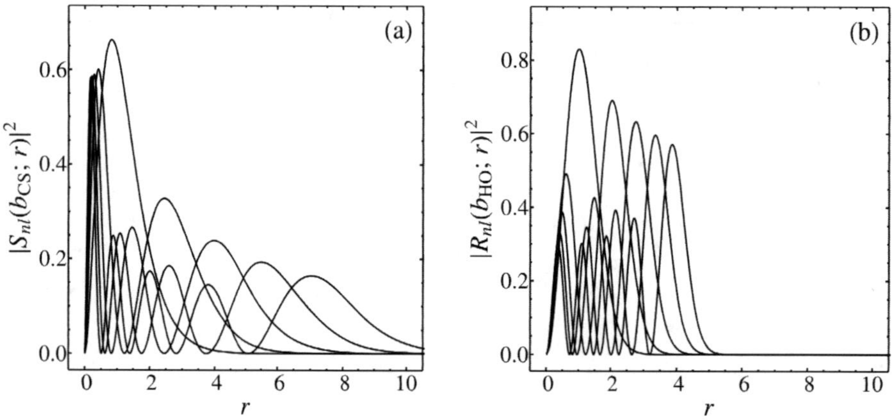

Figure 1. Radial functions for the (a) Coulomb-Sturmian and (b) harmonic oscillator bases, for $l = 0$. The first five functions ($0 \leq n \leq 4$) are shown in each case.

where in this expression we label single-particle orbitals for the oscillator basis by unbarred symbols $a = (n_a l_a j_a)$, $b = (n_b l_b j_b)$, *etc.*, and those for the Coulomb-Sturmian basis by barred symbols $\bar{a} = (\bar{n}_a \bar{l}_a \bar{j}_a)$, $\bar{b} = (\bar{n}_b \bar{l}_b \bar{j}_b)$, *etc.* The coefficients $\langle a|\bar{a}\rangle$ required for the transformation are given by the overlaps of the harmonic oscillator and Coulomb-Sturmian radial wavefunctions, that is, $\langle a|\bar{a}\rangle = \langle R_{n_a l_a}|S_{\bar{n}_a \bar{l}_a}\rangle \delta_{(l_a j_a)(\bar{l}_a \bar{j}_a)}$, where $\langle R_{nl}|S_{\bar{n}l}\rangle = \int_0^\infty dr\, R_{nl}(b_{\mathrm{HO}};r) S_{\bar{n}l}(b_l;r)$. The length parameter for Coulomb-Sturmian and oscillator functions may be chosen independently in this transformation.[2]

For actual calculation of the transformed matrix elements, the infinite sums over single-particle states appearing in the transformation (2) must be truncated, limited in practice by the available set of oscillator-basis matrix elements. In a single-particle space truncated by number of oscillator quanta, to $N \leq N_{\mathrm{cut}}$, the sum appearing in (2) becomes $\sum_{abcd}^{N_a,N_b,N_c,N_d \leq N_{\mathrm{cut}}}$. The question is whether or not we can practically include enough oscillator shells in the sum to achieve sufficient accuracy for the many-body calculation. Notice that the comparatively longer tails of the Coulomb-Sturmian functions [figure 1(a)] imply that they will require oscillator functions of significantly higher n [figure 1(b)] for their expansion. One might therefore ask now what we have gained by changing to an alternative radial basis, if we must still expand the new basis in terms of high-n oscillator basis functions. The essential point is that we need only carry out the expansion at the *two-body* level, where it is still tractable, to encode the information needed for treatment of the *A-body* problem in the new basis.

Although the truncated transformation (2) is found to suffice for the matrix elements of the nucleon-nucleon interaction, with $N_{\mathrm{cut}} \lesssim 13$ (section 3), the matrix elements of the kinetic energy are significantly more sensitive to the shell cutoff in the transformation. These matrix elements may instead be evaluated directly, making use of the particularly simple form of the kinetic energy operator. The NCSM intrinsic Hamiltonian has the form $H = T_{\mathrm{rel}} + V_{NN}$, where

[2] Moreover, it is of practical importance to note that orthogonality of Coulomb-Sturmian wavefunctions $\Lambda_{nlm}(\mathbf{r})$ with *different* l quantum numbers is enforced by the $Y_{lm}(\hat{\mathbf{r}})$ factor, regardless of the radial wavefunction. Therefore, the choice of length parameter may be made independently for each l-space, as b_l, and orthogonality of the basis of single-particle states will still be preserved. The considerations which enter the selection of the b_l values and the prescription used here are discussed further in [10].

$T_{\rm rel}$ is the Galilean-invariant relative kinetic energy operator

$$T_{\rm rel} \equiv \frac{1}{4Am_N} \sum_{ij}{}' (\mathbf{p}_i - \mathbf{p}_j)^2 = \frac{1}{2Am_N} \Bigg[(A-1) \underbrace{\sum_i \mathbf{p}_i^2}_{\text{one-body}} - \underbrace{\sum_{ij}{}' \mathbf{p}_i \cdot \mathbf{p}_j}_{\substack{\textit{separable} \\ \text{two-body}}} \Bigg]. \tag{3}$$

Here the prime on the summation $\sum'_{ij}$ indicates $i \neq j$. This operator is seen to decompose into a one-body part and a separable two-body part. The Coulomb-Sturmian functions have a momentum-space representation $\tilde{S}_{nl}(b;k)$, where $p = \hbar k$, which is given simply in terms of a Jacobi polynomial [9, 15]. Matrix elements of the one-body term are readily evaluated as radial integrals of k^2. Those of the two-body matrix elements factorize by Racah's reduction formula [17], as

$$\langle cd; J|\mathbf{p}_1 \cdot \mathbf{p}_2|ab; J\rangle = (-)^{j_d+j_a+J} \begin{Bmatrix} j_c & j_d & J \\ j_b & j_a & 1 \end{Bmatrix} \langle c\|\mathbf{p}\|a\rangle \langle d\|\mathbf{p}\|b\rangle, \tag{4}$$

where each factor may again be evaluated in terms of a radial integral, more specifically, as $\langle b\|\mathbf{p}\|a\rangle \propto [\int_0^\infty dk\, \tilde{S}_{n_b l_b}(b_{l_b};k)\, k\, \tilde{S}_{n_a l_a}(b_{l_a};k)] \langle l_b j_b \| Y_1 \| l_a j_a\rangle$.

3. Calculations for ^{6}Li

As a basic illustration of the use of the Coulomb-Sturmian basis for NCCI calculations, we now consider the nucleus ^{6}Li. The code MFDn [18, 19] is used for the many-body calculations, taking as its input Hamiltonian two-body matrix elements obtained as outlined in section 2. The calculations are carried out for the JISP16 interaction [20].

The $N_{\rm max}$ truncation for the harmonic oscillator many-body basis is defined by the condition $N_{\rm tot} \leq N_0 + N_{\rm max}$, where $N_{\rm tot} = \sum_i N_i = \sum_i (2n_i + l_i)$ is the total number of oscillator quanta, and N_0 is the minimal number of oscillator quanta for the given number of protons and neutrons. The calculated results depend both on the oscillator length parameter b — which is commonly quoted in terms of the oscillator energy $\hbar\Omega$, where $b = [\hbar/(m_N\Omega)]^{1/2}$, where m_N is the nucleon mass — and the truncation $N_{\rm max}$ for the basis. A natural question is how to truncate the Coulomb-Sturmian many-body basis, where a total number of oscillator quanta is not defined. However, we can *formally* carry over the $N_{\rm max}$ truncation scheme, if, for each Coulomb-Sturmian single-particle state, we simply define $N = 2n + l$ and, for each many-body state, we again define $N_{\rm tot} = \sum_i N_i = \sum_i (2n_i + l_i)$. This choice of many-body truncation is certainly not unique, but it provides a reasonable starting point for further exploration, and it facilitates comparison of convergence rates obtained using the oscillator and Coulomb-Sturmian bases, since the dimensions of the many-body spaces are then the same in both cases.

First, let us review the conventional oscillator results for the energy of the 1^+ ground state, shown in figure 2(a). For any given choice of $N_{\rm max}$, the energy has a variational minimum for some value of $\hbar\Omega$, $\sim 20\,\mathrm{MeV}$ in this example. With increasing $N_{\rm max}$, the ground-state eigenvalue approaches that for the infinite, untruncated space. The horizontal dashed line in figure 2 indicates the best extrapolation from prior calculuations in an $N_{\rm max} = 16$ space [21].

Calculations of the ground-state energy using the Coulomb-Sturmian basis are shown in figure 2(b). At each $N_{\rm max}$ the variational minimum energy obtained with the Coulomb-Sturmian basis in figure 2(b) is substantially higher than that obtained with the oscillator basis in figure 2(a). However, the energies obtained with the Coulomb-Sturmian basis are also falling significantly more rapidly with increasing $N_{\rm max}$, and the quality of extrapolation appears to be comparable. We note that these exploratory results have not yet probed several variational freedoms available with the Coulomb-Sturmian basis, both in the length parameters b_l (see [10]) and in the many-body truncation scheme. The robustness of the many-body calculations against

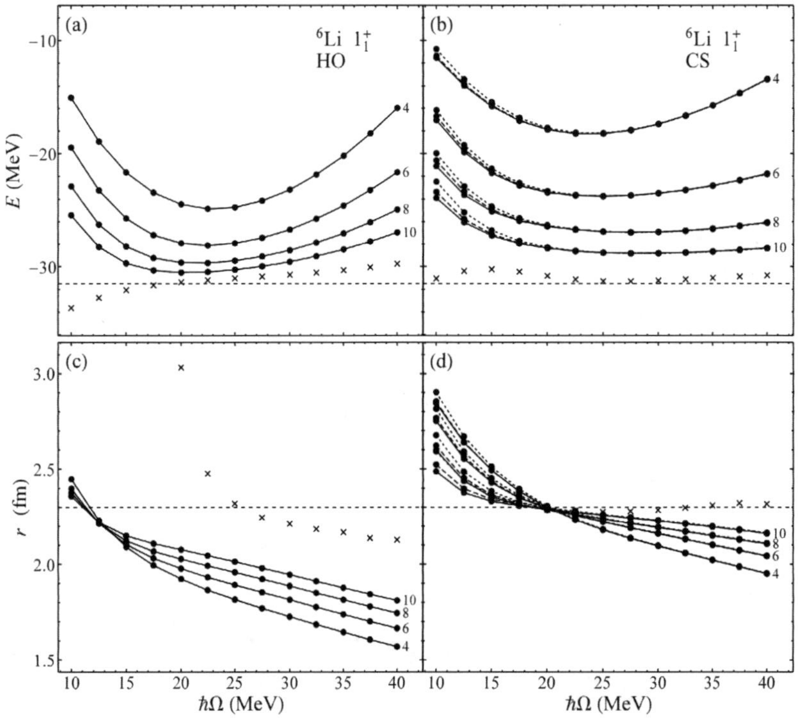

Figure 2. The ^{6}Li 1^+ ground state energy, calculated using the conventional harmonic oscillator basis (left) and the Coulomb-Sturmian basis (right). Calculated energies (top) and RMS radii (bottom) are plotted as a function of the basis $\hbar\Omega$ parameter, for $N_{\text{max}} = 4$, 6, 8, and 10 (successive curves, as labeled). For the Coulomb-Sturmian basis, calculations are shown variously for truncations $N_{\text{cut}} = 9$, 11, and 13 (dotted, dashed, and solid curves, respectively). Exponentially extrapolated values are indicated by crosses. The best extrapolated values from the large-basis calculations of [21] are shown as horizontal dashed lines.

the truncation N_{cut} in the change-of-basis transformation (2) may also be verified from figure 2, where calculations based on two-body matrix elements obtained with $N_{\text{cut}} = 9$, 11, and 13 are overlaid. The results for the ground state energy for $\hbar\Omega \gtrsim 20\,\text{MeV}$ are highly stable with respect to this cutoff. This range safely covers the variational minimum. If necessary, it would also be practicable to carry out transformations in larger oscillator spaces.

The root-mean-square (RMS) radius presents challenges for convergence in NCCI calculations with the conventional oscillator basis [22], since it is more specifically sensitive to the large-r asymptotic properties of the nuclear wavefunction. We compare the results obtained in the oscillator basis, shown in figure 2(c), with those obtained in the Coulomb-Sturmian basis, shown in figure 2(d). The extrapolated values obtained for $\hbar\Omega \gtrsim 20\,\text{MeV}$, *i.e.*, above the crossover point, are reasonably insensitive to $\hbar\Omega$ and are consistent with the best estimate from large oscillator-basis calculations [21]. It would appear that the rate of convergence of the RMS radius obtained with the Coulomb-Sturmian basis is superior to that obtained with the conventional oscillator basis. However, further systematic investigation is required, especially into the stability of extrapolations with increasing N_{max}, before general conclusions may be drawn.

The dominant concern in using any basis other than the harmonic oscillator basis, with

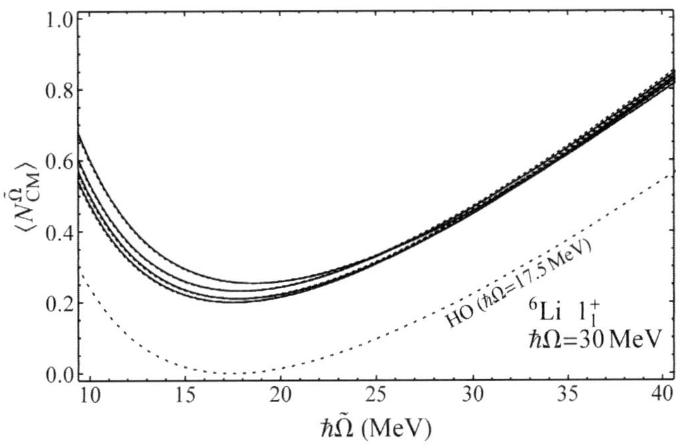

Figure 3. Expectation value of the number operator $N_{\mathrm{c.m.}}^{\tilde{\Omega}}$ for center-of-mass oscillator quanta, as a function of oscillator energy $\hbar\tilde{\Omega}$. These calculations are for the ^{6}Li 1^+ ground state, using the Coulomb-Sturmian basis, with $\hbar\Omega = 30\,\mathrm{MeV}$. Calculations are shown for $N_{\mathrm{max}} = 4$, 6, 8, and 10 (successive curves, top to bottom), and for N_{cut} values as indicated in the caption to figure 2. The analogous curve expected for a pure harmonic oscillator $0s$ function, with $\hbar\Omega = 17.5\,\mathrm{MeV}$, is also shown (dotted curve, labeled).

N_{max} truncation, is incomplete separation of center-of-mass and intrinsic dynamics. There are several aspects to consider: the degree of separation arising in calculations using the Coulomb-Sturmian basis, the spurious state spectrum obtained in such calculations, and the extent to which a Lawson term [23] can be used to purge spurious excitations from the low-lying spectrum. Here we briefly discuss only the first consideration, *i.e.*, separation. However, it is also found that the spurious states are well-separated in the spectrum and are amenable to management with a Lawson term (see [10]).

A first indication of the degree of separation in the many-body eigenstate is provided by the expectation value of the number operator for center-of-mass harmonic oscillator quanta, defined for an arbitrary center-of-mass harmonic oscillator energy $\hbar\tilde{\Omega}$ by

$$N_{\mathrm{c.m.}}^{\tilde{\Omega}} \equiv \frac{1}{\hbar\tilde{\Omega}}\left(\frac{P^2}{2Am_N} + \frac{Am_N\tilde{\Omega}^2R^2}{2} - \frac{3\hbar\tilde{\Omega}}{2}\right), \tag{5}$$

where $\mathbf{R}$ and $\mathbf{P}$ are the center-of-mass coordinate and momentum, respectively. As noted by Hagen *et al.* [24], *if* separation occurs, as $\psi(\mathbf{r}_i;\boldsymbol{\sigma}_i) = \psi_{\mathrm{c.m.}}(\mathbf{R})\psi_{\mathrm{in}}(\mathbf{r}_{ij};\boldsymbol{\sigma}_i)$, and *if* $\psi_{\mathrm{c.m.}}(\mathbf{R})$ happens to be an oscillator $0s$ function, corresponding to some oscillator energy $\hbar\tilde{\Omega}$, then the many-body wavefunction will have $\langle N_{\mathrm{c.m.}}^{\tilde{\Omega}}\rangle = 0$. The expectation value $\langle N_{\mathrm{c.m.}}^{\tilde{\Omega}}\rangle$ is shown as a function of $\hbar\tilde{\Omega}$ for the ^{6}Li ground state in figure 3, for the Coulomb-Sturmian basis calculation with basis $\hbar\Omega = 30\,\mathrm{MeV}$ and no Lawson term. The minimum of $\langle N_{\mathrm{c.m.}}^{\tilde{\Omega}}\rangle$ is obtained at $\hbar\tilde{\Omega} \approx 17.5\,\mathrm{MeV}$. The minimum value decreases with increasing N_{max}, but it appears to be converging towards a *nonzero* $\langle N_{\mathrm{c.m.}}^{\tilde{\Omega}}\rangle \sim 0.2$. The fact that expectation values significantly less than unity are obtained in the calculations indicates that a $0s$ oscillator function dominates the center-of-mass motion, and that an approximate separation of center-of-mass and intrinsic functions is spontaneously arising. However, the nonzero limit indicates that, as the full space is approached, the separated center-of-mass function is *not* strictly taking the form of a $0s$ oscillator function.

4. Summary

Although the conventional oscillator basis for the NCCI approach provides definite benefits, namely, exact center-of-mass factorization in the N_{max} truncation and the simplicity of the Moshinsky transformation for generating two-body matrix elements, it also exhibits nonphysical Gaussian asymptotics at large distances. The Coulomb-Sturmian functions form a complete, discrete set of square-integrable functions with realistic exponential asymptotics. A principal goal in using such a basis is to obtain improved convergence of observables which are sensitive to the asymptotics, such as RMS radii and $E2$ matrix elements. Such a basis might be particularly appropriate to halo nuclei, where the mismatch with the oscillator functions at large distances is particularly severe. The computational framework for the many-body calculation has the standard structure for an nlj single-particle basis, and the interaction matrix elements are transformed from the harmonic-oscillator basis, except that relative kinetic energy matrix elements (and, in fact, those for $N_{\text{c.m.}}$ and r^2, as well) are calculated directly in the Coulomb-Sturmian basis. The initial exploratory calculations considered here indicate that convergence properties are promising. Moreover, the spurious center-of-mass dynamics are found to be tractable.

Acknowledgments

We thank M. Pervin and W. N. Polyzou for pointing out the relevance of the Coulomb-Sturmian basis. This work was supported by the Research Corporation for Science Advancement through the Cottrell Scholar program, by the US Department of Energy under Grants No. DE-FG02-95ER-40934, DE-FC02-09ER41582 (SciDAC/UNEDF), and DE-FG02-87ER40371, and by the US National Science Foundation under Grant No. 0904782. Computational resources were provided by the National Energy Research Supercomputer Center (NERSC), which is supported by the U.S. Department of Energy under Contract No. DE-AC02-05CH11231.

References

[1] Entem D R and Machleidt R 2003 *Phys. Rev.* C **68** 041001
[2] Epelbaum E, Hammer H W and Meißner U G 2009 *Rev. Mod. Phys.* **81** 1773
[3] Navrátil P, Vary J P and Barrett B R 2000 *Phys. Rev. Lett.* **84** 5728
[4] Forssen C, Vary J P, Caurier E and Navratil P 2008 *Phys. Rev.* C **77** 024301
[5] Maris P, Vary J P and Shirokov A M 2009 *Phys. Rev.* C **79** 014308
[6] Moshinsky M and Smirnov Y F 1996 *The Harmonic Oscillator in Modern Physics* (Amsterdam: Harwood Academic Publishers)
[7] Stoitsov M V, Nazarewicz W and Pittel S 1998 *Phys. Rev.* C **58** 2092
[8] Rotenberg M 1962 *Ann. Phys. (N.Y.)* **19** 262
[9] Weniger E J 1985 *J. Math. Phys.* **26** 276
[10] Caprio M A, Maris P and Vary J P 2012 *Phys. Rev.* C **86** 034312
[11] Rotenberg M 1970 *Adv. At. Mol. Phys.* **6** 233
[12] Hylleraas E A 1928 *Z. Phys.* **48** 469
[13] Löwdin P O and Shull H 1956 *Phys. Rev.* **101** 1730
[14] Jacobs S, Olsson M G and Suchyta III C 1986 *Phys. Rev.* D **33** 3338
[15] Keister B D and Polyzou W N 1997 *J. Comput. Phys.* **134** 231
[16] Pervin M 2005 Ph.D. thesis Florida State University
[17] Edmonds A R 1960 *Angular Momentum in Quantum Mechanics* 2nd ed (*Investigations in Physics* no 4) (Princeton, New Jersey: Princeton University Press)
[18] Sternberg P, Ng E G, Yang C, Maris P, Vary J P, Sosonkina M and Le H V 2008 *SC '08: Proceedings of the 2008 ACM/IEEE Conference on Supercomputing* (Piscataway, NJ: IEEE Press) Article No. 15
[19] Maris P, Sosonkina M, Vary J P, Ng E and Yang C 2010 *Procedia Comput. Sci.* **1** 97
[20] Shirokov A M, Vary J P, Mazur A I and Weber T A 2007 *Phys. Lett.* B **644** 33
[21] Cockrell C, Vary J P and Maris P 2012 *Phys. Rev.* C **86** 034325
[22] Bogner S K, Furnstahl R J, Maris P, Perry R J, Schwenk A and Vary J 2008 *Nucl. Phys.* A **801** 21
[23] Gloeckner D H and Lawson R D 1974 *Phys. Lett.* B **53** 313
[24] Hagen G, Papenbrock T and Dean D J 2009 *Phys. Rev. Lett.* **103** 062503

Symmetry-Guided Ab Initio Approach to Light and Medium-Mass Nuclei

T Dytrych[1], **J P Draayer**[1], **K D Launey**[1], **M A Caprio**[2], **D Langr**[3]

[1]Department of Physics and Astronomy, Louisiana State University, Baton Rouge, LA 70803, USA

[2] Department of Physics, University of Notre Dame, Notre Dame, IN 46556-5670, USA

[3] Department of Computer Systems, Czech Technical University, Prague, Czech Republic

E-mail: tdytrych@phys.lsu.edu

Abstract. We use algorithms of computational group theory to perform *ab initio* configuration-interaction calculations in a SU(3)-based coupling scheme for *p*-shell nuclei. Details given for ^{6}Li are reflective of similar results found for ^{8}B, ^{8}Be, ^{12}C, and ^{16}O, all of which exhibit a strong preference for large quadrupole deformations and a narrow set of intrinsic spin quantum numbers. Our results suggest that a small subspace of symmetry-adapted configurations can very closely approximate the exact solutions while allowing for exact factorization of the center-of-mass degrees of freedom. This, in turns, promises to allows us to extend the reach of the ab initio framework for structure and reaction studies towards sd-shell nuclei and beyond.

1. Introduction

Theoretical advances achieved in recent years in the development of realistic nuclear potential models [1–4] along with progress in high performance computing have placed *ab initio* many-particle approaches [5–7] at the frontier of nuclear structure explorations. The *ab initio* methods are built on fundamental principles and therefore hold promise to provide predictive capabilities essential for a description of the structure and reactions of unstable and exotic nuclei, many of which are of high interest, e.g. in nucleosynthesis, but remain inaccessible even to experiment.

The no-core shell model (NCSM) [5] is a prominent *ab initio* method that has achieved a good description of low-lying states and associated spectroscopic properties up through *p*-shell nuclei [8–10]. The NCSM typically employs the Lanczos algorithm to solve the eigenvalue problem for a realistic Hamiltonian. Matrix elements of the Hamiltonian are calculated in a many-particle basis of m-scheme states, which are constructed as an antisymmetrized product of the harmonic oscillator single-particle wave functions, and carry the z-component of the total angular momentum along with the total parity as good quantum numbers. The main limitation of this approach, and the predictive power thereof, is inherently coupled with the combinatorial growth in the size of the many-particle model space with increasing nucleon numbers and expansion in the number of single-particle levels in the model space.

We developed an innovative *ab initio* model, the symmetry-adapted no-core shell model (SA-NCSM), which utilizes a many-particle basis that exploits the physically relevant SU(3)$\supset$SO(3) group-subgroup chain. The significance of the SU(3) group for a microscopic description of the nuclear collective dynamics can be readily seen from the fact that it is the symmetry group of the Elliott model [11], and a subgroup of the Sp$(3, R)$ symplectic model of nuclear collective motion [12,13]. The concept of symmetry-adapted many-particle basis represents a powerful tool that allows one to winnow a model space to correlations indispensable for modeling important modes of nuclear collective dynamics, specifically nuclear deformation and cluster substructures, thereby overcoming the scale explosion bottleneck of *ab initio* nuclear structure computations. Hence, the SA-NCSM framework holds promise to expand dramatically the reach of current *ab initio* approaches toward describing heavier mass nuclei with unprecedented accuracy.

2. *Ab initio* Calculations in a SU(3) Symmetry-Adapted Basis

The basis states of the SA-NCSM are constructed in the proton-neutron formalism and are labeled by the physical SU(3)$\supset$SO(3) subgroup chain quantum numbers $(\lambda\,\mu)\kappa L$, and by proton, neutron, and total intrinsic spins S_p, S_n, and S. The orbital angular momentum L is coupled with S to the total angular momentum J and its projection M_J. Each basis state is thus labeled in the SU(3)-scheme as

$$|\vec{\alpha}\, N(\lambda\,\mu)\kappa L; S_pS_nS; JM_J\rangle, \tag{1}$$

where N signifies the number of harmonic oscillator quanta with respect to the minimal number for a given nucleus. The deformation-related $(\lambda\,\mu)$ set of quantum numbers labels SU(3) irreducible representations (irreps) and bring forward important information about nuclear shapes and deformation. For example, (00), $(\lambda\,0)$ and $(0\,\mu)$ describe spherical, prolate and oblate shapes, respectively. The label κ distinguishes multiple occurrences of the same L value in the parent irrep $(\lambda\,\mu)$. The symbol $\vec{\alpha}$ schematically denotes the additional quantum numbers needed to unambiguously distinguish between irreps carrying the same $N\,(\lambda\,\mu)\,S_pS_nS$ quantum numbers. These irreps compose a well-defined subspace with a unique feature that allows for the complete separation of intrinsic and center-of-mass degrees of freedom [14].

3. Structure of Nuclear Wave Functions

Here we use the SA-NCSM with the bare JISP16 NN interaction [1] to calculate binding energies and determine low-lying eigenstates of ^{6}Li, ^{8}Be, ^{12}C, and ^{16}O nuclei. The resulting wavefunctions are used to determine values or physical observables such as point-particle root-mean-square (rms) matter radii, electric quadrupole moments, magnetic dipole moments, reduced electromagnetic $B(E2)$ and $B(M1)$ transition strengths.

The expansion of calculated wave functions in a physically relevant SU(3)-scheme basis is illuminating salient features that emerge from the complex dynamics of strongly interacting many-particle systems. To explore the nature of the most important correlations, the probability distribution of intrinsic spins $(S_p\,S_n\,S)$ and deformation-related $(\lambda\,\mu)$ quantum numbers of SU(3) for the lowest-lying $T = 0$ states of ^{6}Li were analyzed. Figure 1a shows the probability distribution of intrinsic spins across their Pauli-allowed deformations in the ground state of ^{6}Li. This figure illustrates a facet common to low-energy solutions considered: a highly structured and coherent mix of intrinsic spins and SU(3) spatial quantum numbers that has heretofore gone unrecognized in other *ab initio* studies. These results clearly corroborate the much earlier phenomenological work carried out within the context of the Elliott SU(3) model [11].

Specifically, we found that over 99% of the SA-NCSM eigenstates are accounted for by a small fraction of intrinsic spin combinations. For instance, the lowest-lying eigenstates in ^{6}Li are almost entirely realized in terms of configurations characterized by the following intrinsic spin $(S_p\,S_n\,S)$ quantum numbers: $(\frac{3}{2}\,\frac{3}{2}\,3)$, $(\frac{1}{2}\,\frac{3}{2}\,2)$, $(\frac{3}{2}\,\frac{1}{2}\,2)$, and $(\frac{1}{2}\,\frac{1}{2}\,1)$, with the last one carrying over 90% of the eigenstate. Likewise, the same spin components as in the case of ^{6}Li are found to dominate the ground state and the lowest 1^+, 3^+, and 0^+ excited states of ^{8}B (Table 1). Similarly, the ground state band of ^{8}Be and ^{12}C along with the ground state of ^{16}O are found to be dominated by many-particle configurations carrying total intrinsic spin of the protons

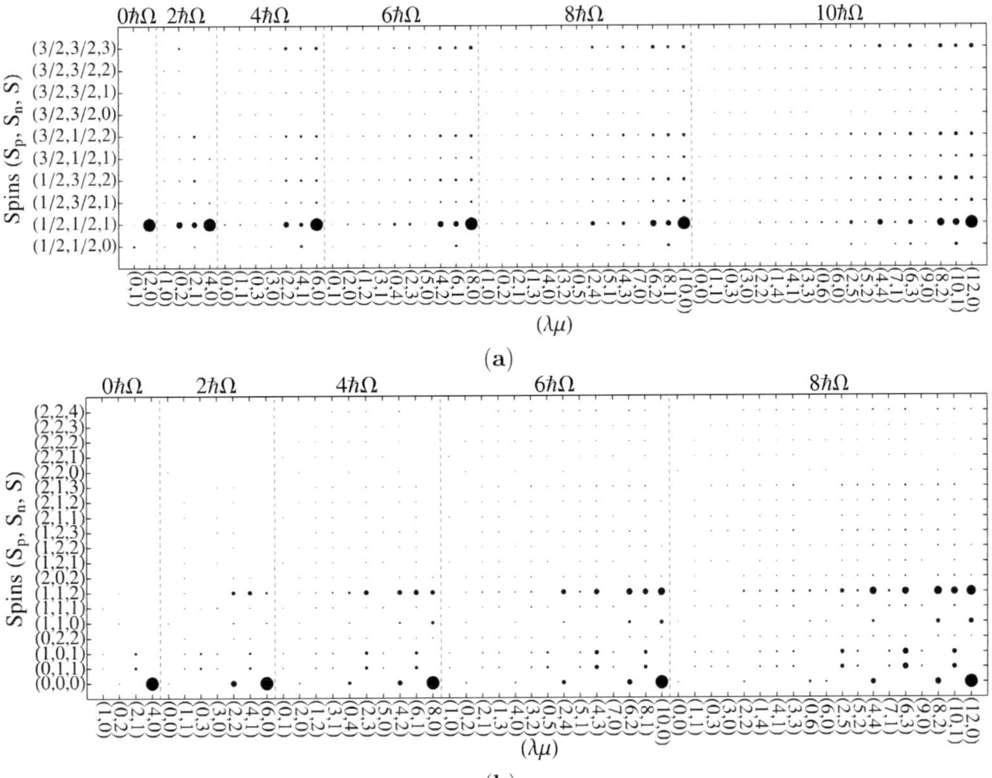

(a)

(b)

Figure 1. Probability distribution of proton, neutron, and total intrinsic spins $(S_p\, S_n\, S)$ across their Pauli-allowed $(\lambda\,\mu)$ deformations (horizontal axis) in the ground 1^+ state of ^{6}Li (a) and the ground 0^+ state of ^{8}Be (b) obtained with JISP16 bare interaction for $\hbar\Omega = 20$ MeV in $N_{\max} = 10$ (a) and $N_{\max} = 8$ (b) full model spaces. The area of each circle is proportional to the total probability of $N\,(\lambda\,\mu)\,S_pS_nS$ states normalized with respect to the total probability of $N\,(2+N\,0)\,\frac{1}{2}\,\frac{1}{2}\,1$ and $N\,(4+N\,0)\,0\,0\,0$ stretched-states, respectively.

Nucleus	$(S_p\, S_n\, S)$	Prob. [%]	$(\lambda_0\, \mu_0)$	Prob. [%]
^{6}Li	$(\frac{1}{2}\,\frac{1}{2}\,1)$	93.26	(2 0)	98.13
^{8}B	$(\frac{1}{2}\,\frac{1}{2}\,1)$	85.17	(2 1)	87.94
^{8}Be	(0 0 0)	85.25	(4 0)	90.03
^{12}C	(0 0 0)	55.19	(0 4)	48.44
^{16}O	(0 0 0)	83.60	(0 0)	89.51

Table 1. Probability amplitude of the dominant $(S_p\, S_n\, S)$ spin configuration and the dominant nuclear shapes according to Eq. (2) for the ground state of p-shell nuclei.

and neutrons equal to zero and one, with the largest contribution due to $(S_p\, S_n\, S) = (0\,0\,0)$ configurations, This is as illustrated, e.g., in Figure 1b for the ground state of ^{8}Be.

The mixing of $(\lambda\,\mu)$ spatial quantum numbers, induced by the SU(3) symmetry breaking terms of realistic interactions, exhibits a remarkably simple coherent pattern. One of its key features is the preponderance of a single $0\hbar\Omega$ SU(3) irrep, the so-called leading irrep, that is, the one characterized by the largest value of the second order SU(3) Casimir invariant, $\hat{C}_2$, and hence corresponding to a large intrinsic quadrupole deformation [15]. For instance, the low-lying $T = 0$ states of ^{6}Li project at 40%-70% level onto the prolate-like $0\hbar\Omega$ SU(3) irrep $(2\,0)$. For the considered states of ^{8}B, ^{8}Be, ^{12}C, and ^{16}O, qualitatively similar dominance of the leading $0\hbar\Omega$ SU(3) irreps is observed – (2 1), (4 0), (0 4), and (0 0) irreps, associated with triaxial, prolate, oblate, and spherical shapes, respectively. Such a clear dominance of the largest $0\hbar\Omega$ deformation within the low-lying states of p-shell nuclei points to the fact that the effective quadrupole-quadrupole interaction of the Elliott SU(3) model of nuclear rotations [11] is realized naturally within the framework of modern realistic interactions.

The analysis reveals that the dominant SU(3)-scheme states at each $N\hbar\Omega$ subspace are typically those with $(\lambda\,\mu)$ quantum numbers such that

$$\lambda + 2\mu = \lambda_0 + 2\mu_0 + N \qquad N = 0, 2, \ldots, \tag{2}$$

where λ_0 and μ_0 denote labels of a leading SU(3) irrep at the $0\hbar\Omega$ ($N = 0$) subspace (Table 1). We conjecture that this coherent pattern of SU(3) quantum numbers reflects the presence of an underlying symplectic $\mathrm{Sp}(3,R)$ symmetry of microscopic nuclear collective motion [12] that governs the low-energy structure of both even-even and odd-odd p-shell nuclei. This can be seen from the fact that configurations with a $(\lambda\,\mu)$ shape that satisfies condition (2) can be determined from the leading SU(3) irrep $(\lambda_0\,\mu_0)$ through a successive application of a specific subset of the $\mathrm{Sp}(3,R)$ symplectic $2\hbar\Omega$ raising operators. This subset is composed of the three operators, $\hat{A}_{zz}$, $\hat{A}_{zx}$, and $\hat{A}_{xx}$, that distribute two oscillator quanta in z and x directions, but none in y direction, thereby inducing SU(3)-scheme configurations with ever-increasing values of the Casimir invariant $\hat{C}_2$. These three operators are the generators of the $\mathrm{Sp}(2,R) \subset \mathrm{Sp}(3,R)$ subgroup [16], and give rise to a hierarchy of deformed shapes that are energetically favored by an attractive quadrupole-quadrupole interaction [13]. Furthermore, there is an apparent hierarchy among states that fulfill condition (2). In particular, the $N\hbar\Omega$ configurations with $(\lambda_0{+}N\;\mu_0)$, the so-called stretched states, carry a noticeably higher intensity than the others. For instance, the $(2{+}N\;0)$ stretched states contribute at the 85% level to the ground state of ^{6}Li. The sequence of the stretched-states, that is, the states with the highest possible deformations, can be formed from many-nucleon correlations of a leading SU(3) irrep by application of the $\hat{A}_{zz}$ operator, which is the generator of $\mathrm{Sp}(1,R) \subset \mathrm{Sp}(2,R) \subset \mathrm{Sp}(3,R)$ subgroup.

The revealed pattern of intrinsic spin and deformation mixing supports a symmetry guided truncation of the $N_{\max}$ model space. Clearly, one can take advantage of the physical relevance of the SU(3)-scheme basis to winnow the full space down to the most relevant configurations that support the strongest many-nucleon correlations of the system using the underlying $\mathrm{Sp}(1,R) \subset \mathrm{Sp}(2,R) \subset \mathrm{Sp}(3,R)$ symmetry considerations. As noted previously, this truncation, while significantly reducing the size of the model space, also preserves the ability to factor out exactly the spurious center-of-mass degrees of freedom.

4. Efficacy of the SU(3) Basis

To probe the adequacy of the symmetry-adapted approach for the *ab initio* modelling of nuclear structure, we used the calculated eigenstates to determine spectroscopic properties of low-lying $T = 0$ states of ^{6}Li using a model space winnowed through symmetry considerations, and compared the outcomes with the corresponding results obtained in the full $N_{\max} = 12$ space.

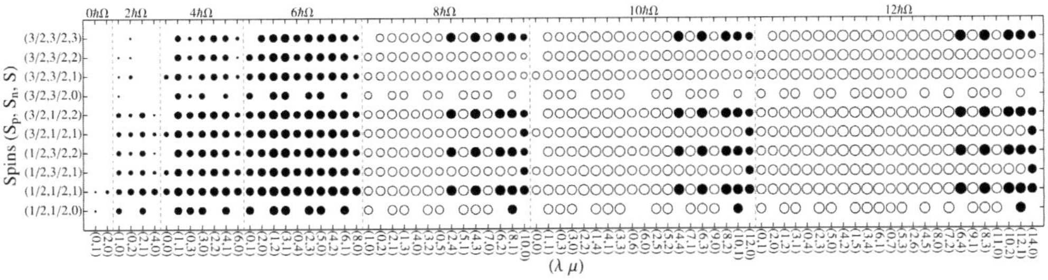

Figure 2. Pauli-allowed $(\lambda\,\mu)$ deformations and their proton, neutron, and total intrinsic spins $(S_p\,S_n\,S)$ for the positive-parity $J = 1$ states of ^{6}Li spanning the full $N_{\rm max} = 12$ model space. Each circle represents basis states carrying the same $N\,(\lambda\,\mu)\,S_pS_nS$ quantum numbers, with the radius being proportional to $\log_{10}$ of the number of such states. Configurations symbolized by the filled circles constitute the symmetry-truncated model space 12[6].

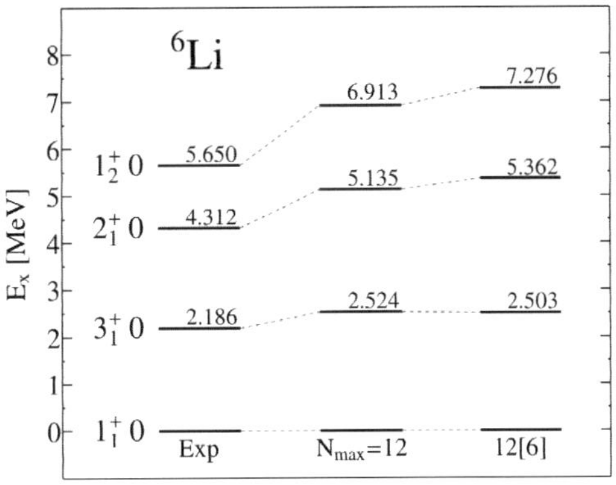

Figure 3. Experimental and theoretical excitation energies of $T = 0$ states of ^{6}Li. Theoretical estimates were obtained at $\hbar\Omega = 22.5$ MeV in the full $N_{\rm max} = 12$ (middle) and the symmetry-truncated 12[6] (right) model spaces.

In the study, Coulomb and bare JISP16 NN interactions were used for $\hbar\Omega$ oscillator strengths ranging from 17.5 up to 25 MeV. The selected model space, depicted in Fig. 3 and denoted by 12[6], incorporates all the configurations carrying up to 6 excitations of oscillator quanta (labeled by [6]) and only a subset of the shapes and a few intrinsic spin components to realize the leading modes of nuclear collective motion for the higher $8\hbar\Omega$, $10\hbar\Omega$ and $12\hbar\Omega$ configurations. The model space 12[6] constitutes a small percentage of the full space. For example, the full $N_{\rm max} = 12$ model space dimension is 4.9×10^7 whereas the dimensions of the 12[6] subspaces with total angular momenta $J = 1$; $J = 2$; and $J = 3$, are 4.3×10^5 (0.87%); 6.5×10^5 (1.32%); and 8.3×10^5 (1.70%), respectively.

The ground state binding energies calculated in 12[6] for oscillator energy $\hbar\Omega$ ranging from 17.5 to 25 MeV represent from 98% up to 98.7% of the full-space binding energy. Furthermore, the excitation energies of 3_1^+, 2_1^+, and 1_2^+ states calculated in 12[6] model space differ only by 20 keV to a few hundreds of keV from the corresponding full-space results, see Fig. 3.

As illustrated in Table 2 for $\hbar\Omega = 17.5$ MeV, the magnetic dipole moments agree to within

	1_1^+0	3_1^+0	2_1^+0	1_2^+0
μ				
Full $N_{\rm max} = 12$	0.838	1.866	0.960	0.336
12[6] SU(3)	0.840	1.866	1.015	0.337
rms				
Full $N_{\rm max} = 12$	2.146	2.092	2.257	2.373
12[6] SU(3)	2.139	2.079	2.236	2.355

Table 2. Magnetic dipole moments μ [μ_N] and point-particle rms matter radii [fm] of $T = 0$ states of ^{6}Li calculated in the full $N_{\rm max} = 12$ and 12[6] model spaces for $\hbar\Omega = 17.5$ MeV. The experimental value for the 1^+ ground state is known to be $\mu = +0.822$ μ_N.

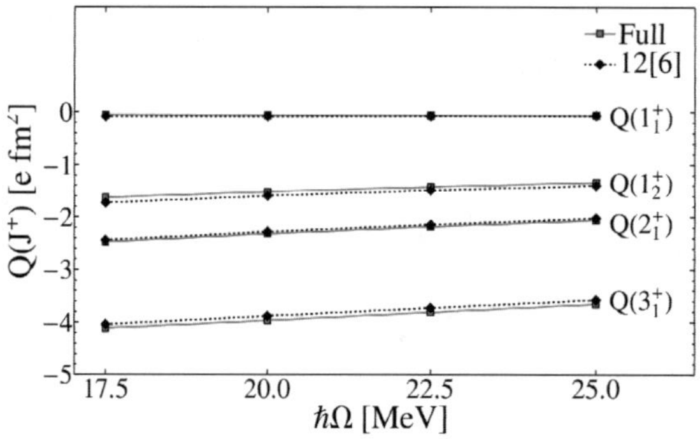

Figure 4. Electric quadrupole moments as a function of $\hbar\Omega$ for T=0 states of ^{6}Li calculated in the full $N_{\rm max} = 12$ (solid red line) and symmetry-truncated 12[6] (dashed black line) model spaces.

less than about 0.3%, or 5% for the $\mu(2_1^+0)$. As the dipole moment is a short-range operator, the results suggest that it may suffice to include all low-lying $\hbar\Omega$ states up to a fixed limit, e.g. $N_{\rm max} = 6$ for ^{6}Li, to account for the most important short-range correlations.

To explore how closely one comes to reproducing the important long-range correlations of the full $N_{\rm max} = 12$ space in terms of nuclear collective excitations within the more restricted 12[6] space, we compared observables that are sensitive to the tails of the wavefunctions; specifically, the point-particle rms matter radii, the electric quadrupole moments and the reduced electromagnetic $B(E2)$ transition strengths. The results for the rms matter radii, listed in Table 2 for $\hbar\Omega = 17.5$ MeV, agree to within 1%. Similarly, the 12[6] eigensolutions yield results for these quantities that track very closely with their full [12] space counterparts for all values of $\hbar\Omega$, as can be seen in Fig. 4 and 5. Also, as the $B(E2)$ strengths almost doubles upon increasing the basis space from $N_{\rm max} = 6$ to $N_{\rm max} = 12$ – a result that suggests that further expansion of the basis space may be needed to reach the experimental value of 21.8(4.8) e^2fm^4, the close correlation between the $N_{\rm max} = 12$ and 12[6] results is even more impressive.

5. Conclusion

We have developed a novel approach that capitalizes on advances being made in *ab initio* methods while exploiting exact and partial symmetries of nuclear many-body system. Using this approach we have demonstrated that the low-lying eigenstates of ^{6}Li, ^{8}Be, ^{12}C, and ^{16}O, which were obtained using the JISP16 NN interaction, exhibit a strong dominance of few intrinsic spin

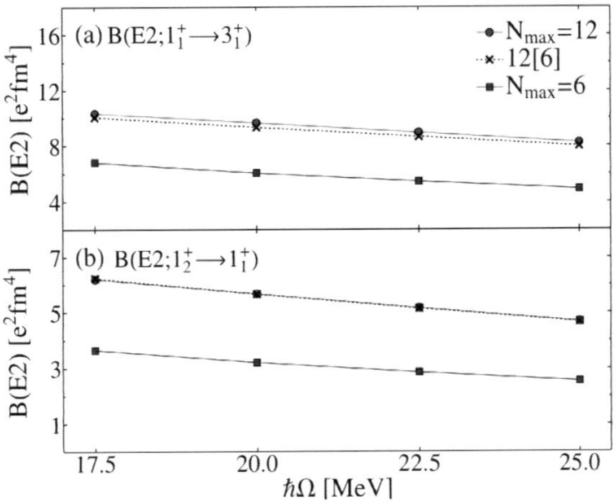

Figure 5. Electric quadrupole transition probabilities as a function of $\hbar\Omega$ for T=0 states of ^{6}Li calculated in the full $N_{\max} = 12$ (solid red line), $N_{\max} = 6$ (solid blue line), and symmetry-truncated 12[6] (dashed black line) model spaces.

components and carry an intriguingly simple pattern of dominant deformations. The results very clearly underscore the significance of the SU(3) scheme, LS-coupling, and underlying symplectic symmetry in enabling an extension, through symmetry-guided model space reductions, of ab initio methods to heavier nuclei beyond ^{16}O.

Acknowledgments
This work was supported by the U.S. NSF (PHY-0500291 & OCI-0904874), the U.S. DOE (DE-SC0005248 & FG02-95ER-40934), SURA, and by the Research Corporation for Science Advancement under a Cottrell Scholar Award. This research used computing resources of the Louisiana Optical Network Initiative, LSU's Center for Computation & Technology, and the National Energy Research Scientific Computing Center, which is supported by the Office of Science of the U.S. Department of Energy under Contract No. DE-AC02-05CH11231.

References

[1] Shirokov A, Vary J, Mazur A and Weber T 2007 *Phys. Lett. B* **644** 33
[2] Wiringa R B, Stoks V G J and Schiavilla R 1995 *Phys. Rev. C* **51** 38
[3] Entem D R and Machleidt R 2003 *Phys. Rev. C* **68** 041001
[4] Epelbaum E 2006 *Prog.Part.Nucl.Phys* **57** 654
[5] Navrátil P, Vary J P and Barrett B R 2000 *Phys. Lett.* **84** 5728
[6] Pieper S C, Varga K and Wiringa R B 2002 *Phys. Rev. C* **66** 044310
[7] Wloch M, Dean D J, Gour J R, Hjorth-Jensen M, Kowalski K, Papenbrock T and Piecuch P 2005 *Phys. Rev. Lett.* **94** 212501
[8] Navrátil P, Quaglioni S, Stetcu I and Barrett B R 2009 *J. Phys. G: Nucl. Part.* **36** 083101
[9] Roth R and Navrátil P 2007 *Phys. Rev. Lett.* **99** 092501
[10] Navratil P, Gueorguiev V, Vary J P, Ormand W and Nogga A 2007 *Phys. Rev. Lett.* **99** 042501 LANL ArXiV: Nucl-Th 0701038
[11] Elliott J P 1958 *Proc. Roy. Sot. A* **245** 128
[12] Rosensteel G and Rowe D 1977 *Phys. Rev. Lett.* **38** 10
[13] Rowe D 1985 *Reports on Progr. in Phys.* **48** 1419
[14] Luo F Q, Caprio M A and Dytrych T *in these proceedings*
[15] Castanos O, Draayer J P and Leschber Y 1988 *Zeit. f. Phys. A* **329** 33
[16] Peteron D R and Hecht K T 1980 *Nucl. Phys. A* **344** 361

Quantum Monte Carlo study of inhomogeneous neutron matter

Stefano Gandolfi

Theoretical Division, Los Alamos National Laboratory Los Alamos, NM 87545, USA

E-mail: stefano@lanl.gov

Abstract. We present an ab-initio study of neutron drops. We use Quantum Monte Carlo techniques to calculate the energy up to 54 neutrons in different external potentials, and we compare the results with Skyrme forces. We also calculate the rms radii and radial densities, and we find that a re-adjustment of the gradient term in Skyrme is needed in order to reproduce the properties of these systems given by the ab-initio calculation. By using the ab-initio results for neutron drops for close- and open-shell configurations, we suggest how to improve Skyrme forces when dealing with systems with large isospin-asymmetries like neutron-rich nuclei.

1. Introduction

Even though neutron drops, neutrons confined by an external field, cannot be studied experimentally, they have nevertheless generated intense theoretical interests for a couple of reasons. Neutron drops provide a very simple model of neutron-rich nuclei, in which the core is modeled as an external potential acting on valence neutrons. For example in Refs. [1, 2, 3] neutron-rich oxygen isotopes have been modeled by neutron drops, and in Ref. [4] the same model has been used to study calcium isotopes. Second, they provide a description of inhomogeneous neutron matter that can be used as data for calibrating model energy density functionals in several conditions [5, 6, 7].

On the experimental side, new facilities plan to study the physics of neutron-rich nuclei close to the drip line. The study of these systems is typically done by using methods based on energy density functionals and effective forces. These forces, like Skyrme or Gogny, are fitted to reproduce properties of nuclear and neutron matter combined with experimental energies of nuclei close to stability. By constraining the density functionals to the equation of state of nuclear and neutron matter, the symmetry energy is implicitly included in these forces. However, the terms dependent on derivatives of proton and neutron densities are constrained only to nuclei close to stability, and then for very small isospin-asymmetries.

The use of these functionals to study nuclei close to the neutron drip line requires then an important extrapolation to large isospin-asymmetries. This extrapolation is even more drastic when the Skyrme forces are used to study the properties of the neutron star crust, where the matter is made by extremely neutron-rich nuclei surrounded by a sea of neutrons. For these reasons, *ab-initio* calculations of these systems starting from accurate nuclear Hamiltonians are important to constrain density functionals.

Important advances have been made in building very accurate Hamiltonians describing nuclear systems. The modern nucleon-nucleon forces fit scattering data with high accuracy [8, 9], and also the three-body forces are quite well constrained to reproduce properties of light nuclei [10, 11].

Furthermore, in recent years new techniques have been developed to solve for the ground-state of nuclear systems. In this paper we present results obtained with Quantum Monte Carlo techniques, i.e. the well known Green's Function Monte Carlo (GFMC) and Auxiliary Field Diffusion Monte Carlo (AFDMC). The two different algorithms use different variational wave functions, but they show similar accuracy in calculating the energy of pure neutron systems.

2. Nuclear Hamiltonian and Quantum Monte Carlo method

In our model, neutrons are non-relativistic point-like particles interacting via two- and three-body forces, that are confined by an external potential:

$$H = \sum_{i=1}^{A} \frac{p_i^2}{2m} + \sum_i V_{ext}(r_i) + \sum_{i<j} v_{ij} + \sum_{i<j<k} v_{ijk} \,, \tag{1}$$

where the external potential is a Harmonic Oscillator (HO)

$$V_{HO}(r) = \frac{1}{2} m\omega^2 r^2 \,, \tag{2}$$

or a Woods-Saxon

$$V_{WS}(r) = -V_0 \frac{1}{1+\exp[(r-r_0)/a]} \,. \tag{3}$$

In our study the parameters are $V_0 = -35.5$ MeV, $r_0 = 3$ fm and $a = 1.1$ fm, and $\hbar^2/m = 41.44$ MeV-fm^2. The parametrization of the WS has been chosen to reproduce the properties of oxygen isotopes [1]; for the HO we consider different values of $\hbar\omega$ to change the density of the system.

The two body-potential that we use is the Argonne AV8' [12], a simplified form of the Argonne AV18 [8]. Although simpler to use in QMC calculations, the AV8' provides almost the same accuracy as AV18 in fitting NN scattering data.

In this work we use the UIX three-body force, that has been originally proposed in combination with the Argonne AV18 and AV8' [13]. Although it slightly underbinds the energy of light nuclei, it has been extensively used to study the equation of state of nuclear and neutron matter [14, 15], and gives an equation of state of pure neutron matter stiff enough to support a neutron star with mass higher than 2 solar masses [14, 15, 16, 17], as suggested by recent observations [18, 19]. The Illinois three-body forces have been introduced to improve the description of both ground- and excited-states of light nuclei with $A \leq 8$ [10], and produce energy spectra for A up to 12 in excellent agreement with data [11], but produce an unphysical overbinding in pure neutron systems [20]. In neutron drops their contribution is also attractive [21].

We solve the many-body ground-state using both the Green's Function Monte Carlo (GFMC) and Auxiliary Field Diffusion Monte Carlo (AFDMC) [22]. The main idea of QMC methods is to evolve a many-body wave function in imaginary-time:

$$\Psi(\tau) = \exp[-H\tau]\Psi_v \,, \tag{4}$$

where Ψ_v is a variational ansatz, and H is the Hamiltonian of the system. In the limit of $\tau \to \infty$, Ψ approaches the ground-state of H. The evolution in imaginary-time is performed by sampling configurations of the system using Monte Carlo techniques, and expectation values are evaluated over the sampled configurations. For more details see for example Refs. [15, 23].

The GFMC has proven to be extremely accurate to study properties of light nuclei. The variational wave function includes all the possible spin/isospin states of the nucleons and provides

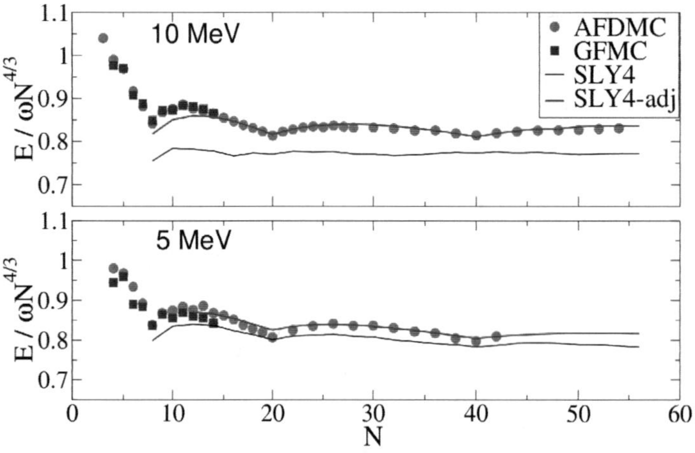

Figure 1. The energy of neutrons in a HO well with $\hbar\omega = 10$ MeV (upper panel) and 5 MeV (lower panel) in units of $\omega N^{4/3}$. The red dots are the results given by AFDMC, blue squares are from GFMC, and the black line is the result obtained using the Skyrme SLY4. The violet line is the adjusted SLY4 where the strength of the gradient, pairing, and spin-orbit terms have been changed. The figure is taken from Ref. [5].

a good variational ansatz to start the projection in imaginary time. The exponential growth of the number of spin-isospin states with particle number limits the calculation to ^{12}C [2] nuclei or drops with 16 neutrons. The AFDMC method does not include all the spin/isospin states in the wave function. The latter are instead sampled using the Hubbard-Stratonovich transformation. Using the AFDMC the calculation can be extended up to many neutrons, making the simulation of large neutron drops with up to several tens of neutrons possible. In the case of homogeneous systems the AFDMC has been used to simulate up to 114 neutrons [15], and up to 70 neutrons in an external well.

3. Results

The energy of neutron drops confined by V_{HO} is shown in Fig. 1 for two different frequencies of the external potential. The red points are the results obtained using the AFDMC method, and the blue ones using the GFMC. The two solid lines are the results given by using the original Skyrme SLY4 force [24], and a modified version. The energy is in units of the Thomas-Fermi energy, that is proportional to $\omega N^{4/3}$, to see the extrapolation to the thermodynamic limit. The two QMC methods agree within 1% for the $\hbar\omega = 10$ MeV trap, and the difference increases up to 4% for $\hbar\omega = 5$ MeV. The reason is that pairing correlations are included in the GFMC wave function, but not in the AFDMC. At low densities neutrons are superfluids, and pairing correlations are quite important to include for open-shell configurations.

The difference between QMC and Skyrme at closed shells is mainly due to two effects, the bulk contribution and the gradient term. Skyrme forces typically give an EOS of pure neutron matter at densities lower than saturation that is more repulsive than microscopic calculations. The equation of state of pure neutron matter is shown in Fig. 2 where we compare the AFDMC results from Refs. [15, 25, 29], the GFMC calculation of Gezerlis and Carlson [26, 27], and the equation of state given by SLY4. We make the reasonable assumption that Skyrme's bulk term cannot explain the difference between QMC and Skyrme energy in neutron drops. Then, since

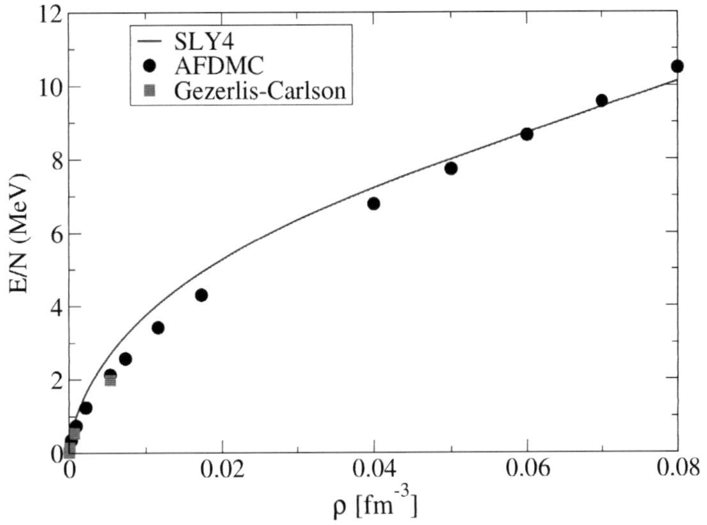

Figure 2. The equation of state of pure neutron matter. The AFDMC results [15, 25] are compared to the QMC results at low densities of Gezerlis and Carlson [26, 27], and to the equation of state of Skyrme SLY4 [28].

the pairing and the spin-orbit terms are expected to be very weak with respect to the gradient term for closed shell configurations, we can use the energy at N=8, 20 and 40 to re-adjust the gradient term of Skyrme. The energy of neutron drops with N near closed shells can be used to adjust the spin-orbit strength because for these configurations the pairing is not important. Finally, by comparing the energy of half-filled shells, we can tune the pairing term.

The ab-initio results can also be used as a benchmark for new Skyrme forces. For example, in a recent paper Kortelainen *et* al. have derived new Skyrme forces, and they have produced neutron drops results as a prediction to be compared with QMC calculations. The comparison is shown in Fig. 3. The agreement between the two new Skyrme forces with QMC results is remarkable considering that the ab-initio energies of neutron drops have not been included as a constraint in producing the UNEDF0 and UNEDF1 forces.

The use of a Woods-Saxon external potential V_{WS} is also interesting because it reflects properties of nuclei. It saturates for some number of neutrons, providing a good test of the surface effects of Skyrme. The comparison between the original and the adjusted Skyrme SLY4 with QMC for neutrons in a WS potential is shown in Fig. 4. We stress that the SLY4 has been adjusted only to reproduce the results in the HO well, and is used here to calculate the neutron drop energies as a prediction. Also in this case the agreement with the QMC calculations is quite good.

Finally, we studied the effect of the adjusted Skyrme to the structure of neutron drops by calculating the rms radii and density. The rms radius $\sqrt{< r^2 >}$ is shown in Fig. 5 where we compare the GFMC estimate with the original and adjusted SLY4 for different numbers of neutrons in the WS well (lower panel), and similarly with HO well for two different frequencies (upper panel). In Fig. 6 we show the radial density calculated with GFMC and Skyrme in the case of HO well for a closed shell configuration with N=8 (upper panel), and for an open-shell with N=14 (lower panel). Both for rms radii and for the density the adjusted Skyrme agree much better with QMC calculations, showing that the energy constraint is sufficient to improve

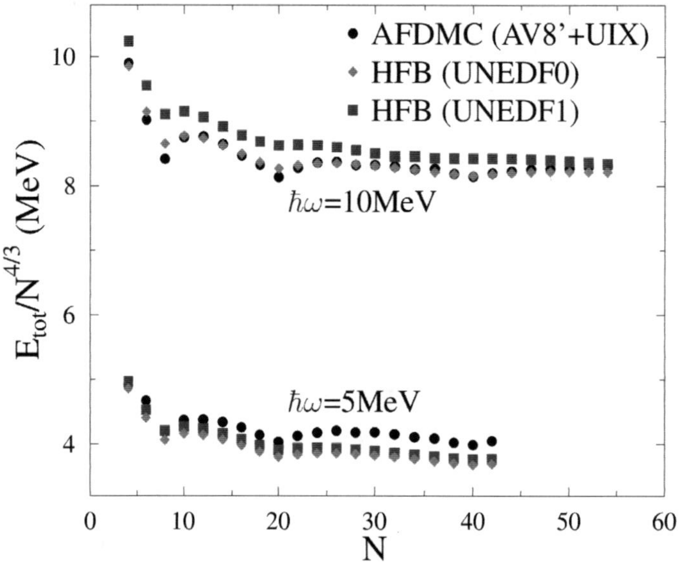

Figure 3. Energy of neutrons in a HO well. The AFDMC results are compared with those given by UNEDF0 and UNEDF1 calculations. Figure taken from Ref. [30] with courtesy of W. Nazarewicz.

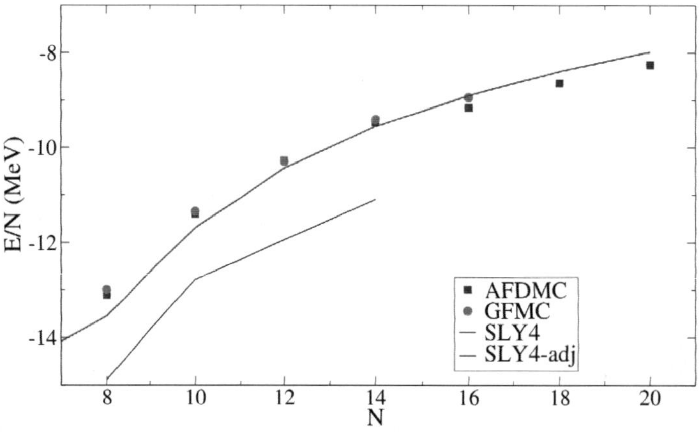

Figure 4. The energy per particle of neutrons in a WS well obtained using AFDMC (blue squares) and GFMC (red points) compared to Skyrme. Figure taken from Ref. [5].

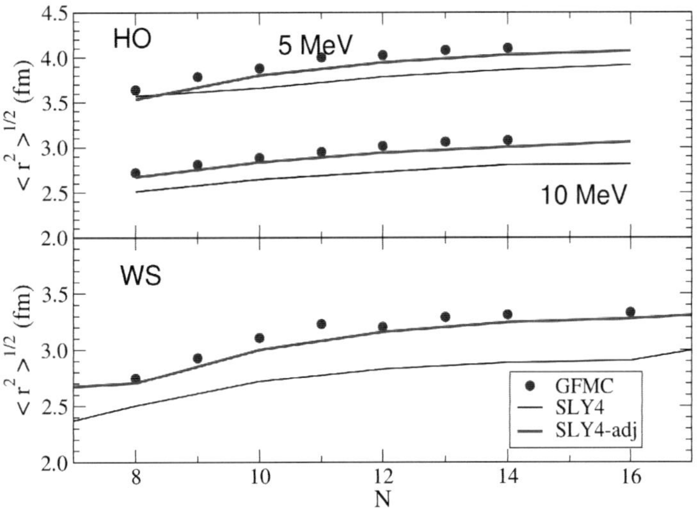

Figure 5. The radii of neutron drops as a function of the number of neutrons calculated using GFMC (blue points) compared with the original (black line) and the adjusted SLY4 (violet line). In the top panel we show the results in the HO external potential, and in the bottom panel those obtained in a WS well. Figure taken from Ref. [5].

the accuracy of density functionals, whose predictive power for other operators would drastically improve.

4. Conclusions

We performed ab-initio calculations of neutrons confined in external potentials. These systems are interesting to study because they provide a simple model to describe neutron-rich nuclei, and can be used to constrain energy density functionals. By using QMC techniques, we calculated the energy of neutrons in different potentials. We have considered both harmonic oscillator and Woods-Saxon wells in order to provide results for systems with rather different geometries. We paid particular attention to the comparison between the GFMC and AFDMC methods. The agreement is quite good, at the few percent level for the total energy of the system.

We compared the energies of neutron drops given by QMC methods with those given by different Skyrme forces, and found that the latter always produce lower energies. Since the density of these systems is lower than nuclear densities, we suggest that the overbinding is mainly due to the (neutron) gradient term. We found that, in order to reproduce our ab-initio results, the pairing and the spin-orbit terms of Skyrme need to be changed too.

We have used AFDMC to calculate the energy of drops up to 54 neutrons, and GFMC to calculate the rms radii and radial densities of drops up to 16 neutrons. After re-adjustment of the SLY4 parametrization, the Skyrme force provides much better agreement for these quantities.

Acknowledgments

The author would like to thank J. Carlson and R. Schiavilla for critical comments on the manuscript, and W. Nazarewicz for the useful discussions and for the permission to show Fig. 3. This work is supported by DOE Grants No. DE-FC02-07ER41457 (UNEDF SciDAC) and No. DE-AC52-06NA25396, and by the LANL LDRD program. Computer time was made available by

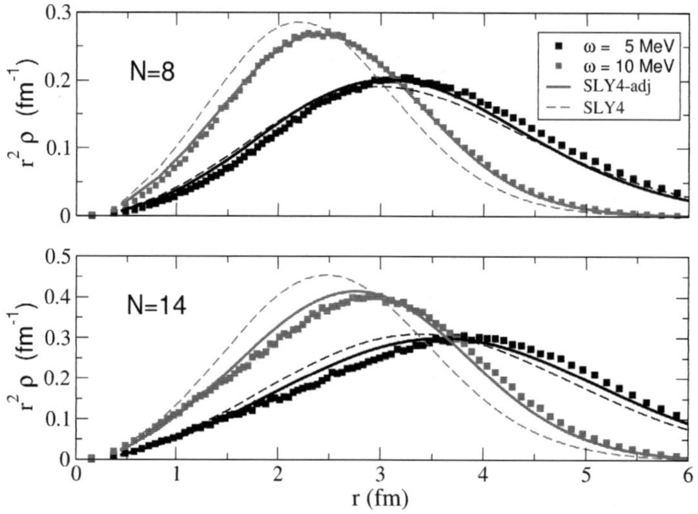

Figure 6. The radial density of neutron drops for different configurations calculated using GFMC (squares) compared with the original (dashed lines) and the adjusted SLY4 (solid lines) in a HO well. In the top panel we show the results of a closed shell configuration with N=8, and in the bottom panel the density of an open shell with N=14. The two colors are for different HO frequencies. Figure taken from Ref. [5].

Los Alamos Open Supercomputing, and by the National Energy Research Scientific Computing Center (NERSC).

References

[1] Chang S Y, Morales J, Jr, Pandharipande V R, Ravenhall D G, Carlson J, Pieper S C, Wiringa R B and Schmidt K E 2004 *Nucl. Phys. A* **746** 215
[2] Pieper S C 2005 *Nuclear Physics A* **751** 516
[3] Gandolfi S, Pederiva F, Fantoni S and Schmidt K E 2006 *Phys. Rev. C* **73** 044304
[4] Gandolfi S, Pederiva F and A Beccara S 2008 *European Physical Journal A* **35** 207–211
[5] Gandolfi S, Carlson J and Pieper S C 2011 *Phys. Rev. Lett.* **106** 012501
[6] Bogner S, Furnstahl R, Hergert H, Kortelainen M, Maris P *et al.* 2011 *Phys. Rev. C* **84** 044306
[7] Drut J E and Platter L 2011 *Phys. Rev. C* **84**(1) 014318
[8] Wiringa R B, Stoks V G J and Schiavilla R 1995 *Phys. Rev. C* **51** 38
[9] Entem D R and Machleidt R 2003 *Phys. Rev. C* **68** 041001
[10] Pieper S C, Pandharipande V R, Wiringa R B and Carlson J 2001 *Phys. Rev. C* **64** 014001
[11] Pieper S C 2008 *AIP Conf. Proc.* **1011** 143
[12] Wiringa R B and Pieper S C 2002 *Phys. Rev. Lett.* **89** 182501
[13] Pudliner B S, Pandharipande V R, Carlson J and Wiringa R B 1995 *Phys. Rev. Lett.* **74** 4396
[14] Akmal A, Pandharipande V R and Ravenhall D G 1998 *Phys. Rev. C* **58** 1804
[15] Gandolfi S, Illarionov A Y, Schmidt K E, Pederiva F and Fantoni S 2009 *Phys. Rev. C* **79** 054005
[16] Gandolfi S, Illarionov A Y, Fantoni S, Miller J, Pederiva F and Schmidt K 2010 *Mon. Not. R. Astron. Soc.* **404** L35
[17] Gandolfi S, Carlson J and Reddy S 2012 *Phys. Rev. C* **85** 032801
[18] Demorest P B, Pennucci T, Ransom S M, Roberts M S E and Hessels J W T 2010 *Nature* **467** 1081
[19] Steiner A W and Gandolfi S 2012 *Phys. Rev. Lett.* **108** 081102
[20] Sarsa A, Fantoni S, Schmidt K E and Pederiva F 2003 *Phys. Rev. C* **68** 024308
[21] Carlson J, Gandolfi S, Maris P, Pieper S C and Vary J P In preparation
[22] Schmidt K E and Fantoni S 1999 *Phys. Lett. B* **446** 99

[23] Pudliner B S, Pandharipande V R, Carlson J, Pieper S C and Wiringa R B 1997 *Phys. Rev. C* **56** 1720
[24] Chabanat E, Bonche P, Haensel P, Meyer J and Schaeffer R 1995 *Phys. Scr.* **T56** 231
[25] Gandolfi S, Illarionov A Y, Fantoni S, Pederiva F and Schmidt K E 2008 *Phys. Rev. Lett.* **101**(13) 132501
[26] Gezerlis A and Carlson J 2008 *Phys. Rev. C* **77** 032801
[27] Gezerlis A and Carlson J 2010 *Phys. Rev. C* **81**(2) 025803
[28] Chabanat E, Bonche P, Haensel P, Meyer J and Schaeffer R 1998 *Nuclear Physics A* **635** 231–256
[29] Gandolfi S, Illarionov A Y, Pederiva F, Schmidt K E and Fantoni S 2009 *Phys. Rev. C* **80**(4) 045802
[30] Kortelainen M, McDonnell J, Nazarewicz W, Reinhard P G, Sarich J, Schunck N, Stoitsov M V and Wild S M 2012 *Phys. Rev. C* **85**(2) 024304

HITES 2012 IOP Publishing
Journal of Physics: Conference Series **403** (2012) 012017 doi:10.1088/1742-6596/403/1/012017

Low-energy Electro-weak Reactions

Doron Gazit

Racah Institute of Physics and The Hebrew University Center for Nanoscience and Nanotechnology, The Hebrew University, 91904 Jerusalem, Israel

E-mail: doron.gazit@mail.huji.ac.il

Abstract. Chiral effective field theory (EFT) provides a systematic and controlled approach to low-energy nuclear physics. Here, we use chiral EFT to calculate low-energy weak Gamow-Teller transitions. We put special emphasis on the role of two-body (2b) weak currents within the nucleus, and discuss their applications in predicting physical observables.

1. Introduction

Electro-weak nuclear reactions at low energies, which are the subject of this contribution, have an important role in many physical scenarios. They are the microscopic motor of astrophysical phenomena, such as solar fusion, core collapse supernovae and the nucleosynthesis within. In addition, these reactions are used to study nuclear structure and dynamics. Moreover, low-energy β-decays and weak transitions provide a window into high-energy physics. For example, superallowed decays allow high precision tests of the Standard Model, and $0\nu\beta\beta$ decays probe the nature of neutrinos, their hierarchy and mass. High accuracy theoretical, as well as experimental understanding of the nuclear response is a key ingredient needed to accomplish these tasks.

However, the non-perturbative character of QCD at low energies makes a theoretical calculation of the nuclear response extremely challenging. In the last two decades, chiral effective field theory of QCD at low energies was developed, enabling a construction of a nuclear Lagrangian that is consistent with QCD. Chiral EFT provides a systematic basis for nuclear forces and consistent electroweak currents [1, 2], where pion couplings contribute both to the electroweak axial current and to nuclear interactions. This is already seen at leading order: the axial constant, g_A, determines the axial one-body (1b) current and the one-pion-exchange potential. Two-body (2b) currents, also known as meson-exchange currents, enter at higher order, just like three-nucleon (3N) forces [1]. As shown in Fig. 1, the leading axial contributions are due to long-range one-pion-exchange and short-range parts [2], with couplings c_3, c_4 and c_D, which also enter the leading 3N (and subleading NN) forces [1, 3].

In this contribution we present some recent studies of chiral currents and their consistency with nuclear forces. We emphasize the importance of using chiral 2b currents in weak Gamow-Teller transitions in nuclei.

2. Gamow-Teller transitions

At low energies, the coupling of weak probes to a nucleus is given by the current-current interaction, $H_W = \frac{G_F}{\sqrt{2}} \int d^3\mathbf{r}\, e^{-i\mathbf{p}\cdot\mathbf{r}} j_{L\mu} J_L^{\mu\dagger}$ + h.c., where G_F is the Fermi constant, $\mathbf{p}$ the momentum transferred from nucleons to leptons, and $j_{L\mu}$ the leptonic current of an electron

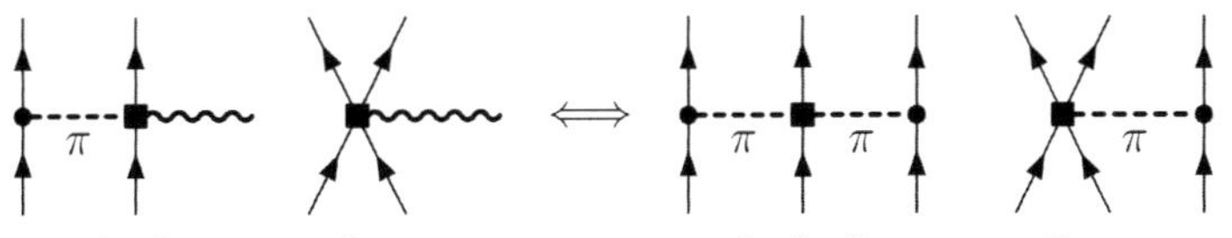

Figure 1. In chiral EFT, Feynman diagrams of chiral 2b currents (left) correspond to 3N force contributions (right) that depend on the same low energy constants.

coupled to a left-handed electron neutrino. This approximation creates a separation between the leptons and baryons, which is a great simplification of the problem. It suggests that only the symmetry of the interaction is needed to determine the structure of the current. For example, any axial probe, regardless of its fundamental origin, would couple to the same current within the nucleus, which is the Nöther current derived from the axial symmetry of the chiral Lagrangian, which we will consider up to next-to-next-to-next-to leading order (N^3LO) [2].

The nuclear current $J_L^{\mu\dagger}$ is organized in an expansion in powers of momentum $Q \sim m_\pi$ over a breakdown scale $\Lambda_b \sim 500\,\text{MeV}$. To order Q^2 (and also Q^3 in this counting), the 1b current, $J_L^{\mu\dagger}(\mathbf{r}) = \sum_{i=1}^{A} \tau_i^- \left[\delta^{\mu 0} J_{i,1b}^0 - \delta^{\mu k} J_{i,1b}^k\right] \delta(\mathbf{r}-\mathbf{r}_i)$, has temporal and spatial parts in momentum space [2]:

$$J_{i,1b}^0(p^2) = g_V(p^2) - g_A \frac{\mathbf{P}\cdot\boldsymbol{\sigma}_i}{2m} + g_P(p^2)\frac{E\,(\mathbf{p}\cdot\boldsymbol{\sigma}_i)}{2m}, \tag{1}$$

$$\mathbf{J}_{i,1b}(p^2) = g_A(p^2)\,\boldsymbol{\sigma}_i - g_P(p^2)\frac{\mathbf{p}\,(\mathbf{p}\cdot\boldsymbol{\sigma}_i)}{2m} + i(g_M+g_V)\frac{\boldsymbol{\sigma}_i\times\mathbf{p}}{2m} - g_V\frac{\mathbf{P}}{2m}, \tag{2}$$

where $E = E_i - E_i'$, $\mathbf{p} = \mathbf{p}_i - \mathbf{p}_i'$, and $\mathbf{P} = \mathbf{p}_i + \mathbf{p}_i'$. $g_V(p^2)$, $g_A(p^2)$, $g_P(p^2)$, and $g_M(p^2)$ are the vector (V), axial (A), pseudo-scalar (P), and magnetic (M) couplings. In chiral EFT, the p dependence is due to loop corrections and pion propagators, to order Q^2: $g_{V,A}(p^2) = g_{V,A}\,(1 - 2\frac{p^2}{\Lambda_{V,A}^2})$, with $g_V = 1$, $\Lambda_V = 850\,\text{MeV}$, $\Lambda_A = 2\sqrt{3}/r_A = 1040\,\text{MeV}$; $g_P(p^2) = \frac{2g_{\pi pn}F_\pi}{m_\pi^2+\mathbf{p}^2} - 4\,g_A(p^2)\frac{m}{\Lambda_A^2}$ and $g_M = \mu_p - \mu_n = 3.70$, with pion decay constant $F_\pi = 92.4\,\text{MeV}$, $m_\pi = 138.04\,\text{MeV}$, and $g_{\pi pn} = 13.05$ [4].

At leading order Q^0, only the momentum-independent g_A and g_V terms contribute. They give rise to low-momentum-transfer Gamow-Teller and Fermi decays. The former has the known form, $E_1^A|_{\text{LO}} = \frac{i\,g_A}{\sqrt{6\pi}} \sum_{i=1}^{A} \sigma_i \tau_i^-$.

Among the Q^2 terms, form-factor-type (ff) contributions and the g_P part of $\mathbf{J}_{i,1b}$ dominate. The remaining Q^2 terms are odd under parity, so they require either a P-wave lepton or another odd-parity term to connect to 0^+ states, common for ground states. In such cases, the $\mathbf{P}$ and E terms in Eqs. (1) and (2) can be neglected, and only the term with the large $g_M + g_V = 4.70$ is kept.

At order Q^3, 2b currents enter in chiral EFT [2]. These include vector spatial, axial temporal, and axial spatial parts[1]. The first two are odd under parity, and therefore can be neglected in couplings to 0^+ ground states. In addition, within chiral EFT the vector spatial current is conserved, satisfying the CVC hypothesis, thus at low-energies their contribution can be calculated via the Siegert theorem. As a result, the dominant weak 2b currents include an axial

[1] Vector temporal parts do not contribute at this order [2].

Table 1. Calculated ^{3}H, ^{3}He and ^{4}He g.s. energies (MeV) and point-proton root-mean-squared radii (fm)

	^{3}H		^{3}He		^{4}He	
	$E_{\text{g.s.}}$	$\langle r_p^2\rangle^{1/2}$	$E_{\text{g.s.}}$	$\langle r_p^2\rangle^{1/2}$	$E_{\text{g.s.}}$	$\langle r_p^2\rangle^{1/2}$
NN	−7.852(4)	1.651(5)	−7.124(4)	1.847(5)	−25.39(1)	1.515(2)
$NN+NNN$	−8.473(4)	1.605(5)	−7.727(4)	1.786(5)	−28.50(2)	1.461(2)
Expt.	−8.482	1.60	−7.718	1.77	−28.296	1.467(13)

spatial component, $\mathbf{J}_{2b}^{\text{axial}} = \sum_{i<j}^{A} \mathbf{J}_{ij}$, with [2]

$$\begin{aligned}\mathbf{J}_{12} = & -\frac{g_A}{F_\pi^2}\Big[2d_1(\boldsymbol{\sigma}_1\tau_1^- + \boldsymbol{\sigma}_2\tau_2^-) + d_2\,\boldsymbol{\sigma}_\times\tau_\times^-\Big] \\ & -\frac{g_A}{2F_\pi^2}\frac{1}{m_\pi^2+\mathbf{k}^2}\Big[\Big(c_4+\frac{1}{4m}\Big)\mathbf{k}\times(\boldsymbol{\sigma}_\times\times\mathbf{k})\,\tau_\times^- \\ & +4c_3\mathbf{k}\cdot(\boldsymbol{\sigma}_1\tau_1^- + \boldsymbol{\sigma}_2\tau_2^-)\mathbf{k} - \frac{i}{2m}\mathbf{k}\cdot(\boldsymbol{\sigma}_1-\boldsymbol{\sigma}_2)\mathbf{q}\,\tau_\times^-\Big], \end{aligned} \quad (3)$$

where $\tau_\times^- = (\tau_1\times\tau_2)^-$ and the same for $\boldsymbol{\sigma}_\times$, $\mathbf{k} = \frac{1}{2}(\mathbf{p}_2' - \mathbf{p}_2 - \mathbf{p}_1' + \mathbf{p}_1)$ and $\mathbf{q} = \frac{1}{4}(\mathbf{p}_1 + \mathbf{p}_1' - \mathbf{p}_2 - \mathbf{p}_2')$. Equation (3) includes contributions from the one-pion-exchange c_3, c_4 parts and from the short-range couplings d_1, d_2, where due to the Pauli principle only the combination $d_1 + 2d_2 = c_D/(g_A\Lambda_\chi)$ enters [with $\Lambda_\chi = 700\,\text{MeV}$], see Fig 1. This result is extremely interesting as it connects the low-energy constants (LECs) determining the strength of the meson exchange current (MEC) contact term to c_D, which is one of the two unknown low-energy constants, c_D and c_E, that determine the 3N force up to N^{3}LO, and do not contribute to 2N force. This relation shows that one can use a weak observable, e.g., an empirical extraction of $\langle E_1^A\rangle|_{emp}$ from a decay rate, as a constraint for the determination of c_D and c_E.

3. Using triton β-decay to determine the 3N force

We use the relation between the short range contribution to the current and the 3N force to fully constrain the Lagrangian in systems with up to 3-body, and predict ^{4}He properties. The calculation steps are as follows: (i) calculate the ^{3}H and ^{3}He g.s. wave functions by solving the Schrödinger equation for three nucleons interacting via the chiral NN potential at N^{3}LO of Ref. [5] and the NNN interaction at N^{2}LO [6]; (ii) determine for which c_D values along the trajectory the calculated reduced matrix element of the E_1^A operator (at N^{3}LO) reproduces $\langle E_1^A\rangle_{emp}$ extracted from ^{3}H β-decay. We find that these two constraints are uncorrelated, producing a stringent determination of the LECs: $c_D = -0.2 \pm 0.1$ and $c_E = -0.205 \pm 0.015$ [3].

With this calibration of c_D and c_E, for this potential, in principle, any other calculation is a prediction! In Table 1 a collection of $A = 3$ and 4 data is given, obtained with and without inclusion of the NNN force for $c_D = -0.2$ ($c_E = -0.205$). The $NN+NNN$ predictions for the ^{4}He are in good agreement with measurement.

To this important conclusion, one can add the results of few additional tests, appearing in Fig. 2, that are aimed to probe the axial correlations in the nucleus, by analyzing the sensitivity of the triton half life to NNN force and/or MEC:

- *The fundamental importance of the axial two-body currents.* By suppressing the MEC, in the whole investigated c_D-c_E range, the calculations underpredict $\langle E_1^A\rangle_{emp}$ by about 2%. Within the MEC, one finds a large cancellation between the long-range one-pion-exchange term (corresponding to the left diagram in Fig. 1), and the contact term (second diagram from the left Fig. 1).

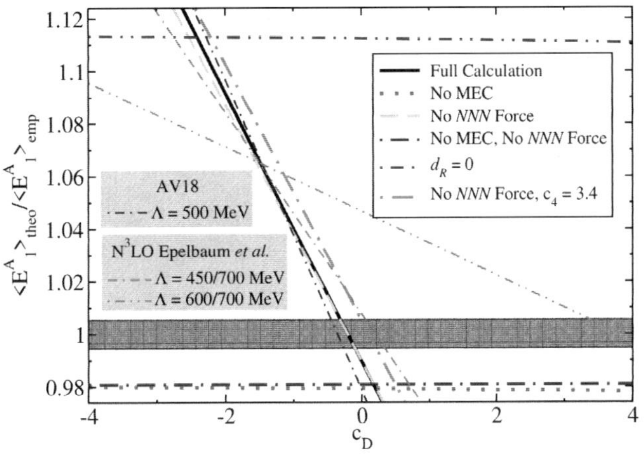

Figure 2. The ratio $\langle E_1^A\rangle_{th}/\langle E_1^A\rangle_{emp}$ using the N^{3}LO NN potential [5] with and without the N^{2}LO NNN interaction [6], and the axial current with and without MEC. The shaded area is twice the experimental uncertainty. Also shown: results for the phenomenological AV18 potential (with $\Lambda = 500$ MeV imposed in the current), and for the N^{3}LO NN potential of Epelbaum et al. [7] (with $\Lambda = 450, 600$ MeV, and a 700 MeV spectral-function cutoff in the two-pion exchange), which should be compared with the full calculation (black solid line). The line denoted by $d_R = 0$ includes only the long range contribution to the MEC, neglecting the contact term contribution. Figure taken from [3].

- *A negligible effect of the NNN force on this observable.*
- *Weak dependence on the specific character of the force.*

These features, which might be unique to s-shell nuclei, entail that the determination of c_D and c_E obtained in this work is robust.

4. Extracting the weak structure of the nucleon from μ^- capture on ^{3}He [8]

In the previous section it was shown that for a weak transition in the 3-nucleon sector, the specific character of the force had a minor effect on the reaction rate. We extrapolate this understanding to the muon capture on ^{3}He which results in a triton, $\mu^- +^3\text{He} \to \nu_\mu +^3\text{H}$. We study this reaction in a hybrid approach: the calculation uses the chiral EFT based weak current, combined with the phenomenological nucleon-nucleon potential Argonne v_{18} (AV18) [9] augmented by the Urbana IX (UIX) [10] three nucleon force. This allows a study of the cutoff variation of the observable.

μ capture on ^{3}He has an accurately measured rate $\Gamma(\mu^- +^3\text{He} \to \nu_\mu +^3\text{H})^{\text{exp}} = 1496(4)$ Hz, i.e. a $\pm 0.3\%$ precision [11]. Combining with the sizable momentum transfer in ordinary muon capture (OMC), that enhances the effect of the induced-pseudoscalar form factor, allows its study. In addition, one can use OMC to study second class currents [12].

The calculation process demands fixing the single contact unknown LEC, which determines the weak axial current. We do this by reproducing the experimental triton half-life, for various cutoff values for the chiral current. The results for the nuclear matrix element of the muon capture process show a 9% effect due to the MEC contribution, and a small 0.3% effect due to the cutoff dependence. It is worthwhile noting that the relative contribution of the MEC to this process is almost three times bigger than the MEC contribution to the triton half life. An extremely weak cutoff dependence shows that the essential physics is captured in the chiral EFT operators.

The final prediction for the capture rate is

$$\Gamma = 1499(2)_\Lambda(3)_{\rm NM}(5)_t(6)_{\rm RC}\,{\rm Hz}, \tag{4}$$

where the first error is due to the Heavy Baryon chiral perturbation theory (HBχPT) cutoff, the second is due to uncertainties in the extrapolation of the form factors to finite momentum transfer, and in the choice of the specific nuclear model, the third error is related to the uncertainty in the triton half life, and the last error is due to theoretical uncertainty in the electroweak radiative corrections calculated for nuclei [13]. This sums to a total error estimate of about 1%, which overlaps with the uncertainty

The conservative error estimation still allows rather interesting conclusions. First, one notices that the calculated capture rate agrees with the experimental measurement $\Gamma(\mu^- + {}^3\,{\rm He} \rightarrow \nu_\mu + {}^3\,{\rm H})^{\rm exp}_{\rm stat} = 1496(4)\,{\rm Hz}$. Thus, one concludes that EFT accurately predicts the capture rate.

However, the most interesting result concerns the weak form factors of the nucleon. In order to constrain the induced pseudoscalar and second class form factors, we take the following approach. In each case, we set all the other form factors to their nominal value, and change this form factor in a way which keeps an overlap between the experimental rate and the theoretically allowed rate. The nominal value of the form factor is set to reproduce the experimental measurement.

The resulting constraint on the induced pseudoscalar form factor is:

$$g_P(q^2 = -0.954 m_\mu^2) = 8.13 \pm 0.6\,, \tag{5}$$

in very good agreement with the HBχPT prediction of Ref. [14].

A second conclusion concerns the contribution of second class currents. While the current constraint on the axial G- parity breaking term is rather weak, this calculation puts the tightest limit on CVC, constraining it to $m_e F_S/F_V = (0.5 \pm 2) \times 10^{-4}$, which is consistent with CVC. One notes that this is already close to the regime of CVC breaking, predicted by χPT [14].

A recent calculation in the framework of the previous section, i.e., taking currents and forces from the same chiral EFT Lagrangian, has reached similar results [15].

5. The in-medium behavior of the axial constant

Surprisingly, key aspects of β decays remain a puzzle. In particular, when calculations of Gamow-Teller (GT) transitions of the spin–isospin-lowering operator $g_A\boldsymbol{\sigma}\tau^-$ are confronted with experiment, some degree of renormalization, or "quenching" q, of the axial coupling $g_A^{\rm eff} = q g_A$ is needed. Compared to the single-nucleon value $g_A = 1.2695(29)$, the GT term seems to be weaker in nuclei. This was first conjectured in studies of β-decay rates, with a typical $q \approx 0.75$ in many-body calculations [16, 17]. In view of the significant effect on weak reaction rates, it is no surprise that this suppression has been the target of many theoretical works. It is also a major uncertainty for $0\nu\beta\beta$ decay nuclear matrix elements (NMEs), which probe larger momentum transfers of the order of the pion mass, $p \sim m_\pi$, where the renormalization could be different. Here we revisit this puzzle based on chiral EFT currents.

The lightest nucleus that undergoes a β-decay is the triton. However, the theory cannot be checked in the triton since its half-life is used to calibrate the strength of the MEC. The lightest nucleus that can provide a test to the theory is thus ^{6}He ($J^\pi = 0^+$), an unstable nucleus, which undergoes a β decay with a half-life $\tau_{1/2} = 806.7 \pm 1.5\,$msec to the ground state of ^{6}Li ($J^\pi = 1^+$) [18]. As this is a 6-body problem, we use a soft nuclear potential, JISP16. This potential leads to an underbinding of about 80keV for the ^{3}He and 120keV for the triton, where we calibrate the one unknown LEC determining the current. One calculates binding energies of 28.70MeV for ^{6}He and 31.46 MeV for ^{6}Li. These results underestimate the experimental energies by about

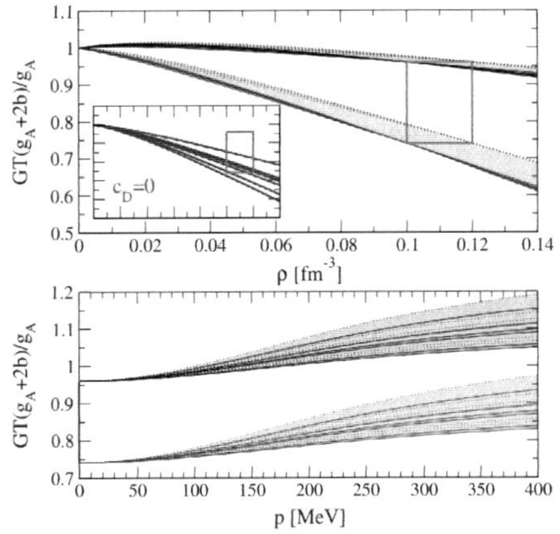

Figure 3. Top panel: g_A plus 2b-current contributions for $p = 0$ GT transitions normalized to g_A as a function of density ρ. The boundaries of the box are given by $q = 0.74/0.96$ and $\rho = 0.10, ..., 0.12\,\mathrm{fm}^{-3}$. The different curves correspond to possible calibrations of c_D (see Ref. [19] for details), with shaded regions for the density range. Intermediate quenching values would lie between these regions. The inset shows the quenching predicted by the long-range parts of 2b currents only ($c_D = 0$). Bottom panel: Same as top, but as a function of momentum transfer p for empirical/smaller quenching $q = 0.74/0.96$. Figure taken from Ref. [19].

0.5 MeV, though the difference $\Delta E = 2.76$MeV differs by merely 34keV from the experimental value.

Incorporating the χPT based contributions to the weak-current one can calculate the full ^{6}He-^{6}Li GT matrix-element, as a function of the cutoff imposed in the weak current:

$$|\mathrm{GT}(^6\mathrm{He})|_{\mathrm{theo}} = 2.198(1)_\Lambda(2)_\mathrm{N}(4)_\mathrm{t}(5)_{\mathrm{g_A}} = 2.198 \pm 0.007. \qquad (6)$$

The first error is the cutoff variation dependence, the second is numerical, the third is due to uncertainties in the triton half-life, and the last is due to uncertainties in g_A. This should be compared to the experimental matrix-element $|\mathrm{GT}(^6\mathrm{He})|_{\mathrm{expt}} = 2.161 \pm 0.005$. Thus, the theory overpredicts GT by about 1.7%, which is reasonable considering the approximation in using a phenomenological, pure 2-body, potential.

For heavier nuclei a microscopic calculation is still out of reach. We thus study the impact of chiral 2b currents in nuclei at the normal-ordered 1b level by summing the second nucleon over occupied states in a spin and isospin symmetric reference state or core. Taking a Fermi gas approximation for the core and neglecting tensor-like terms $(\mathbf{k}\cdot\boldsymbol{\sigma}\,\mathbf{k} - \frac{1}{3}k^2\boldsymbol{\sigma})\tau^-$, we obtain the normal-ordered 1b current [19]:

$$\mathbf{J}^{\mathrm{eff}}_{i,2\mathrm{b}} = -g_A\boldsymbol{\sigma}_i\tau_i^- \frac{\rho}{F_\pi^2}\left[\frac{c_D}{g_A\Lambda_\chi} + \frac{2}{3}c_3\frac{\mathbf{p}^2}{4m_\pi^2 + \mathbf{p}^2} + I(\rho, P)\left(\frac{1}{3}(2c_4 - c_3) + \frac{1}{6m}\right)\right], \qquad (7)$$

where $\rho = 2k_\mathrm{F}^3/(3\pi^2)$ is the density of the reference state, k_F the corresponding Fermi momentum, and $I(\rho, P)$ is due to the summation in the exchange term. The effective 1b current $\mathbf{J}^{\mathrm{eff}}_{i,2\mathrm{b}}$ only contributes to the GT operator and can be included as a correction to the $g_A(p^2)$ part of the 1b current, Eq. (2). This demonstrates that chiral 2b currents naturally contribute to the quenching of GT transitions. Compared to light nuclei, their contributions are amplified because of the larger nucleon momenta. A quenching of $p \approx 0$ GT transitions is predicted, as well as the momentum transfer dependence, as seen in Fig. 3.

This formalism can be used to evolve the chiral EFT understanding various phenomena, including the prediction of $0\nu\beta\beta$ decay rates. Indeed, such a calculation of the $0\nu\beta\beta$ decay operator based on chiral EFT currents shows that 2b contributions to the nuclear matrix elements are significant, and range from -35% to 10% [19]. Another recent application, which will not be covered in this contribution, is to the scattering of a weakly interacting massive particle (WIMP) candidate on nuclei [20].

6. Summary

We showed that chiral EFT enables parameter free calculations of electro-weak Gamow Teller transitions, that can be used in various fields: from constraining the nuclear forces, through a deeper understanding of in-medium nuclear effects, and up to predicting weak reactions for the use of experiments studying the limits of the Standard Model.

7. Acknowledgments

It is a great pleasure to thank my collaborators in the works presented in this contribution. This work was supported by the German Federal Ministry of Education and Research (BMBF) ARCHES – Award for Research Cooperation and High Excellence in Science.

References

[1] E. Epelbaum, H.-W. Hammer and U.-G. Meißner 2009 *Rev. Mod. Phys.* **81** 1773
[2] T. S. Park *et al.* 2003 *Phys. Rev. C* **67** 055206; D. Gazit 2007, Ph.D. thesis, Hebrew University, *arXiv* 0807.0216
[3] D. Gazit, S. Quaglioni and P. Navratil 2009 *Phys. Rev. Lett.* **103** 102502
[4] V. Bernard, L. Elouadrhiri and U.-G. Meißner 2002 *J. Phys. G* **28** R1
[5] D. R. Entem and R. Machleidt 2003 *Phys. Rev. C* **68** 041001(R)
[6] U. van Kolck 1994 *Phys. Rev. C* **49** 2932; E. Epelbaum *et al.* 2002 *Phys. Rev. C* **66** 064010; P. Navrátil 2007 *Few Body Syst.* **41** 117
[7] E. Epelbaum, W. Glöckle, and U.-G. Meißner 2005 *Nucl. Phys. A* **747** 362
[8] D. Gazit 2008 *Phys. Lett. B* **666** 472 (2008)
[9] R.B. Wiringa, V.G.J. Stoks, R. Schiavilla 1995 *Phys. Rev. C* **51** 38
[10] B.S. Pudliner, et al. 1997 *Phys. Rev. C* **56** 1720
[11] P. Ackerbauer, *et al.* 1998 *Phys. Lett. B* **417** 224
[12] N. Severijns, M. Beck, and O. Naviliat-Cuncic 2006 *Rev. Mod. Phys.* **78** 991
[13] A. Czarnecki, W.J. Marciano, A. Sirlin 2007 *Phys. Rev. Lett.* **99** 032003
[14] N. Kaiser 2001 *Phys. Rev. C* **64** 028201
[15] L. E. Marcucci *et al.* 2012 *Phys. Rev. Lett.* **108** 052502
[16] B. H. Wildenthal, M. S. Curtin and B. A. Brown 1983 *Phys. Rev. C* **28** 1343; G. Martínez-Pinedo *et al.* 1996 *Phys. Rev. C* **53** R2602
[17] M. Bender *et al.* 2002 *Phys. Rev. C* **65** 054322; T. R. Rodríguez and G. Martínez-Pinedo 2010 *Phys. Rev. Lett.* **105** 252503
[18] F. Ajzenberg-Selove 1998 *Nucl. Phys. A* **490** 1
[19] J. Menéndez, D. Gazit and A. Schwenk 2011 *Phys. Rev. Lett.* **107** 062501
[20] J. Menéndez, D. Gazit and A. Schwenk 2012 *Phys. Rev. D* accepted for publication, *arXiv* 1208.1094

Search for center-of-mass excitation free states in the SU(3) no-core shell model space

F Q Luo[1], M A Caprio[1] and T Dytrych[2]
[1]Department of Physics, University of Notre Dame, Notre Dame, IN 46556-5670, USA
[2]Department of Physics and Astronomy, Louisiana State University, Baton Rouge, LA 70803-4001, USA
E-mail: fluo@nd.edu

Abstract. We address the removal of states with center-of-mass excitation from the SU(3) no-core shell model [SU(3)-NCSM] space. After these spurious states have been removed, the remaining center-of-mass excitation free (CMF) states form a suitable model space for describing the intrinsic dynamics of nuclei. We formulate the problem of finding the CMF states as a null-space problem for the center-of-mass harmonic oscillator lowering operator $\tilde{B}^{(01)}_{\mathrm{c.m.}}$, acting on the SU(3)-NCSM many-body space. We outline the construction process for the matrix representation of $\tilde{B}^{(01)}_{\mathrm{c.m.}}$ and provide a simple application of the method to the deuteron.

1. Introduction

The ability to carry out no-core configuration interaction calculations of light nuclei, in the no-core shell model (NCSM) [1–3], has made a significant contribution to recent progress in the *ab initio* description of nuclei. However, the dimensionality of the nuclear model space becomes computationally prohibitive as the number of active nucleons and orbitals increases. Symmetry can play a significant role in addressing this problem, by assisting in the selection of the physically relevant portions of the model space. Elliott [4–6] explored the $\mathrm{U}(3) \supset \mathrm{SU}(3)$ symmetry of the harmonic oscillator, which serves as an organizational scheme for quadrupole deformation and rotation in the nuclear shell model. The U(3) algebra of the oscillator is contained in a larger $\mathrm{Sp}(3,\mathbb{R})$ algebra, which is found to have a close connection both to the dominant components of the nuclear Hamiltonian and to nuclear collective motion [7–9]. Following the discovery of evidence for SU(3) and $\mathrm{Sp}(3,\mathbb{R})$ symmetry in conventional NCSM calculations for light nuclei [10, 11], the NCSM has been reformulated in terms of an SU(3)-based model space, in the SU(3) no-core shell model [SU(3)-NCSM] [12]. This development provides the foundations for future realization of a symplectic no-core shell model (Sp-NCSM) [13], making full use of $\mathrm{Sp}(3,\mathbb{R})$ symmetry.

In predicting physical properties of the nucleus, only the intrinsic dynamics of the nucleus is of interest, not the center-of-mass motion. In principle, the center-of-mass motion may be eliminated from the problem by explicitly changing variables to relative coordinates. However, the nuclear many-body state must be antisymmetrized, and with increasing nucleon number this process rapidly becomes prohibitive in relative coordinates [3]. On the other hand, for a many-body basis constructed from antisymmetrized products of single-particle states, as in the NCSM, antisymmetrization is straightforward, but both center-of-mass (spurious) and intrinsic excitations are included in the model space [14, 15]. Consequently many of the nuclear eigenstates obtained by diagonalizing the Hamiltonian will carry center-of-mass excitation, presenting challenges for the study of the intrinsic excitations.

The usual approach to addressing the problem of spurious states consists of modifying the nuclear *Hamiltonian*, by adding a Lawson term [16] proportional the harmonic oscillator Hamiltonian $H_{\mathrm{c.m.}}$ for the center-of-mass coordinate. After diagonalization, the spurious states remain in the spectrum but are shifted to an excitation energy above the low-lying intrinsic states of interest. However, the other possibility is to modify the *model space*, so as to explicitly remove the spurious states, before the Hamiltonian is diagonalized. This approach may be made feasible through the use of SU(3) symmetry [17–19]. After the spurious states are removed, the remaining center-of-mass excitation free (CMF) states form a suitable model space for describing the intrinsic dynamics of nuclei.

In this contribution, we establish an algorithm for identifying the CMF subspace within the SU(3)-NCSM space. The CMF states are associated with the lowest eigenvalue of the center-of-mass Hamiltonian $H_{\mathrm{c.m.}}$. However, they also constitute the null space of the center-of-mass harmonic oscillator lowering operator $\tilde{B}^{(01)}_{\mathrm{c.m.}}$, which serves as the basis for a simpler numerical approach developed here. It should also be noted that the SU(3)-NCSM model space is the starting point for constructing the Sp-NCSM model space [13]. The present results are intended to provide the foundation for obtaining Sp$(3,\mathbb{R})$ irreducible representations (irreps) which are free of spurious excitations.[1]

We first outline the construction of the SU(3)-NCSM basis (section 2). We then introduce an algorithm for identification of the CMF subspace within an SU(3)-NCSM model space, based on constructing the matrix representation of $\tilde{B}^{(01)}_{\mathrm{c.m.}}$ with respect to the SU(3)-NCSM basis and solving for the null space (section 3). Finally, we provide a simple application of the present method to the SU(3)-NCSM model space for the deuteron (section 4).

2. SU(3)-NCSM basis states

The many-body basis states for the SU(3)-NCSM have good SU(3) $\times$ SU(2) quantum numbers, where the SU(3) symmetry label $(\lambda\mu)$ characterizes the spatial degrees of freedom, according to the Elliott classification, while the SU(2) label S describes the total intrinsic spin angular momentum. The creation operator for a nucleon in a given major oscillator shell η comprises an SU(3) $\times$ SU(2) tensor $a^{\dagger}_{(\eta 0)1/2}$, where the labels denote $(\lambda\mu)S = (\eta 0)\frac{1}{2}$ [20]. The operators $a^{\dagger}_{(\eta 0)1/2}$ are then used as the fundamental units in building up an SU(3) $\times$ SU(2)-coupled nuclear state.

Specifically, each SU(3)-NCSM basis state is characterized by a definite distribution of nucleons over the major shells. First, all nucleons in the each major shell η are combined to form a configuration of the type $[a^{\dagger}_{(\eta 0)1/2} \times a^{\dagger}_{(\eta 0)1/2} \times \cdots]^{(\lambda_\eta \mu_\eta) S_\eta}$, with good SU(3) $\times$ SU(2) coupling [21], separately for the protons and neutrons. Then, such configurations for individual major shells are coupled successively to form a total proton state and total neutron state carrying good SU(3) $\times$ SU(2) quantum numbers. Finally, the proton state and neutron state are coupled, to give the SU(3)-NCSM basis state. Since the major shells have definite occupations, the state may be classified, as usual in the oscillator basis for the NCSM [3], by the number N_{ex} of harmonic oscillator excitations, taken relative to the minimal number of oscillator quanta possible for the given number of protons and neutrons.

[1] Briefly, an Sp$(3,\mathbb{R})$ irrep is labeled by its extremal SU(3) state [13]. Other Sp-NCSM states within this irrep are constructed by acting on the extremal state with an SU(3)-coupled product $[A^{(20)} \times A^{(20)} \ldots \times A^{(20)}]^{(\lambda\mu)}$, where $A^{(20)}$ is the Sp$(3,\mathbb{R})$ raising operator. Since the action of $A^{(20)}$ does not introduce any center-of-mass excitation, an Sp$(3,\mathbb{R})$ irrep is CMF as long as its extremal SU(3) state is CMF.

The resulting SU(3)-NCSM basis state has the form, with all coupling labels shown explicitly,

$$\Big(| ((((\gamma_{p,0} \times \gamma_{p,1})^{\rho_{p,0}\omega_{p,0}} \times \gamma_{p,2})^{\rho_{p,1}\omega_{p,1}} \times \gamma_{p,3})^{\rho_{p,2}\omega_{p,2}} \ldots \times \gamma_{p,\eta_{\max}})^{\rho_p\omega_p} \rangle$$
$$\times | ((((\gamma_{n,0} \times \gamma_{n,1})^{\rho_{n,0}\omega_{n,0}} \times \gamma_{n,2})^{\rho_{n,1}\omega_{n,1}} \times \gamma_{n,3})^{\rho_{n,2}\omega_{n,2}} \ldots \times \gamma_{n,\eta_{\max}})^{\rho_n\omega_n} \rangle \Big)^{\rho\omega}. \quad (1)$$

Here the symbol $\gamma_{s,\eta}$ represents the labels $\gamma = [f_1, f_2, \ldots, f_N]\alpha(\lambda\mu)S$ needed to completely specify the coupling of nucleons of type s (*i.e.*, protons or neutrons) within major shell η [21]. Specifically, each major shell has associated with it a U(N) algebra [$N = (\eta + 1)(\eta + 2)$] consisting of bilinears of creation and annihilation operators, for which the irreps are labeled by $[f_1, f_2, \ldots, f_N]$, where we consider only antisymmetric irreps, and $\sum_{i=1}^{N} f_i$ equals the occupation of the shell. Within a U(N) irrep, a multiplicity index α is required to distinguish SU(3) $\times$ SU(2) irreps with the same quantum numbers $(\lambda\mu)S$, yielding the labeling scheme

$$\underset{[f_1,f_2,\ldots,f_N]}{\mathrm{U}(N)} \quad \underset{\alpha}{\supset} \quad \underset{(\lambda\mu)}{\mathrm{SU}(3)} \times \underset{S}{\mathrm{SU}(2)} . \quad (2)$$

The symbols $\omega_{s,\eta}$ in (1) then indicate the SU(3) $\times$ SU(2) coupling labels [$\omega \equiv (\lambda\mu)S$] of successive shells, and $\rho_{s,\eta}$ denotes the multiplicity index for this coupling. Finally we have total couplings $\omega_p \equiv (\lambda_p\mu_p)S_p$ for the protons, $\omega_n \equiv (\lambda_n\mu_p)S_n$ for the neutrons, and $\omega \equiv (\lambda\mu)S$ for the entire basis state, with corresponding multiplicity indices ρ_p, ρ_n, and ρ, respectively.

The expression in (1) represents not just a single state but an entire set of states, with various values for the quantum numbers associated with the branching of SU(3) $\times$ SU(2) into angular momentum subalgebras: the orbital (spatial) angular momentum L, the inner multiplicity κ for this L within the SU(3) irrep $(\lambda\mu)$, and the total angular momentum J, as well as its z-projection M. However, these states share the same "internal" microscopic structure, given by the same couplings of the particles at the level of SU(3) $\times$ SU(2). Therefore, they may be thought of as a single *reduced state* for certain purposes, in particular, evaluation of reduced matrix elements under the SU(3) $\times$ SU(2) Wigner-Eckart theorem and, as we shall see, identification of CMF linear combinations. The analogy is to angular momentum theory, where one may consider the states $|JM\rangle$, for different M, to be substates of a single state $\|J\rangle$, more formally, a tensorial set or SU(2) irrep.

3. Construction of CMF states

The separation of the many-body space into CMF and spurious parts simplifies in the context of an SU(3) $\times$ SU(2)-coupled basis, since the process may be carried out independently within subspaces characterized by definite SU(3) $\times$ SU(2) quantum numbers. To start with, the center-of-mass Hamiltonian $H_{\mathrm{c.m.}}$ does not connect states involving different numbers N_{ex} of oscillator excitations.[2] Thus, the separation of CMF states may be carried out separately within the space of states with each specific number of oscillator quanta, which we denote by $W_{N_{\mathrm{ex}}}$. Furthermore, $H_{\mathrm{c.m.}}$ commutes with the SU(3) generators and may therefore be diagonalized within a subspace with good SU(3) quantum numbers [18]. As an operator acting only on spatial degrees of freedom, $H_{\mathrm{c.m.}}$ also commutes with the total spin operators for protons and neutrons, as well as their combined spin. Thus, collecting these properties, the separation of CMF states may be carried out separately within subspaces of given N_{ex}, $(\lambda\mu)$, S_p, S_n, and S, which we denote by $W_{N_{\mathrm{ex}}}[(\lambda\mu)S_pS_nS]$.

[2] It is due to this property that the usual $N_{\max}$ truncation scheme for the NCSM [3], based on the total number of oscillator quanta for the many-body state, permits an exact factorization of center-of-mass and intrinsic wavefunctions.

In considering how to extract CMF states, we note that these states are defined by their relation to the center-of-mass harmonic oscillator raising and lowering operators. The center-of-mass raising operator $B^{\dagger}_{\text{c.m.}}$, which is an $L = 1$ operator, furthermore constitutes an SU(3) (10) tensor, with components $B^{\dagger(10)}_{\text{c.m.},L=1,M}$. This operator may be written in terms of single-particle harmonic-oscillator raising operators $b^{\dagger(10)}$ as [13, 20]

$$B^{\dagger(10)}_{\text{c.m.}} = \frac{1}{\sqrt{A}} \sum_{i=1}^{A} b^{\dagger(10)}(i). \tag{3}$$

The corresponding lowering operator $\tilde{B}^{(01)}_{\text{c.m.}}$ similarly has the form

$$\tilde{B}^{(01)}_{\text{c.m.}} = \frac{1}{\sqrt{A}} \sum_{i=1}^{A} \tilde{b}^{(01)}(i). \tag{4}$$

The center-of-mass harmonic oscillator Hamiltonian is built from these operators as

$$H_{\text{c.m.}} = \hbar\omega \Big(B^{\dagger(10)}_{\text{c.m.}} \cdot \tilde{B}^{(01)}_{\text{c.m.}} + \frac{3}{2} \Big), \tag{5}$$

where the dot indicates a spherical tensor scalar product. The CMF states are then defined by the property that they have no center-of-mass excitations, *i.e.*, they have zero eigenvalue for the center-of-mass number operator $N_{\text{c.m.}} = B^{\dagger(10)}_{\text{c.m.}} \cdot \tilde{B}^{(01)}_{\text{c.m.}}$. Equivalently, however, they are identified by the property that they are annihilated by the center-of-mass lowering operator $\tilde{B}^{(01)}_{\text{c.m.}}$.

Either of these criteria allow the problem of identifying CMF states to be formulated as a null-space problem. We seek the subspace $W^{\text{CMF}}_{N_{\text{ex}}}[(\lambda\mu)S_pS_nS]$ of $W_{N_{\text{ex}}}[(\lambda\mu)S_pS_nS]$ consisting of states $|\Psi^{\text{CMF}}\rangle$ such that $N_{\text{c.m.}}|\Psi^{\text{CMF}}\rangle = 0$ or, equivalently, $\tilde{B}^{(01)}_{\text{c.m.},L=1,M}|\Psi^{\text{CMF}}\rangle = 0$. In practice, this means first representing the operator as a matrix with respect to the SU(3)-NCSM basis and then solving for a complete set of null vectors of this matrix. The result yields new basis states, for $W^{\text{CMF}}_{N_{\text{ex}}}[(\lambda\mu)S_pS_nS]$, as linear combinations of the original basis states, for $W_{N_{\text{ex}}}[(\lambda\mu)S_pS_nS]$.

Although we could in principle search for the null space of either $N_{\text{c.m.}}$ or $\tilde{B}^{(01)}_{\text{c.m.}}$, there is a practical advantage to working with $\tilde{B}^{(01)}_{\text{c.m.}}$. While $N_{\text{c.m.}}$ is a two-body operator, $\tilde{B}^{(01)}_{\text{c.m.}}$ is simply a one-body operator. Consequently, evaluation of matrix elements is computationally less involved. Note that $N_{\text{c.m.}}$ acts within the space $W_{N_{\text{ex}}}$, conserving the number of oscillator excitations, while $\tilde{B}^{(01)}_{\text{c.m.}}$ instead connects the space $W_{N_{\text{ex}}}$ to the next lower space $W_{N_{\text{ex}}-1}$.

The problem of identifying CMF states is simplified by the realization that it may be formulated entirely at the level of reduced states. An SU(3) × SU(2)-reduced state is CMF if and only if its substates, labeled by κLJ, are all CMF. [Recall $H_{\text{c.m.}}$ commutes with the SU(3) generators, which connect states within an irrep.] We need thus only find a basis of SU(3) × SU(2)-*reduced* states, for $W^{\text{CMF}}_{N_{\text{ex}}}[(\lambda\mu)S_pS_nS]$, independent of κLJ. We also need only consider the SU(3)×SU(2)-*reduced* matrix elements of the operator $\tilde{B}^{(01)}_{\text{c.m.}}$, rather than the matrix elements among individual κLJ substates, which are much greater in number.

To explicitly relate the null space problem to SU(3) × SU(2)-reduced matrix elements, let us return for a moment to κLJ states and observe that any CMF state $|\Psi^{\text{CMF}}_{N_{\text{ex}}}[(\lambda\mu)S_pS_nS;\kappa LJM]\rangle$ within the $W_{N_{\text{ex}}}[(\lambda\mu)S_pS_nS]$ subspace must satisfy

$$\langle \Psi_{N_{\text{ex}}-1}[(\lambda'\mu')S'_pS'_nS';\kappa'L'J'M'] | \tilde{B}^{(01)}_{\text{c.m.},L_0=1,M_0} | \Psi^{\text{CMF}}_{N_{\text{ex}}}[(\lambda\mu)S_pS_nS;\kappa LJM] \rangle = 0, \tag{6}$$

for every state $|\Psi_{N_{\text{ex}}-1}[(\lambda'\mu')S'_pS'_nS';\kappa'L'J'M']\rangle \in W_{N_{\text{ex}}-1}$. Since we are working with states of good angular momentum, we can immediately rewrite the condition in terms of a reduced matrix element as

$$\langle \Psi_{N_{\text{ex}}-1}[(\lambda'\mu')S'_pS'_nS';\kappa'L'J'] \| \tilde{B}^{(01)}_{\text{c.m.},L_0=1} \| \Psi^{\text{CMF}}_{N_{\text{ex}}}[(\lambda\mu)S_pS_nS;\kappa LJ] \rangle = 0. \tag{7}$$

Note that, since $\tilde{B}^{(01)}_{\text{c.m.}}$ acts only on spatial degrees of freedom, we actually need only consider the case $(S'_p S'_n S') = (S_p S_n S)$. The ordinary SU(2)-reduced matrix element in (7) is related to the SU(3) $\times$ SU(2)-reduced matrix element of $\tilde{B}^{(01)}_{\text{c.m.}}$ by the SU(3) Wigner-Eckart theorem (and LS-coupling relations), as

$$\begin{aligned}\langle \Psi_{N_{\text{ex}}-1}[(\lambda'\mu')S_pS_nS;\kappa'L'J'] \| \tilde{B}^{(01)}_{\text{c.m.},L_0=1} \| \Psi^{\text{CMF}}_{N_{\text{ex}}}[(\lambda\mu)S_pS_nS;\kappa LJ]\rangle \\ = (-)^{J+L'+1+S}\hat{L}'\hat{J}'\begin{Bmatrix} L & J & S \\ J' & L' & 1\end{Bmatrix}\left(\begin{matrix}(\lambda\mu) & (01) \\ \kappa L & 1\end{matrix}\middle\| \begin{matrix}(\lambda'\mu') \\ \kappa' L'\end{matrix}\right) \\ \times \langle \Psi_{N_{\text{ex}}-1}[(\lambda'\mu')S_pS_nS] \| \tilde{B}^{(01)}_{\text{c.m.}} \| \Psi^{\text{CMF}}_{N_{\text{ex}}}[(\lambda\mu)S_pS_nS]\rangle, \qquad (8)\end{aligned}$$

where $\hat{J} \equiv (2J+1)^{1/2}$, and the quantity in parentheses is an SU(3) Clebsch-Gordan coefficient [22]. The condition that the complete set of SU(2)-reduced matrix elements appearing on the left hand side of (8) vanish is equivalent to the condition that the single SU(3) $\times$ SU(2)-reduced matrix element on the right hand side vanish.

Thus, for the CMF states, we seek SU(3) $\times$ SU(2)-reduced states $\|\Psi^{\text{CMF}}_{N_{\text{ex}}}[(\lambda\mu);S_pS_nS]\rangle$ such that

$$\langle \Psi_{N_{\text{ex}}-1}[(\lambda'\mu')S_pS_nS] \| \tilde{B}^{(01)}_{\text{c.m.}} \| \Psi^{\text{CMF}}_{N_{\text{ex}}}[(\lambda\mu)S_pS_nS]\rangle = 0, \qquad (9)$$

for all possible reduced states $\|\Psi_{N_{\text{ex}}-1}[(\lambda'\mu')S_pS_nS]\rangle \in W_{N_{\text{ex}}-1}[(\lambda'\mu')S_pS_nS]$. The subspaces $W_{N_{\text{ex}}-1}[(\lambda'\mu')S_pS_nS]$ which may be linked with $W_{N_{\text{ex}}}[(\lambda\mu)S_pS_nS]$ through $\tilde{B}^{(01)}_{\text{c.m.}}$ are restricted by the SU(3) tensor character of $\tilde{B}^{(01)}_{\text{c.m.}}$. Specifically, $(\lambda'\mu')$ must be contained in the product $(\lambda\mu)\times(01)$ which, from the general rules of SU(3) coupling [23], may be seen to consist of

$$(\lambda\mu)\times(01) = \begin{cases}(01) & \lambda=\mu=0 \\ (0\,\mu+1)\oplus(1\,\mu-1) & \lambda=0,\mu\geq 1 \\ (\lambda 1)\oplus(\lambda-1\,0) & \lambda\geq 1,\mu=0 \\ (\lambda\,\mu+1)\oplus(\lambda+1\,\mu-1)\oplus(\lambda-1\,\mu) & \lambda\geq 1,\mu\geq 1.\end{cases} \qquad (10)$$

The problem of finding reduced states which satisfy (9) can be converted into searching for the null space of a matrix, the entries of which are the SU(3) $\times$ SU(2)-reduced matrix elements of $\tilde{B}^{(01)}_{\text{c.m.}}$ between the SU(3) $\times$ SU(2)-reduced basis states for the $W_{N_{\text{ex}}}[(\lambda\mu)S_pS_nS]$ subspace and the SU(3) $\times$ SU(2)-reduced basis states for each of the possible subspaces $W_{N_{\text{ex}}-1}[(\lambda'\mu')S_pS_nS]$. The resulting matrix has the form illustrated in figure 1, where the horizontal dashed lines delimit submatrices corresponding to the different final spaces with $(\lambda'\mu') = (\lambda'_1\mu'_1)$, $(\lambda'_2\mu'_2)$, ..., as allowed by the selection rule (10). (There will be at most three such submatrices.) The entries of the null vectors then give the expansion coefficients for the basis states for the CMF space $W^{\text{CMF}}_{N_{\text{ex}}}[(\lambda\mu)S_pS_nS]$ in terms of the original basis states of $W_{N_{\text{ex}}}[(\lambda\mu)S_pS_nS]$.

The SU(3)-reduced matrix elements of $\tilde{B}^{(01)}_{\text{c.m.}}$ entering into the matrix of figure 1 can be calculated numerically using methods from [24]. In particular, the SU(3)$\times$SU(2) matrix elements of a one-body operator such as $\tilde{B}^{(01)}_{\text{c.m.}}$ (or an n-body operator in general) can readily be computed, once the operator is expressed in second-quantized form in terms of SU(3) $\times$ SU(2)-coupled products of creation operators $a^\dagger_{(\eta 0)1/2}$ and annihilation operators $\tilde{a}_{(0n)1/2}$. For the center-of-mass annihilation operator $\tilde{B}^{(01)}_{\text{c.m.}}$, the second-quantized form is obtained as

$$\tilde{B}^{(01)}_{\text{c.m.}} = \frac{1}{\sqrt{A}}\sum_\eta \sqrt{\frac{(\eta+1)(\eta+2)}{3}}\langle \eta \,\|\, \tilde{b}^{(01)} \,\|\, \eta+1\rangle \left[a^\dagger_{(\eta 0)1/2}\times \tilde{a}_{(0,\eta+1)1/2}\right]^{(01)0}, \qquad (11)$$

where $\langle \eta \,\|\, \tilde{b}^{(01)} \,\|\, \eta+1\rangle = \sqrt{\eta+3}$.

$$
\begin{bmatrix}
\langle \Psi^{(\lambda'_1\mu'_1)}_{N_{\mathrm{ex}}-1,1} \| \tilde{B}^{(01)}_{\mathrm{c.m.}} \| \Psi^{(\lambda\mu)}_{N_{\mathrm{ex}},1} \rangle & \langle \Psi^{(\lambda'_1\mu'_1)}_{N_{\mathrm{ex}}-1,1} \| \tilde{B}^{(01)}_{\mathrm{c.m.}} \| \Psi^{(\lambda\mu)}_{N_{\mathrm{ex}},2} \rangle & \cdots \\
\langle \Psi^{(\lambda'_1\mu'_1)}_{N_{\mathrm{ex}}-1,2} \| \tilde{B}^{(01)}_{\mathrm{c.m.}} \| \Psi^{(\lambda\mu)}_{N_{\mathrm{ex}},1} \rangle & \langle \Psi^{(\lambda'_1\mu'_1)}_{N_{\mathrm{ex}}-1,2} \| \tilde{B}^{(01)}_{\mathrm{c.m.}} \| \Psi^{(\lambda\mu)}_{N_{\mathrm{ex}},2} \rangle & \cdots \\
\vdots & \vdots & \\
\hdashline
\langle \Psi^{(\lambda'_2\mu'_2)}_{N_{\mathrm{ex}}-1,1} \| \tilde{B}^{(01)}_{\mathrm{c.m.}} \| \Psi^{(\lambda\mu)}_{N_{\mathrm{ex}},1} \rangle & \langle \Psi^{(\lambda'_2\mu'_2)}_{N_{\mathrm{ex}}-1,1} \| \tilde{B}^{(01)}_{\mathrm{c.m.}} \| \Psi^{(\lambda\mu)}_{N_{\mathrm{ex}},2} \rangle & \cdots \\
\langle \Psi^{(\lambda'_2\mu'_2)}_{N_{\mathrm{ex}}-1,2} \| \tilde{B}^{(01)}_{\mathrm{c.m.}} \| \Psi^{(\lambda\mu)}_{N_{\mathrm{ex}},1} \rangle & \langle \Psi^{(\lambda'_2\mu'_2)}_{N_{\mathrm{ex}}-1,2} \| \tilde{B}^{(01)}_{\mathrm{c.m.}} \| \Psi^{(\lambda\mu)}_{N_{\mathrm{ex}},2} \rangle & \cdots \\
\vdots & \vdots & \\
\hdashline
\vdots & \vdots &
\end{bmatrix}
$$

Figure 1. Form of the matrix representation of $\tilde{B}^{(01)}_{\mathrm{c.m.}}$, as reduced matrix elements between the SU(3)-NCSM basis states for the $W_{N_{\mathrm{ex}}}[(\lambda\mu)S_pS_nS]$ subspace (identified with the columns) and the SU(3)-NCSM basis states for each of the possible subspaces $W_{N_{\mathrm{ex}}-1}[(\lambda'\mu')S_pS_nS]$ (identified with the rows).

4. Application to the deuteron

The case of deuterium provides a simple illustration of the methods described in section 3. The full SU(3)-NCSM space for the deuteron may be broken into subspaces $W_{N_{\mathrm{ex}}}[(\lambda\mu)\frac{1}{2}\frac{1}{2}S]$, with $S = 0$ and 1. The dimensions of these subspaces — specifically, the number of SU(3) $\times$ SU(2)-reduced basis states, not the total number of κLJ basis states, which would be much higher — are indicated by the areas of the outer, light circles in figure 2. Then, for each value of N_{ex}, it is found that solution of the null space problem for the center-of-mass annihilation operator yields two CMF reduced states. Both have $(\lambda\mu) = (N_{\mathrm{ex}}0)$, one with $S = 0$ and one with 1. These CMF subspaces are indicated by the inner, dark circles in figure 2.

The deuteron provides a particularly illuminating test case, since the quantum numbers of the CMF spaces obtained in figure 2 may be understood through simple arguments. Let $N_{\mathrm{c.m.}}$ and N_{rel} denote the number of oscillator excitations of the center-of-mass and relative degrees of freedom, respectively, so $N_{\mathrm{ex}} = N_{\mathrm{c.m.}} + N_{\mathrm{rel}}$. The CMF condition imposes $N_{\mathrm{c.m.}} = 0$ and thus $N_{\mathrm{rel}} = N_{\mathrm{ex}}$. For the two-particle system, the transformation between single-particle and relative coordinates is straightforward. There is only a single relative coordinate vector, and the harmonic oscillator in this coordinate is equivalent to the harmonic oscillator for a single particle in three dimensions. For a given N_{ex}, it will therefore carry SU(3) quantum numbers $(N_{\mathrm{ex}}0)$. The center-of-mass oscillator carries (00) for a CMF state. Therefore the CMF state as a whole will have $(\lambda\mu) = (N_{\mathrm{ex}}0) \times (00) = (N_{\mathrm{ex}}0)$, as well. Since the deuteron consists of distinguishable particles, the coupling of spins is independent of the spatial degrees of freedom, and both $S = 0$ and 1 are obtained.

5. Conclusions

The elimination of spurious center-of-mass excitations is essential to the problem of determining the intrinsic structure of nuclei. If an SU(3)-coupled harmonic oscillator basis is used for the many-body problem, as in the SU(3)-NCSM, the separation may be carried out at the level of SU(3) irreps, in particular, within subspaces of fixed number of oscillator quanta and $(\lambda\mu)S_pS_nS$ labels. We have formulated the problem of finding the CMF subspace as a matrix null-space problem, based on the SU(3) $\times$ SU(2)-reduced matrix elements of the center-of-mass annihilation operator. It is therefore possible to remove spurious contributions from the SU(3)-NCSM model space *prior* to diagonalization of the Hamiltonian, rather than through a Lawson term. In the

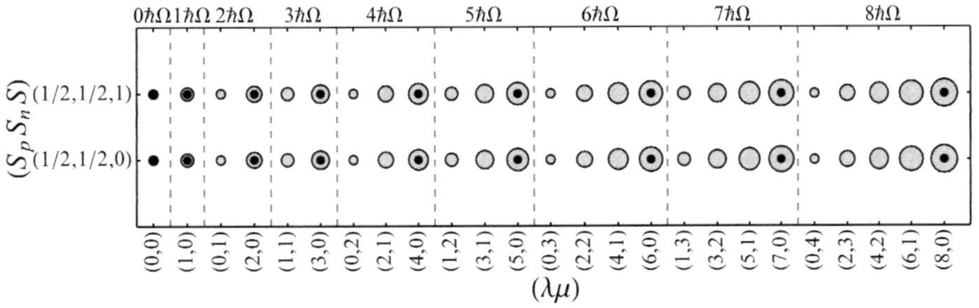

Figure 2. Subspaces of the SU(3)-NCSM model space for the deuteron, arranged by N_{ex} (labeled by $N_{\text{ex}}\hbar\Omega$ at top), SU(3) quantum numbers (at bottom), and spin quantum numbers $(S_pS_nS) = (\frac{1}{2}\frac{1}{2}S)$ (at left). The area of the light (outer) circle indicates the dimension of $W_{N_{\text{ex}}}[(\lambda\mu)S_pS_nS]$, and the area of the dark (inner) circle, when present, indicates the dimension of its CMF subspace $W_{N_{\text{ex}}}^{\text{CMF}}[(\lambda\mu)S_pS_nS]$. Here, dimension refers to the number of SU(3)×SU(2)-reduced basis states.

context of the SU(3)-NCSM, this raises the possibility of significant reductions in dimensionality. However, this work is more broadly motivated by the goal of constructing nonspurious irreps for an Sp(3, $\mathbb{R}$) symmetry-adapted no-core shell model.

Acknowledgments

Discussions with C. Bahri are gratefully acknowledged. This work was supported by the Research Corporation for Science Advancement under a Cottrell Scholar Award, by the US Department of Energy under Grants No. DE-FG02-95ER-40934 and DE-SC0005248, and by the US National Science Foundation under Grant No. OCI-0904874. Computational resources were provided by the University of Notre Dame Center for Research Computing.

References

[1] Navrátil P, Vary J P and Barrett B R 2000 *Phys. Rev. Lett.* **84** 5728
[2] Navrátil P, Vary J P and Barrett B R 2000 *Phys. Rev.* C **62** 054311
[3] Navrátil P, Quaglioni S, Stetcu I and Barrett B R 2009 *J. Phys.* G **36** 083101
[4] Elliott J P 1958 *Proc. R. Soc. London* A **245** 128
[5] Elliott J P 1958 *Proc. R. Soc. London* A **245** 562
[6] Elliott J P and Harvey M 1963 *Proc. R. Soc. London* A **272** 557
[7] Rosensteel G and Rowe D J 1977 *Phys. Rev. Lett.* **38** 10
[8] Rosensteel G and Rowe D J 1980 *Ann. Phys. (N.Y.)* **126** 343
[9] Draayer J P, Weeks K J and Rosensteel G 1984 *Nucl. Phys.* A **413** 215
[10] Dytrych T, Sviratcheva K D, Bahri C, Draayer J P and Vary J P 2007 *Phys. Rev.* C **76** 014315
[11] Dytrych T, Sviratcheva K D, Bahri C, Draayer J P and Vary J P 2007 *Phys. Rev. Lett.* **98** 162503
[12] Dytrych T, Draayer J P, Launey K D, Caprio M A and Langr D in these proceedings
[13] Dytrych T, Sviratcheva K D, Draayer J P, Bahri C and Vary J P 2008 *J. Phys.* G **35** 123101
[14] Elliott J P and Skyrme T H R 1955 *Proc. R. Soc. London* A **232** 561
[15] Baranger E and Lee C W 1961 *Nucl. Phys.* **22** 157
[16] Gloeckner D H and Lawson R D 1974 *Phys. Lett.* B **53** 313
[17] Kretzschmar M 1960 *Z. Phys.* **158** 284
[18] Verhaar B J 1960 *Nucl. Phys.* **21** 508
[19] Hecht K T 1971 *Nucl. Phys.* A **170** 34
[20] Escher J and Draayer J P 1998 *J. Math. Phys.* **39** 5123
[21] Draayer J P, Leschber Y, Park S C and Lopez R 1989 *Comput. Phys. Commun.* **56** 279
[22] Akiyama Y and Draayer J P 1973 *Comput. Phys. Commun.* **5** 405
[23] O'Reilly M F 1982 *J. Math. Phys.* **23** 2022
[24] Bahri C and Draayer J P 1994 *Comput. Phys. Commun.* **83** 60

Large-scale ab initio configuration interaction calculations for light nuclei

Pieter Maris[1], H Metin Aktulga[2], Mark A Caprio[3], Ümit V Çatalyürek[4,5], Esmond G Ng[2], Dossay Oryspayev[6], Hugh Potter[1], Erik Saule[4], Masha Sosonkina[7], James P Vary[1], Chao Yang[2] and Zheng Zhou[8]

[1] Department of Physics and Astronomy, Iowa State University, Ames, IA 50011, USA
[2] Computational Research Div, Lawrence Berkeley National Lab, Berkeley, CA 94720, USA
[3] Department of Physics, University of Notre Dame, Notre Dame, IN 46556, USA
[4] Dept. of Biomedical Informatics, The Ohio State University, Columbus, OH 43210, USA
[5] Dept. of Electrical and Computer Engineering, The Ohio State University, Columbus, OH 43210, USA
[6] Dept. of Electrical and Computer Engineering, Iowa State University, Ames, IA 50011, USA
[7] Dept. of Modeling, Simulation and Visualization Eng., Old Dominion University, Norfolk, VA 23529, USA and Ames Laboratory, Ames, IA, 50011, USA
[8] School of Computer, Wuhan University, P. R. China

Abstract. In ab-initio Configuration Interaction calculations, the nuclear wavefunction is expanded in Slater determinants of single-nucleon wavefunctions, and the many-body Schrodinger equation becomes a large sparse matrix problem. The challenge is to reach numerical convergence to within quantified numerical uncertainties for physical observables using finite truncations of the infinite-dimensional basis space. We discuss strategies for constructing and solving the resulting large sparse matrix eigenvalue problems on current multicore computer architectures. Several of these strategies have been implemented in the code MFDn, a hybrid MPI/OpenMP Fortran code for ab-initio nuclear structure calculations that can scale to 100,000 cores and more. Finally, we will conclude with some recent results for ^{12}C including emerging collective phenomena such as rotational band structures using SRG evolved chiral N3LO interactions.

1. No-Core Configuration Interaction calculations

Solving for nuclear properties with the best available nucleon-nucleon (2NF) potentials [1, 2], supplemented by three-nucleon forces (3NF) as needed [3, 4, 5], using a quantum many-particle framework that respects all the known symmetries of the potentials is referred to as an "ab initio" problem and is recognized to be computationally hard.

A commonly used approach in nuclear physics is the Configuration Interaction (CI) method for solving the many-body nuclear Hamiltonian in a (sufficiently large) single-particle basis space. In this approach, the many-body Schrödinger equation becomes a large sparse matrix eigenvalue problem. In a No-Core CI (NCCI) approach [6], the wavefunction Ψ of a nucleus consisting of A nucleons (protons and neutrons) is expanded in an A-body basis of Slater determinants Φ_k of single-particle wavefunctions ϕ_i

$$\Psi(\vec{r_1}, \ldots, \vec{r_A}) = \sum c_k \Phi_k(\vec{r_1}, \ldots, \vec{r_A}), \tag{1}$$

Published under licence by IOP Publishing Ltd

with $\Phi_k(\vec{r_1}, \ldots, \vec{r_A}) = \mathcal{A}[\phi_{n_1 l_1 j_1 m_1}(r_1)\, \phi_{n_2 l_2 j_2 m_2}(r_2) \ldots \phi_{n_A l_A j_A m_A}(r_A)]$. Conventionally, one uses a harmonic oscillator (HO) basis for the single-particle wavefunctions, but it is straightforward to extend this approach to a more general single-particle basis [7, 8]. The single-particle wavefunctions are labelled by the quantum numbers n, l, j, and m, where n and l are the radial and orbital HO quantum numbers, with $N = 2n + l$ the number of HO quanta; j is the total single-particle spin, and m its projection along the z-axis. The many-body basis states have well-defined total spin-projection, which is simply the sum of m of the single-particle states, $M = \sum m$, hence the name M-scheme. However, the many-body basis states do not have a well-defined total spin J. Some of the benefits of this scheme is that it is very simple to implement, and that in two runs (one for positive and one for negative parity), we get the complete low-lying spectrum, including the ground state, even if the spin of the ground state is unknown.

In a complete basis, this method would give exact results for a given input interaction V. However, practical calculations can only be done in a finite-dimensional truncation of a complete basis. Different truncation schemes have different convergence rates: Full Configuration Interaction or FCI employs a widely-know truncation in which all many-body basis states are retained that can be constructed from a finite set of single-particle states. This method is commonly used in atomic and molecular physics, but has a rather slow convergence rate in typical ab initio nuclear structure calculations [9, 10]. The so-called $N_{\max}$ truncation, in which the total number of HO quanta in a basis state is limited: $\sum N_k \leq N_0 + N_{\max}$, is generally much more efficient for NCCI calculations. Here, N_k is the number of quanta of each of the single-particle states in the many-body basis state; N_0 is the minimal number of quanta for that nucleus; and $N_{\max}$ is the truncation parameter. For HO single-particle states, or more general, for any basis in which the single-particle states have radial and orbital quantum numbers n and l, we have $N_k = 2n_k + l_k$. Furthermore, in combination with a HO basis for the single-particle basis, this truncation leads to an exact factorization of the center-of-mass wavefunction and the relative wavefunction.

For a very light nucleus like ^{4}He or ^{6}Li one can achieve convergence of the ground state energy and other observables for many modern 2NFs by simply going to a sufficiently large basis, but for larger nuclei and/or with 3NFs that is not practical. In order to reliably extrapolate to the complete basis, we need results in finite bases up to at least $N_{\max} = 8$, and preferably $N_{\max} = 10$ or even higher [11, 12], depending on the interaction. This leads to a computational challenge, because the basis space dimension increases rather rapidly with $N_{\max}$ and with the number of particles. In the left panel of Fig. 1 we show the basis space dimensions for a number of $N = Z$ nuclei in the p-shell (solid curves), as well as a few nuclei in the sd-shell. The basis space dimension for even the first couple of nuclei in the sd-shell (more than 8 protons and 8 neutrons) is of the order of 10 billion or more at $N_{\max} = 8$. Furthermore, the sparsity of the matrix depends on the 'rank' of the potential as shown in the right panel of Fig. 1: a two-body potential leads to a much sparser many-body matrix than a three-body potential [13]. As the matrix size increases, there is a clear need for high-performance computing: obtaining the lowest 10 to 20 eigenvalues and eigenvectors of matrices with a dimension of the order of several billion is a nontrivial task, even if the matrix is extremely sparse.

2. Computational Challenges and Solutions

For our calculations we use the code MFDn, which is a hybrid MPI/OpenMP parallel Fortran 90 code for nuclear structure calculations, that has been in development for over two decades. Significant improvements in its performance have been made over the last five years [14, 15, 16] under the UNEDF SciDAC program. MFDn constructs the many-body basis states in M-scheme and the corresponding many-body matrix, and solves for the lowest eigenstates using an iterative Lanczos algorithm. Each Lanczos iteration consists of a matrix-vector multiplication followed by orthogonalization. At the end of a run MFDn writes the eigenvectors to file, and evaluates

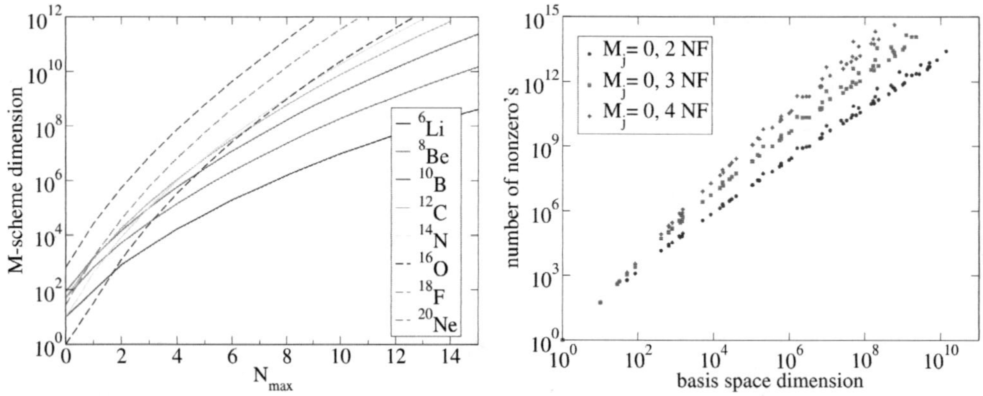

Figure 1. Left: Basis space dimension as function of $N_{\rm max}$ for $N = Z$ nuclei in the p-shell (solid), as well as a few nuclei in the sd-shell (dashed); Right: number of nonzero matrix elements as function of dimension, with 2-body forces (2NF, blue circles,), 2- and 3-body forces (3NF, red squares) and up to 4-body forces (4NF, green diamonds).

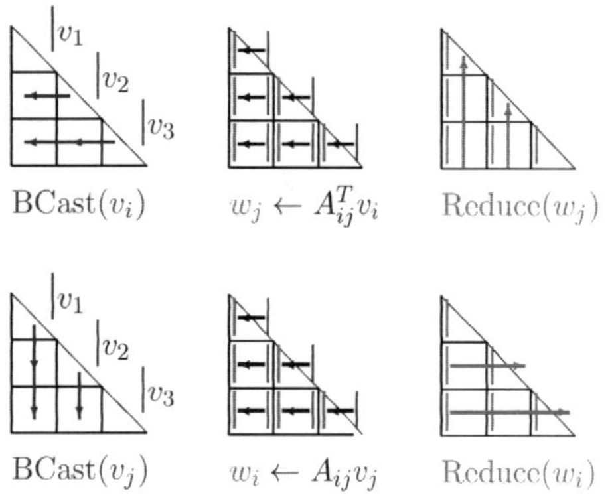

Figure 2. Communication patterns for SpMV and transpose SpMV.

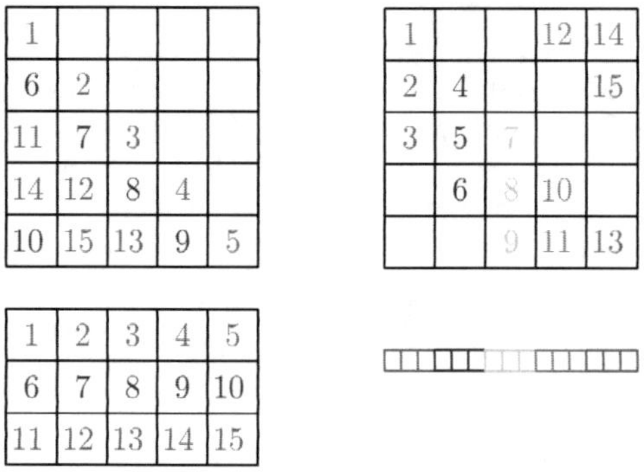

Figure 3. Left: Diagonal-Major distribution of lower matrix triangle over $n(n+1)/2$ processors, with cyclic distribution of Lanczos vectors over n processors; Right: Load-Balanced Column-Major distribution with hierarchical 1-dimensional distribution of Lanczos vectors over $n(n+1)/2$ processors.

selected physical observables which can be compared to experimental data (the wavefunctions themselves are not experimentally observable). It also writes out the one-body density matrix elements (in the underlying single-particle basis), which can be used for further analysis.

Each vector is distributed over n processors, and the matrix itself is distributed in a 2-dimensional fashion over all available processors. We work with the lower triangle only, because the matrix is symmetric, so we need $n(n+1)/2$ processors. For a sparse matrix-vector multiplication (SpMV) we need to broadcast the input vector along the columns and along the rows, then perform a local SpMV and transpose SpMV on each processor, and finally reduce the local output vectors along both the rows and the columns to get the global output vector, as

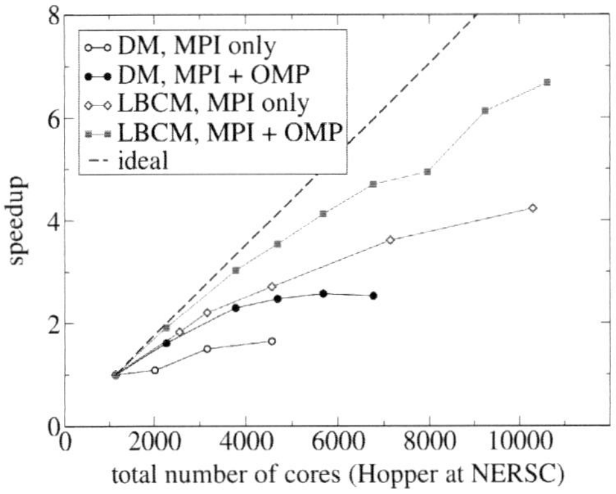

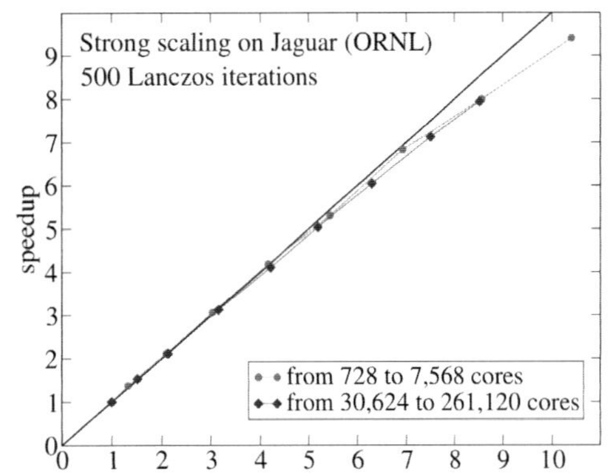

Figure 4. Speedup of the Lanczos iterations for different processors layout for ^{10}Be at $N_{\max} = 8$ using NN-only interactions (left) and with LBCM MPI+OMP for ^{7}Li (red, D= $6.2 \cdot 10^{6}$ and NNZ= 10^{11}) and ^{10}Be (blue, D= $1.6 \cdot 10^{8}$ and NNZ=$5.2 \cdot 10^{12}$) at $N_{\max} = 8$ using 3NF (right), where D is the basis dimension, and NNZ the number of nonzero's in half the matrix.

illustrated in Fig. 2. The local work load of the SpMV is proportional to the number of nonzero matrix elements on each processor. With a round-robin distribution of the M-scheme basis states over the n processors, we achieve almost perfect local load balancing, both in memory usage and the (local) CPU usage.

However, the overall performance also depends on the communication overhead, and it turns out that the mapping of the matrix blocks onto the MPI processors has a large influence on the communication time [16]. The layout that was initially implemented in MFDn was Diagonal-Major (DM), as indicated in the left panel of Fig. 3. However, this mapping leads to poor scaling and increasing communication time as the number of processors increases: the communication volume is not balanced among different communicator groups (e.g. the first column communicator group contains n processors whereas the last column communicator group contains only one processor). Furthermore, on modern large-scale high performance computer systems, job-schedulers often use as default a space-filling Z curve (or a similar) algorithm for ranking processing units in order to ensure that physically nearby processing units are ranked consecutively. With the DM mapping of the left panel of Fig. 3, this means that the 'diagonal' processors are physically close to each other, creating local communication hot-spots, whereas members of the same column communicator group, as well as members of the same row communicator group, tend to be spread out over the system.

On the other hand, with a Load-Balanced Column-Major (LBCM) mapping as indicated in the right panel of Fig. 3, the communication volume is balanced, and members of the same column communicator group are physically close together, leading to much lower communication times. The overall performance of this mapping is indeed significantly improved over the original DM mapping [16], see the left panel of Fig. 4. We may be able to further optimize these mappings for specific hardware and interconnection networks.

For large runs on multi-core architectures such as the Cray XT4/5, it is significantly more efficient to run the code in a hybrid mode, using OpenMP threading within a node, and MPI between nodes [15]. As the number of threads per node increases on modern multiprocessor systems, the main memory system on a node is often split into multiple memory banks with each bank associated with a set of cores. This is referred to as Nonuniform Memory Access (NUMA), and accesses to remote memory banks (but still within the same node) reduce scalability when using one MPI processor node, e.g. using one MPI processor per node with 24 threads on Hopper (Cray XE6) at NERSC, or one MPI processor per node with 16 threads on Jaguar (Cray XK6) at ORNL; performance is significantly better using one MPI processor per NUMA node instead, which is what we used in Fig. 4. We are currently working on improving the performance on NUMA architectures by focussing on the data locality [17].

The Lanczos vectors are stored in memory, distributed over all processors; this allows for fast and efficient re-orthoganalization after every SpMV. The mapping of the Lanczos vectors on the processors also influences the communication overhead significantly. The hierarchical 1-dimensional mapping of the right panel of Fig. 3 seems to be most efficient [16]. Note that the re-orthogonalization is done in double precision, even though the matrix and vectors themselves are stored in single precision. This turns out to be essential in order to maintain numerical accuracy over hundreds of Lanczos iterations with matrix dimensions in the billions.

With these recent improvements, we obtain very good scaling characteristics. With 3NF we can scale on Jaguar at ORNL up to the full machine (300k cores), as illustrated in the right panel of Fig. 4. The minimal number of processors needed for a run is determined by the amount of memory needed to store all nonzero matrix elements in compressed-column format (CCF), or alternatively, in compressed-row format (CRF). On currently available machines, the largest many-body basis dimension that MFDn can handle is about 10 to 15 billion with 2NFs only, and about one billion with 2NFs plus 2NFs.

Despite the significant performance improvement, for the current algorithm and implementation the main challenge remains the amount of memory needed to store the matrix in core. An obvious alternative would be to use an 'out-of-core' implementation for the SpMV, reading the matrix from disk as needed. Indeed, such a simple out-of-core implementation for MFDn exists, and has been quite useful on clusters with a local disk associated with each processor, e.g. for the calculations presented in Ref. [12]. However, on modern large-scale high performance computer systems with a parallel filesystem, this simple out-of-core implementation performs rather poorly.

On the other hand, the emergence of high performance computing platforms equipped with solid state drives (SSD) presents an opportunity to dramatically increase the efficiency of out-of-core calculations. In Ref. [18], we explore the advantages and challenges associated with performing the SpMV on a small SSD testbed. We use a novel distributed out-of-core linear algebra framework, called Distributed Out-of-Core (DOoC) plus Linear Algebra Frontend (LAF). This DOoC+LAF framework provides an easy-to-use high-level application interface for linear algebra operations such as SpMV, while performing efficient execution by orchestrating pipelined execution of computation, communication and I/O. Preliminary experiments indicate that the out-of-core implementation on the SSD testbed can compete with an in-core implementation in terms of total CPU hours and energy efficiency.

Another recent hardware development is the advent of GPUs and accelerators, which are now employed in several of the top-ten of the fastest supercomputers. However, it is far from trivial to achieve the speedup potentially provided by GPUs. In particular for a heavily memory-bound application such as MFDn, the bottleneck is not floating point operations, but aggregate memory and memory access. Nevertheless, there may be some parts of the MFDn that could benefit from GPUs. We are currently exploring the use of GPUs for the decoupling of the 3NF matrix elements in coupled JT-format. For a simple test we find a speedup of a factor of 20 to 50 on a GPU using CUDA, as compared to a serial code on a CPU. These speedup values include the transfer times between the host CPU and the GPU. Integrating this in a production code for nuclear structure, however, is still a challenge.

In addition, there may be opportunities to use GPUs in combination with a 'matrix-free' SpMV algorithm, in which the necessary matrix elements are constructed as needed, rather than all stored in memory. Obviously, such an implementation would have a strongly reduced

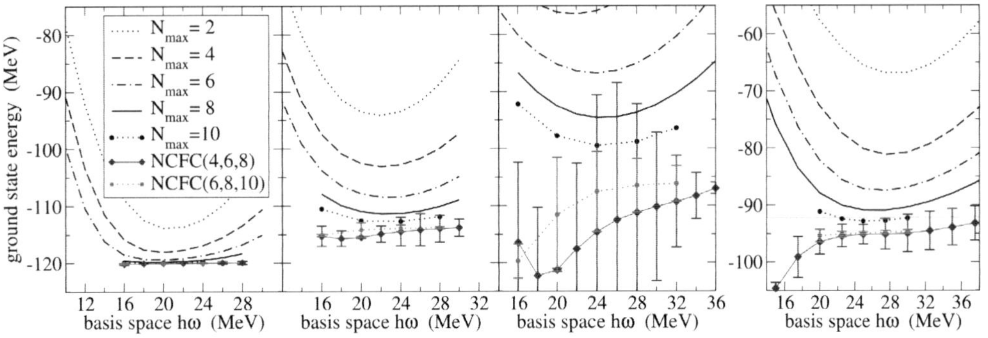

Figure 5. Ground state energy of ^{12}C with SRG evolved chiral N3LO at, from left to right, $\lambda = 1.5$, 2.0, and 2.5 fm^{-1}; and with JISP16 (rightmost), with the experimental value indicated by a green line. (The Coulomb interaction is not included in the first three panels.)

memory footprint, at the cost of additional computation, which may be advantageous if this can be done efficiently on the GPUs. Rather than a true 'matrix-free' SpMV algorithm, we are exploring the possibilities to combine storing part of the matrix in core (on the CPUs) and using a 'construction on-the-fly' for the other part of the matrix (on the GPUs). Such a hybrid algorithm, using CPUs only, has already been used for the largest calculations to date with 2NFs plus 3NFs [19].

3. Recent results for ^{12}C with chiral 2-body interactions

In Fig. 5, we show new results for ^{12}C using different 2NFs: JISP16 and chiral N3LO interactions, SRG renormalized [12] to three different values of the evolution parameter λ. We did not include the Coulomb interaction in the chiral N3LO calculations, though it is included in the JISP16 calculations. Fig. 5 clearly shows the difference in convergence; furthermore, the actual (converged or extrapolated) binding energy depends strongly on the SRG parameter λ. This effect is also seen in lighter nuclei and arises because we have neglected induced many-body forces forces in the SRG evolution. Inclusion of induced 3NFs reduces this λ dependence to less than a percent over this λ range [20].

For the extrapolation we use the empirical behavior [11, 12] that the nuclear binding energies seem to converge exponentially with $N_{\max}$

$$E^N_{\text{binding}} = E^{\infty}_{\text{binding}} + a_1 \exp(-a_2 N_{\max}) . \tag{2}$$

Here we use three consecutive $N_{\max}$ values at a fixed basis space parameter $\hbar\omega$ in order to estimate a converged binding energy at that $\hbar\omega$. The error estimate is based on the difference with the extrapolated value obtained from the next-smaller basis spaces at the same $\hbar\omega$ value. An extrapolated result that is within the error estimates of the previous $N_{\max}$ extrapolation, together with a consistent decrease and reduced dependence on $\hbar\omega$ as $N_{\max}$ increases, indicates the optimal $\hbar\omega$ range for this extrapolation method. For $\lambda = 1.5$ fm^{-1} the ground state is extremely well converged, and there is no real need to extrapolate to the infinite basis limit, and we find a binding energy of 120.03(5) MeV. For $\lambda = 2.0$ fm^{-1} the extrapolation seems to be reasonably independent of $\hbar\omega$ and with decreasing error bars over a region of 20 MeV $\leq \hbar\omega \leq$ 30 MeV, and an extrapolated result of 113.9(6) MeV. At $\lambda = 2.5$ fm^{-1} the convergence is much slower. Nevertheless, the extrapolation gives consistent results, with only a very small dependence on $\hbar\omega$, in the region region 24 MeV $\leq \hbar\omega \leq$ 36 MeV, which suggests an extrapolated

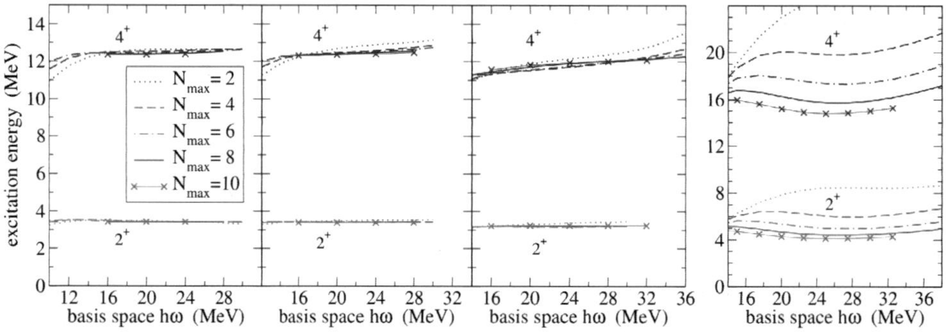

Figure 6. Excitation energy of the 2^+ (red) and 4^+ states (blue) of ^{12}C with SRG evolved chiral N3LO at, from left to right, $\lambda = 1.5$, $\lambda = 2.0$, and $\lambda = 2.5$ fm^{-1}; and with JISP16 (right).

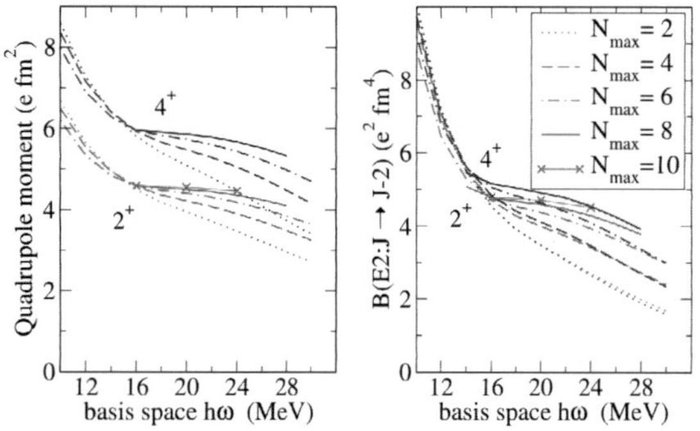

Figure 7. Quadrupole moments of the 2^+ and 4^+ states of ^{12}C (left) and B(E2)'s for the 2^+ to 0^+ and the 4^+ to 2^+ transitions (right), all at $\lambda = 1.5$.

result of 106.5±5.0 MeV. Finally, for JISP16 the extrapolation works very well, with extrapolated results almost independent of $\hbar\omega$, and a binding energy of 94.9(4) MeV, that is, JISP16 overbinds ^{12}C by about 2.5 MeV.

In addition to the ground state energy, we also looked at the lowest 2^+ and 4^+ states. With the chiral N3LO interactions, these are the two lowest excited states, and their excitation energies are extremely well converged. These three states have energies that are indicative of a rotational band: the ratio $E^{(4)}_{\rm exc}/E^{(2)}_{\rm exc}$ is 3.6, independent of λ, compared to 3.33 for a pure rotational band [21]. Note the different scale for the JISP16 results in Fig. 6; furthermore, with JISP16 there is a 0^+ and a 1^+ state with slightly smaller excitation energies than the 4^+ state which are not shown in this figure.

Also the quadrupole moments and B(E2) transitions of the 2^+ and 4^+ states are in reasonable agreement with a rotational model. At $\lambda = 1.5$ fm^{-1} both the quadrupole moments and the B(E2) transitions show convergence, and we find $\mathcal{Q}^{(4)}/\mathcal{Q}^{(2)} = 1.30$, compared to 1.27 for a rotor. Our calculations suggest an intrinsic quadrupole of about $Q_0 = -16.1$ e fm^2. This in turn would lead us to expect B(E2) values of 5.16 e^2 fm^4 and 7.4 e^2 fm^4 for the $2^+ \to 0^+$ and $4^+ \to 2^+$ transitions respectively, if the rotational relations for E2 matrix elements hold. Our NCCI calculations indicate a B(E2) for the $2^+ \to 0^+$ transition that is about 10% below this value, and that for the $4^+ \to 2^+$ transition is about 25% below that of the rotational model. However, our results for that last transition are not quite converged yet. It would be very interesting to look in more detail at these states with the Symmetry-Adapted No-Core Shell Model [22].

Of course, these numbers are not (yet) to be compared to experimental numbers, since we incorporated neither induced nor explicit 3NFs in these calculations. However, such calculations

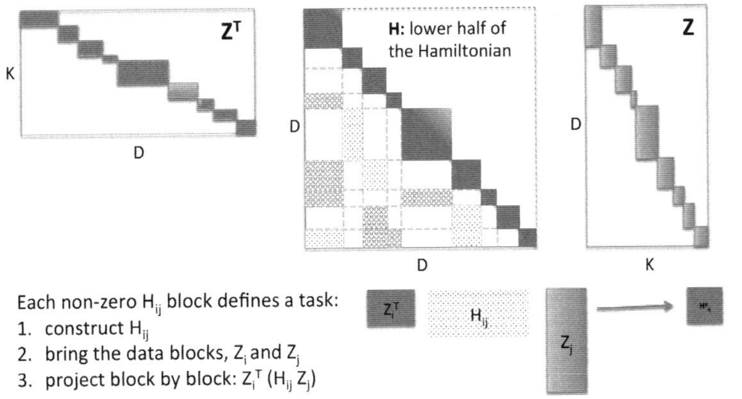

Figure 8. Projection of an n by n many-body matrix in M-scheme to a k by k many-body matrix in coupled-J scheme (adapted from Ref. [24]).

are currently underway on Jaguar at Oak Ridge National Laboratory, and would not have been possible without the recent improvements in the performance of our codes [16].

4. Taming the scale explosion

Despite the tremendous increase in available computing resources and the significant improvements in the performance of our codes, the limiting factor for our calculations remains the combinatorial scale explosion of the dimension, see Fig. 1. As we have just seen for the calculations of ^{12}C, $N_{\text{max}} = 10$ results are highly desirable – however, the corresponding basis dimension is 8 billion. Hence there is an ongoing effort to reduce the basis dimension while keeping the relevant physics.

One way to reduce the basis dimension is by exploiting additional symmetries. Instead of using an M-scheme, one could construct many-body basis states with well-defined total spin J, that is, construct a basis out of eigenstates of the J^2 operator. This can be done by starting out with an M-scheme basis, and diagonalizing the J^2 operator in this M-scheme basis numerically. It is easy to see that the J^2 operator becomes a block-diagonal matrix in an M-scheme basis, provided that basis states with the same quantum numbers n_i, l_i, j_i (but different m_i) are grouped together. Thus diagonalizing the J^2 operator reduces to diagonalizing a block-diagonal matrix, that is, repeatedly diagonalizing much smaller (sparse) matrices. The caveat is that some of the blocks that need to be diagonalized become rather large. Nevertheless, this method has been implemented and has been demonstrated to work [23, 24]. In order to achieve good parallel performance and scaling, the largest blocks are diagonalized first, using all processors; next, blocks of medium size are diagonalized, using a subset of processors; and finally the smallest blocks are diagonalized on single processors [23].

Once the J^2 operator is diagonalized, we can project the many-body Hamiltonian onto the invariant subspace for a given value of J, and obtain the spectrum for that J-value, see Fig. 8. Note, however, that the projected many-body matrix is not as sparse as the original matrix in M-scheme. The typical sparsity of the many-body matrix in coupled-J scheme with 2NF only is therefore similar to that of a many-body matrix in M-scheme with 3NF, see Fig. 9. In fact, although the dimension of this subspace is much smaller than that of the corresponding M-scheme basis, the number of nonzero matrix elements in the projected many-body matrix can be (significantly) larger than that of the many-body matrix in M-scheme. Furthermore, the construction of the many-body matrix, as well as the construction of the basis itself, is much more involved than in M-scheme. Nevertheless, for certain applications, in particular when a lot of states with the same spin (and parity) are required, a basis with well-defined total spin J

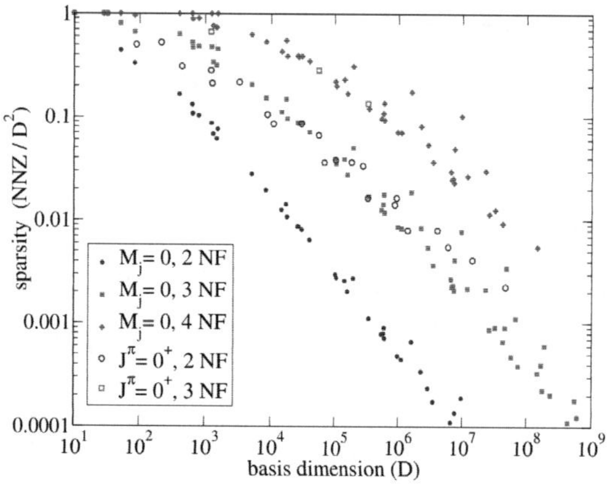

Figure 9. Comparison of the sparsity of the many-body matrix in M-scheme (solid symbols) and in coupled-J scheme (open symbols).

is advantageous over an M-scheme basis.

Instead of, or in addition to, exploiting symmetries, one could also reduce the basis dimension by keeping only the most important basis states. The question then becomes: How to select the most important basis states? At least three different methods are currently being pursued, namely the Importance-Truncated No-Core Shell Model (IT-NCSM) [25], the no-core Monte-Carlo Shell Model (MCSM) [9], and the Symmetry-Adapted No-Core Shell Model (SA-NCSM) [22, 27].

The IT-NCSM works with an M-scheme basis, and typically works with basis dimensions that are an order of magnitude (or more) smaller than the corresponding complete M-scheme basis. The sparsity of the resulting matrix is similar to that of the complete M-scheme basis and thus the number of nonzero matrix elements is typically several orders of magnitude smaller than of the complete M-scheme basis. One can make an estimate of the numerical error introduced by truncating the complete M-scheme basis [25].

The no-core MCSM [9] is an extension of the original MCSM [26], and is based on a truncation of the single-particle states, rather than on the many-body states. The many-body basis is constructed from linear combinations of non-orthogonal angular momentum and parity projected deformed Slater determinants with good total angular momentum projection M using a stochastic Monte-Carlo algorithm. Although the actual many-body basis in the no-core MCSM consists often of only a (few) hundred Monte-Carlo basis states, in contrast to the millions or billions of basis states in the corresponding FCI basis, the results from the no-core MCSM are in very good agreement with the corresponding FCI results [9].

The SA-NCSM is based on the observation that nuclear wavefunctions are often dominated by many-body configurations of specific shape, spatial, and spin deformations [22]. By restricting the many-body basis to the dominant symplectic or SU(3) structures up to an $N_{\max}$ cutoff, one obtains almost the same results as with the complete M-scheme basis, but in a much smaller basis [27]. Note, however, that the number of nonzero matrix elements could be larger than in the corresponding complete M-scheme calculation. In other words, the resulting many-body matrices are less sparse than the corresponding matrices in coupled-J scheme, which in turn are less sparse than the corresponding matrices in M-scheme.

Last but not least, is the use of a more general single-particle wavefunctions than the HO basis. Although there are certain advantages to using a HO basis, one is not restricted to this basis. A drawback of the HO basis is that the falloff (Gaussian asymptotics) of the HO functions at large radius makes them rather poorly suited for the description of the asymptotic properties of the nuclear wavefunction. In Ref. [7] we have established the foundations for carrying out NCCI calculations with an alternative many-body basis built from Coulomb–Sturmian functions. They provide a complete, discrete set of functions with a realistic exponential falloff. Initial results for ^{6}Li in this basis are encouraging. In particular, the convergence properties of observables such as the root-mean-square radius are promising.

Note that neither the IT-NCSM nor the no-core MCSM lead to an exact factorization of the center-of-mass motion and the relative motion. Also with the Coulomb–Sturmian basis there is no exact factorization. However, in practice, the spurious center-of-mass dynamics seems to be tractable [7, 28]. On the other hand, the SA-NCSM truncation, in combination with a HO single-particle basis, does lead to an exact factorization of the center-of-mass wavefunction and the relative wavefunction.

Acknowledgments

This work was supported in part by U.S. Department of Energy Grants DE-FC02-09ER41582 (SciDAC/UNEDF), DE-FG02-87ER40371, DE-FG02-95ER-40934, DE-FC02-06ER2775, and by the U.S. NSF grants 0904782, CNS-0643969, OCI-0904809, OCI-0904802, and by the Research Corporation for Science Advancement under a Cottrell Scholar Award. Computational resources were provided by the National Energy Research Supercomputer Center (NERSC), which is supported by the DOE Office of Science, and by an INCITE award, "Nuclear Structure and Nuclear Reactions", from the DOE Office of Advanced Scientific Computing. This research used resources of the Oak Ridge Leadership Computing Facility at ORNL, which is supported by the DOE Office of Science under Contract DE-AC05-00OR22725.

References

[1] Wiringa R B, Stoks V G J and Schiavilla R 1995 *Phys. Rev.* **C51** 38
[2] Entem D R and Machleidt R 2003 *Phys. Rev.* **C68** 041001
[3] Pieper S C, Pandharipande V R, Wiringa R B and Carlson J 2001 *Phys. Rev.* **C64** 014001
[4] Epelbaum E, Hammer H-W and Meissner U G 2009 *Rev. Mod. Phys.* **81** 1773
[5] Navratil P, Gueorguiev V G, Vary J P, Ormand W E and Nogga A 2007 *Phys. Rev. Lett.* **99** 042501
[6] Navratil P, Vary J P and Barrett B R 2000 *Phys. Rev. Lett.* **84** 5728
[7] Caprio M, Maris P and Vary J P 2012 *Phys. Rev.* **C86** 034312
[8] Caprio M, Maris P and Vary J P 2012 these proceedings (*Preprint* 1208.6055)
[9] Abe T, Maris P, Otsuka T, Shimizu N, Utsuno Y and Vary J 2012 *Phys. Rev.* **C** in press
[10] Maris P 2012 accepted for publication in CCP2011
[11] Maris P, Vary J P and Shirokov A M 2009 *Phys. Rev.* **C79** 014308
[12] Bogner S K, Furnstahl R J, Maris P, Perry R J, Schwenk A, and Vary J P 2008 *Nucl. Phys.* **A801** 21
[13] Vary, J P, Maris, P, Ng, E G, Yang, C and Sosonkina, M 2009 *J. Phys.: Conf. Ser.* **180** 012083
[14] Sternberg P, Ng E G, Yang C, Maris P, Vary J P, Sosonkina M and Le H V 2008 in *Proc. of the 2008 ACM/IEEE conf. on Supercomputing* SC '08 (Piscataway, NJ: IEEE Press) 15:1
[15] Maris P, Sosonkina M, Vary J P, Ng E G and Yang C 2010 *Procedia Computer Science* **1** 97
[16] Aktulga H M, Yang C, Ng E G, Maris P and Vary J P 2012 in *Lecture Notes in Computer Science* 7484 (Springer) 830
[17] Srinivasa S, Sosonkina M, Maris P, and Vary J P 2012 *Procedia Computer Science* **9** 256
[18] Zhou Z, Saule E, Aktulga H M, Yang C, Ng E G, Maris P, Vary J P and Catalyurek U V 2012 in *Proc. of IEEE Cluster* to be published
[19] Maris P Vary J P, Navratil P, Ormand W E, Nam H, and Dean D J 2011 *Phys. Rev. Lett.* **106** 202502
[20] Roth R, Langhammer J, Calci A, Binder S and Navratil P 2012 *Phys. Rev. Lett.* **107** 072501
[21] Rowe D J 1970 *Nuclear Collective Motion* (London: Methuen, London)
[22] Dytrych T, Sviratcheva K D, Bahri C, Draayer J P and Vary J P 2008 *J. Phys.* **G35** 095101
[23] Aktulga H M, Yang C, Ng E G, Maris P and Vary J P 2011 in *proceedings of HPCS2011* 176
[24] Aktulga H M, Yang C, Catalyurek U V, Maris P, Vary J P and Ng E G 2011 in *Lecture Notes in Computer Science* **7155** Euro-Par workshops (1) (Springer) 305
[25] Roth R 2009 *Phys. Rev.* **C79**, 064324
[26] Otsuka T, Honma M, Mizusaki T, Shimizu N and Utsuno Y 2001 *Prog. Part. Nucl. Phys.* **47** 319
[27] Dytrych T, Draayer J P, Launey K D, Caprio M A, Langr D 2012 these proceedings
[28] Hagen, G, Papenbrock, T and Dean, D J 2009 *Phys. Rev. Lett.* **103** 062503

New Horizons in Ab Initio Nuclear Structure Theory

Robert Roth, Joachim Langhammer, Sven Binder, Angelo Calci
Institut für Kernphysik, Technische Universität Darmstadt, 64289 Darmstadt, Germany

Abstract.
Nuclear interactions derived within chiral effective field theory enable nuclear structure and reaction calculations solidly rooted in QCD. Using chiral two- and three-body interactions in ab initio studies of a variety of nuclear observables is, both, promising and challenging. Particularly the inclusion of chiral 3N interactions into exact and approximate many-body calculations is demanding and computationally expensive. We present recent key developments that facilitate ab initio calculations of ground and low-lying excited states of p- and sd-shell nuclei with full 3N interactions in the Importance-Truncated No-Core Shell Model using consistent Similarity Renormalization Group transformations of the NN+3N Hamiltonian. The treatment of 3N interactions without approximations allows for first studies of the dependence of nuclear-structure observables on the details of chiral interactions. Moreover, we present results for heavy nuclei obtained in Coupled Cluster calculations using a normal-ordered two-body approximation of the 3N interaction.

1. Introduction

Chiral effective field theory (EFT) provides the most systematic approach to QCD-based Hamiltonians for nuclear structure calculations today. Starting from nucleons and pions as relevant degrees of freedom and the symmetries of low-energy QCD, chiral EFT provides nuclear two-, three- and higher-body interactions in a consistent way. A remaining challenge is to incorporate these QCD-based Hamiltonians in ab initio nuclear structure and reaction calculations, in order to make precise predictions and compare to experiment to provide feedback for the improvement of chiral EFT.

The most advanced nuclear Hamiltonian from chiral EFT available in terms of matrix elements at the moment consists of a high-precision two-body (NN) interaction at N^3LO [1, 2] along with a chiral three-body (3N) force at N^2LO [3, 4]. Computations are underway to provide matrix elements of the 3N force consistently at order N^3LO [5, 6]. Of course, the direct inclusion of these Hamiltonians into nuclear structure calculations is desirable, however this would require huge, computationally intractable model spaces to account for the strong short-range correlations induced by the bare interaction. One possibility to tame these correlations is to transform the Hamiltonian by means of a unitary transformation in the framework of the so called Similarity Renormalization Group (SRG) [7, 8]. As a result, the transformed Hamiltonians exhibit a improved convergence behavior and facilitate the subsequent many-body calculations. The SRG transformation can be carried out consistently, i.e. without refitting the low-energy constants afterwards. Moreover, it does not depend on the investigated nuclei or the many-body method and model space. In summary, we obtain a universal transformed Hamiltonian and observables with favorable convergence behavior useful for a variety of many-body methods.

Published under licence by IOP Publishing Ltd

In the following we present nuclear structure calculations including the transformed NN+3N interactions for two different many-body methods. One of them is the No-Core Shell Model (NCSM) [9] and its extension the Importance-Truncated No-Core Shell Model (IT-NCSM) [10, 11]. These ab initio methods are applicable for nuclei up to the lower-sd shell and yield ground states as well as low-lying excited states including all relevant observables with exact inclusion of full NN+3N interactions. Since the inclusion of 3N forces is computationally demanding, we have developed new approaches for handling the 3N matrix elements. Together with the importance truncation, this technique permits ab initio calculations for heavier nuclei in much larger model spaces than before.

Beyond the sd shell we use a different many-body method, the Coupled Cluster approach, which is the method of choice especially if one is interested in ground-states of nuclei near closed shells. Instead of the inclusion of the full 3N interaction, we employ the so called Normal-Ordered Two-Body approximation of the 3N interaction. We have benchmarked this approximation extensively and found it to be very accurate, while it reduces the computational cost tremendously. In combination with the coupled cluster approach we can compute nuclear properties including 3N interaction effects even for heavy nuclei.

2. Similarity Renormalization Group with 3N Interactions

The Similarity Renormalization Group (SRG) is one of the most flexible and still simple methods to unitarily evolve the Hamiltonian into band-diagonal form with respect to a chosen uncorrelated many-body basis. The unitary transformation of the Hamiltonian is governed by the so-called flow equation

$$\frac{\mathrm{d}}{\mathrm{d}\alpha} H_\alpha = [\eta_\alpha, H_\alpha]\,, \qquad \eta_\alpha = (2\mu)^2 [T_{\mathrm{int}}, H_\alpha] = -\eta_\alpha^\dagger\,, \tag{1}$$

with η_α as generator of the transformation. Note that this generator itself depends on the continuous flow parameter α. In order to obtain the transformed Hamiltonian (or other consistently transformed observables, e.g. multipole operators or radii), one has to solve a first-order differential equation, which is a rather simple task. The flexibility of the SRG is rooted in the fact that one is completely free in choosing an appropriate generator of the transformation as long as it is antihermitian. In the following we restrict ourselves to the definition of the generator as commutator of the intrinsic kinetic energy T_{int} with the transformed Hamiltonian, which is common in context of nuclear physics, see for example [12]. However, other choices of the generator are possible and under discussion.

We solve the operator equation (1) by converting it into matrix-element representation with respect to a n-body basis. Technically, this results in a system of coupled ordinary differential equations for the matrix elements of the transformed Hamiltonian H_α, which can be solved using standard Runge-Kutta methods. The effect of the SRG transformation on the matrix elements of the NN+3N interaction in the harmonic oscillator Jacobi basis is depicted in Fig. 1. In the left-hand panel we show the initial interaction, i.e. $\alpha = 0\,\mathrm{fm}^4$. The matrix consists of matrix elements with large absolute values even in far off-diagonal regions. The right-hand panel shows the SRG transformed matrix for flow parameter $\alpha = 0.16\,\mathrm{fm}^4$. Clearly, the SRG yields the desired prediagonalization of the matrix, i.e. a decoupling of the high- and low-energy degrees of freedom.

The introduction of the n-body states projects the Fock-space operator equation on the n-body space, i.e. flow-induced operator contributions beyond the n-body level are discarded. Currently, we can account for all many-body contributions up to the three-body level, i.e. we are able to include all two- and three-body terms of the evolved Hamiltonian. In consequence, we can define three different SRG-evolved Hamiltonians which we will distinguish in the discussion of our many-body calculations later on: *(i) NN-only*: we start with an initial NN interaction

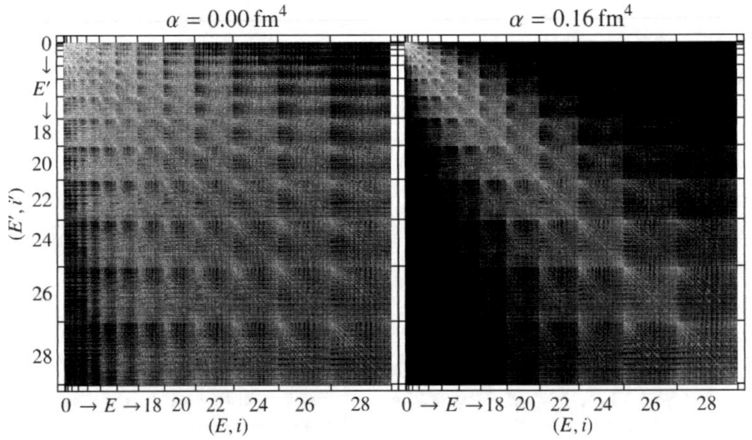

Figure 1. Effect of the SRG evolution on the matrix elements of the NN+3N-full interaction in a three-body harmonic oscillator Jacobi basis. Plotted are the absolute values of the matrix elements, where light colors represent large values and dark colors represent values near zero.

and keep the two-body terms only; *(ii) NN+3N-induced*: we start with an initial NN interaction and keep two- and induced three-body terms; *(iii) NN+3N-full*: we start with an initial NN+3N interaction and keep the two- and all three-body terms.

However, we have to omit all four- and higher-order many-body forces, which are induced during the unitary flow. This truncation formally violates the unitary character of the SRG. To assess the relevance of these neglected contributions we can investigate the dependence of the eigenvalues of the Hamiltonian (or any other observable) on a variation of the SRG flow parameter α. If omitted many-body forces are important, we will detect it after exact many-body calculations by an α-dependence of the energy eigenvalues.

3. Importance Truncated No-Core Shell Model

As a first example we discuss the application of consistently SRG-transformed chiral NN+3N interactions in nuclear-structure calculations for nuclei throughout the p- and lower-sd-shell. To solve the eigenvalue problem of the Hamiltonian, we use the Importance Truncated No-Core Shell Model (IT-NCSM), which was developed in our group [10, 11]. The underlying idea is that a significant part of the NCSM Slater-determinant basis is not required for a proper description of most of the low-lying spectrum of the nuclear Hamiltonian, which is what we are interested in. With help of an *a priori* importance measure, one can quantify the importance of each individual Slater determinant. All states with importance measure below a specific threshold are discarded, and consequently the dimension of the matrix eigenvalue problem is reduced. Nevertheless, after a sequence of IT calculations for a set of threshold values we are able to recover the full NCSM results with an *a posteriori* extrapolation of the desired observables to vanishing threshold [11]. Therefore, we can compute via the IT-NCSM all observables which can be calculated in the NCSM framework. Furthermore, we developed a new handling of the 3N matrix elements, essentially by storing J, T-coupled matrix elements and taking advantage of an efficient on-the-fly decoupling of these matrix elements. It turned out that this new scheme is the key for the inclusion of 3N forces in huge model spaces in particular for nuclei in the upper p-shell and beyond [13].

In Figure 2 we show ground-state energies obtained from IT-NCSM calculations for ^{12}C

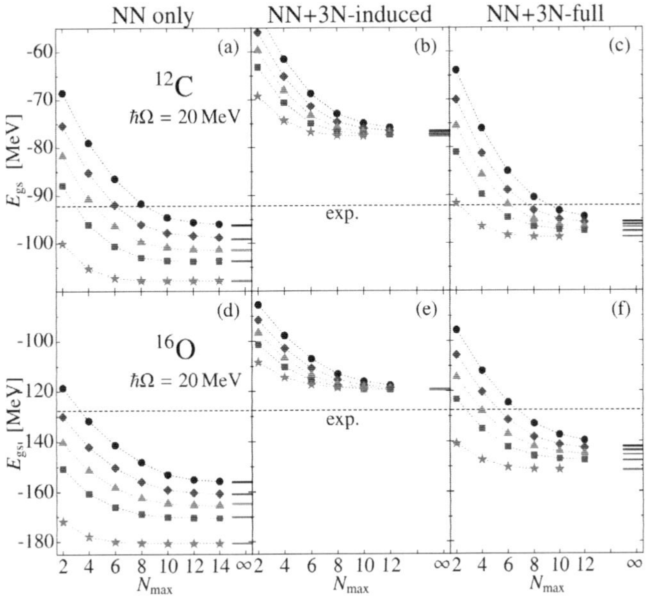

Figure 2. IT-NCSM ground-state energies for ^{12}C and ^{16}O as function of N_{max} for the three types of Hamiltonians (see column headings) for a set of flow parameters: $\alpha = 0.04\,\text{fm}^4$ (●), $0.05\,\text{fm}^4$ (◆), $0.0625\,\text{fm}^4$ (▲), $0.08\,\text{fm}^4$ (■), and $0.16\,\text{fm}^4$ (★). Taken from Ref. [14].

and ^{16}O for a set of flow parameters α as function of the model space size N_{max}. Due to the importance truncation and the new handling of 3N matrix elements we are able to reach $N_{\text{max}} = 12$ model spaces, which is sufficient to obtain converged ground-state energies for the soft interactions. The left panel shows our results using the NN-only Hamiltonian, i.e. without any 3N force, which show a strong dependence on the SRG flow parameter. This behavior hints at relevant, omitted induced 3N forces. If we include them consistently using the NN+3N-induced Hamiltonian as shown in the middle panel, the α-dependence vanishes. This means the induced higher-order many-body forces coming from the initial two-body force are irrelevant. The resulting ground-state energies are shifted towards less binding and can be seen as the true answer starting from an initial two-body force. The right panel shows the results from the NN+3N-full Hamiltonian, i.e. including the initial 3N force, with a re-emergence of the α-dependence of the ground-state energies. Thus, omitted 4N contributions that are induced during the SRG transformation with the initial 3N interaction have sizeable effects on the ground-state energies of nuclei beyond the lower p-shell [14, 15].

The impact of induced 4N interactions depend on the observables — it is much weaker e.g. for excitation energies. We show in Figure 3 the spectrum of ^{12}C computed with the three different Hamiltonians for two different SRG parameters. Once the spectra are converged with respect to model space size, they do not show a relevant dependence on α. Therefore, benchmarks of the chiral NN+3N forces against experiment are possible through spectroscopic studies in the upper p- and sd-shell, even without the inclusion of SRG-induced 4N forces. Two examples are shown in Figure 4, where we compare the spectrum of ^{12}C obtained with the standard NN+3N-full Hamiltonian to spectra obtained with modified 3N interactions: in the middle panel we use a new set of low-energy constants c_1, c_3 and c_4, which we get by shifting their original values. These

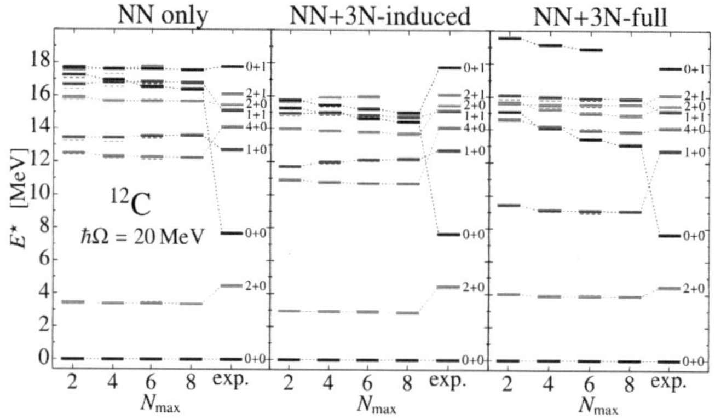

Figure 3. Excitation spectrum for the lowest positive-parity states (labelled $J\pi T$) in ^{12}C for the NN-only, the NN+3N-induced, and the NN+3N-full Hamiltonian with $\alpha = 0.08\,\text{fm}^4$. The thin dashed bars (mostly hidden) show results for $\alpha = 0.0625\,\text{fm}^4$. Taken from Ref. [14].

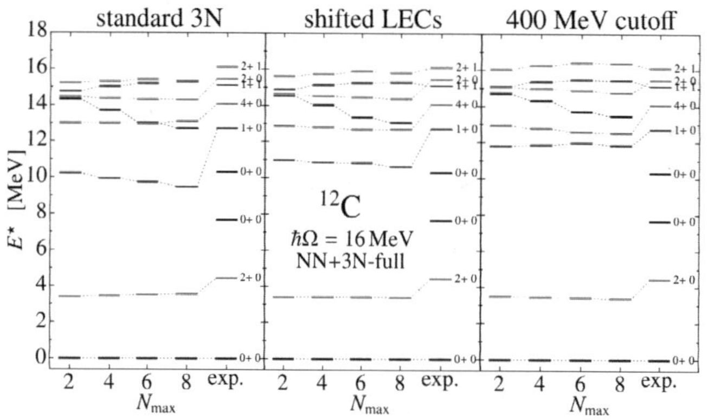

Figure 4. Excitation spectrum for the lowest positive-parity states (labelled $J\pi T$) in ^{12}C using NN+3N-full Hamiltonians with different chiral 3N interactions at $\alpha = 0.08\,\text{fm}^4$. In the left panel using the standard initial 3N interaction, in the middle with shifted low-energy constants c_i for the initial 3N interaction and in the right panel using a modified 3N interaction with cutoff 400 MeV.

shifts are motivated and specified in Ref. [16]. We observe only minor changes in the excitation energies of most of the states, except for the first 1^+ state. In the right panel we show the ^{12}C spectrum resulting from an initial 3N interaction with reduced cutoff $\Lambda = 400\,\text{MeV}$. Again the first 1^+ state is most sensitive to this cutoff variation. Note that we refit the low-energy constant c_E to the ^{4}He binding energy. These studies motivate a more detailed sensitivity analysis of nuclear spectra in this mass range, e.g. with respect to the low-energy constants and the cutoff of the initial 3N force [17].

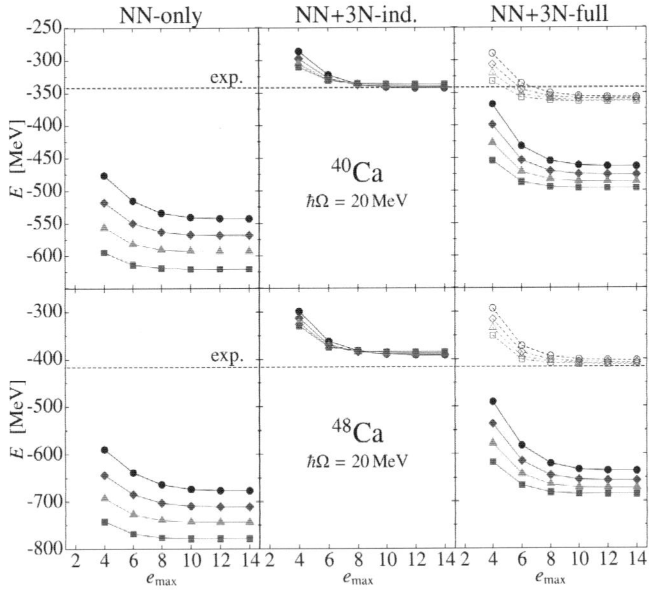

Figure 5. CCSD ground-state energies for ^{40}Ca and ^{48}Ca computed for the three types of Hamiltonians using the NO2B approximation for a set of flow parameters: $\alpha = 0.04\,\text{fm}^4$ (●), $0.05\,\text{fm}^4$ (◆), $0.0625\,\text{fm}^4$ (▲), and $0.08\,\text{fm}^4$ (■). Filled symbols for the NN+3N-full Hamiltonian are for the standard chiral 3N interaction with cutoff 500 MeV, the open symbols for a modified 3N interaction with cutoff 400 MeV. Taken from Ref. [15].

4. Coupled Cluster Theory

Beyond nuclei of the sd-shell our method of choice for studies of the SRG-evolved chiral NN+3N interactions is Coupled Cluster Theory, which is a standard technique in quantum chemistry. For our nuclear studies we employ the Coupled Cluster Singles and Doubles (CCSD) approximation [18], which eventually can be improved using the non-iterative ΛCCSD(T) correction for the energy [20, ?]. We include the 3N interactions using the so-called Normal-Ordered Two-Body (NO2B) approximation. We benchmarked this approximation extensively by comparing it to results obtained using the exact 3N force and found remarkably good agreement at the order of 1% for ground-state energies [15]. Computationally, the benefit of this NO2B approximation is that we have to handle two-body terms only, which decreases the computational cost significantly.

Exemplarily, we show in Figure 5 ground-state energies for ^{40}Ca and ^{48}Ca obtained from CCSD calculations, for the first time including effects of chiral 3N forces. The flow-parameter dependence of the energies computed with the NN+3N-full Hamiltonian confirms the relevance of induced 4N forces for heavier nuclei. Moreover, the open symbols correspond to calculations using the initial 3N force with lowered cutoff of 400 MeV leading to a much weaker α dependence. This offers the possibility to test chiral NN+3N Hamiltonians in ab initio calculations throughout the nuclear chart using the NO2B approximation alongside with CCSD.

5. Conclusions

We have discussed our recent progress regarding the consistent SRG-transformation of NN+3N Hamiltonians, the handling of matrix-elements of 3N forces, and their use in IT-NCSM and CCSD approaches for ab initio studies with chiral NN+3N interactions. This paves the way for extensive studies of the spectroscopy of p- and sd-shell nuclei and their sensitivity to details of the chiral forces, e.g. low-energy constants or cutoffs. Furthermore, ab initio predictions of binding energies including the effects of 3N forces are now possible even for medium mass nuclei, which is crucial in particular in the region of drip lines. In parallel we are working on the implementation of 3N forces into the formalism of the NCSM combined with the Resonating Group Method, which enables ab initio predictions of scattering and reactions of light nuclei. Together, these advances establish new horizons in ab initio nuclear structure and reaction theory and lead into an era of precise QCD-based low-energy nuclear theory.

Acknowledgements

Numerical calculations have been performed at the Jülich Supercomputing Centre, the LOEWE-CSC, and the National Energy Research Scientific Computing Center (NERSC), which is supported by the U.S. DOE under Contract No. DE-AC02-05CH11231. Supported by the Deutsche Forschungsgemeinschaft through SFB 634, the Helmholtz International Centre for FAIR within the LOEWE initiative of the state of Hesse, and by the BMBF (06DA9040I). We thank Petr Navrátil for providing the initial 3N matrix elements and for intense and productive discussions.

References

[1] Entem D R and Machleidt R 2003 *Phys. Rev. C* **68** 041001(R)
[2] Epelbaum E, Glöckle W and Meißner U G 2005 *Nucl. Phys. A* **747** 362
[3] Epelbaum E, Nogga A, Glöckle W, Kamada H, Meißner U G and Witala H 2002 *Phys. Rev. C* **66** 064001
[4] Navratil P 2007 *Few Body Syst.* **41** 117
[5] Bernard V, Epelbaum E, Krebs H and Meißner U G 2008 *Phys. Rev. C* **77** 064004
[6] Bernard V, Epelbaum E, Krebs H and Meissner U G 2011 *Phys. Rev. C* **84** 054001
[7] Roth R, Neff T and Feldmeier H 2010 *Prog. Part. Nucl. Phys.* **65** 50
[8] Bogner S K, Furnstahl R J and Schwenk A 2010 *Prog. Part. Nucl. Phys.* **65** 94
[9] Navrátil P, Quaglioni S, Stetcu I and Barrett B 2009 *J. Phys. G: Nucl. Part. Phys.* **36** 083101
[10] Roth R and Navrátil P 2007 *Phys. Rev. Lett.* **99** 092501
[11] Roth R 2009 *Phys. Rev. C* **79** 064324
[12] Bogner S K, Furnstahl R J and Perry R J 2007 *Phys. Rev. C* **75** 061001(R)
[13] Roth R, Calci A, Langhammer J and Binder S, in prep.
[14] Roth R, Langhammer J, Calci A, Binder S and Navrátil P 2011 *Phys. Rev. Lett.* **107** 072501
[15] Roth R, Binder S, Vobig K, Calci A, Langhammer J and Navrátil P 2012 *Phys. Rev. Lett.* **109** 052501
[16] Skibiński R, Golak J, Topolnicki K, Witała H, Epelbaum E, Glöckle W, Krebs H, Nogga A and Kamada H 2011 *Phys. Rev. C* **84** 054005
[17] Roth R, Langhammer J, Calci A and Binder S, in prep.
[18] Hagen G, Papenbrock T, Dean D J and Hjorth-Jensen M 2010 *Phys. Rev. C* **82** 034330
[19] Taube A G and Bartlett R J 2008 *J. Chem. Phys.* **128** 044110
[20] Taube A G and Bartlett R J 2008 *J. Chem. Phys.* **128** 044111

Oscillator basis, scattering and nuclear structure

A M Shirokov[1,2,3], A I Mazur[2], J P Vary[3] and I A Mazur[2]

[1] Skobeltsyn Institute of Nuclear Physics, Lomonosov Moscow State University, Moscow 119234, Russia

[2] Pacific National University, 136 Tikhookeanskaya st., Khabarovsk 680035, Russia

[3] Department of Physics and Astronomy, Iowa State University, Ames, Ia 50011-3160, USA

Abstract. We note that a Lanczos algorithm can be used to generate a harmonic oscillator basis. We use it to formulate quantum scattering theory utilizing an expansion of scattering wave functions in series of oscillator functions. The continuum spectrum solutions of the Schrödinger equation are found by means of Lanczos iterations. This formalism provides a possibility to extend a nuclear shell model, in particular, an *ab initio* no-core shell model, on a scattering domain.

Impressive progress in the *ab initio* description of nuclear structure has been achieved during the last decade. In addition, research programs aimed at an *ab initio* description of nuclear reactions have advanced. First should be mentioned a combined no-core shell model/resonating group method (NCSM/RGM) approach developed by Navrátil, Quaglioni et al [1]. Various reactions with light nuclei were successfully calculated by means of the NCSM/RGM method. Unfortunately, the RGM approach involves some model assumptions regarding the reaction mechanism which limit the predictive power usually associated with *ab initio* methods. The first attempt to calculate nucleon-nucleus scattering based on quantum Monte Carlo calculations was performed in Ref. [2]. The Gamow shell model [3] provides a possibility to calculate widths of nuclear resonant states. A Lorentz Integral Transform method [4] makes it possible to calculate observables in photonuclear and electroweak reactions within various *ab initio* approaches. This method was successfully used in calculations of cross sections of various photodisintegration, electrodisintegration and electroweak reactions on light nuclei within the hyperspherical harmonics approach [5]. A generalized Lanczos technique was suggested in Ref. [6] for calculations of nuclear response for any multipole operator and general electroweak response functions and electromagnetic responses in particular within theoretical approaches utilizing Slater determinants built on harmonic oscillator basis functions.

We suggest here a formalism in scattering theory based on the Lanczos algorithm generating the harmonic oscillator basis. Using this formalism, we propose a new approach to nuclear reactions. Our approach utilizes the many-body harmonic oscillator basis and Lanczos algorithm in calculations of scattering observables.

A harmonic oscillator provides a natural basis for nuclear structure studies. It is widely used in various shell model applications, in particular, in *ab initio* NCSM [7, 8]. A diagonalization of the shell model Hamiltonian matrix in the many-body oscillator basis is conventionally performed using the Lanczos algorithm. We note that the suggested formalism in quantum scattering theory is equivalent to the J-matrix formalism with oscillator basis [9, 10]. However the explicit use of Lanczos iterations within the proposed approach to nuclear reactions makes it possible to integrate it naturally into modern *ab initio* methods for nuclear structure studies.

Quantum scattering theory compares asymptotics of wave functions Ψ_E fitting the Schrödinger equation with the Hamiltonian H describing the system of interest,

$$H\Psi_E = E\Psi_E, \tag{1}$$

with asymptotics of the wave function Ψ_E^0 of the reference Hamiltonian H^0,

$$H^0\Psi_E^0 = E\Psi_E^0. \tag{2}$$

We discuss here the case of H^0 being a free Hamiltonian, i. e., it includes the operator of kinetic energy of relative motion of colliding particles T only,

$$H^0 = T. \tag{3}$$

We denote by ϕ_i the oscillator state of angular momentum J with i excitation quanta, i. e. the state with i quanta counted from the lowest number of oscillator quanta for the particular value of J. The kinetic energy operator is known to have a tridiagonal matrix in the basis of states ϕ_i, i. e., matrix elements $T_{ij} = \langle\phi_i|T|\phi_j\rangle$ differ from zero only if $i = j$ or $i = j \pm 2$. Therefore applying operator T to the state ϕ_i we obtain

$$T\phi_i = T_{i,i-2}\phi_{i-2} + T_{ii}\phi_i + T_{i,i+2}\phi_{i+2}. \tag{4}$$

We note that the Lanczos algorithm with kinetic energy operator T generates the oscillator basis of states with any given angular momentum. ϕ_0 is the lowest oscillator state with the angular momentum J. Let us use ϕ_0 as a pivot vector in the Lanczos procedure. By applying T to ϕ_0, we get

$$T\phi_0 = T_{00}\phi_0 + T_{02}\phi_2. \tag{5}$$

After orthogonalization to ϕ_0 and normalization we obtain ϕ_2 as a first Lanczos vector. The second Lanczos vector we obtain by applying T to ϕ_2,

$$T\phi_2 = T_{20}\phi_0 + T_{22}\phi_2 + T_{24}\phi_4, \tag{6}$$

by orthogonalizing the result to ϕ_0 and ϕ_2 and normalization. Clearly ϕ_4 appears to be this second Lanczos vector. In the same manner we obtain ϕ_6 as the third Lanczos vector, etc. Thus the oscillator basis appears to be a Lanczos basis generated by the kinetic energy operator with ϕ_0 as a pivot vector.

We can construct the oscillator basis by Lanczos procedure not only from below, i. e. starting from the lowest oscillator state, but also from above starting from oscillator states with arbitrary large oscillator quanta. Suppose we have oscillator functions ϕ_N and ϕ_{N+2}. Applying T to ϕ_N, orthogonalizing the result to ϕ_N and ϕ_{N+2} and normalizing, we obtain ϕ_{N-2} as the next Lanczos basis state. At the next step we obtain ϕ_{N-4}, etc.

Expanding the wave function Ψ_E in series of oscillator functions,

$$\Psi_E = \sum_{n=0}^{\infty} a_{2n}(E)\,\phi_{2n}, \tag{7}$$

we reduce the free Schrödinger equation (2) to an infinite set of three-term recurrent relations (TRR)

$$T_{N,N-2}\,a_{N-2}(E) + \big(T_{NN} - E\big)a_N(E) + T_{N,N+2}\,a_{N+2}(E) = 0. \tag{8}$$

For positive energies E, these TRR have two linearly-independent solutions, $S_N(E)$ and $C_N(E)$, which analytical expressions are well-known [9, 10]. Only $S_N(E)$ fits physical TRR boundary condition,

$$T_{00}\, S_0(E) + \big(T_{02} - E\big) S_2(E) = 0, \tag{9}$$

while $C_N(E)$ does not,

$$T_{00}\, C_0(E) + \big(T_{02} - E\big) C_2(E) \neq 0. \tag{10}$$

The solution $S_N(E)$ is related to the physical solution Ψ_E^0 of the free Schrödinger equation (2) with sine-like asymptotics:

$$\sum_{N=0}^{\infty} S_N(E)\, \phi_N = \Psi_E^0 = \sqrt{\frac{2}{\pi}}\, kr\, j_l(kr) \underset{r\to\infty}{\longrightarrow} \sqrt{\frac{2}{\pi}} \sin\left(kr - \frac{\pi l}{2}\right), \tag{11}$$

where $k = \sqrt{2mE}/\hbar$ is the momentum and m is the reduced mass. The solution $C_N(E)$ defines the function

$$\Psi_E^C = \sum_{N=0}^{\infty} C_N(E)\, \phi_N \tag{12}$$

which is regular at the origin and asymptotically coincides with the irregular solution

$$\Psi_E^{0irreg} = -\sqrt{\frac{2}{\pi}}\, kr\, n_l(kr) \tag{13}$$

of the free Schrödinger equation (2):

$$\Psi_E^C \underset{r\to\infty}{\longrightarrow} \Psi_E^{0irreg} \underset{r\to\infty}{\longrightarrow} \sqrt{\frac{2}{\pi}} \cos\left(kr - \frac{\pi l}{2}\right). \tag{14}$$

An arbitrary solution of the TRR (8) is a linear combination of solutions $S_N(E)$ and $C_N(E)$. Properly normalized solutions can be expressed as

$$a_N(E) = \cos\delta\, S_N(E) + \sin\delta\, C_N(E). \tag{15}$$

The respective wave function defined through Eq. (7) behaves asymptotically as

$$\Psi_E \underset{r\to\infty}{\longrightarrow} \sqrt{\frac{2}{\pi}} \sin\left(kr - \frac{\pi l}{2} + \delta\right) \tag{16}$$

where δ is a scattering phase shift.

Instead of the TRR solutions $S_N(E)$ and $C_N(E)$, one can use another pair of linearly independent solutions, $C_N^+(E)$ and $C_N^-(E)$:

$$C_N^{\pm}(E) = C_N(E) \pm i S_N(E). \tag{17}$$

With the help of $C_N^+(E)$ and $C_N^-(E)$ we can define wave functions Ψ_E^+ and Ψ_E^-,

$$\Psi_E^{\pm} = \sum_{N=0}^{\infty} C_N^{\pm}(E)\, \phi_N, \tag{18}$$

with asymptotic behavior

$$\Psi_E^{\pm} \underset{r\to\infty}{\longrightarrow} \sqrt{\frac{2}{\pi}}\,(\mp i)^l\,e^{\pm ikr}. \tag{19}$$

The functions Ψ_E^+ are of a particular importance in the case of negative energies E when $k = i\sqrt{2m|E|}/\hbar$ since they decrease asymptotically as bound state wave functions.

We suppose that the Hamiltonian H can be accurately enough approximated by

$$H = T + V^{N_{\max}}, \tag{20}$$

where T is a kinetic energy operator and $V^{N_{\max}}$ is a potential energy in the P space, i. e. an infinite potential energy matrix in the oscillator basis is truncated to a finite matrix which involves only oscillator states with excitation quanta $N \le N_{\max}$:

$$V^{N_{\max}} = \sum_{n,m=0}^{N_{\max}/2} |\phi_{2n}\rangle\langle\phi_{2n}|V|\phi_{2m}\rangle\langle\phi_{2m}|. \tag{21}$$

The kinetic energy operator T is not truncated in Eq. (20). We note that in conventional shell model applications not only the potential energy V but the kinetic energy operator T is also truncated to the P space. Therefore the suggested approach extends the nuclear shell model Hamiltonian and is expected to improve the results obtained in conventional nuclear structure calculations. Note also that the potential energy matrix elements $\langle\phi_N|V|\phi_M\rangle$ decrease with N and M when excitation quanta N and M are large enough. On the other hand, the diagonal T_{NN} and off-diagonal $T_{N,N\pm2}$ kinetic energy matrix elements increase linearly with N for large excitation quanta N. This different behavior of T and V matrix elements with excitation quanta justifies the approximation (20).

We use H to produce the Lanczos basis from above. When applied to the oscillator states ϕ_N and ϕ_{N+2} with large enough quanta N and $N+2$, $N > N_{\max}$, H has the same effect as the pure kinetic energy operator T. Therefore we reproduce with H the oscillator basis states ϕ_{N-2}, ϕ_{N-4}, ..., $\phi_{N_{\max}+2}$ in the Q space and the highest oscillator state in the P space $\phi_{N_{\max}}$. The matrix elements of H in our Lanczos basis in the Q space and the only off-diagonal matrix element of H relating the P and Q spaces are equivalent to the kinetic energy matrix elements:

$$H_{NN} = T_{NN}, \qquad N > N_{\max} \tag{22}$$

$$H_{N,N+2} = H_{N+2,N} = T_{N,N+2}, \qquad N \ge N_{\max} \tag{23}$$

$$H_{N,N+m} = H_{N+m,N} = 0, \qquad m > 2 \tag{24}$$

Starting from $\phi_{N_{\max}}$, the potential energy enters the Lanczos procedure. The matrix elements H_{NN} with $N \le N_{\max}$ and $H_{N,N+2} = H_{N+2,N}$ with $N < N_{\max}$ differ from T_{NN} and $T_{N,N+2}$ respectively, all matrix elements $H_{N,N+m} = H_{N+m,N}$ with any N and $m > 2$ vanish. As a result, for the set of coefficients $a_N(E)$ of the oscillator expansion (7) of the wave function Ψ_E we obtain a TRR

$$H_{N,N-2}\,a_{N-2}(E) + \big(H_{NN} - E\big)a_N(E) + H_{N,N+2}\,a_{N+2}(E) = 0. \tag{25}$$

This TRR should be supplemented by the boundary condition at $N = 0$:

$$H_{00}\,a_0(E) + \big(H_{02} - E\big)a_2(E) = 0. \tag{26}$$

The TRR (25) is infinite, i. e. N can take any even positive value.

To find the solutions of TRR (25), we assign any value to $a_0(E)$, obtain $a_2(E)$ using the boundary condition (26), with known $a_0(E)$ and $a_2(E)$ we calculate $a_4(E)$ using TRR (25), next we calculate $a_6(E)$ by means of TRR (25), etc.

In the case of scattering states ($E > 0$), the expansion coefficients $a_N(E)$ with $N \geq N_{\max}$ should be proportional to the r.h.s. of Eq. (15). Knowing the values of two expansion coefficients in the Q space, say, $a_N(E)$ and $a_M(E)$ with $M \neq N$, $M \geq N_{\max}$, $N \geq N_{\max}$, we obtain the scattering phase shift

$$\tan\delta = -\frac{S_N(E) - \frac{a_N(E)}{a_M(E)} S_M(E)}{C_N(E) - \frac{a_N(E)}{a_M(E)} C_M(E)}. \tag{27}$$

Expression (27) can be easily derived from Eq. (15).

In the case of bound states ($E < 0$), the coefficients $a_N(E)$ in the Q space are linear combinations of $C_N^+(E)$ and $C_N^-(E)$. The solution $C_N^-(E)$ exponentially increases with N while $C_N^+(E)$ exponentially decreases with N in the limit of $N \to \infty$. We need to find numerically a negative energy E which provides the decreasing behavior of $a_N(E)$ with N. This is the bound state energy associated with the S-matrix pole which improves the pure variational calculation in the finite oscillator basis.

We use the neutron-nucleus elastic scattering as an example to demonstrate how this Lanczos formalism can be used for a description of nuclear reactions. In this case, we construct a conventional shell model Hamiltonian H^{A+1} of the $(A+1)$-body system which includes all oscillator states with excitation quanta $N \leq N_{\max}^{A+1}$. The Hamiltonian H^{A+1} acts in the P space and describes the relative kinetic energy of all $A+1$ particles and interactions between them.

We need to define also a channel Hamiltonian

$$h^{ch} = T^{n-A} + H^A \tag{28}$$

which includes a kinetic energy operator of the relative n–A motion T^{n-A} and a truncated Hamiltonian of an A-body subsystem

$$H^A = T^A + V^A. \tag{29}$$

The operator T^A describes the relative kinetic energy of A particles and interaction V^A between them. The operator H^A is defined in the A-body relative motion oscillator states with excitation quanta $N \leq N_{\max}^A$. The wave function Ψ^A describes the internal motion of the A-body subsystem in the ground state,

$$H^A\Psi^A = E^A\Psi^A. \tag{30}$$

The A-body system is supposed to be bound, and hence $E^A < 0$.

The channel Hamiltonian generally acts in the Q space. However it should have non-zero matrix elements between some states in the Q space and the P space to provide coupling of P and Q spaces. Therefore we can define the channel Hamiltonian H^{ch} as

$$H^{ch} = Ph^{ch}Q + Qh^{ch}P + Qh^{ch}Q. \tag{31}$$

where P is a projector on the P space, Q is a projector on the Q space and

$$Q + P = 1. \tag{32}$$

The Hamiltonian describing our system is

$$H = H^{A+1} + H^{ch}. \tag{33}$$

We construct the Lanczos basis from above starting from channel states $\phi_N\Psi^A$ and $\phi_{N+2}\Psi^A$. Here ϕ_N and ϕ_{N+2} are oscillator functions of relative $n{-}A$ motion with N and $N+2$ quanta respectively. It is supposed that $N > \mathcal{N}^{A+1}_{\max}$ where $\mathcal{N}^{A+1}_{\max} = N^{A+1}_{\max} + N^{A+1}_{tot} - N^A_{tot}$ and N^A_{tot} and N^{A+1}_{tot} are the *total* oscillator quanta in the lowest configuration of the A-body and $(A+1)$-body systems respectively. Clearly our states belong to the Q space, and applying to them the Hamiltonian H is equivalent to the application of the channel Hamiltonian H^{ch} to them. Applying H^{ch} to $\phi_N\Psi^A$ we obtain successively Lanczos states $\phi_{N-2}\Psi^A$, $\phi_{N-4}\Psi^A$, ..., $\phi_{\mathcal{N}^{A+1}_{\max}}\Psi^A$. The respective TRR is

$$T^{n-A}_{N,N-2}\,a_{N-2}(E) + \big(T^{n-A}_{NN} - E^{A+1} + E^A\big)a_N(E) + T^{n-A}_{N,N+2}\,a_{N+2}(E) = 0, \quad N > \mathcal{N}^{A+1}_{\max}. \tag{34}$$

Here E^{A+1} is the total energy of the $(A+1)$-body system, and

$$E^{n-A} = E^{A+1} - E^A \tag{35}$$

is a kinetic energy of the relative $n{-}A$ motion.

Up to this point all Lanczos states $\phi_{N-2}\Psi^A$, $\phi_{N-4}\Psi^A$, ..., $\phi_{\mathcal{N}^{A+1}_{\max}}\Psi^A$ were obtained analytically. Starting from the Lanczos vector $\phi_{\mathcal{N}^{A+1}_{\max}}\Psi^A$, the complete $(A+1)$-particle P-space Hamiltonian H^{A+1} is involved in the Lanczos procedure. All the remaining Lanczos iterations look like the conventional Lanczos iterations used in standard shell model codes with a specific pivot vector $\phi_{\mathcal{N}^{A+1}_{\max}}\Psi^A$. Note however that the Hamiltonian involves an additional term H^{ch} [see Eq. (31)] and the pivot vector $\phi_{\mathcal{N}^{A+1}_{\max}}\Psi^A$ generally includes few terms with excitation quanta $N > N^{A+1}_{\max}$ when the A-body wave function Ψ^A is a mixture of A-body components Ψ^A_M with excitation oscillator quanta $M = 0,\, 2,\, 4,\, ...,\, N^A_{\max}$:

$$\Psi^A = \alpha_0\Psi^A_0 + \alpha_2\Psi^A_2 + \alpha_4\Psi^A_4 + ... + \alpha_{N^A_{\max}}\Psi^A_{N^A_{\max}}. \tag{36}$$

By means of the Lanczos algorithm we obtain TRR

$$H^{A+1}_{N,N-2}\,a_{N-2}(E) + \big(H^{A+1}_{NN} - E^{A+1}\big)a_N(E) + H^{A+1}_{N,N+2}\,a_{N+2}(E) = 0, \quad N \le \mathcal{N}^{A+1}_{\max}, \tag{37}$$

where

$$H^{A+1}_{\mathcal{N}^{A+1}_{\max},\mathcal{N}^{A+1}_{\max}+2} = T^{n-A}_{\mathcal{N}^{A+1}_{\max},\mathcal{N}^{A+1}_{\max}+2}. \tag{38}$$

The Lanczos iterations will mix all many-body states in the $(A+1)$-body system. Generally the number of Lanczos iterations can be as large as the dimensionality of the $(A+1)$-body P space. N in the TRR (37) does not have a meaning of oscillator quanta, it is used only to distinguish various Lanczos basis states and can take negative values. After some reasonable number of Lanczos iterations we should stop and solve the combined set of TRR (34) and (37) for bound or scattering states by the methods discussed above.

The suggested approach can be extended to describe the scattering of charged particles. In this case we include the Coulomb interaction $V^{Coulomb}$ in the channel Hamiltonian H^{ch}. At large excitation quanta, the operator $T^{n-A} + V^{Coulomb}$ can be accurately approximated by a tridiagonal matrix [11]. Therefore at large excitation quanta we can still use TRR (34) with matrix elements T^{n-A}_{NN} corrected by the Coulomb interaction. However the Coulomb interaction starts mixing P and Q space states at larger values of excitation quanta than the strong nucleon-nucleon interaction. Therefore Lanczos iterations will involve the P space states at larger values of excitation quanta than in the case of uncharged particles. This is equivalent to adding of a few additional vectors from the Q space to a huge number of basis states in the P space in the conventional diagonalization of the shell model Hamiltonian by means of Lanczos iterations.

We hope that the suggested Lanczos approach will be efficient in the *ab initio* description of various nuclear reactions with light nuclei and for improving results for bound state energies obtained in shell model calculations.

This work was supported in part by the US Department of Energy Grants No. DE-FG02-ER40371 and DE-FC02-09ER41582 (SciDAC/UNEDF), by the Ministry of Education and Science of the Russian Federation Grant 14.V37.21.1297 and by the Russian Foundation of Basic Research.

References

[1] Navrátil P, Quaglioni S, Stetcu I and Barrett B R 2009 *J. Phys.* G **36** 083101
Navrátil P, Quaglioni S and Roth R 2011 *J. Phys.: Conf. Series* **312** 082002
[2] Nollett K M, Pieper S C, Wiringa R B, Carlson J and Hale G M 2007 *Phys. Rev. Lett.* **99** 022502
[3] Michel N, Nazarewicz W, Ploszajczak M and Bennaceur K 2002 *Phys. Rev. Lett.* **89** 042502
[4] Efros V D 1999 *Phys. At. Nucl.* **62** 1833
[5] Efros V D, Leidemann W, Orlandini G and Barnea N 2007 *J. Phys.* G **34** R459
Leidemann W 2007 *Nucl. Phys.* A **790** 24
Leidemann W 2009 *Int. J. Mod. Phys.* E **18** 1339
[6] Haxton W C, Nollett K M and Zurek K M 2005 *Phys. Rev.* C **72** 065501
[7] Zheng D C, Vary J P and Barrett B R 1994 *Phys. Rev.* C **50** 2841
Zheng D C, Barrett B R, Vary J P, Haxton W C and Song C L 1995 *Phys. Rev.* C **52** 2488
[8] Navrátil P, Vary J P and Barrett B R 2000 *Phys. Rev. Lett.* **84** 5728
Navrátil P, Vary J P and Barrett B R 2000 *Phys. Rev.* C **62** 054311
[9] Yamani H A and Fishman L 1975 *J. Math. Phys.* **16** 410
[10] Bang J M, Mazur A I, Shirokov A M, Smirnov Yu F and Zaytsev S A 2000 *Ann. Phys. (NY)* **280** 299
[11] Okhrimenko I P 1984 *Nucl. Phys.* A **424** 121

Weakly bound systems, continuum effects, and reactions

Y Jaganathen[1], **N Michel**[2,3] **and M Płoszajczak**[1]

[1] Grand Accélérateur National d'Ions Lourds (GANIL), CEA/DSM - CNRS/IN2P3, BP 55027, F-14076 Caen Cedex, France

[2] Department of Physics and Astronomy, University of Tennessee, Knoxville, Tennessee 37996, USA

[3] Physics Division, Oak Ridge National Laboratory, Oak Ridge, Tennessee 37831, USA

E-mail: ploszajczak@ganil.fr

Abstract. Structure of weakly bound/unbound nuclei close to particle drip lines is different from that around the valley of beta stability. A comprehensive description of these systems goes beyond standard Shell Model (SM) and demands an open quantum system description of the nuclear many-body system. We approach this problem using the Gamow Shell Model (GSM) which provides a fully microscopic description of bound and unbound nuclear states, nuclear decays, and reactions. We present in this paper the first application of the GSM for a description of the elastic and inelastic scattering of protons on ^{6}He.

1. Introduction

The nuclear SM was proposed almost sixty years ago [1, 2]. Soon after, the interacting SM was developed and used extensively to understand a wealth of data on nuclear levels, moments, collective excitations, electromagnetic and β decays, and various particle decays. Interacting SM describes the nucleus as a closed quantum system: nucleons occupy bound, hence well localized, single-particle orbits of a harmonic oscillator potential and are isolated from the environment of unbound scattering states. Since the scattering continuum is not considered explicitly, the presence of decay thresholds and the interplay of Hermitian (internal) and anti-Hermitian (external, via the continuum) mixing of SM configurations is neglected. This competition yields a complicated interference pattern [3] and is a source of many collective features such as, e.g., the resonance trapping [4, 5, 6] and super-radiance phenomenon [7, 8], the multichannel coupling effects in reaction cross-sections [9] and shell occupancies [10], the clustering [11], the modification of spectral fluctuations [12], the deviations from Porter-Thomas resonance widths distribution [6, 13], and so on.

It was clear in the fifties that the SM is a phenomenological tool which employs the strongly renormalized bare nuclear interaction between nucleons [14] and neglects coupling to the continuum [15]. The overwhelming success of the SM, its elegance and simplicity resulted in neglecting most of these questionable assumptions. Unfortunately, the divide between discrete and continuum states in SM has led to an artificial separation of nuclear structure from nuclear reactions, and hindered a deeper understanding of nuclear properties. It is often believed that the understanding why the simple interacting SM works so well will be advanced when the goal of *ab initio* many-body approaches including the continuum coupling will be achieved.

Many structural properties of the nucleus are determined by means of nuclear collisions and this calls for a unified theoretical framework. Feshbach at the beginning of the sixties formulated a unified theory of nuclear reactions using the effective Hamiltonian and the projection operator method to select the open channel components of the wave function [16]. This development led to various formulations of the real-energy continuum shell model (CSM) [17, 18, 19, 20, 21, 22]. Nowadays, the real-energy CSM provides a unified description of the structure and reactions with up to two nucleons in the scattering continuum [20, 22].

At the same time, Fano noticed that the exact coincidence of different configurations above the lowest particle emission threshold makes the perturbation theory inadequate and calls for a generalization of the standard SM [23]. The achievement of Fano's goal took almost forty years and required the development of new mathematical approach of the rigged Hilbert space [24], new methods to deal with diverging integrals (matrix elements of one- and two-body operators) involving resonance and scattering states [28, 29, 30, 31, 32], the formulation of the generalized completeness relation including single-particle (s.p.) bound states, resonances and scattering states [25]. These different and independent developments in mathematics and physics enabled finally a satisfactory formulation of the new many-body theory, the GSM [33, 34, 35, 36], which offers a fully symmetric treatment of bound, resonance and scattering states in the multiparticle framework. In this formulation, the maximum number of particles in the scattering continuum is not *a priori* prescribed as in the real-energy CSM, but follows from the Schrödinger variational principle for the many-body Hamiltonian.

Until now, GSM has been primarily used in the context of nuclear structure. (For a recent review, see Ref. [26].) In this paper, we shall extend GSM to reaction problems by the coupled-channel (CC) formulation of the scattering problem. The very first application of the GSM-CC formalism will be presented in this paper for the proton scattering on ^{6}He target. The proposed GSM-CC formalism can be easily straightforwardly generalized for the description of nuclear reactions in the *ab initio* framework of the No-Core Gamow Shell Model [27].

In Sect. 2 we present essential features of the GSM. The reaction wave function and the derivation of GSM-CC equations in the coordinate space representation are discussed in Sect. 3. The method for solving the CC equations is presented in Sect. 4. The first application of the GSM-CC formalism for a description of ^{6}He+p reaction is discussed in Sect. 5. Finally, main conclusions of this work are summarized in Sect. 6.

2. The Gamow Shell Model

Resonance phenomena are generic intrinsic properties of many mesoscopic systems. What is specific to atomic nuclei are strong nucleon-nucleon correlations which impose a simultaneous description of the configuration mixing and the coupling to decay channels. The fundamental difficulty in the CSM formulation and the reason why the Fano's program [23] had to wait forty years to find a comprehensive solution is the fact that resonances do not belong to the Hilbert space. Therefore, whatever strategy is chosen to formulate a configuration interaction approach in open quantum systems, a key points are always: (i) the treatment of s.p. resonances in a many-body framework, and (ii) the definition of complete s.p. and many-body bases.

For the GSM, a s.p. basis is given by the Berggren ensemble [25] which consists of Gamow (resonant) states and the non-resonant continuum (see Fig. 1). (For a detailed description of the GSM see Ref. [26].) The GSM Hamiltonian is Hermitian. However, since the s.p. vectors have either outgoing or scattering asymptotics, the Hamiltonian matrix in GSM is complex symmetric and its eigenvalues are complex above the first particle emission threshold. Hence, both real-energy and complex-energy CSM formulations lead to a non-Hermitian eigenvalue problem above the threshold and contain all salient features of an interplay between opposite effects of Hermitian and anti-Hermitian couplings.

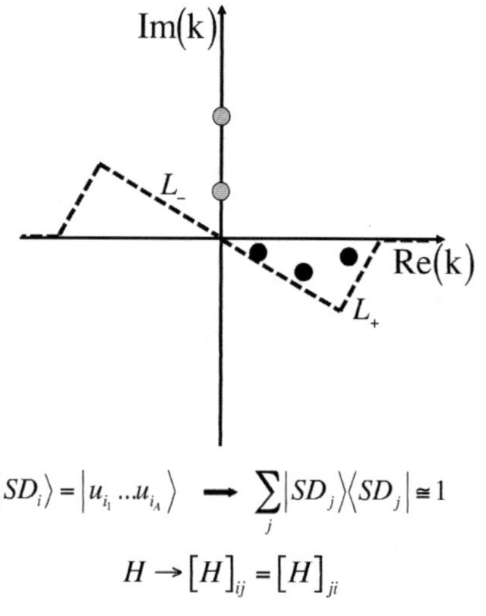

Figure 1. The schematic illustration of a complete s.p. basis in k-plane for GSM calculations. The complete many-body basis is spanned by Slater determinants involving nucleons on bound, resonant and continuum s.p. states along the contour L_+ in the complex k-plane. GSM Hamiltonian matrix is complex symmetric and yields complex energy eigenvalues above the first particle emission threshold.

2.1. The Hamiltonian

The translationally invariant GSM Hamiltonian in intrinsic nucleon-core coordinates of the cluster-orbital shell model[37], can be written as:

$$H = \sum_{i=1}^{A_{val}} \left[\frac{p_i^2}{2\mu_i} + U_i \right] + \sum_{i<j}^{A_{val}} \left[V_{ij} + \frac{1}{M_c} \mathbf{p_i p_j} \right], \tag{1}$$

where M_c is the mass of the core, μ_i is the reduced mass of either the proton or neutron ($1/\mu_i = 1/m_i + 1/M_c$), U is the s.p. potential describing the field of the core, V is the two-body residual interaction between valence nucleons. The last term in (1) represents the recoil term.

The particle-core interaction is a sum of nuclear and Coulomb terms: $U = U^N + U^C$. The nuclear potential U^N is approximated either by a Woods-Saxon (WS) field with a spin-orbit term [35] or by the Gamow-Hartree-Fock (GHF) potential [36]. The Coulomb field U^C is generated by a Gaussian density of Z_c core protons [38]. Similarly, the residual interaction can split into nuclear and Coulomb parts: $V = V^N + V^C$, where V^N is the modified surface Gaussian interaction (MSGI) [38] and V^C is the two-body Coulomb interaction.

V^C can be rewritten as: $U^C_{Z_{val}-1} + \left[V^C - U^C_{Z_{val}-1} \right]^{HO}$, where $U^C_{Z_{val}-1}$ takes care of the asymptotic behavior of the Coulomb interaction. The second term in this equation and the two-body recoil term can be expanded in the harmonic oscillator basis [39, 38] which provides an accurate treatment of the long-range physics of the Coulomb potential. In this work, we took 9 harmonic oscillator shells with the oscillator length $b = 2\,\mathrm{fm}$.

3. N-body GSM reaction wave functions

The CC framework is a convenient to formulate the GSM description of reactions involving one proton (neutron) scattering processes. The CC equations are obtained from the following A-body matrix elements:

$$\langle \mathcal{A}\{\langle \Psi_f|^{J_f} \otimes \langle r_f\ \ell_f\ j_f\ \tau_f|\}^{J_A}_{M_A}|H|\mathcal{A}\{|\Psi_i\rangle^{J_i} \otimes |r_i\ \ell_i\ j_i\ \tau_i\rangle\}^{J_A}_{M_A}\rangle \quad , \tag{2}$$

where $|\Psi_i\rangle^{J_i}_{M_i}$ ($|\Psi_f\rangle^{J_f}_{M_f}$) are initial (final) GSM eigenvectors of $(A-1)$-body system, r_i, r_f are radial coordinates, τ_i, τ_f are isospin quantum numbers (proton or neutron), and ℓ_i, j_i, ℓ_f, j_f are angular quantum numbers. All A-body wave functions are fully antisymmetrized, as emphasized by the $\mathcal{A}$ symbol.

In order to express the antisymmetry in a convenient way, the $|r\ell j\tau\rangle$ channel is expanded in a s.p. basis of GSM wave functions $u_n(r)$ generated by the s.p. potential U_{basis}. This implies for the associated creation operator:

$$a^\dagger_{r\ell j\tau} = \sum_n u_n(r)\ a^\dagger_{n\ell j\tau} \quad . \tag{3}$$

Hence, the expression of considered A-body wave function becomes:

$$\mathcal{A}|\{|\Psi\rangle^J \otimes |r\ell j\tau\rangle\}^{J_A}_{M_A}\rangle \quad = \quad \sum_n u_n(r)\{a^\dagger_{n\ell j\tau}|\Psi\rangle^J\}^{J_A}_{M_A} \quad . \tag{4}$$

States of the target $(A-1)$-nucleus are eigenstates of the Hamiltonian and can be expressed in the basis of Slater determinant. It is then convenient to express channel states in (A)-nucleus in the same Slater determinant basis. Consequently, all formal operations involving many-body operators and channel/target states become straightforward using the second quantization.

3.1. GSM-CC equations in the coordinate space

In order to evaluate Eq. (2), we separate the Hamiltonian $H = T + U_{core} + V_{res}$ into basis and residual parts:

$$H = T + U_{basis} + (V_{res} - U_0) \quad , \tag{5}$$

where U_{basis} is the optimal potential of A-particle system and $U_0 = U_{basis} - U_{core}$. U_{core} is the potential generated by the core. The advantage of this decomposition is that $V_{res} - U_0$ is finite-range and $T + U_{basis}$ is diagonal in the basis of Slater determinants used.

Let us consider first: $|\ell_i\ j_i\ \tau_i\rangle = |\ell_f\ j_f\ \tau_f\rangle = |\ell j\tau\rangle$ in Eq. (2), as H in this case contains infinite-range components leading to Dirac delta's which have to be calculated analytically. In order to derive these expressions, we suppose that only a finite number of Slater determinants appear in target many-body states. This assumption is always valid because GSM eigenvectors expansion coefficients decrease exponentially with the energy of basis scattering Slater determinants and, therefore, can be approximated with any arbitrary precision by a finite expansion of Slater determinants. As a consequence, the antisymmetry between $|u_n\ell j\tau\rangle$ and $|\Psi\rangle^J$ in Eq. (4) no longer plays any role for n larger than a given n_{max}. This implies that the creation operators in Eq. (4) can be replaced by the tensor products.

It is convenient to have matrix elements $\langle\alpha\beta|V_{res} - U_0|\gamma\delta\rangle$ vanished when $n_\alpha > n_{max}$ (same for β, γ or δ). This is always the case in GSM calculations as one uses a finite model space. It is thus convenient to rewrite the Hamiltonian H of Eq. (5) introducing an operator acting only on the target:

$$H = T + U_{basis} + (V_{res} - U_0)^{A-1} + [(V_{res} - U_0) - (V_{res} - U_0)^{A-1}] \ , \tag{6}$$

where one defines $(V_{res} - U_0)^{A-1}$ by its action on the non-antisymmetrized A-body states $|\Psi\rangle^J_M \otimes |u_n \ell j m \tau\rangle$:

$$(V_{res} - U_0)^{A-1}(|\Psi\rangle^J_M \otimes |u_n \ell j m \tau\rangle) = [(V_{res} - U_0)|\Psi\rangle^J_M] \otimes |u_n \ell j m \tau\rangle \ , \tag{7}$$

i.e., $(V_{res} - U_0)^{A-1}$ is the finite-range part of $V_{res} - U_0$ acting on $(A-1)$-body states only. Inserting (6), (7) in (4), one can show that it is only the sum involving $T + U_{basis} + (V_{res} - U_0)^{A-1}$ and $n \geq 0$ which generates Dirac delta's in the matrix element (2). Consequently, one can rewrite the matrix element as a sum of two terms, one which is finite and the other which is infinite:

$$\begin{aligned}
&\langle \mathcal{A}\{\langle\Psi_f|^{J_f} \otimes \langle r_f \ell j \tau|\}^{J_A}_{M_A}|H|\mathcal{A}\{|\Psi_i\rangle^{J_i} \otimes |r_i \ell j \tau\rangle\}^{J_A}_{M_A}\rangle \\
&= \left[\sum_{n_i n_f}^{n_{max}} u_{n_i}(r_i)\, u_{n_f}(r_f)\, \langle \mathcal{A}\{\langle\Psi_f|^{J_f} \otimes \langle u_{n_f} \ell j \tau|\}^{J_A}_{M_A}|H|\mathcal{A}\{|\Psi_i\rangle^{J_i} \otimes |u_{n_i} \ell j \tau\rangle\}^{J_A}_{M_A}\rangle - \right. \\
&- \sum_{n \leq n_{max}} u_n(r_i)\, u_n(r_f)\, \langle\{\langle\Psi_f|^{J_f} \otimes \langle u_n \ell j \tau|\}^{J_A}_{M_A}|T + U_{basis} + \\
&\left. + (V_{res} - U_0)^{A-1}|\{|\Psi_i\rangle^{J_i} \otimes |u_n \ell j \tau\rangle\}^{J_A}_{M_A}\rangle \right] + \sum_{n \geq 0} u_n(r_i)\, u_n(r_f) \\
&\langle\{\langle\Psi_f|^{J_f} \otimes \langle u_n \ell j \tau|\}^{J_A}_{M_A}|T + U_{basis} + (V_{res} - U_0)^{A-1}|\{|\Psi_i\rangle^{J_i} \otimes |u_n \ell j \tau\rangle\}^{J_A}_{M_A}\rangle \ .
\end{aligned} \tag{8}$$

The term in between square bracket is finite and can be calculated using Slater determinant expansion of considered many-body states and employing standard SM formulas. As $|\Psi_i\rangle^{J_i}_{M_i}$ and $|\Psi_f\rangle^{J_f}_{M_f}$ are eigenvectors of H, the second term of Eq. (8) does not vanish if:

$$|\Psi_i\rangle^{J_i}_{M_i} = |\Psi_f\rangle^{J_f}_{M_f} = |\Psi\rangle^J_M \ .$$

If $|\ell_i\ j_i\ \tau_i\rangle \neq |\ell_f\ j_f\ \tau_f\rangle$, one can show using the orthogonality of $|\ell_i\ j_i\ \tau_i\rangle$ and $|\ell_f\ j_f\ \tau_f\rangle$ that only the sum over n_i, n_f in Eq. (8) is nonzero and can be calculated straightforwardly from the Slater determinant expansions of $|\Psi_i\rangle^{J_i}_{M_i}$ and $|\Psi_f\rangle^{J_f}_{M_f}$.

4. Resolution of the CC equations

Let us consider the scattering A-body state decomposed in reaction channels:

$$|\Phi\rangle = \sum_c \int_0^{+\infty} \mathcal{A}|\{|\Psi_c\rangle^{J_c} \otimes |r\ \ell_c j_c \tau_c\rangle\}^{J_A}_{M_A}\rangle u_c(r) r^2\, dr \ , \tag{9}$$

where c is the reaction channel defined by the $(A-1)$-body state $|\Psi_c\rangle^{J_c}_{M_c}$ and the one-body quantum numbers (ℓ_c, j_c, τ_c). $u_c(r)$ is the radial amplitude of the c channel to be determined. The CC equations in this basis follow from the Schrödinger equation $H|\Phi\rangle = E|\Phi\rangle$:

$$\sum_{c'} \int_0^{+\infty} \langle \mathcal{A}\{\langle\Psi_c|^{J_c} \otimes \langle r\ \ell_c j_c \tau|\}^{J_A}|H - E|\mathcal{A}\{|\Psi_{c'}\rangle^{J_{c'}} \otimes |r' \ell_{c'} j_{c'} \tau_{c'}\rangle\}^{J_A}\rangle u_{c'}(r')(r')^2\, dr' = 0 \ . \tag{10}$$

Eq. (10) is a generalized eigenvalue problem. Indeed, different channels are mutually non-orthogonal because target and projectile are antisymmetrized. In order to deal with a standard eigenvalue problem, one introduces the channel overlap matrix:

$$O(n,c,n',c') = \langle \mathcal{A}\{\langle\Psi_c|^{J_c} \otimes \langle n\ \ell_c j_c \tau|\}^{J_A} | \mathcal{A}\{|\Psi_{c'}\rangle^{J_{c'}} \otimes |n'\ell_{c'}j_{c'}\tau_{c'}\rangle\}^{J_A}\rangle \ . \tag{11}$$

Eq. (10) can then be written in a matrix form: $HU = EOU$, where $U = \{u_c(r)\}_c$ is the vector of considered channels. Introducing: $W = O^{\frac{1}{2}}U$, and the modified Hamiltonian: $H_m = O^{-\frac{1}{2}}HO^{-\frac{1}{2}}$, one obtains the standard eigenvalue problem: $H_mW = EW$.

The overlap matrix O is defined in the Berggren basis and can be calculated using the Slater determinant expansion of channels. As the antisymmetry acts locally, it is convenient to introduce the harmonic oscillator expansion for the finite-range part of $O^{-\frac{1}{2}}$. For this, O is expanded in the basis of harmonic oscillator channels to obtain the finite-range part of $O^{-\frac{1}{2}}$: $\Delta = O^{-\frac{1}{2}} - I_d$. Then, H_m can be separated into long- and short-range parts:

$$H_m = H + H\Delta + \Delta H + \Delta H \Delta \ ,$$

where all terms involving Δ are expanded in the harmonic oscillator basis. Thus, the added part of H in H_m can be treated similarly to the short-range residual interaction.

Using results of Sec. (3.1) and the transformation described above, one can write Eq. (10) as a system of non-local differential equation with respectively local ($V_{cc'}^{(loc)}(r)$) and non-local ($V_{cc'}^{(non-loc)}(r,r')\ \forall c$) optical potentials:

$$\frac{\hbar^2}{2\mu}\left(-w_c''(r) + \frac{\ell_c(\ell_c+1)}{r^2}w_c(r)\right) + \sum_{c'}\left(V_{cc'}^{(loc)}(r)w_{c'}(r) + \int_0^{+\infty} V_{cc'}^{(non-loc)}(r,r')w_{c'}(r')\ dr'\right) =$$
$$= (E - E_{T_c})\ w_c(r) \ , \tag{12}$$

where μ is the reduced mass of the particle and E_{T_c} is the energy of $|\Psi_c\rangle_{M_c}^{J_c}$. This set of non-local differential equations is then solved numerically using the modified equivalent potential method which yields equations local without singularities in the potentials [40]. The initial vector $U = \{u_c(r)\}_c$ of channel functions is obtained by multiplying solutions of Eq. (12): $W = \{w_c(r)\}_c$, by $O^{-\frac{1}{2}}$: $U = W + \Delta W$.

5. Discussion of results for ^{6}He+p reaction

Results presented in this section correspond to ^{6}He target nucleus in 2 states: $J^\pi = 0_1^+$ and 2_1^+. The configuration space for neutrons correspond to $0p_{3/2}$ resonance and 17 states of a discretized $p_{3/2}$ contour in the complex k plane. The configuration space for protons includes $0p_{1/2}$ and $0p_{3/2}$ resonances and 17 states of a discretized contour for each resonance. Moreover, for protons we include partial waves s, d, f, g, h which are decomposed using a real-energy contour as in this case the resonant poles are high in energy and very broad.

The s.p. basis in ^{6}He and ^{7}Li is generated by the same GHF potential, produced by the WS potential of the core and the MSGI two-body interaction between valence nucleons. Parameters of the core potential: the radius R = 1.993 fm (1.954 fm), the depth of the central part V_0 = 50.188 MeV (49.268 MeV), the diffuseness d = 0.497 fm (0.657 fm), and the spin-orbit strength V_{so} = 7.488 MeV (8.23 MeV) for protons (neutrons) have been chosen to fit $p_{1/2}$ and $s_{1/2}$ phase shifts in ^{4}He+p and ^{4}He+n, and energies/widths of $3/2_1^-$ and $1/2_1^-$ resonances in ^{5}He and ^{5}Li. The radius of the Coulomb potential in this calculation is R_C = 1.952 fm. Parameters of the MSGI interaction have been chosen to reproduce energies of the states in ^{6}He (0_1^+ and 2_1^+) and in ^{7}Li ($3/2_1^-$, $1/2_1^-$, $7/2_1^-$ and $5/2_1^-$) relative to the ^{4}He core. Eq. (12) are then solved using the multidimensional variant of one-dimensional iterative procedure [40] where the input functions for the first iteration come from the diagonalization of H_m in the Berggren basis.

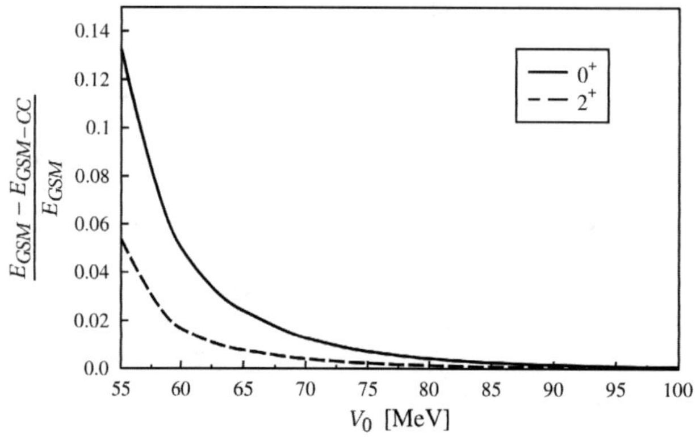

Figure 2. Relative energy difference between GSM and GSM-CC calculations for ^{6}He in $J^\pi = 0^+$ (the solid line) and 2^+ (the dashed line) states as a function of the depth of the WS potential generated by ^{4}He core. The model space is spanned by $0p_{3/2}$ resonance and 30 $p_{3/2}$ states of the discretized continuum. As a two-body interaction, the MSGI interaction with coupling strengths: $V(J=0, T=1) = -20\text{MeV}\times\text{fm}^3$ and $V(J=2, T=1) = -15\text{MeV}\times\text{fm}^3$ is used.

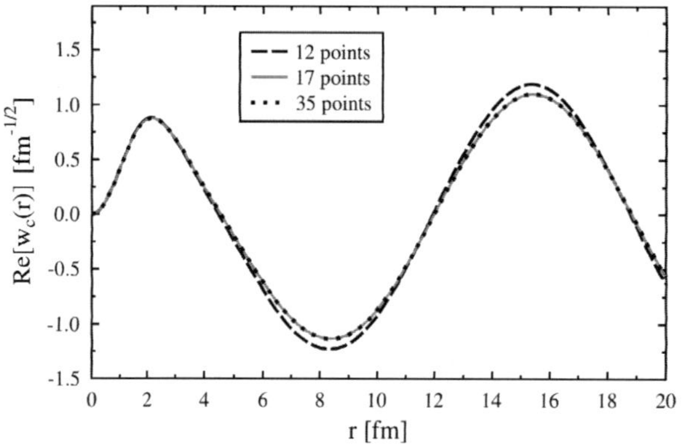

Figure 3. The real part of $3/2^-$ scattering wave function at the CM energy E_{CM}=5 MeV in the entrance channel $[^6\text{He}(0_1^+) \otimes \pi p_{3/2}]$ is plotted for different discretization densities of the $p_{3/2}$ contour in the complex k-plane.

5.1. Numerical tests

Before showing results for ^{6}He+p scattering, it is instructive to compare the spectra of GSM-CC calculations with those obtained in GSM by a direct diagonalization in Berggren basis. One should remember that the configuration space in both calculations is in general not the same.

Fig. 2 compares results of the GSM calculation for ^{6}He with the GSM-CC results for a system

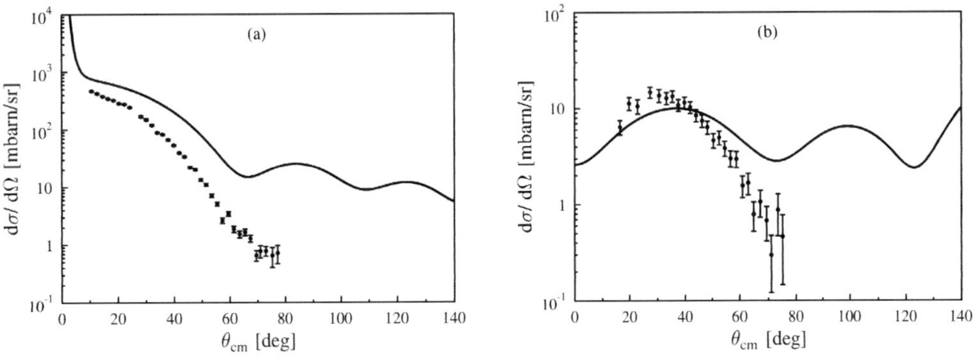

Figure 4. Elastic (left panel) and inelastic (right panel) GSM-CC angular cross sections for the reaction ^{6}He+p at the CM energy E_{CM}=35.1 MeV. The experimental data are taken from Ref. [41].

^{5}He($3/2^-_1$)+n in a model space including $0p_{3/2}$ s.p. resonance and $p_{3/2}$ continuum discretized with 30 points. One can see that if the continuum coupling is weak, *i.e.* for deep WS potentials, then the relative difference between GSM and GSM-CC calculations tends to zero. On the contrary, for weakly bound systems the GSM-CC approach with a limited number of reaction channels can be a poor approximation of the GSM calculation in the complete many-body space.

The discretization density of the Berggren basis is an essential ingredient of the GSM-CC calculations. Fig. 3 shows a convergence of the $3/2^-$ scattering wave function in ^{7}Li as a function of the number of $p_{3/2}$ states on the discretized contour. One can see that the scattering wave function is fully converged with 17 continuum states.

5.2. Elastic and inelastic ^{6}He+p cross section

Angular cross-sections calculated in GSM-CC for ^{6}He+p reaction at E_{CM} = 35.1 MeV are in the qualitative agreement with experimental data [41] (see Fig. 4). Quantitative discrepancies between theory and experiment may have several origins. Firstly, the excitation energy in Fig. 4 is such that the contribution of ^{4}He core excitations in this reaction cannot be neglected. Secondly, the chosen effective interaction is rather schematic. Thirdly and most importantly, the target is a weakly bound halo state whereas the reaction product (^{7}Li) in its ground state is well bound with strong ^{3}H cluster correlations due to the proximity of ^{3}H decay threshold. This implies that the GSM-CC configuration space generated by adding one proton to ^{6}He in 2 discrete states (0^+_1 ground state and 2^+_1 resonance) only, may by insufficient to produce configurations obtained by a direct diagonalization of ^{7}Li in the Berggren basis. Indeed, the configuration spaces in GSM-CC and GSM become equivalent if *all* discrete and continuum states of ^{6}He are used to generate ^{7}Li configurations. Including only few states of ^{6}He in GSM-CC approach requires a strong renormalization of the effective nucleon-nucleon interaction (the optical potentials) to compensate for the missing configurations. By construction, such a renormalization will generate identical energy spectra for both full space and restricted space GSM-CC calculations. However, the reaction cross-sections in these two settings will be different. The comparison between energy spectra obtained for the same Hamiltonian either by a direct diagonalization in Berggren basis or by solving GSM-CC equations for a limited number of many-body target states gives indication of how strong should be the renormalization of effective two-body interaction (the optical potentials) due to the neglected channels. In Table 1, we compare energy spectra

Table 1. GSM and GSM-CC spectra for low-lying states of ^{7}Li are compared with the experimental data. For the details of these calculations, see the discussion in the text.

J^{π}	E_{GSM} (MeV)	E_{GSM-CC} (MeV)	E_{exp} (MeV)
$3/2^-$	-17.83	-10.946	-10.949
$1/2^-$	-21.18	-10.469	-10.471
$7/2^-$	-21.01	-6.307	-6.297
$5/2^-$	-31.26	-4.509	-4.345

obtained in GSM and GSM-CC approaches using the same MSGI interaction which was fitted in the GSM-CC approach to reproduce the low-lying states of ^{6}He and ^{7}Li. One may see that ^{7}Li configurations involving ^{6}He scattering states which are missing in the GSM-CC calculation, lead to a dramatic lowering of the GSM eigenenergies and, hence, are essential to understand the dynamics in the ^{6}He+p reaction.

Results shown in Table 1 demonstrate that none of the CC approaches which neglects continuum states of the target can provide a realistic description of the proton scattering on weakly-bound neutron-rich nuclei. Similar conclusion about the importance of high-lying states in the continuum of a target $(A-1)$-nucleus on low-lying properties of the (A)-nucleus have been made recently in the Shell Model Embedded in the Continuum [42].

6. Conclusions

We have proposed the unified description of nuclear structure and nuclear reactions in the framework of GSM which includes continuum couplings in the many-body framework. This CC formulation of the GSM can be employed further in the No-Core GSM framework [27] with the bare interaction between free nucleons. The great advantage of the GSM-CC formalism is that the approximation of neglecting high-lying target states can be checked by comparing the calculated GSM-CC energy spectrum with those obtained in (complete) GSM calculation. In this way, one can quantify the role of these neglected configurations in the (A)-particle wave function and estimate the scale of renormalization corrections in the microscopic optical potentials.

In spite of recent developments in *ab initio* description of nuclear states and progress in open quantum system formulation of the nuclear many-body problem, the microscopic description of nuclear reactions with weakly bound targets continues to be a formidable challenge as it requires including large number of target states to obtain convergent results. In this respect, the task of a unified microscopic description of nuclear structure and nuclear reactions with weakly bound exotic nuclei remains still a distant perspective.

Acknowledgments

This paper is written to honor important and numerous contributions of Jerry P. Draayer to the nuclear many-body theory. This work has been supported in part by the by the MNiSW grant No. N N202 033837; the Collaboration COPIN-GANIL; and U.S. Department of Energy under Contract Nos. DE-FG02-96ER40963 (University of Tennessee) and DE-FG02-10ER41700 (French-U.S. Theory Institute for Physics with Exotic Nuclei).

References

[1] Göppert-Mayer M 1949 *Phys. Rev.* **75** 1969
[2] Haxel O, Jensen J.H.D. and Süss H.E. 1949 *Phys. Rev.* **75** 1766
[3] Rotter I, Persson E, Pichugin K and Seba P 2000 *Phys. Rev.* A **62** 450
[4] Kleinwächter P and Rotter I 1985 *Phys. Rev.* C **32** 1742
Rotter I 1991 *Rep. Prog. Phys.* **54** 635
[5] Sokolov V.V. and Zelevinsky V.G. 1988 *Phys. Lett.* B **202** 10
Sokolov V.V. and Zelevinsky V.G. 1989 *Nucl. Phys.* A **504** 562
[6] Drożdż S, Okołowicz J, Płoszajczak M and Rotter I 2000 *Phys. Rev.* C **62** 24313
[7] Dicke R.H. 1954 *Phys. Rev.* **93** 99
[8] Auerbach N and Zelevinsky V.G. 2011 *Rep. Prog. Phys.* **74** 106301
[9] Baz A.I. (1957) *Soviet Phys.-JETP* **6** 709
Newton R.G. 1959 *Phys. Rev.* **114** 1611
Hategan C 1978 *Ann. Phys. (NY)* **116** 77
[10] Michel N, Nazarewicz W and Płoszajczak M 2007 *Phys. Rev.* C **75** 031301
Michel N, Nazarewicz W and Płoszajczak M 2007 *Phys. Rev.* C **82** 044315
[11] Okołowicz J, Płoszajczak M and Nazarewicz W 2012 On the origin of nuclear clustering *Preprint* arXiv:1202.6290
Okołowicz J, Nazarewicz W and Płoszajczak M 2012 Toward understanding the microscopic origin of nuclear clustering *Preprint* arXiv:1207.6225
[12] Fyodorov Y.V. and Khoruzhenko B.A. 1999 *Phys. Rev. Lett.* **83** 65
[13] Kohler P.E., Bečvař F, Krtička M, Harvey J.A. and Guber K.H. 2010 *Phys. Rev. Lett.* **105** 072502
Celardo G.L., Auerbach N, Izrailev F.M. and Zelevinsky V.G. 2011 *Phys. Rev. Lett.* **106** 042501
[14] Brueckner K.A. 1954 *Phys. Rev.* **96** 508
[15] Ehrman J.B. 1951 *Phys. Rev.* **81** 412
Thomas R.G. 1952 *Phys. Rev.* **88** 1109
[16] Feshbach H 1958 *Ann. Phys. (NY)* **5** 357
Feshbach H 1962 *Ann. Phys. (NY)* **19** 287
[17] Mahaux C and Weidenmüller H.A. 1969 *Shell Model Approach to Nuclear Reactions* (North-Holland, Amsterdam)
[18] Barz H.W., Rotter I and Höhn J 1977 *Nucl. Phys.* **A 275** 111
[19] Philpott R.J. 1977 *Fizika* **9** 109
[20] Bennaceur K, Nowacki F, Okołowicz J and Płoszajczak M 1999 *Nucl. Phys.* **A 651** 289
[21] Volya A and Zelevinsky V 2006 *Phys. Rev.* C **74** 064314
[22] Rotureau J, Okołowicz J and Płoszajczak M 2006 *Nucl. Phys.* **A 767** 13
[23] Fano U 1961 *Phys. Rev.* **124** 1866
[24] Gel'fand I.M. and Vilenkin N.Ya. 1961 *Generalized Functions, Vol. 4*, Academic Press, New York
Maurin K 1968 *Generalized Eigenfunction Expansions and Unitary Representations of Topological Groups*, Polish Scientific Publishers, Warsaw
[25] Berggren T 1968 *Nucl. Phys.* **A 109** 265
[26] Michel N, Nazarewicz W, Płoszajczak M and Vertse T 2009 *J. Phys.* G **36** 013101
[27] Papadimitriou G, Rotureau J, Michel N, Płoszajczak M and Barrett B.R. in preparation
[28] Zel'dovich Ya.B. 1960 *Zh. Eksp. i Theor. Fiz.* **39** 776
[29] Hokkyo N 1965 *Prog. Theor. Phys.* **33** 1116
[30] Romo W.J. 1968 *Nucl. Phys.* **A 116** 617
[31] Zimányi J, Zimányi M, Gyarmati B, and Vertse T 1970 *Acta Phys. Hung.* **28** 251
[32] Gyarmati B and Vertse T 1971 *Nucl. Phys.* **A 160** 523
[33] Michel N, Nazarewicz W, Płoszajczak M and Bennaceur K 2002 *Phys. Rev. Lett.* **89** 042502
[34] Id Betan R, Liotta R.J., Sandulescu N and Vertse T 2002 *Phys. Rev. Lett.* **89** 042501
[35] Michel N, Nazarewicz W, Płoszajczak M and Okołowicz J 2003 *Phys. Rev.* C **67** 054311
[36] Michel N, Nazarewicz W and Płoszajczak M 2004 *Phys. Rev.* C **70** 064313
[37] Suzuki Y and Ikeda K 1988 *Phys. Rev.* C **38** 410
[38] Michel N, Nazarewicz W, and Płoszajczak M 2010 *Phys. Rev.* C **82** 044315
[39] Hagen G, Hjorth-Jensen M and Michel N 2006 *Phys. Rev.* C **73** 064307
[40] Michel N 2009 *Eur. Phys. Journal* A **42** 523
[41] Lagoyannis A et al 2001 *Phys. Lett.* B **157** 18
[42] Okołowicz J, Płoszajczak M and Luo Y.A. 2008 *Acta Physica Polonica* B **39** 389

The Super-Radiant Mechanism and the Widths of Compound Nuclear States

N Auerbach
School of Physics and Astronomy, Tel Aviv University, Tel Aviv, 69978, Israel

E-mail: auerbach@post.tau.ac.il

Abstract. In the introduction I will present the theory of the super-radiant mechanism as applied to various phenomena. I will then discuss the statistics of resonance widths in a many-body Fermi system with open decay channels. Depending on the strength of the coupling to the continuum such systems show deviations from the standard Porter-Thomas distribution. The deviations result from the process of increasing interaction of the intrinsic states through the common decay channels. In the limit of very strong coupling this leads to super-radiance. The results I will present are important for the understanding of recent experimental data concerning the width distribution of compound neutron resonances in nuclei.

1. Introduction

The possibility of forming a ''super-radiant'' (SR) state in a gas of atoms confined to a volume of a size smaller than the wave length of radiation was proposed by Dicke [1]. In the absence of a direct interaction, the atoms are coupled through their common radiation field. This indirect interaction through the continuum leads to a redistribution of lifetimes among intrinsic states. A rapidly decaying SR state is created at the expense of the rest of the states of the system that are ''robbed'' of their decay width and become narrow. This mechanism has a general origin and analogous phenomena should appear in many quantum systems when quasi-bound states are strongly coupled through common decay channels. The SR approach has been used in many different fields. For example, it was applied in chemistry [2], atomic physics [3], condensed matter physics [4, 5], intermediate energy nuclear physics [6, 7], in particle physics [8], in the theory of nuclear reactions [5, 9, 10]. In the next section we will develop the SR formalism for some simple cases.

2. The projection formalism.

Following the projection formalism [6,11], we divide the Hilbert space of nuclear states into two parts, the $\{Q\}$-subspace involving complicated many-body states $|q\rangle$, and the subspace $\{P\}$ of open channels $|c\rangle$. We use the notations Q and P for the corresponding projection operators onto the above subspaces. The wave function of the total system,

$$|\Psi\rangle = Q|\Psi\rangle + P|\Psi\rangle \qquad (1)$$

is the solution of the Schrödinger equation

$$H|\Psi\rangle = E|\Psi\rangle \qquad (2)$$

that can be written as a set of coupled equations,

$$(E - H_{QQ})Q|\Psi\rangle = H_{QP}\,P|\Psi\rangle \qquad (3)$$

and

$$\left(E - H_{PP}\right)P\left|\Psi\right\rangle = H_{PQ}Q\left|\Psi\right\rangle, \tag{4}$$

where we use the notations $H_{AB} = AHB$. Eliminating the part $P\left|\Psi\right\rangle$, we obtain an equation in the Q-space:

$$\left(E - \widetilde{H}_{QQ}\right)Q\left|\Psi\right\rangle = 0 \tag{5}$$

with the effective Hamiltonian

$$\widetilde{H}_{QQ} = H_{QQ} + H_{QP}\frac{1}{E^{(+)} - H_{PP}}H_{PQ}. \tag{6}$$

Here, $E^{(+)} \equiv E + i0$ contains the infinitesimal imaginary term $+i0$ ensuring correct asymptotic behaviour of the continuum wave functions. The second term of the effective Hamiltonian contains a real and imaginary part of the propagator

$$G^{(+)}\left(E\right) = \frac{1}{E^{(+)} - H_{PP}} \tag{7}$$

arising from the principal value and the delta-function $\delta\left(E - H_{PP}\right)$ (on-shell contributions from channels c open at energy E), respectively. The imaginary part of the effective Hamiltonian is $-\left(i/2\right)W$, with

$$W = 2\pi\sum_{c} H_{QP}\left|c\right\rangle\left\langle c\right|H_{PQ}. \tag{8}$$

Thus, the effective Hamiltonian (7) in Q-space is anti-Hermitian,

$$\widetilde{H}_{QQ} = \overline{H}_{QQ} - \frac{i}{2}W, \tag{9}$$

where $\overline{H}_{QQ}$ is a symmetric and real matrix that includes, apart from the original Hamiltonian in the Q-space-H_{QQ}, the principal value contribution of the PQ-coupling. The second part is anti-Hermitian. The eigenvalues of $\widetilde{H}$, $\tilde{E} = E - \left(i/2\right)\Gamma$ are complex poles of the scattering matrix corresponding to the resonances in the cross sections.

To demonstrate in a simple way the role of the anti-Hermitian term we assume that only one channel is open. Then the matrix W, Eq. (8), has a completely separable form,

$$\left\langle q\right|W\left|q'\right\rangle = 2\pi A_q^c A_{q'}^{c*}, \tag{10}$$

where:

$$A_q^c = \left\langle q\right|H_{QP}\left|c\right\rangle. \tag{11}$$

The rank of the matrix W is 1, so that all the eigenvalues of this matrix are zero except one that has the value equal to the trace of the matrix:

$$\Gamma_s = \sum_q \left\langle q\right|W\left|q\right\rangle = 2\pi\sum_q \left|A_q^c\right|^2 \equiv \sum_q \Gamma_q^{\uparrow} \tag{12}$$

with $\Gamma_q^{\uparrow}$ denoting the escape width of the individual levels before the W-matrix is diagonalized. The special unstable state with width Γ_s is often referred to as the *super-radiant* (SR), in analogy to the Dicke coherent state [1] of a set of two-level atoms coupled through the common radiation field. Here the coherence is generated by the *common decay channel.* The stable states are trapped and decoupled from the continuum.

In the more general case of N intrinsic states and N_c open channels with $N_c << N$ the super-radiant mechanism survives if the mean level spacing D of internal states and their decay widths $\Gamma_q^{\uparrow}$ satisfy:

$$\kappa^c = \frac{\Gamma_q^{\uparrow}}{D} > 1 \tag{13}$$

In this case we have N_c broad states, while the rest of states $N - N_c$ become very narrow.

3. Porter-Thomas Distribution and the Super-Radiant Mechanism.

We will now discuss an example that illustrates the use of the super-radiant approach to a complicated problem in nuclear physics. One the best examples in which the interplay between intrinsic dynamics and decay channels is manifested are the low-energy neutron resonances. These resonances are interpreted as quasi-stationary levels of the compound nucleus formed after the neuron is captured. These resonances were described, with certain degree of success, by the theory of random matrices [12]. With exceedingly complicated wave functions of the compound states their components are Gaussian distributed. This applies also to the specific component related to the channel, thus the neutron in the continuum and the residual nucleus in the ground state. The decay width of the neutron resonance is proportional to the amplitude squared of this component and the width distribution is therefore given by the χ_ν^2 function with $\nu = 1$, that is, the Porter-Thomas distribution (PTD). Recently experiments found significant departures from the PTD [13]. Attempts to still use the χ_ν^2 distribution in order to fit the data require $\nu < 1$. This result has been interpreted as a breakdown of the random matrix theory for the case when the components of compound states are the same as for a closed system. This might be valid only in the limit of very weak continuum coupling. The standard PTD does not account for the openness of the system. The openness of the system inevitably brings in the elements of coherence through the common decay channel even in the case of intrinsic chaos. This is expected naturally to introduce deviations from the PTD.

In a paper published recently [14] it was shown that a proper description of unstable quantum states using the super-radiant mechanism can in fact produce deviations from the PTD of the same type as observed in experiments. The coupling to open channels is essential in the redistribution of the decay widths of the compound states. The details of the calculations are described in reference [14]. When the strength of the coupling between compound states and the open channel(s) (denoted by κ) is very small ($\kappa << 1$) one finds that the distribution of the calculated widths follows the PTD. With increasing coupling the deviations become more pronounced. At $\kappa \sim 1$ the super-radiant stage is reached and the distribution of widths becomes singular. In Figures 1 and 2 some of the results from reference [14] are shown.

We see in Figure 1 that as κ increases, the fits to the numerical results involve χ_ν^2 with decreasing values of ν. In Figure 2 we present the deviations from the PTD curve when κ increases. The observed deviations from the PTD seem to exceed than the predicted by our new results for rather small values of the parameter κ. This might suggest that there are other factors related to nuclear structure that go beyond statistical considerations. The intrinsic dynamics can have some non-chaotic

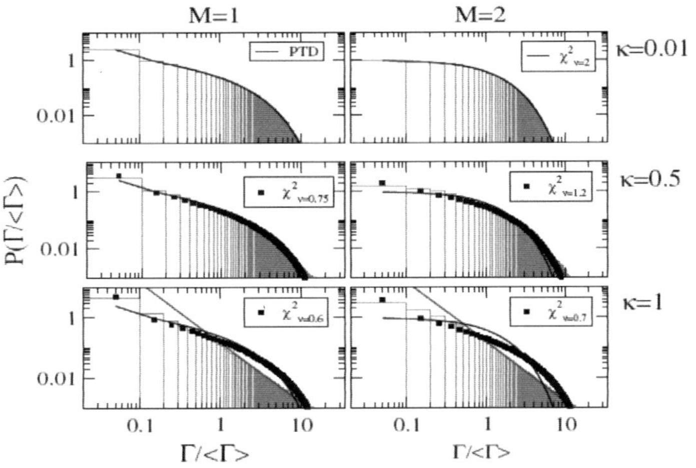

Figure 1. Distribution of widths for one open channel (M=1) and for two channels (M=2) for an intrinsic Gaussian Orthogonal Ensemble Hamiltonian and for different continuum coupling strength κ. The numerical results are given by histograms; the PTD for M=1 and the $\chi^2_{\nu=2}$ distribution for M=2 are shown as a smooth curve. The full squares stand for the best fit to a χ^2_ν distribution. Taken from Ref. [14].

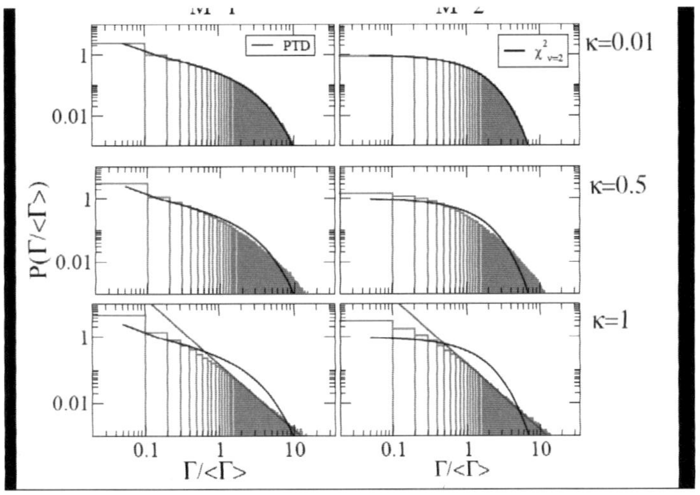

Figure 2. [illegible]istribution of widths for one open channel (M=1) and for two channels (M[illegible] for an intrinsic [illegible]ussian Orthogonal Ensemble Hamiltonian and for different continuum couplir[illegible] trength κ. The [illegible] ion for M=2 are shown as a smooth curve in each case indicating clearly the deviations from PTD (or $\chi^2_{\nu=2}$) as κ increases.

components, the energy dependence of neutron resonances might not be the one that is usually assumed (for s states $\sqrt{E}$), especially due to single-particle resonances in the vicinity of the energies studied in the experiment [15].

In summary, the interpretation of the widths of compound resonances as a strength of the pure neutron component in the wave function fails due to the coupling to the continuum that has to be taken into account in a proper statistical description. This phenomenon is of general nature and might influence many other quantum systems.

I wish to thank L.Celardo, F. Izrailev and V. Zelevinsky for their collaboration.

References

[1] Dicke R H 1954 *Phys. Rev.* **93** 99
[2] Pavlov-Verevkin V B 1988 *Phys. Lett.* A **129** 168
[3] Flambaum V V, Gribakina A A, Gribakin G F 1996 *Phys. Rev.* A **54** 2066
[4] Pichugin K, Schanz H, Šeba P, 2001 *Phys. Rev.* E **64** 056227
[5] Volya A and Zelevinsky V 2005 *Nuclei and Mesoscopic Physics* (Ed. Zelevinsky V), vol. 777 *(*AIP Conference Proceedings*)* p 229
[6] Auerbach N 1994 *Phys. Rev. C* **50** 1606
[7] Auerbach N, Zelevinsky V 2002 *Phys. Rev.* C **65** 034601
[8] Auerbach N, Zelevinsky V, Volya A 2004 *Phys. Lett.* B **590** 45
[9] Sokolov V V, Zelevinsky V 1989 *Nucl. Phys.* A **504** 562
[10] Rotter I 1991 *Rep. Prog. Phys.* **54** 635
[11] Feshbach H 1958 *Ann. Phys.* **5** 357; Feshbach H 1962 *Ann. Phys.* **19** 287
[12] Porter C E 1965 *Statistical Theories of Spectra: Fluctuations* (New York: Academic)
[13] Koehler P E, Becvar F, Krticka M, Harvey J and Gruber K H 2010 *Phys. Rev. Lett.* **105** 072502
[14] Celardo G L, Auerbach N, Izrailev F, and Zelevinsky V G 2011 *Phys. Rev. Lett.* **106** 042501
[15] Weidenmuller H A 2010 *Phys. Rev. Lett.* **105** 232501

HITES 2012 IOP Publishing
Journal of Physics: Conference Series **403** (2012) 012024 doi:10.1088/1742-6596/403/1/012024

Microscopic description of the elusive Hoyle state

Alison Dreyfuss[1], Kristina D Launey[2], Tomáš Dytrych[2], Jerry P Draayer[2], and Chairul Bahri[3]

[1]Keene State College, Keene, NH 03435, USA
[2]Department of Physics and Astronomy, Louisiana State University, Baton Rouge, LA 70803, USA
[3]Department of Physics, University of Notre Dame, Notre Dame, Indiana 46556-5670, USA

Abstract. Within a symmetry-guided shell-model framework and using a fraction of the model space extended beyond current no-core shell-model limits along with a schematic effective many-nucleon interaction, we gain new insights into the many-body dynamics that give rise to the ground state and low-lying 0^+ states of ^{12}C and ^{16}O. In particular, we gain further understanding of the alpha-clustering nature of the challenging Hoyle state and its first 2+ excitation in ^{12}C, as well as the corresponding states in ^{16}O. This provides guidance for *ab initio* shell models by informing key features of the underlying nuclear structure and interaction.

1. Introduction

^{12}C is produced in hot stars through a series of nuclear fusion reactions called the triple-alpha process. The triple-alpha process begins with the statistically unlikely collision and subsequent fusion of two alpha particles into a highly unstable ^{8}Be nucleus, which, in another highly improbable event, then fuses with a third alpha particle through collision into a ^{12}C nucleus. In order for the ^{12}C nucleus formed during this process to reach a stable state, there must exist a low-lying, spin-zero, even-parity state of ^{12}C, as postulated by Hoyle [1]. The Hoyle state and its first 2^+ excitation remain a difficult challenge for no-core shell model studies of ^{12}C.

2. NCSpM basis

The fully microscopic no-core symplectic shell model (NCSpM) uses a symplectic $Sp(3, \mathbb{R})$ basis and $Sp(3, \mathbb{R})$-preserving interactions. When applied to the full model space, the model coincides exactly with the NCSM for a given $N_{\rm max}$ cutoff, which is the maximum allowed harmonic oscillator (HO) excitations above the lowest energy configuration for a given nucleus. The symplectic irreps of NCSpM divide the basis space into 'vertical slices', each of which is comprised of basis states of a definite $(\lambda\,\mu)$ deformation. By considering only those symplectic configurations of all $Sp(3, \mathbb{R})$ irreps in the $N_{\rm max}$ model space that contribute significantly to the shape of the nucleus in question, we can greatly reduce the model space. With this reduction in model space, the NCSpM is able to provide shell model calculations beyond current NCSM limits; namely, up through $N_{\rm max} = 20$ for ^{12}C and $N_{\rm max} = 16$ for ^{16}O, the model spaces we found sufficient for the convergence of results.

3. NCSpM interaction

We use a very simple and elegant Hamiltonian with an effective interaction derived from the long-range expansion of the two-body central nuclear force:

$$H_{\text{eff}} = H_0 - \frac{\chi}{2}\frac{1}{\gamma}\left(e^{\gamma(Q.Q - \langle Q.Q\rangle_n)} - 1\right). \tag{1}$$

This includes both the spherical HO potential, which, together with the kinetic energy yields the HO Hamiltonian H_0, and the $Q.Q$ quadrupole-quadrupole interaction not restricted to a single shell. We remove the average contribution, $\langle Q.Q\rangle_n$, of $Q.Q$ within a subspace of n HO excitations [2], which helps eliminate the considerable renormalization of the zero-point energy, while retaining the $Q.Q$-driven behavior of the wavefunctions.

In its zeroth-order approximation (for parameter $\gamma \rightarrow 0$) for a valence shell, this Hamiltonian simplifies to the established Elliott model [3, 4, 5]. The coupling constant χ is taken to be proportional to $\hbar\Omega$, and to decrease, at leading order, with the total number of HO excitations, as shown by Rowe [6]. Hence, in the Hamiltonian (1), χ is fixed by the value of $\hbar\Omega$ (not varied) and the eigenstates are thus $\hbar\Omega$-independent.

Table 1. Experimental and theoretical NCSpM ^{12}C and ^{16}O energy spectra (E_{exp} and E_{theo} in MeV, respectively). The NCSpM calculations are performed for ^{12}C for $\gamma = -1.71 \times 10^{-4}$ and three Sp(3, $\mathbb{R}$) irreps, $0\hbar\Omega$ 0p-0h (0 4), $2\hbar\Omega$ 2p-2h (6 2), and $4\hbar\Omega$ 4p-4h (12 0), in the $N_{\max} = 20$ model space; and for ^{16}O for $\gamma = -0.50 \times 10^{-4}$ and three Sp(3, $\mathbb{R}$) irreps, $0\hbar\Omega$ 0p-0h (0 0), $2\hbar\Omega$ (4 2), and $4\hbar\Omega$ (8 4), in the $N_{\max} = 16$ model space. Experimental energies for all states are taken from Ref. [7] unless otherwise specified.

Sp(3, $\mathbb{R}$) irrep	J^π	E_{theo}, MeV	E_{exp}, MeV
		^{12}C	
(0 4)	0^+	0.0	0.0
	2^+	4.44	4.44 5
	4^+	14.74	14.08
(12 0)	0^+	7.49	7.65
	2^+	5.58	9.84 Ref. [8]
(6 2)	0^+	9.90	9.93 Ref. [8]
		^{16}O	
(0 0)	0^+	0.0	0.0
(8 4)	0^+	4.66	6.05
	2^+	6.38	6.92
	4^+	10.68	10.36
(4 2)	0^+	12.29	12.05

4. Results

Table 1 shows the NCSpM results in considerable agreement with the experimental values for low-lying states in both ^{12}C and ^{16}O. NCSpM results were calculated for ^{12}C using $\hbar\Omega = 18$ MeV given by the empirical estimate $\approx 41/A^{1/3} = 17.9$ MeV and $N_{\max} = 20$, which we found sufficient to yield convergence. The space was further reduced by selecting only the most relevant symplectic irreps: the spin-zero ($S = 0$) $0\hbar\Omega$ 0p-0h (0 4), $2\hbar\Omega$ 2p-2h (6 2), and $4\hbar\Omega$ 4p-4h (12 0) symplectic bandheads, and all multiples thereof up through $N_{\max} = 20$. Note that the NCSpM energies for ^{12}C given in Table 1 are rescaled by an overall factor of ~ 2, which was found by fixing the lowest 2^+ by its experimental value. This scaling factor has no implications on the underlying physics, as an overall factor for H does not change its properties and eigenstates, nor any associated observables.

The low-lying energy spectrum and eigenstates for ^{16}O were also calculated using the NCSpM with $\hbar\Omega = 16$ MeV (following the $41/A^{1/3}$ estimation). The $N_{\max} = 16$ model space used is further reduced by selecting the spin-zero ($S = 0$) $0\hbar\Omega$ (0 0), $2\hbar\Omega$ (4 2), and $4\hbar\Omega$ (8 4) symplectic bandheads, and all multiples thereof up through $N_{\max} = 16$. The values for ^{16}O in Table 1 are not rescaled.

These results indicate that the lowest 0^+, 2^+, and 4^+ states of the $0\hbar\Omega$ 0p-0h (0 4) symplectic irrep calculated for $\gamma = -1.71 \times 10^{-4}$ in ^{12}C closely reproduce the *g.st.* rotational band, the lowest 0^+ states of the $4\hbar\Omega$ 4p-4h (12 0) slice is found to lie close to the Hoyle state, and the lowest 0^+ state of the $2\hbar\Omega$ 2p-2h (6 2) slice lies close to the 10-MeV 0^+ resonance (third 0^+ state). Results for ^{16}O indicate that, relative to the lowest 0^+ state of the $0\hbar\Omega$ (0 0), the $4\hbar\Omega$ (8 4) slice reproduces a 0^+-2^+-4^+ rotational band over the second 0^+ state in ^{16}O, while the lowest 0^+ state of the $2\hbar\Omega$ (4 2) is found to lie very close to the third 0^+ state.

Table 2. Probability distribution for ^{12}C as a function of the n total excitations for the lowest 0^+ and 2^+ states as calculated by the NCSpM (independent of $\hbar\Omega$) and the SA-NCSM with the bare JISP16 interaction for $\hbar\Omega$=20 MeV and $N_{\max} = 8$.

n	NCSpM	SA-NCSM (with bare JISP16)
	$J = 0^+$	
0	64.49	69.25
2	18.80	17.24
4	11.11	9.37
6	3.69	2.93
8	1.31	0.84
	$J = 2^+$	
0	64.82	68.17
2	0.68	17.88
4	10.90	9.59
6	3.68	2.79
8	1.09	0.76

A comparison of the probability distributions for the *g.st.* rotational band as calculated using the NCSpM with *ab initio* results when only configurations of zero proton and neutron spins ($S_{\rm p,n} = 0$) are selected (for $0^+_{g.st.}$ and 2^+_1 for ^{12}C, see Table 2) shows encouraging similarities. In particular, we compare NCSpM eigenstates, which are $\hbar\Omega$-independent, to SA-NCSM calculations [9, 10] with the bare JISP16 realistic interaction for $\hbar\Omega = 20$ MeV (around the minimum of the calculated binding energy for ^{12}C) and a $N_{\max} = 8$ model space, which appears to be sufficient for convergence for the ^{12}C *g.st.* rotational band for both models. This agreement indicates that the schematic interaction used in NCSpM captures most of the underlying physics in the realistic interaction for low-energy nuclear dynamics.

The model also successfully reproduces a number of other observables in both ^{12}C and ^{16}O, including, for example, mass rms radii, electric quadrupole moments and $B(E2)$ transition strengths, all of which give additional information about the structure of the nuclei. For example, for ^{12}C, the $B(E2;\, 2_1^+ \to 0_{g.st.}^+)$ is calculated to be 5.12 W.u., close to the 4.65 W.u. experimental measurement.

While the model includes an adjustable parameter, γ, this parameter only controls the decrease rate of the $Q.Q$ interaction toward higher-lying shells. The many-body apparatus itself is fully microscopic and no adjustments are possible. Hence, as γ varies, there is only a small window of possible γ values that, for large enough $N_{\max}$, closely reproduces the relative positions of the three lowest 0^+ states (see Fig. 1 for ^{12}C). $B(E2)$ transition strengths are also found to be sensitive to the γ values, as shown in Fig. 1 for $B(E2;\, 2_1^+ \to 0_{g.st.}^+)$, and the small γ window indeed corresponds to $B(E2)$ values that are in agreement with experiment. In addition, Figure 1 shows the dependence of the energy (not rescaled) of the 2^+ and 4^+ states of (0 4) on γ. Notice that while the 2^+ and 4^+ energies are almost entirely independent of γ, the $B(E2)$ values vary within a range of ~ 5 W.u. A more dramatic variation with γ is observed for the 0^+ states of ^{12}C, e.g., the (12 0) 0^+ energy (corresponding to the Hoyle-state energy) and the (6 2) 0^+ energy vary within ~ 100 MeV and 50 MeV, respectively, for γ from -4×10^{-5} to -2×10^{-2}.

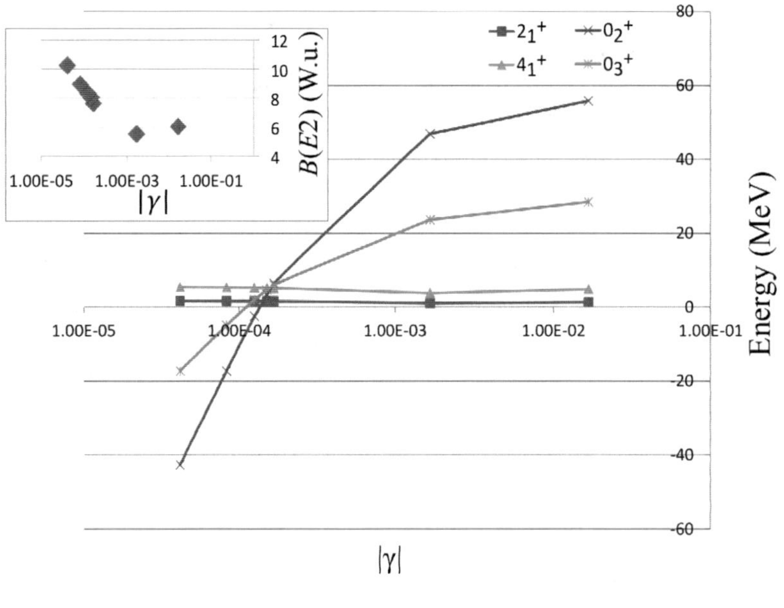

Figure 1. Dependence of the energy of the 2^+ and 4^+ states of (0 4) (labeled as 2_1^+ and 4_1^+), the 0^+ state of (12 0) (labeled as "0_2^+") and the 0^+ state of (6 2) (labeled as "0_3^+"), as well as of the $B(E2;\, 2_1^+ \to 0_{g.st.}^+)$ transition strength in W.u. (insert), on γ for ^{12}C.

5. Conclusion

Calculations using the symplectic no-core shell model within a reduced model space and employing a schematic effective many-nucleon interaction were presented. We showed that this symmetry-guided framework is not limited to a specific nucleus but effectively describes low-lying eigenstates, e.g., of ^{12}C and ^{16}O. This model can be used as a guide to help *ab initio* models move toward heavier nuclear systems, and points to important underlying collective and cluster phenomena in nuclei.

Acknowledgments
This work was supported by the U.S. NSF (0904874), the U.S. DOE (DE-SC0005248 & FG02-95ER-40934), and SURA. ACD further acknowledges support by the U.S. NSF (1004822) through the REU Site in the Department of Physics & Astronomy at LSU. We also acknowledge DOE/NERSC and LSU/LONI for providing HPC resources.

References

[1] Hoyle F 1954 *Astrophys. J.* **1** 121
[2] Rosensteel G and Draayer J P 1985 *Nucl. Phys. A* **436** 445
[3] Elliott J P 1958 *Proc. Roy. Sot. A* **245** 128
[4] Elliott J P 1958 *Proc. Roy. Sot. A* **245** 562
[5] Elliott J P and Harvey M 1962 *Proc. Roy. Sot. A* **272** 557
[6] Rowe D J 1967 *Phys. Rev.* **162** 866
[7] Ajzenberg-Selove F 1990 *Nucl. Phys. A* **506** 1
[8] Itoh M et al. 2011 *Phys. Rev. C* **84** 054308
[9] Dytrych T et al. 2012 to be submitted to Phys. Rev. Lett.
[10] Draayer J P, Dytrych T, Launey K D, and Langr D 2012 *Prog. Part. Nucl. Phys.* **67** 516

Nuclear Reactions: A Challenge for Few- and Many-Body Theory

Ch Elster, L Hlophe

Department of Physics and Astronomy and Institute of Nuclear and Particle Physics, Ohio University, Athens, OH 45701, USA

E-mail: elster@ohio.edu

Abstract. A current interest in nuclear reactions, specifically with rare isotopes concentrates on their reaction with neutrons, in particular neutron capture. In order to facilitate reactions with neutrons one must use indirect methods using deuterons as beam or target of choice. For adding neutrons, the most common reaction is the (d,p) reaction, in which the deuteron breaks up and the neutron is captured by the nucleus. Those (d,p) reactions may be viewed as a three-body problem in a many-body context. This contribution reports on a feasibility study for describing phenomenological nucleon-nucleus optical potentials in momentum space in a separable form, so that they may be used for Faddeev calculations of (d,p) reactions.

1. Introduction

Today the non-relativistic nuclear three-body system can numerically be solved with high precision in momentum [1] as well as coordinate space [2], including three-nucleon forces. Beyond this, three-body models can play an important role in the description of many-body problems.

Since the full many-nucleon problem presently can only be solved for light nuclei [3, 4, 5, 6], a common approach to treating reactions involving heavier nuclei is concentrate on fewer, relevant degrees of freedom and employ few-body methods. Such an approach is mostly justified for direct reactions, and may be expected to be reasonable if either the projectile fails to significantly penetrate the nucleus (i.e. it is peripheral) or if it penetrates, it moves very fast with respect to the nucleons bound in the nucleus. When a reaction is peripheral, it can be expected that only very few degrees of freedom of the nucleus may be excited due to the relatively low density at the nuclear surface. However, one can argue that in case of attractive optical potentials a reaction will also be mostly peripheral. The projectile has a higher kinetic energy and a shorter wave length inside the nucleus, so that increased oscillations of the wave function inside the nucleus tend do suppress the contribution of the nuclear interior.

2. (d,p) Reactions as Three-Body Problem

Scattering and reaction processes induced by deuterons as projectile are perhaps the most natural three-body problem in the realm of nuclear reactions. The binding energy of the deuteron is so small that its root-mean-square radius (rms) is significantly larger than the range of the nuclear force, meaning that most of the time the neutron and the proton inside the deuteron can be viewed as being outside the range of the interaction. Thus, when a deuteron interacts with a

Journal of Physics: Conference Series **403** (2012) 012025 doi:10.1088/1742-6596/403/1/012025

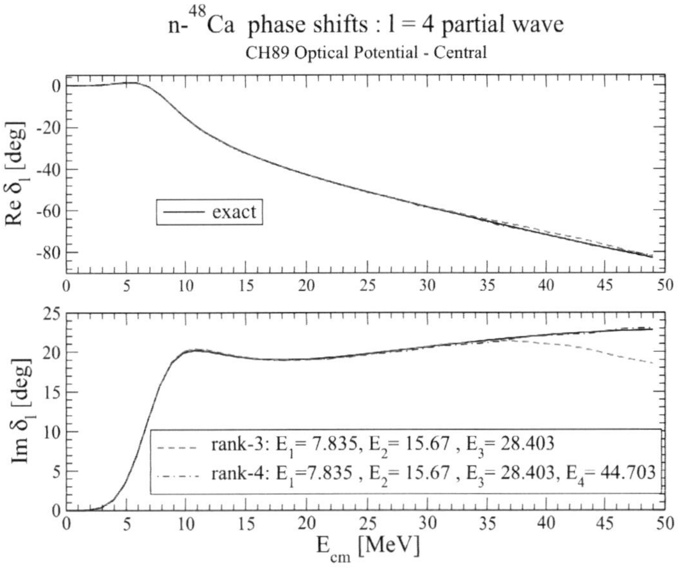

Figure 1. The $l = 4$ phase shift calculated from the central part of the CH89 optical potential [8] for n+^{48}Ca (solid line) together with a rank-3 (dashed line) and rank-4 (dash-dotted line) separable representation as function of the c.m. energy. The support points are c.m. energies.

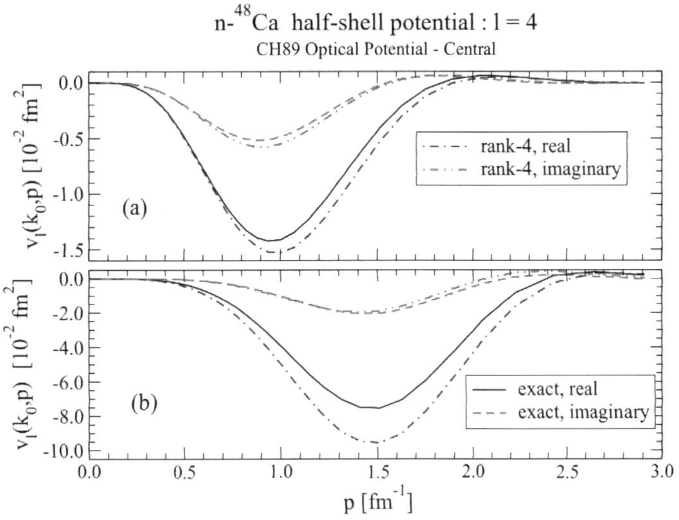

Figure 2. The half-shell potentials $v_{l=4}(k_0, p)$ corresponding to the CH89 central optical potential (exact) together with the rank-4 separable representation at fixed momentum k_0 corresponding to $E_{c.m.} = 4.897$ MeV (a) and $E_{c.m.} = 48.97$ MeV (b).

compact nucleus, one may expect that it will behave like a three-body system consisting of a proton p, a neutron n, and a nucleus A.

The most obvious three-body effects are rearrangement (stripping followed by pickup) and breakup processes. In order to treat those processes as well as elastic deuteron scattering on the same footing, deuteron-nucleus scattering should be treated as three-body problem making assuming that the system can be described by a three-body Hamiltonian. It should be noted that the extraction of an effective three-body Hamiltonian from the many-body problem is non-trivial. A common approach is using as forces between the neutron and proton (both interacting via the nucleon-nucleon force) neutron and proton-nucleus optical potentials, which in turn are fitted to a large body of elastic scattering data (e.g. [7, 8]). Since these optical potentials are in general energy dependent, a common procedure is to take the nucleon-nucleus optical potentials at half the deuteron incident energy. With this Hamiltonian Faddeev type equations for (d,p) reactions can be set up and solved [9]. Treating nuclear reactions within a Faddeev three-body formulation is at the forefront of developments in nuclear reactions. Existing formulations need to be extended, since the original Faddeev description of a three-body problem does not consider possible internal excitations of the particles active in the scattering process. In a (d,p) reaction with a nucleus, it is definitely possible that the nucleus can be excited during the scattering process. When considering reactions involving exotic nuclei, this may become even more urgent when the exotic nuclei under consideration are loosely bound.

A Faddeev formulation of (d,p) reactions including excitations of the nuclear target has been proposed in Refs. [10, 11]. Solving Faddeev equations is most conveniently carried out in momentum space as coupled integral equations, since those automatically contain the boundary conditions. Thus all interactions in the three-body Hamiltonian need to be given in momentum space. Specifically, a momentum space Faddeev formulation requires as input from the sub-systems, which are here either the deuteron of the neutron-nucleus or proton-nucleus pair, two-body transition matrices (i.e. the interactions summed to all orders) for those sub-systems. It is well known that the numerical effort in solving momentum space Faddeev integral equations is reduced when separable representations of the sub-system transition amplitudes are used. In the case of exited states of the target, separable interactions have the additional important feature that for any resonance as well as for a bound state any two-body transition amplitude is separable. For this reason the formulation in Ref. [10] is based on using separable transition amplitudes for the sub-systems.

3. Separabilization of Nucleon-Nucleus Potentials

Separable representations of nucleon-nucleon (NN) t-matrices are available since the 1980s (see e.g. Refs. [12, 13]). However, as far as neutron or proton-nucleus optical potentials are concerned, there only exist very few separable potentials [11, 14] for light nuclei, which are based on Yamaguchi-type form factors. Those form factors may not be well suited for parameterizing optical potentials for heavier nuclei, for which excellent phenomenological descriptions in terms of Wood-Saxon functions exist (see e.g. [8, 7]). Considering the variety of nuclei for which there is or will be experimental information available from (d,p) reactions, one will need a separabilization procedure that is sufficiently general so that it can be applied to a variety of nucleon-nucleus optical potentials over a wide range of nuclei. In addition, transition amplitudes calculated from phenomenological optical potentials need to be well represented over a wider range of energies.

The separable representation of two-body interactions suggested by Ernst-Shakin-Thaler [15] (EST) looks well suited to achieve this goal. The basic idea of the EST separabilization of a two-body transition amplitude is that one picks a fixed number of energies E_{k_E} for which the transition amplitude is exact, and all other energies are described by interpolation between those points E_{k_E} by the EST separable transition amplitude in which the half-shell amplitudes

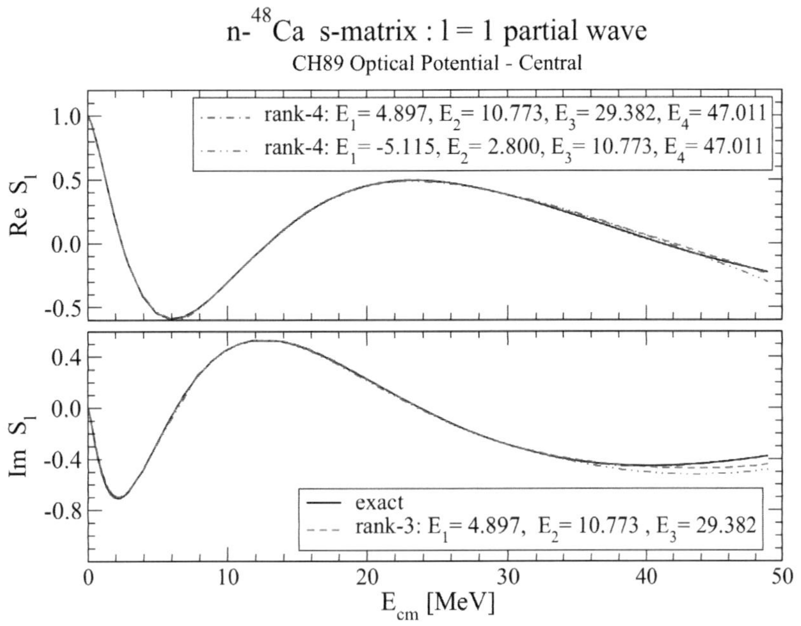

Figure 3. The real (upper panel) and imaginary (lower panel) partial wave s-matrix for $l = 1$ calculated from the central part of the CH89 optical potential [8] for n+^{48}Ca (solid line) as function the c.m. energy. A rank-3 (dashed) and two different rank-4 (dash-dotted and dash-double-dotted) are compared to the exact calculation. The energies of the support points are c.m. energies.

evaluated at E_{k_E} serve as form-factors. The number of energy points E_{k_E} determines the rank of the separable t-matrix. The description of the original t-matrix can in principle be made successively more accurate with increasing support points E_{k_E}.

Though the EST scheme was extensively used to represent NN t-matrices [12, 13], it has to our knowledge not yet been applied to optical potentials for nucleon-nucleus scattering. We report here on a study in which we use the EST scheme to represent the partial wave projected t-matrices for neutron scattering from ^{48}Ca based on the Chapel-Hill CH89 phenomenological global optical potential [8]. For this study we concentrate on the central part of the optical potential and neglect the spin-orbit terms.

According to Ref. [15] a separable potential of arbitrary rank is given as

$$\mathbf{V} = \sum_{i,j} v|\Psi_i\rangle\langle\Psi_i|M|\Psi_j\rangle\langle\Psi_j|v, \tag{1}$$

where $|\Psi_i\rangle$ stands for either a bound-state $|\Psi_B\rangle$ or a scattering state $|\Phi^{(+)}_{k_E}\rangle$. If only bound states are considered, the standard unitary pole approximation (UPA) is recovered. The matrix M is defined and constrained by

$$\delta_{ik} = \sum_j \langle\Psi_i|M|\Psi_j\rangle\langle\Psi_j|v|\Psi_k\rangle = \sum_j \langle\Psi_i|v|\Psi_j\rangle\langle\Psi_j|M|\Psi_k\rangle. \tag{2}$$

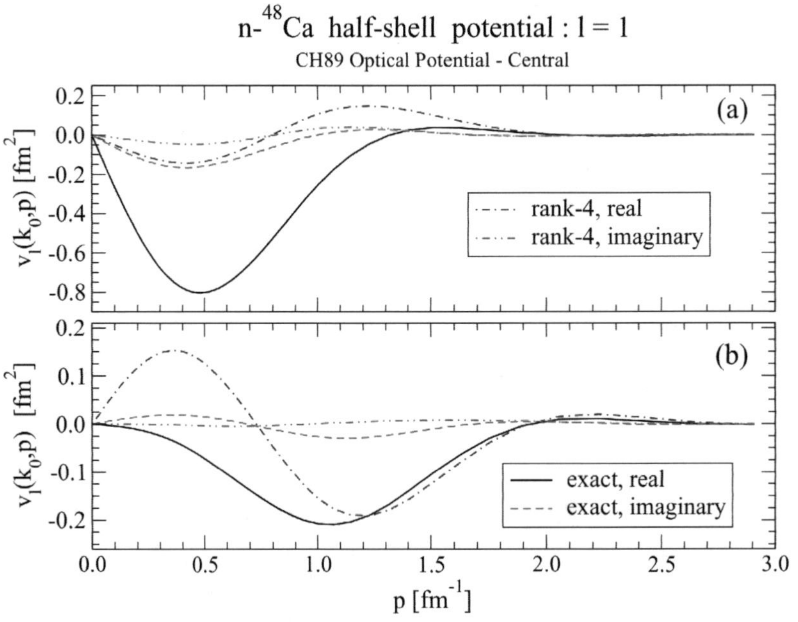

Figure 4. The real (solid) and imaginary (dashed) partial wave projected half-shell CH89 [8] optical potentials for n+^{48}Ca scattering are shown for fixed on-shell momentum k_0 as function of the half-shell momentum p. In (a) the on-shell momentum k_0 corresponds to $E_{c.m.} = 4.897$ MeV, and in (b) to $E_{c.m.} = 48.97$ MeV. A potential leading to a rank-4 separable representation of the corresponding s-matrix is shown by the dashed-dotted line (real part) and dashed-double-dotted line (imaginary part).

In order for the t-matrix elements to have the same structure as the Born term of the Lippmann-Schwinger integral equation, they must be of the form

$$\hat{t}(E) = \sum_{i,j} v|\Psi_i\rangle \tau_{ij}(E) \langle \Psi_j| v \; . \tag{3}$$

It can be shown [15] that the quantities $\tau_{ij}(E)$ fulfill

$$\sum_j \tau_{ij}(E) \; \langle \Psi_j | v - v g_0(E) v | \Psi_k \rangle = \delta_{ik} \; , \tag{4}$$

and thus can be calculated by inverting the matrix $\langle \Psi_j | v - v g_0(E) v | \Psi_k \rangle$ for given $|\Psi_{j,k}\rangle$, so that finally the momentum space t-matrix elements are given as

$$\hat{t}(p, p', E) = \sum_{ij} \langle p | v_l | \psi_i \rangle \tau_{ij}(E) \langle \psi_j | v_l | p' \rangle \tag{5}$$

The central part of the CH89 optical potential for n+^{48}Ca scattering is attractive and supports either one ($l = 2, 3$) or two ($l = 0, 1$) bound states in the lower partial waves. Partial wave beyond $l = 4$ do not support bound states. For those partial waves the EST support points at with the exactly calculated half-shell t-matrix is used to construct the form factors for the separable expansion are all chosen to be at positive c.m. energies. In Fig. 1 the real and imaginary part of the $l = 4$ phase shift are shown for a rank-3 and a rank-4 approximation and compared to the exactly calculated phase shift. The most prominent structure of the phase shift occurs roughly between 5 and 15 MeV c.m. energy. Thus two support points are needed to fit this structure and one additional at around 30 MeV to describe the flat behavior in that region. If one is interested to describe the phase shift at higher energies, then additional support points can be added without problem.

In Fig. 2 the corresponding half-shell potentials $v_{l=4}(k_0, p)$, calculated using Eq. (1), are shown for two different values of k_0 and compared to the original CH89 optical potential. For the upper panel (a) k_0 is fixed corresponding to $E_{c.m.} = 4.897$ MeV and for the lower panel (b) to $E_{c.m.} = 48.97$ MeV. Since the CH89 optical potential is energy dependent, so are the separable expansions. In the latter, the form factors are energy dependent, which is formally allowed [16]. The original, local Wood-Saxon potential and the non-local rank-4 separable representation are amazingly similar half-shell.

As example for a lower partial wave we consider $l = 1$. This partial wave supports two bound states in its Wood-Saxon well, a ground state at $E_{B_1} = 28.3$ MeV and an excited state at $E_{B_2} = 5.1$ MeV. For a (d,p) reaction only the latter is relevant, since a neutron may be captured into this state. In Fig. 3 the partial wave $l = 1$ s-matrix calculated from the CH89 cental optical potential is shown together with a selection of separable representations. If one is only interested in a representation of the s-matrix for scattering energies, then one can proceed in exactly the same fashion as in the higher partial waves that do not support bound states. A rank-3 and a rank-4 separable representation constructed in this fashion are shown in Fig. 3 as dashed and dash-dotted lines. The rank-4 expansion describes the s-matrix essentially perfectly up to 50 MeV. In a three-body calculation, which takes into account the capture of a neutron into the $l = 1$ bound state close to threshold, the pole structure given by this bound state must be taken into account exactly. In a separable expansion this can be easily accomplished by using the bound state wave function as form factor (as done in UPA approximations) for one of the expansion terms. The dash-double-dotted line in Fig. 3 represents a separable expansion in which one of the terms contains the bound state wave function of the excited state as form factor. Obviously, the positive energy EST support points need to be readjusted. Their values are indicated in Fig. 3. We also tested if the description of the s-matrix depends on introducing the ground state wave function as an additional form factor. However, since its energy is quite far below threshold, the effect on the separable representation for positive energies is negligible.

In Fig. 4 the corresponding partial wave half-shell potentials $v_{l=1}(k_0, p)$ are shown for the central, local CH89 optical potential as function of the half-shell momentum together a non-local, separable rank-4 potential obtained via Eq. (1). As in Fig. 2, an on-shell point k_0 corresponding to $E_{c.m.} = 4.897$ MeV is chosen for panel (a) and to $E_{c.m.} = 48.97$ MeV for panel (b). In this case the both half-shell potentials differ considerably from each other, despite describing the same s-matrix elements. The phenomenon that potentials may show considerable off-shell differences but still lead to very similar t-matrices (and thus s-matrices) has been studied for various NN potentials [17]. Since the information from two-body subsystems which enters Faddeev calculations only enters via t-matrices, differences in potentials do not matter. If the fully off-shell t-matrices show substantial differences in momentum regions important to a Faddeev calculations will need to be investigated further.

4. Summary and Outlook

For describing all aspects of (d,p) reactions on nuclei a three-body Faddeev formulation is currently the most desirable ansatz, since in a Faddeev formulation elastic scattering, breakup and rearrangement processes are treated on the same footing. Considerable progress has been made in solving Faddeev equations in momentum space for (d,p) reactions [9], however there are still major open issues. One of them is the inclusion of target excitations, i.e. the intrinsic excitation of one of the three-body constituents, which is not inherent in the original Faddeev formulation for three-body scattering. A formulation for (d,p) reactions including target excitations has recently been proposed in Ref. [10]. which is based on a separable representation of the transition amplitudes of the two-body sub-systems. While separable representations for the NN sub-system have been developed some time ago, for the neutron- and proton-nucleus sub-system corresponding representations do not exist. In a feasibility study based on neutron scattering off ^{48}Ca and using only the central piece of a Wood-Saxon based phenomenological optical potential fit [8], we show that for c.m. energies below 50 MeV the s-matrix can be well represented by rank-4 separable transition amplitudes. Future work will include a closer inspection of the fully off-shell behavior of the transition amplitudes derived from local optical potentials compared to their separable representations in momentum regions relevant for solving (d,p) Faddeev equations as well as considering representations for the proton-nucleus optical potential.

Acknowledgments
This work was performed in part under the auspices of the U. S. Department of Energy under contract No. DE-FG02-93ER40756 with Ohio University and under contract No. DE-SC0004084 (TORUS Collaboration).

References

[1] Glöckle W, et. al. 1996 *Phys. Rep.* **274** 107
[2] Barletta P, Kievsky A 2009 *Few Body Syst.* **45** 25
[3] Pieper S C, Varga K, and Wiringa R B 2002, *Phys. Rev. C* **66** 044310
[4] Pieper S C 2008 arXiv:0711.1500 and *Riv. Nuovo Cim.* **31** 709
[5] Navratil P and Quaglioni S 2011 *Phys. Rev. C* **83** 044609
[6] Navratil P, Roth R, and Quaglioni S 2010 *Phys. Rev. C* **82** 034609
[7] Weppner S P, Penney R B, Diffendale G W, and Vittorini G 2009 *Phys. Rev. C* **80** 034608
[8] Varner R L, Thompson W J, McAbee T L, Ludwig E J, Clegg T B 1991 *Phys. Rept.* **201** 57
[9] Deltuva A and Fonseca A C 2009 *Phys. Rev. C* **79** 014606
[10] Mukhamedzhanov A M, Eremenko V, Sattarov A I 2012 *Phys. Rev. C* **86** 034001
[11] Alt E O, Blokhintsev L D, Mukhamedzhanov A M, Sattarov A I 2007 *Phys. Rev. C* **75** 054003
[12] Haidenbauer J and Plessas W 1984 *Phys. Rev. C* **30** 1822
[13] Haidenbauer J , Koike Y and Plessas W 1986 *Phys. Rev. C* **33** 439
[14] Miyagawa K and Koike Y 1989 *Prog. Theor. Phys.* **82** 329
[15] Ernst D J , Shakin C M and Thaler R M 1973 *Phys. Rev. C* **8** 46
[16] Ernst D J , Londergan J T , Moniz E J and Thaler R M 1974 *Phys. Rev. C* **10** 1708
[17] Redish E F and Stricker-Bauer K 1987 *Phys. Rev. C* **36** 513

Exploring R-matrix ideas for the description of one-nucleon transfer reactions

J E Escher[1], **A M Mukhamedzhanov**[2] **and I J Thompson**[1]
[1]Lawrence Livermore National Laboratory, Livermore, CA 94551, U.S.A.
[2]Cyclotron Institute, Texas A&M University, College Station, TX 77843, U.S.A.

E-mail: escher1@llnl.gov

Abstract. Deuteron-induced reactions, in particular (d,p) one-neutron transfer reactions, have been used for decades to investigate the structure of nuclei. These reactions, carried out in inverse kinematics, are expected to play a central role in the study of weakly-bound systems at modern radioactive beam facilities. While the theoretical framework and its computational implementation for describing (d,p) reactions have seen much progress over the decades, open questions remain and need to be addressed, including the proper treatment of transfers to resonance states. Recently, a new formalism was proposed [1] that, in principle, describes transfers to both bound and resonance states. The new formalism is summarized here and illustrated; implications are discussed.

1. Introduction

The (d,p) reaction provides the simplest mechanism for creating a more neutron-rich isotope in specific final states by adding a neutron to a target nucleus. The reaction has been extensively used to study the properties of single-particle orbitals: their spins and parities, their energies and evolutions with neutron number. With the move to radioactive beam facilities, it becomes possible to study a much wider range of isotopes, in particular those short-lived species that will provide new insights into the nuclear force and that play an important role in astrophysical environments. As experimental techniques have been greatly improved over the decades, the theoretical descriptions of the reactions have also evolved. Butler's simple theory has made way to distorted-wave descriptions, coupled-channels treatments have been introduced to account for processes beyond the one-step mechanisms, and state-of-the-art approaches include contributions from the breakup of the deuteron [2, 3, 4]. More work is needed, though, to ensure that proper techniques are available and tested for applications in the new neutron and proton-rich regimes of interest.

An area in need of improvement is the treatment of reactions that populate resonance states. Resonances, long-lived but unstable states, occur in light, medium, and heavy nuclear systems. Their properties, in particular their energies and widths, are of great interest to nuclear astrophysics, as many reaction that occur in astrophysical environments proceed through resonances. Furthermore, resonances provide both challenges and stringent tests for nuclear structure models, as predicting their properties involves a proper treatment of the nuclear many-body system including the continuum. Current theories of transfer reactions that populate resonance states suffer from two major shortcomings. On the practical side, one has to deal

Published under licence by IOP Publishing Ltd

with convergence issues, as the matrix element that describes the transition to the final state includes contributions from very large distances outside the nucleus. Conceptually, it is also not clear what spectroscopic information can be extracted from a transfer to a resonance; a solid connection between experimental observables and typical resonance parameters remains to be established. Recent formal work by Mukhamedzhanov [1] has laid the foundation for going beyond these limitations. In Ref. [1], cross sections for (d,p) transfer reactions are expressed in terms of parameters that are typically employed in R-matrix theory. R-matrix theory is a standard tool for extracting resonance properties such as energies and widths from nucleon capture and scattering experiments [5, 6], and the new approach establishes a similar link between resonance properties and transfer reactions.

In the next section, we briefly summarize the formalism developed in Ref. [1], in particular the introduction of 'exterior' and 'interior' contributions to the transfer matrix element and the emergence of a surface term. In Section 3, we will illustrate the underlying ideas with examples and discuss the implications.

2. R-matrix concepts in binary and transfer reactions

Experimentally, resonance structures are most often studied in elastic and inelastic scattering reactions. For almost all nuclei resonance properties cannot be theoretically predicted, but the phenomenological R-matrix approach has been extremely useful for the interpretation of experiments and for extracting resonance energies and widths from measured cross sections [5, 6]. An extension of the approach, applicable to (d,p) transfer reactions, was recently developed [1] with the goal to provide a practical and sound way for extracting structure information from transfer experiments.

2.1. R-matrix approach to binary reactions

In the R-matrix approach for a binary reaction, the configuration space is separated into two regions: an exterior region, in which the reaction partners are well-separated and interact only via the Coulomb interaction (for charged particles), and an interior region, where a confined compound system exists that is governed by the nuclear and Coulomb forces. Formally, the nuclear wave function in the interior is expanded in some suitable set of basis functions, while in the exterior, it takes the form of a scattering function, and matching conditions are imposed at the surface. In typical applications, the parameters are adjusted to reproduce measured cross sections. This approach makes it possible to parameterize the collision matrix, and thus the calculated cross section, in terms of a few formal R-matrix parameters, which can then be related to observed quantities.

2.2. R-matrix ideas in the description of (d,p) reactions

Even though R-matrix theory [5] is typically employed in the description of binary reactions, such as elastic or inelastic scattering, or capture reactions, the underlying ideas can be applied more generally. For (d,p) reactions, an extension of the approach was worked out by Mukhamedzhanov [1], who introduced a surface that defines interior and exterior regions. The separation into the different regions is based on the distance r_{nA} of the deposited neutron from the center of the target nucleus, in analogy to what is done in descriptions of binary reactions. The standard distorted-wave Born Approximation (DWBA) transition matrix element can be written in post or prior form:

$$M^{(post)} = \left\langle \varphi_F \chi_{pF}^{(-)} | \Delta V_{pF} | \varphi_d \varphi_A \chi_{dA}^{(+)} \right\rangle = \left\langle I_A^F \chi_{pF}^{(-)} | \Delta V_{pF} | \varphi_d \chi_{dA}^{(+)} \right\rangle \quad (1)$$

$$M^{(prior)} = \left\langle \varphi_F \chi_{pF}^{(-)} | \Delta V_{dA} | \varphi_d \varphi_A \chi_{dA}^{(+)} \right\rangle = \left\langle I_A^F \chi_{pF}^{(-)} | \Delta V_{dA} | \varphi_d \chi_{dA}^{(+)} \right\rangle \quad (2)$$

where φ_A and φ_F denote the wave functions of the initial (A) and final ($F = A + n$) nuclei, respectively, and $I_A^F(r_{nA}) = \langle \varphi_A | \varphi_F \rangle$ is the associated overlap function, which depends on the coordinate of interest, r_{nA}. The distorted waves in the entrance and exit channels are given by χ_{dA} and χ_{pF}, respectively, and φ_d is the deuteron wave function. Post and prior forms require different transition operators

$$\Delta V_{pF} = V_{pn} + V_{pA} - U_{pF}, \text{ (post)} \tag{3}$$

$$\Delta V_{dA} = V_{pA} + V_{nA} - U_{dA}, \text{ (prior)} \tag{4}$$

respectively. In the post form, ΔV_{pF} contains the interaction between the proton and the target nucleus (V_{pA}), the proton-neutron interaction (V_{pn}), and the optical potential for the exit channel (U_{pF}), while in the prior form, ΔV_{dA} contains the interaction between the neutron and the target (V_{nA}), the proton-neutron interaction (V_{pn}), and the optical potential in the entrance channel (U_{dA}). In principle, post and prior formulations give numerically the same matrix element, but in practice small differences arise from ambiguities in the choice of the potentials used and the introduction of additional approximations.

In both formulations, the transition matrix element can be separated into two parts,

$$M^{(DWBA)} = M_{int}(0, R_{cut}) + M_{ext}(R_{cut}, \infty), \tag{5}$$

where R_{cut} refers to the specific value chosen for the coordinate r_{nA}, for which to carry out the separation. The interior term involves an integration from $r_{nA} = 0$ to $r_{nA} = R_{cut}$, while the exterior term involves an integration from $r_{nA} = R_{cut}$ to very large radii, $r_{nA} \to \infty$. Below, in Section 3, we will illustrate that for typical (d,p) transfer reactions and cutoff radii roughly equal to the size of the target nucleus, the main contributions to the post form matrix element come from the exterior region, while the prior form matrix element is dominated by interior contributions. This is significant, as in Ref. [1], it was further demonstrated that the post-form exterior matrix element can be expressed in terms of a surface term plus a prior-form exterior matrix element, so that the final expression for the matrix element is:

$$M^{(DWBA)} = M_{int}^{(post)}(0, R_{cut}) + M_{surf}(R_{cut}) + M_{ext}^{(prior)}(R_{cut}, \infty). \tag{6}$$

The same result is obtained if one starts with a consideration of the prior form and expresses the interior part in terms of a surface integral and remaining contributions from the interior. A dominant surface matrix element would provide several advantages: a) It would reduce the dependence of the cross section calculations on the model used for the interior portion of the overlap function; b) it would reduce the impact of the slow convergence of calculations of the exterior term when resonances are considered; and c) it would establish a useful link between resonance properties and transfer observables, since the surface term $M_{surf}(R_{cut})$, which is evaluated at a specific radius (R_{cut}), can be parameterized in terms of quantities that are familiar from traditional R-matrix approaches, namely a channel radius (here the cutoff radius R_{cut}), logarithmic boundary conditions (here logarithmic derivatives of known Hankel functions), and reduced-width amplitudes (here related to the asymptotic normalization of the overlap function). The contributions to typical cross sections arising from the three different terms in Eq. 6 will be discussed in Section 3 below.

3. Insights and implications

The separation into interior and exterior regions, and subsequent introduction of a surface term, can lead to improved descriptions of (d,p) reactions, in particular in cases where the final state is a resonance.

3.1. Interior and exterior contributions to the transition matrix element

Transfer experiments are often carried out under conditions that presumably make the reaction peripheral, *i.e.* they probe the overlap function in the region of the nuclear surface. The primary contributions to the transfer matrix elements should therefore come from r_{nA} values that are roughly equal to the size of the target nucleus. Figure 1, however, shows that the importance of a particular nuclear region depends on the form (post or prior) chosen. Shown are the contributions to the transfer cross section arising from the interior term in Eq. 5 only. Both post and prior forms are considered. The calculated peak cross section, normalized to the peak cross section obtained in the full calculation, is given as a function of the cutoff radius R_{cut}. We observe that the contributions for radii below 4 fm are small. In particular in the post formalism, the integration has to be carried out to fairly large radii in order to account for the majority of the cross section. This illustrates that the post form is more sensitive to the overlap function at larger radii (*i.e.* dominated by exterior contributions), while the prior formalism is more sensitive to the overlap at smaller radii (*i.e.* dominated by interior contributions). The reason for this can be found in the difference in the structure of the transition operators, Eqs. 3 and 4. Of course, as long as no further approximations are made, and both interior and exterior contributions are taken into account, both forms give the same result. Additional calculations, for (d,p) reactions on other nuclei (C, Ca, Pb), with various deuteron beam energies, give similar results. The findings also apply to reactions that populate resonances, which are found to be sensitive to larger values of r_{nA}, *i.e.* more peripheral regions of the nucleus. The interior contributions from the post-form matrix element were found to be small for all cases considered.

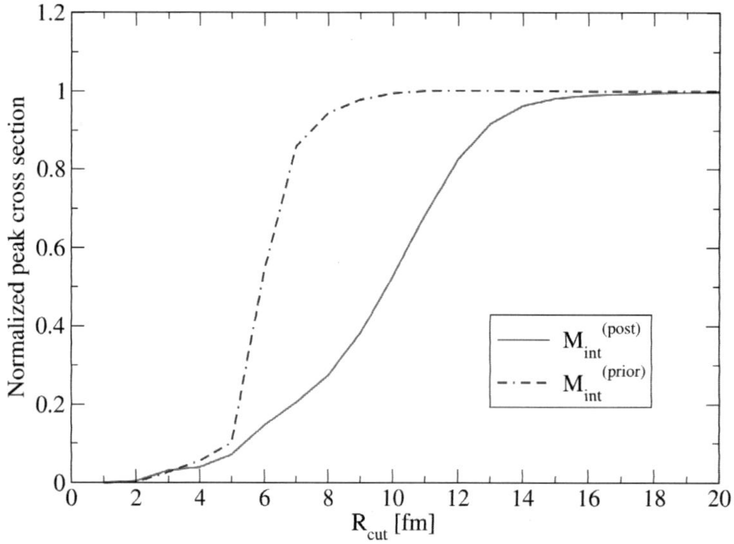

Figure 1. Examination of the role of interior contributions, for the ^{90}Zr(d,p) transfer reaction to the $5/2^+$ ground state in ^{91}Zr. Shown is the peak cross section obtained by including only the interior term (M_{int}) in the transition matrix element, as a function of the cutoff radius R_{cut}. The cross sections are normalized relative to the peak cross sections obtained in the full calculation. Results are given for both the post and prior formalisms.

3.2. Role of the surface term

The smallness of the interior contribution to post-form transition matrix element supports the introduction of the surface term, as shown in Eq. 6. The replacement $M_{ext}^{(post)}(R_{cut},\infty) \rightarrow M_{surf}(R_{cut}) + M_{ext}^{(prior)}(R_{cut},\infty)$ is illustrated in Figure 2. Shown are the contributions to the cross section arising from the three terms in Eq. 6, as function of the surface radius R_{cut}. A region (6-9 fm) can be identified in which the contributions from the interior post and exterior prior terms are (relatively) small and the surface term dominates. Also shown, for reference purposes, is the exterior post contribution that was replaced (dotted light-colored curve). Similar results have been obtained for other target nuclei and bombarding energies. Reactions leading to ground states, low-lying excited states, as well as resonances have been considered, with similar findings. For resonance reactions, it was observed that the surface term tends to be peaked at larger radii, *i.e.* transfers to resonance states are more peripheral.

While the surface contribution shown in Fig. 2 is dominant around 7-8 fm, it can be observed that there are non-negligible contributions from both the interior (post) and exterior (prior) terms. It was shown in Ref. [1] that the exterior prior term does no longer appear when deuteron breakup effects are explicitly included via the continuum-discretized coupled-channels (CDCC) formalism [3], *i.e.* $M^{CDCC} = M_{int}^{CDCC}(0, R_{cut}) + M_{surf}^{CDCC}(R_{cut})$. The interior post contribution, on the other hand, requires further consideration. In the DWBA approach, this contribution can be minimized by appropriately optimizing the potentials in the transition operator, Eq. 3. This has to be done in a manner consistent both with the spirit of the DWBA and with the determination of the optical-model wave function. This matter is presently under investigation.

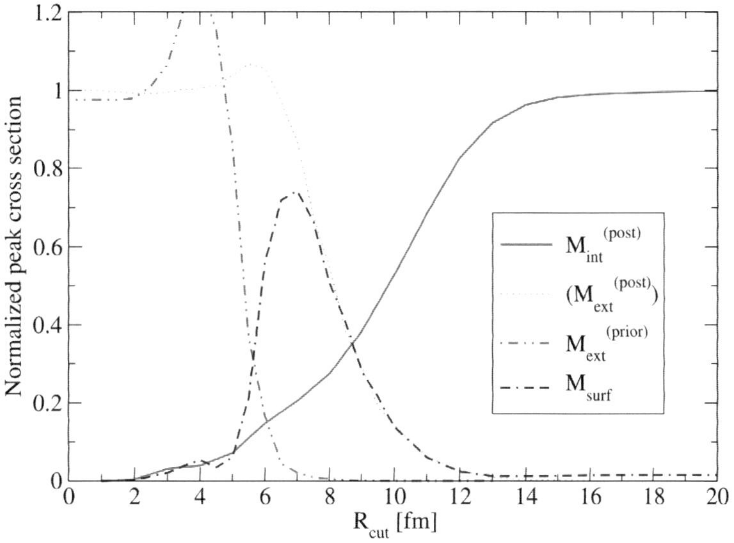

Figure 2. Contributions from the $M_{int}^{(post)}$, M_{surf}, $M_{ext}^{(prior)}$ terms to the ^{90}Zr(d,p)^{91}Zr$_{gs}$ transfer cross section. Shown are the peak cross sections, normalized as in Fig. 1, as a function of the surface radius (R_{cut}). Also given is the exterior post contribution (dotted light-colored curve) that was replaced with the introduction of the surface term. The calculations were carried out using a modified version of the reaction code FRESCO [7].

4. Conclusions and outlook

One-nucleon transfer reactions, in particular (d,p) reactions, have played a central role in nuclear structure studies for many decades. Present theoretical descriptions of the underlying reaction mechanisms are insufficient for addressing the challenges and opportunities that are opening up with new radioactive beam facilities. We have discussed a new theoretical approach that was recently introduced specifically to address shortcomings in the description of transfers to resonance states. The method builds on ideas from the very successful R-matrix theory; in particular it uses a similar separation of the parameter space into interior and exterior regions, and introduces a parameterization that can be related to physical observables, which, in principle, makes it possible to extract meaningful spectroscopic information from experiments. We have illustrated some aspects of the approach, in particular the role of the surface term. The calculations were carried out in the DWBA approximation. Extensions to include deuteron break-up contributions in a CDCC version of the approach are underway.

Acknowledgments

This work, performed under the auspices of the U.S. Department of Energy by Lawrence Livermore National Laboratory under Contract DE-AC52-07NA27344, is supported by the DOE through the Topical Collaboration TORUS.

References

[1] Mukhamedzhanov A M 2011 *Phys. Rev. C* **84**(4) 044616
[2] Satchler G R 1983 *Direct Nuclear Reactions* (Oxford University Press)
[3] Austern N, Iseri Y, Kamimura M, Kawai M, Rawitscher G and Yahiro M 1987 *Phys. Rep.* **154** 125–204
[4] Thompson I J and Nunes F 2009 *Nuclear Reactions for Astrophysics: Principles, Calculation and Applications of Low-Energy Reactions* (Cambridge: Cambridge University Press)
[5] Lane A M and Thomas R G 1958 *Rev. Mod. Phys.* **30** 257
[6] Descouvemont P and Baye D 2010 *Reports on Progress in Physics* **73** 036301
[7] Thompson I J 1988 *Computer Physics Reports* **7** 167–212

Why are light nuclei so important?

Martin Freer
School of Physics and Astronomy University of Birmingham, Birmingham, B15 2TT, UK

E-mail: M.Freer@bham.ac.uk

Abstract. This contribution explores the structure of light nuclei and how the harmonic oscillator may be used as a simple basis for their understanding. There is a particular focus on the nuclei ^{12}C and ^{9}Be.

1. Context

The quest to understand the structure of nuclei from stable to the drip-lines and from light to heavy, has driven the subject for the last several decades. Over that period nuclear theory has evolved considerably with the appearance of density functional approaches which have an ambition to be universal, the Greens Function Monte Carlo (GFMC) with an *ab initio* motivated nucleon-nucleon interaction with the inclusion of three-body forces, antisymmetrised molecular dynamics and fermionic molecular dynamics approaches which use an effective nucleon-nucleon interaction with nucleons simulated by Gaussian wave-packets and finally the no-core shell model (which featured strongly in the present conference). Many of these approaches are limited to the domain of light nuclei and hence it is here that their overlap is maximal. From an experimental perspective, the documentation of the structural properties of the light nuclei is most complete and hence provides the most stringent test of nuclear theory.

In many cases the structure of nuclei is driven by the behavior of, and correlations among, the valence nucleons. Deformation, pairing, evolution of shell structure and collective modes are all induced by the valence particles. In heavy nuclei the number of valence nucleons is relatively small (relative to the core nucleons), but this is not the case in light systems. Here the effect of the correlations becomes amplified and in certain systems distinct cluster-like structures appear. These cluster structures display particular spatial symmetries which are reminiscent of close packing of spheres. Remarkably, many of these symmetries can be found in the deformed harmonic oscillator, which in principle simulates the independent particle motion of the nucleons within the mean-field. In fact, the symmetries displayed by the deformed harmonic oscillator also reveal the emergence of cluster like structures in neutron-rich nuclei, such as ^{9}Be.

Even though the spectroscopy of the light nuclei has been subjected to the greatest scrutiny, it remains clear that there are substantial gaps in our knowledge – witness the missing first excited $1/2^{+}$ state in ^{9}B (the analogue is known in ^{9}Be) and the lack of clarity regarding the structure of the excitation energy spectrum of ^{12}C. Pinning down some of these open experimental questions and constraining the experimental situation is critical as this provides the foundation for further refining models of nuclei.

This contribution explores both the simple underlying ideas connected with the evolution of cluster-like structures and some of the developments in determining the structure of light nuclei, especially ^{12}C. Section 2 describes the symmetries of the deformed harmonic oscillator, Section 3 the

Published under licence by IOP Publishing Ltd

symmetries and structure of ^{12}C, Section 4 recent experimental measurements of the structure of ^{12}C and finally Section 5 explores the nature of clustering in neutron-rich systems, e.g. ^{9}Be.

2. The deformed harmonic oscillator and symmetries in symmetric and asymmetric nuclear matter

It is well-known that the solutions of the Schrödinger equation for the deformed harmonic oscillator (DHO) potential provides a reasonable approximation to the more realistic Woods-Saxon potential. With the addition of the spin-orbit interaction a reasonable description of the shapes and single-particle structure of light stable nuclei can be found. The solutions of the deformed harmonic oscillator are shown in Fig. 1.

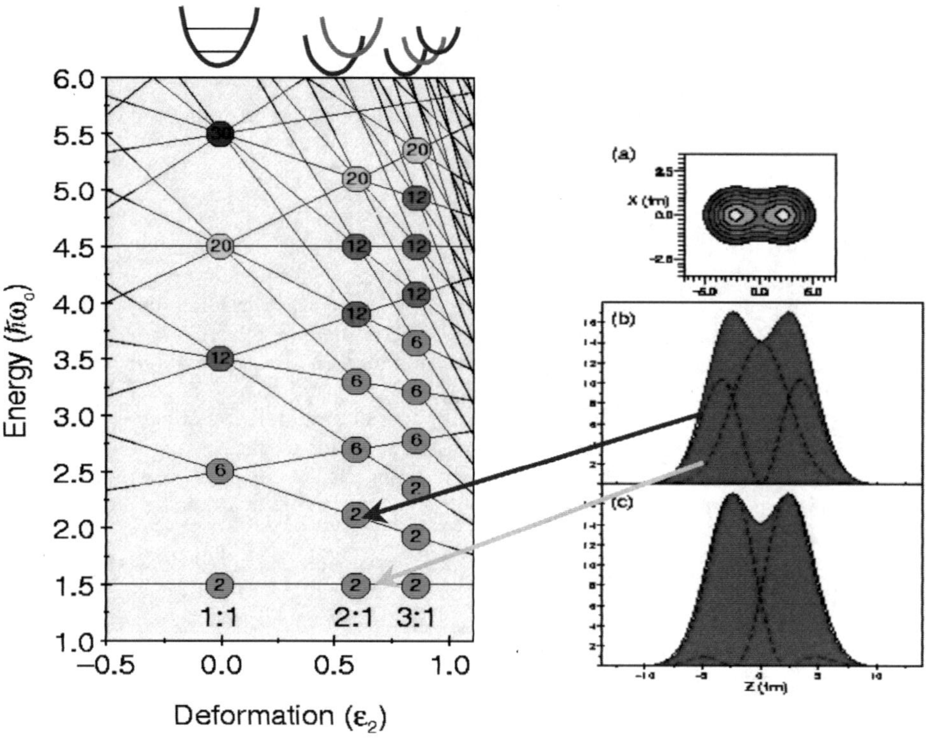

Figure 1. Energy levels of the deformed harmonic oscillator (left). The numbers at the level crossings at deformations of 1:1, 2:1, 3:1 etc, correspond to the degeneracy. The repeating of the spherical pattern twice at 2:1 indicates that two degenerate harmonic potentials co-exist. The right hand plots show the densities of the simplest 2:1 deformed configuration, which would correspond to ^{8}Be. The density is formed from the $[n_x, n_y, n_z] = [0,0,0]$ and $[0,0,1]$ harmonic oscillator levels (b), which may be decomposed into two alpha-particles (c). (a) shows the density calculated in the *x*-*z* plane.

The symmetries are revealed in the degeneracies of the energy levels as the potential is prolate deformed ($\varepsilon_2>0$). At deformations of 2:1 and 3:1 the spherical degeneracies (2, 6, 12, 20,...) are repeated two and three times, respectively. These symmetries extend to the behaviour of the densities calculated for systems of equal, and even, numbers of protons and neutrons. For example, in the formation of ^{8}Be within this framework, the density would correspond to the sum of the densities associated with the $[n_x, n_y, n_z]$ = [0,0,0] and [0,0,1] orbitals. These are shown in Fig. 1 (right hand side). The two centred symmetry is evident. These densities can be decomposed into two centres:

$$\phi_\alpha(\pm) = \frac{1}{\sqrt{2}}\left([0,0,0] \pm [0,0,1]\right). \quad (1)$$

The resulting densities which may be associated with alpha-particles are also shown in Fig. 1 (part c). These ideas were explored at length in Ref. [1]. The case of ^{8}Be is rather trivial, consisting of only two centres and two alpha-particles. The ^{24}Mg nucleus is much more complex, being composed of 6 alpha-particles. This point is worth emphasising...... even though the clusters themselves do not exist explicitly as quasi-free alpha-particles, the symmetries reflect the underlying geometric arrangements

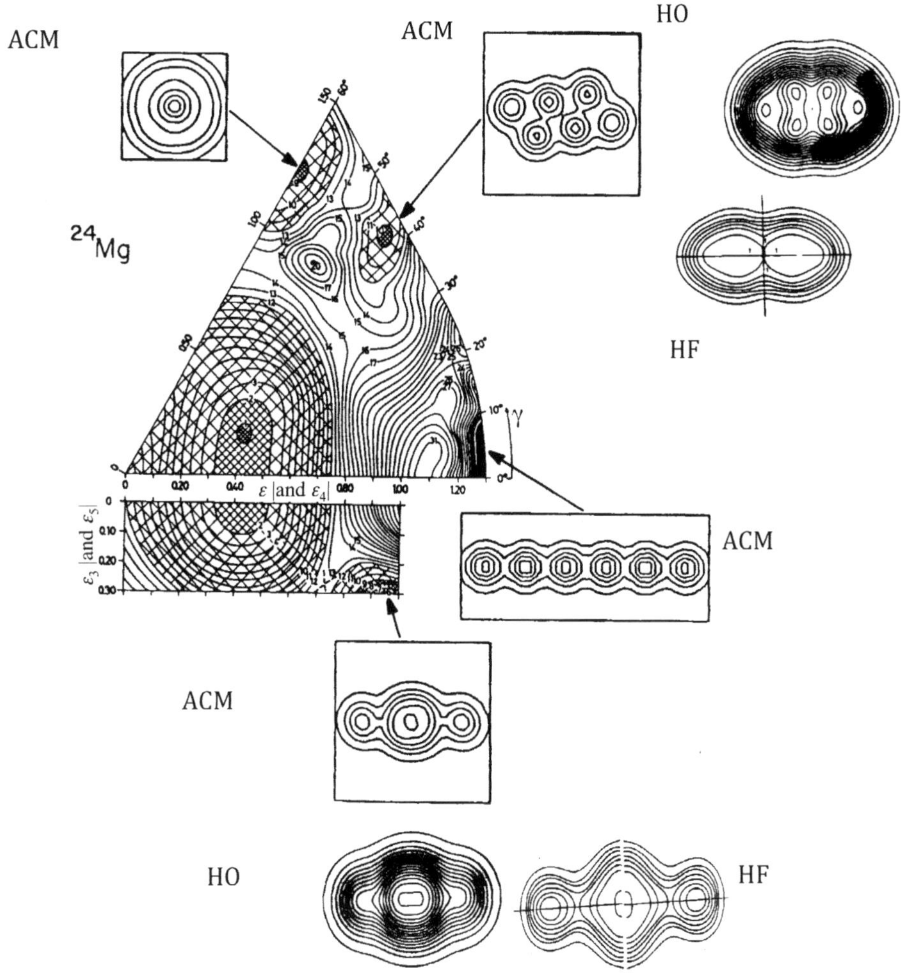

Figure 2. Comparison between the Nilsson-Strutinsky calculations of the potential energy surface of ^{24}Mg [2] (center) and the Alpha Cluster Model structures, ACM, (density plots around the outside) [3], harmonic oscillator, HO, and Hartree-Fock, HF [4].

of clusters. Experimentally, the nucleus ^{8}Be has a prolate deformation (two touching alpha-particles), ^{12}C is oblate (triangular arrangement of 3 alpha-particles), ^{16}O spherical (tetrahedral arrangement of 4 alpha-particles), ^{20}Ne is prolate with an octupole deformation (^{16}O+alpha structure)...... Again, this does not imply the existence of free alpha-particles but that the symmetry is retained in the ground-state. For the clustering to appear explicitly an excitation which is equivalent to the associated *N*-alpha decay threshold is required – the Ikeda rule [6]. For these reasons the alpha-cluster structure in the ^{8}Be ground-state is explicitly developed – it is unbound to two alpha-decay by 92 keV.

Fig. 2 illustrates a comparison between four different types of calculation. In the centre is the potential energy surface associated with a Nilsson-Strutinsky (NS) type calculation [2]. Here the local minima in the potential energy surface are associated with shell-structure in the deformed energy level scheme, which echo the shell structure in the DHO. Around the outside, and with a one-to-one correspondence, are calculations from an alpha cluster model (ACM) [3] and from the mean-field Hartree Fock (HF) approach [4]. In each case it is possible to identify the different shapes in the ACM and HF calculations uniquely with the minima in the potential energy surface. Moreover, from knowledge of the underlying configuration associated with the potential energy minima of the NS potential energy surface, it is possible to deduce also the harmonic oscillator structure. Correspondingly, the harmonic oscillator (HO) densities are shown next to those of the ACM and HF. Remarkably, the symmetries that are found in the HO are repeated in the ACM and HF. This underlines the fact that these symmetries are a necessary component of models of light nuclei and are a reflection of the underlying cluster structure.

3. ^{12}C and dynamical symmetries

In the case of ^{12}C, the system can be constructed from a variety of geometric arrangements of three alpha-particles. It might be expected that the compact equilateral-triangle arrangement is the lowest energy configuration. Such an arrangement possesses a D_{3h} point group symmetry. The corresponding rotational and vibrational spectrum is described by a form [6]

$$E = E_0 + Av_1 + Bv_2 + CL(L+1) + D(K \pm 2l)^2 , \qquad (2)$$

where $v_{1,2}$ are vibrational quantum numbers, and v_2 is doubly degenerate; $l = v_2, v_2 - 2, \ldots.1$ or 0, L is angular momentum, M would be its projection on a laboratory fixed axis and K a body-fixed axis [6]. A, B, C and D are adjustable parameters. The spectrum of states predicted by the choice $A = 7.0$, $B = 9.0$, $C = 0.8$ and $D = 0.0$ MeV is shown in Fig. 3 [6].

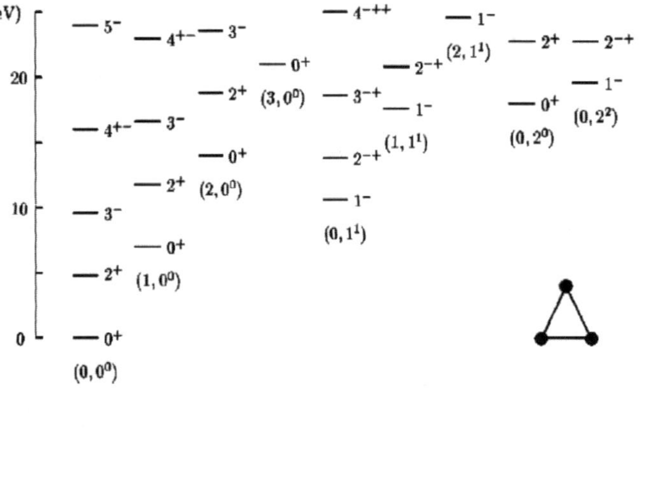

Figure 3. Spectrum of the energy levels of an equilateral triangle configuration. The bands are labeled by $(v_1;v^l_2)$. See Ref. [6] for further details.

The ground state band, $(v_1;v^l_2) = (0,0^0)$, contains no vibrational modes and coincides well with the observed experimental spectrum. Here the states correspond to different values of K ($K = 3n$, n = 0,1,2...) and L. For K=0, L=0, 2, 4 etc....., which is a rotation of the plane of the triangle about a line of symmetry, whereas for $K \neq 0$ $L = K, K+1, K+2$..... In the present case, K=0 or 3 is plotted with the parity being given by $(-1)^K$. The K=0 states coincide well with the well-known 0^+ (ground-state), 2^+ (4.4 MeV) and 4^+ (14.1 MeV) states. The K=3 states correspond to a rotation about an axis which passes through the centre of the triangle, the first of which has spin and parity 3^- and coincides with the 9.6 MeV, 3^-, excited state. This state may be thought of in terms of three alpha-particles rotating around the centre-of-mass each with one unit of angular momentum. Correspondingly, the next such state would be K=6, $J^\pi=6^+$. A prediction of this model is that there should be a 4^- state almost degenerate with the 4^+ state. A recent measurement involving studies of the alpha-decay correlations indicated that the 13.35 MeV unnatural-parity state possessed $J^\pi=4^-$ [7]. The close degeneracy with the 14.1 MeV 4^+ state would appear to confirm the D_{3h} symmetry. In this picture the 0^+ state at 7.65 MeV would correspond to a vibrational mode (v_1=1). The coupling of rotational modes would then produce a corresponding 2^+ state at 4.4 MeV above 7.65 MeV, i.e. 12.05 MeV; there is no known 2^+ state at this energy, pointing to the more complex structure of this state.

4. ^{12}C recent experimental developments

If the 7.65 MeV state in ^{12}C has a structure similar to that of the ground-state then a 2^+ state close to 12 MeV is expected. The closest state which has been tabulated with these characteristics is at 11.16 MeV [8]. This state was observed in the $^{11}B(^3He,d)^{12}C$ reaction, but has not been observed in measurements subsequently.

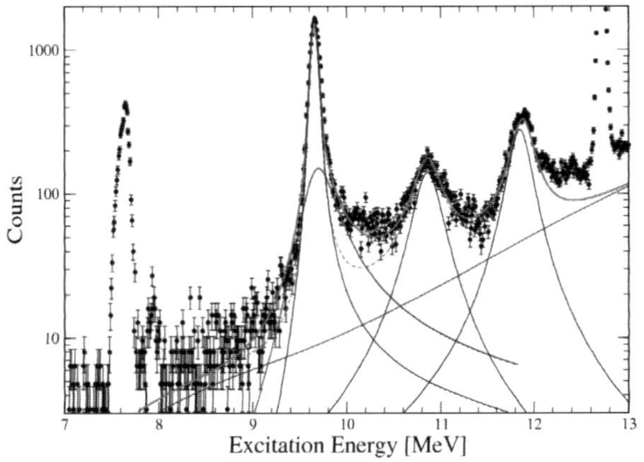

Figure 4. ^{12}C excitation energy spectrum from the $^{11}B(^3He,d)$ reaction. The measurement clearly demonstrates the absence of the 11.16 MeV state in ^{12}C [9].

A recent re-measurement of this reaction using the K600 spectrometer at iThemba (the previous measurements were made using photographic plates) demonstrates that the earlier observation of a state at 11.16 MeV was an experimental artifact and no such state exists [9].

Recent studies of the $^{12}C(\alpha,\alpha')$ [10] and $^{12}C(p,p')$ [11,12] reactions indicate the presence of a 2^+ state close to 9.6-9.7 MeV with a width of 0.5 to 1 MeV. The state is only weakly populated in these reactions, presumably due to its underlying cluster structure, and is broad. Consequently, its distinction from other broad-states and dominant collective excitations (e.g. the 9.6 MeV, 3^-) makes its unambiguous identification challenging. Further evidence for such an excitation comes from measurements of the $^{12}C(\gamma,3\alpha)$ reaction performed at the HIGS facility, TUNL [13]. Here a measurable cross section for this process was observed in the region of 9-10 MeV which cannot be attributed to known states in this region. Furthermore, the angular distributions of the alpha-particles are consistent with an L=2 pattern, indicating a dominant 2^+ component. Based on a rather simple description of this state in terms of three alpha-particles with radii given by the experimental charge radius, it is possible to use the 2 MeV separation between the Hoyle-state and the proposed 2^+ excitation to draw some conclusions as to the arrangements of the clusters. This would indicate that rather than a linear arrangement of the three clusters that a more appropriate description would be a loose arrangement of the alpha-particles in something approaching a triangular structure.

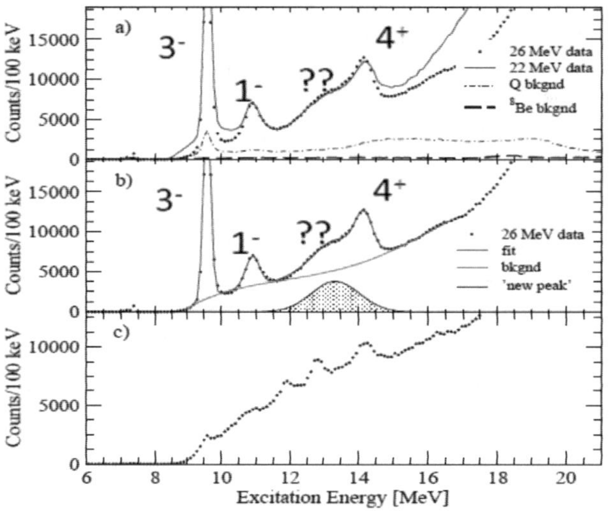

Figure 5. [Top] Arrangement of the 4 silicon strip detectors used in the measurements in Ref. [11]. The beam passes from bottom to top in this picture. [Bottom] Carbon-12 excitation energy spectra. a) The blue line shows the measurement at 22 MeV. The spectrum corresponding to measurements at 26 MeV is shown by the dots. The backgrounds obtained by gating above the ^{8}Be peak (bold dashed line) and both above and below the Q-value peak (dot-dashed line) are both illustrated. b) Fit to the 26 MeV data is given by the blue solid line. The polynomial background (red-line) and line-shape for the new peak (shaded area) is shown. c) Excitation energy spectrum for events not proceeding via the decay to the ^{8}Be ground-state. The proposed new state is indicated by "??".

A natural extension of such a conclusion is that there should also be a collective 4^+ state. Using the simple $j(j + 1)$ scaling, a 4^+excitation close to $E_x(^{12}C)$ = 14 MeV would be expected. Recent measurements of the two reactions $^9Be(\alpha,3\alpha)n$ and $^{12}C(\alpha,3\alpha)^4He$ have been performed [14]. In these measurements three alpha-particles were detected in an array of four silicon strip detectors (shown in

Fig 5). The analysis required that two of the three alpha-particles came from the decay of the ground-state of ^{8}Be. For the decay of ^{12}C to ^{8}Be+α this ensures that the decay process can proceed through only natural parity states (i.e. 0^+,1^-, 2^+ ....). This restricts the complexity of the excitation energy spectrum. The measurements for both the ^{9}Be and ^{12}C targets reveal the known 3^-, 1^- and 4^+ states at 9.64, 10.84 and 14.08 MeV, respectively. However, there is an additional component to the spectrum close to 13.3 MeV (marked by '??') with a width estimated to be 1.7 MeV (Fig. 5). It is believed that this is not a contaminant and is observed with similar properties in all spectra. Angular correlation measurements made using the ^{12}C target are not definitive, but indicate a 4^+ assignment.

5. Neutron-rich clusters

It has long been documented by von Oertzen and others [15] that when neutrons are added to systems in which there is a well established alpha-cluster structure, the neutrons inhabit a multicentre system and are in fact exchanged between the centres. In this way the orbits of the neutrons are analogues of those occupied by electrons in atomic molecules, with the one exception that the Pauli Exclusion Principle must be observed with respect to the cores. In the case of ^{9}Be, molecule is formed from two alpha-particles and a neutron. At each centre the neutron should occupy a *p*-orbital and as a consequence the molecular orbital is a linear combination of *p*-orbitals. These linear combinations give rise to sigma and pi-type molecular orbitals which are characterized by the projection of the angular momentum onto the axis of symmetry. The pi-configuration (l=1) has been linked to the ^{9}Be ground-state ($3/2^-$) and the sigma-orbit (l=0) the $1/2^+$ first excited state.

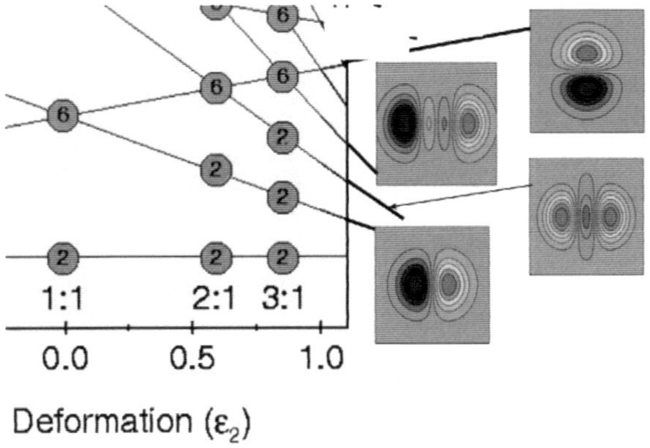

Figure 6. A section of the energy level scheme of the deformed harmonic oscillator and the associated wave-functions.

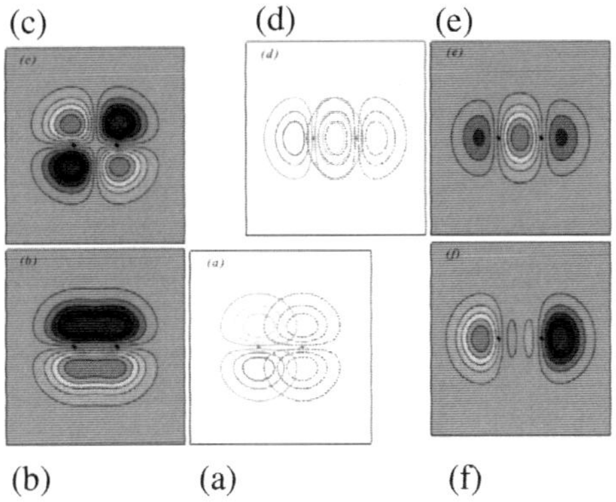

Figure 7. Molecular orbitals associated with linear combinations of HO orbitals orbits associated with one oscillator quantum (a) shows the overlap of the two individual wave-functions. Diagrams (b) and (c) the result of forming linear combinations: (b) corresponds to the binding pi-state, and (c) to the anti-binding state. Diagram (d) shows the overlap of the two orbits with the orientation of the orbits aligned with the separation axis, forming the sigma-configurations, and (e) and (f) the two linear combinations.

These configurations can be linked to the deformed harmonic oscillator and hence to other nuclear models, e.g. the Nilsson model. Fig. 6 shows the deformed harmonic oscillator and the wave-functions associated with each level. Fig. 7 shows the types of molecular wave-functions that can be formed from the superposition of two p-type orbitals. There is an obvious connection between the molecular and harmonic oscillator orbits. In the formation of the nucleus ^{9}Be in the deformed harmonic oscillator basis, the valence neutron would occupy the series of levels associated with the degeneracy '6' at a deformation of 2:1. These orbits have a structure which overlaps strongly with the pi (Fig. 7b) and sigma (Fig. 7e) molecular orbitals. Within the Nilsson scheme these levels have angular momenta projections $3/2^-$ and $1/2^+$ respectively – precisely as observed experimentally.

6. Summary

The structure of light nuclei is sensitive to nucleon-nucleon correlations which can manifest themselves as clusterisation. In the ground states of nuclei the explicit clusterisation is suppressed but the symmetries are retained. These are seen in the fingerprint of the excited states which can be described by dynamical symmetries which reflect those of the underlying structure. The spectrum of ^{12}C is a good example of this. The exchange of neutrons between alpha-particle cores can be described in terms of molecular orbitals. Both cluster symmetries and molecular orbitals may be understood within the framework of the deformed harmonic oscillator.

Acknowledgements
I would like to acknowledge the very significant contributions the work described here from collaborators at iThemba Labs South Africa, Notre Dame USA and the University of Birmingham UK.

References
[1] Freer M, Betts R R and Wuosmaa A H 1995 *Nucl. Phys.* **A587** 36
[2] Leander G and Larsson S E 1975 *Nucl. Phys.* **A239** 93
[3] Marsh S and Rae W D M 1986 *Phys. Lett.* **B180** 185
[4] Flocard H, Heenen P H, Krieger S J, Weiss M S 1984 *Prog. Theor. Phys.* **72** 1000
[5] Ikeda K, *et al.* 1968 *Prog. Theor. Phys. Suppl.* Extra Number 464
[6] Bijker R and Iachello F 2000 *Phys. Rev.* C **61** 067305
[7] Freer M, *et al.* 2007 *Phys. Rev.* C **76** 034320
[8] Reynolds G M, Rundquist D E, Poichar M 1971 *Phys. Rev.* C **3** 442
[9] Smit F D, *et al.* 2012 *Phys. Rev.* C **86** 037301
[10] Itoh M, *et al.* 2004 *Nucl. Phys.* A **738** 268
[11] Freer M, *et al.* 2009 *Phys. Rev.* C **80** 041303
[12] Zimmerman W R, Destefano N E, Freer M, Gai M, Smit F D 2011 *Phys. Rev.* C **84** 027304
[13] Gai M for the UConn-Yale-Duke-Weizmann-PTB-UCL Collaboration 2011 *APH Pol* B **42** 0775
[14] Freer M *et al.* 2011 *Phys. Rev.* C **83** 034314
[15] von Oertzen W, Freer M and Kanada-En'yo Y 2006 *Physics Reports* **432** 43

Journal of Physics: Conference Series **403** (2012) 012028 doi:10.1088/1742-6596/403/1/012028

Clusters and halos in light nuclei

Thomas Neff
GSI Helmholtzzentrum für Schwerionenforschung GmbH, Planckstraße 1, 64291 Darmstadt, Germany
E-mail: t.neff@gsi.de

Abstract. The fermionic molecular dynamics approach uses Gaussian wave packets as single-particle basis states. Many-body basis states are Slater determinants projected on parity, angular momentum and total linear momentum. The wave-packet basis is very flexible – FMD contains harmonic oscillator shell model and Brink-type cluster states as special cases. The parameters of the wave packets are obtained by variation. A realistic effective interaction derived from the Argonne V18 interaction by means of the unitary correlation operator method is employed. We discuss the fully microscopic calculation of the ^{3}He(α,γ)^{7}Be capture reaction within the FMD approach. The model space contains frozen cluster configurations at large distances and polarized configurations in the interaction region. The polarized configurations are essential for a successful description of the ^{7}Be bound state properties and for the S- and D-wave scattering states. The calculated cross section agrees well with recent measurements regarding both the absolute normalization and the energy dependence. We also discuss the structure of the cluster states, including the famous Hoyle state, in ^{12}C. From the two-body densities we conclude that the Hoyle state has a spatially extended triangular α-cluster structure, whereas the third 0^+ state features a chain-like obtuse triangle structure. We also calculate the $N\hbar\Omega$ decomposition of our wave functions to illuminate the challenges of no-core shell model calculations for these cluster states.

1. Introduction
Cluster and halo states present a challenge for nuclear structure calculations. They are typically found close to thresholds and feature extended tails in the wave functions. In many-body approaches like the no-core shell model where the wave functions are expanded in a harmonic oscillator basis the description of such spatially extended wave functions require huge model spaces. Cluster models on the other hand are by construction well suited to describe the asymptotics of such states. However cluster models are an oversimplification. With realistic interactions it is not possible to obtain reasonable results as polarization effects of the clusters are missing. With the fermionic molecular dynamics approach we attempt to combine the advantages of a microscopic model with those of the cluster model.

2. Fermionic Molecular Dynamics
In the fermionic molecular dynamics (FMD) approach we employ Gaussian wave packets

$$\langle \mathbf{x} | q \rangle = \frac{(\mathbf{x}-\mathbf{b})^2}{2a} \otimes | \chi^{\uparrow}, \chi^{\downarrow} \rangle \otimes | \xi \rangle \tag{1}$$

as single-particle basis states. The complex parameters $\mathbf{b}$ encode the mean positions and momenta of the wave packets and a the widths of the wave packets. The spins can assume any direction, isospin is ±1 denoting a proton or a neutron. Intrinsic many-body basis states are Slater determinants

$$\left| Q \right\rangle = \mathcal{A}\left\{ \left| q_1 \right\rangle \otimes \ldots \otimes \left| q_A \right\rangle \right\} . \tag{2}$$

The intrinsic states $\left| Q \right\rangle$ reflect deformation or clustering and break the symmetries of the Hamiltonian with respect to parity, rotation and translation. To restore the symmetries the intrinsic basis states are projected on parity, angular momentum and total linear momentum

$$\left| Q; J^\pi MK; \mathbf{P}=0 \right\rangle = \hat{P}^\pi \hat{P}^J_{MK} \hat{P}^{\mathbf{P}=0} \left| Q \right\rangle. \tag{3}$$

In a full FMD calculation the many-body Hilbert space is spanned by a set of N intrinsic basis states $\left\{ \left| Q^{(a)} \right\rangle, a = 1, \ldots, N \right\}$. In the end the full wave many-body state is obtained by diagonalizing the Hamiltonian in this set of non-orthogonal basis states.

3. Unitary Correlation Operator Method

Starting from the realistic Argonne V18 interaction [1] we derive an effective low-momentum interaction using the unitary correlation operator method (UCOM). The basic idea of the UCOM approach is to explicitly include short-range central and tensor correlations into the many-body state by means of a unitary operator [2, 3, 4]. The correlation operators can also be mapped onto the Hamiltonian to define a correlated interaction which not only contains two-body but also higher order contributions. However, the UCOM interaction is defined as the two-body part the correlated Hamiltonian and we neglect higher order terms. The UCOM interaction provides the same phase shifts as the original interaction but behaves differently in many-body systems. No-core shell model calculations show that the two-body UCOM interaction gives a good description of *s*- and light *p*-shell nuclei [4].

4. Radiative capture reaction ^{3}He(α,γ)^{7}Be

A recent application of the FMD approach is the calculation of the ^{3}He(α,γ)^{7}Be radiative capture reaction [5]. This reaction plays an important role in the solar proton-proton chains and determines the production of ^{7}Be and ^{8}B neutrinos [6, 7]. This reaction has been studied extensively from the experimental side in recent years [8, 9, 10, 11, 12]. However it is still not possible to reach the low energies relevant for solar burning in experiment. From the theory side this reaction has been investigated using simple potential models, where ^{3}He and ^{4}He are treated as point-like particles interacting via an effective nucleus-nucleus potential, e.g., [13] or microscopic cluster models, e.g., [14, 15] where the ^{7}Be bound and scattering states are constructed from microscopic ^{3}He and ^{4}He clusters interacting via an effective nucleon-nucleon interaction. *Ab-initio* calculations using variational Monte Carlo [16] and no-core shell model wave functions [17] were used to calculate asymptotic normalization coefficients for the bound states but relied on potential models for the scattering phase shifts.

In the FMD calculation for the ^{3}He(α,γ)^{7}Be reaction we divide the model space into two regions. In the external region bound and scattering states are described by ^{3}He and ^{4}He clusters in their FMD ground states. These microscopic wave functions can also be rewritten as resonating group (RGM) wave functions. The RGM representation allows us to include boundary conditions for bound and scattering states by matching to Whittaker and Coulomb functions at the channel radius (a=12 fm). For this we employ the microscopic R-matrix method developed by the Brussels group [18]. In the interaction region the model space is enlarged by additional FMD many-body configurations obtained by variation after parity and angular momentum projection on spin-parity $1/2^+$, $3/2^+$, $5/2^+$ and $3/2^-$, $1/2^-$, $7/2^-$, $5/2^-$. A constraint on the radius of the intrinsic states is used to vary the distance between the clusters.

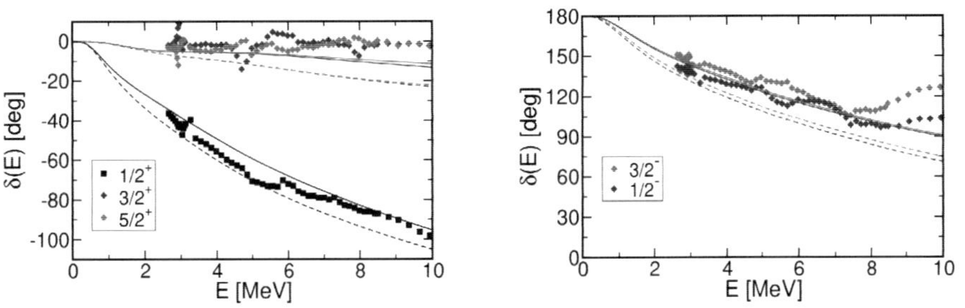

Figure 1. ^{4}He-^{3}He scattering phase shifts. Dashed lines show results using only frozen configurations, solid lines show results with full FMD model space. Left: S- and D-wave phase shifts. Right: P-wave phase shifts. Experimental results are from [20] and [21].

4.1. Capture Cross Section

For energies up to 2.5 MeV only the capture from S- and D-wave scattering states into the $3/2^-$ and $1/2^-$ bound states has to be considered. The addition of polarized configurations is essential in the FMD framework — using the frozen configurations only, ^{7}Be is only bound by 200 keV. Including the polarized configurations the $3/2^-$ state is bound by 1.49 MeV and the $1/2^-$ state by 1.31 MeV with respect to the cluster threshold. The calculated splitting between the two states is too small compared to the experimental value of 430 keV. This is related to a deficiency of the two-body interaction – additional spin-orbit strength is assumed to be coming from three-body forces. Fortunately it turns out that the capture cross section depends strongly on the centroid energy, which is reproduced nicely, but only very weakly on the splitting between the two bound states. An important test of the ground state wave function is provided by the charge radius. The calculated value 2.67 fm agrees well with the experimental value of 2.647(17) fm [19]. This is important as the dipole matrix element and therefore the capture cross section depends on an accurate description of the tail of the bound state wave functions. In Fig. 1 the calculated phase shifts for the S-, D- and P-waves are compared with the experimental phase shift analyses [20, 21]. There is a noticeably change in the results going from the model space containing only the frozen configurations to the full FMD model space. In case of the S- and D-wave phase shifts we find a good agreement with the experimental data. Also the P-wave phase shifts are essentially in good agreement with experiment, with the exception of the splitting.

The capture cross section is calculated using the microscopic wave functions of the bound and scattering states. The result for the total cross section for the ^{3}He(α,γ)^{7}Be capture is shown in the left part of Fig. 2 in form of the astrophysical S-factor. It agrees very well with the recent experimental data both in absolute normalization and in the energy dependence. Our result for the isospin mirror reaction ^{3}H(α,γ)^{7}Li is shown on the right hand side of Fig. 2. Whereas the energy dependence of the calculated S-factor agrees very well with the data the absolute cross section is larger then the data by Brune *et al.* by about 15%. This is surprising as the FMD results for the ^{7}Li bound states and the scattering phase shifts are of similar quality than those for ^{7}Be.

The total cross section can be decomposed into S- and D-wave contributions as shown in the left part of Fig. 3. If we compare our results with microscopic cluster model calculations, for example to those by Kajino [15] we observe that the biggest differences are found in the S-wave contribution. Our results also deviate from the empirical correlation between radius or quadrupole moment of the ^{7}Be ground state with the S-factor at zero energy that was found in the microscopic cluster model using different phenomenological interactions.

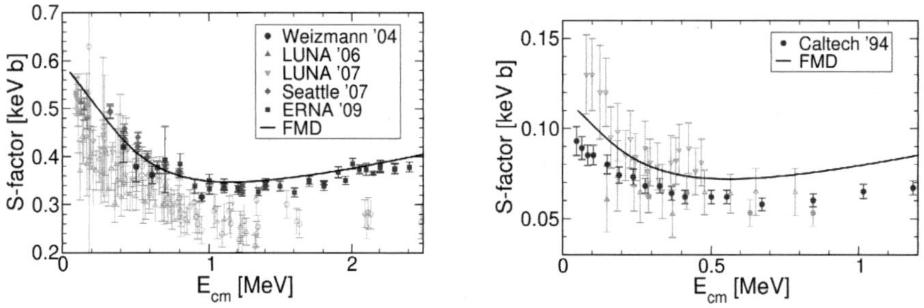

Figure 2. Left: S-factor for the ^{3}He(α,γ)^{7}Be reaction. Recent experimental data [8, 9, 10, 11, 12] are shown as colored symbols, older data as gray symbols. Right: S-factor for the ^{3}H(α,γ)^{7}Li reaction. Most recent data [22] is shown as colored symbols, older data as gray symbols.

For a comparison with the cluster model picture we can analyze the A-body bound and scattering state wave functions in terms of overlap functions. Here the full microscopic wave function is projected on the subspace of cluster configurations. If one takes antisymmetrization between the clusters into account properly (by folding with the square-root of the RGM norm kernel) these overlap functions can be interpreted as the relative wave functions of point-like ^{3}He and ^{4}He clusters. It is interesting that the overlap functions deviate from the Whittaker and Coulomb functions, that describe the asymptotic behavior of the cluster motion for bound and scattering states, up to distances of about 9 fm. These differences from the asymptotic behavior manifest themselves also in the dipole strength. The dipole matrix elements calculated from the overlap functions as shown in the right part of Fig. 3 agree with the matrix elements from the microscopic wave functions within 2%. They have sizable contributions already for distances as small as 3 fm and deviate significantly from matrix elements calculated from the asymptotic Whittaker and Coulomb functions to distances of up to 10 fm. This differences indicate that considering this reaction as a simple external capture process is not possible.

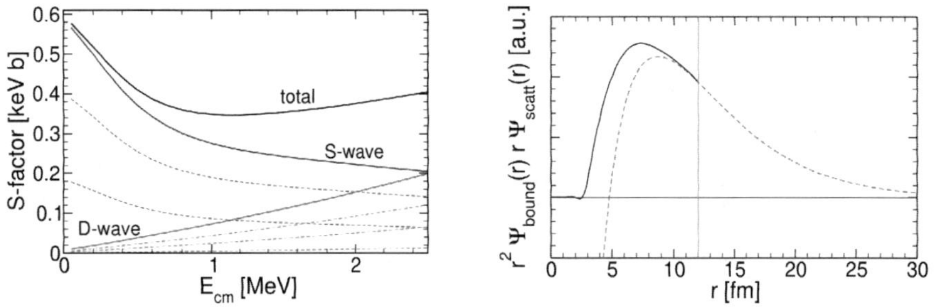

Figure 3. Left: S- and D-wave contributions to the S-factor for the ^{3}He(α,γ)^{7}Be reaction. Dashed lines show contributions of individual transitions. Right: Dipole strength between the $3/2^-$ bound state and the 50 keV $1/2^+$ scattering state calculated from the overlap functions (solid line), from the Whittaker and Coulomb functions matched at the channel radius (dashed line).

5. Cluster States in ^{12}C

The structure of the second 0^+ state in ^{12}C, the famous Hoyle state, is still one of the hottest topics in nuclear structure. In [23] we investigated the structure of the Hoyle state using the FMD approach. The model space consisted of configurations obtained by variation and a full set of three-α configurations. A UCOM interaction with some phenomenological modifications regarding the strength of the spin-orbit force and the saturation properties of the two-body interaction was used in that calculation. We compared the results with a microscopic cluster model using a phenomenological Volkov interaction. These cluster model calculations reproduced previous results by Kamimura [24] and are also very close to those obtained by Funaki et al. [25]. We found for both models that the Hoyle state has a very dilute, extended three-α structure. This is illustrated in Fig. 4 where we show the intrinsic FMD basis states that have the largest overlap with the ground state and the Hoyle state.

We used these wave functions also to calculate the transition form factor from the ground state to the Hoyle state. This transition form factor can be directly compared to electron scattering data [23, 26]. The good agreement of calculation and experiment is a strong confirmation for a spatially extended structure for the Hoyle state.

5.1. Two-body densities

Observables like radii and form factors are scalar quantities that provide information about the size of the states but they do not provide direct information about the structure of the states. The old question whether the Hoyle state should be interpreted as a linear chain of α-particles, a triangular structure or a gas-like structure can therefore not be answered directly by these experimental observables. There is also the questions of how we should compare the wave functions obtained in different many-body approaches like the cluster model, the no-core shell model or as obtained on the lattice [27].

In case of FMD or the cluster model the individual basis states can be easily interpreted in terms of the intrinsic structure as shown in Fig. 4. However the eigenstates are linear combinations of many basis states and the non-orthogonality of the basis states might question the validity of the obtained picture.

To remedy this situation we propose to use two-body densities to analyze the structure of the ^{12}C eigenstates. In Fig. 5 we show the diagonal part of the two-body density integrated over the center-of-mass coordinates and summed over all spin-isospin channels which can be expressed as

$$\rho^{(2)}(\mathbf{r}) = \langle \Psi \,|\, \sum_{i<j} \delta(\hat{\mathbf{r}}_i - \hat{\mathbf{r}}_j - \mathbf{r}) \,|\, \Psi \rangle \,. \tag{4}$$

The two-body density $\rho^{(2)}(\mathbf{r})$ tells us about the probability to find a pair of nucleons at a given distance $\mathbf{r}$. In the case of ^{12}C where we expect an intrinsic α-cluster structure the two-body density should directly reflect the correlations between the α-clusters. The two-body density

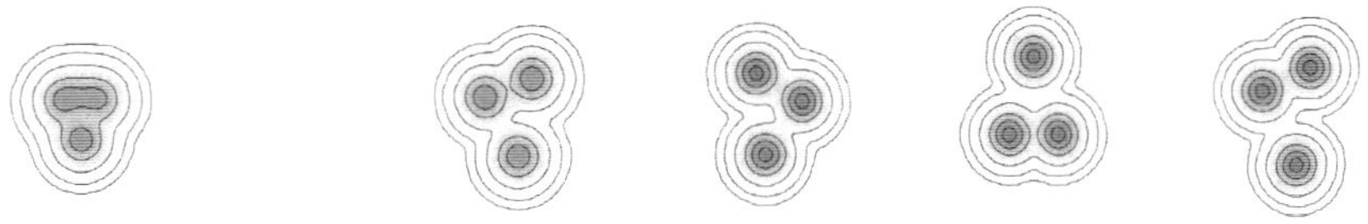

Figure 4. (Left) Intrinsic FMD basis state that has the largest overlap with the ground state. (Right) The four intrinsic FMD basis states that have the largest overlaps with the Hoyle state. The basis states are not orthogonal.

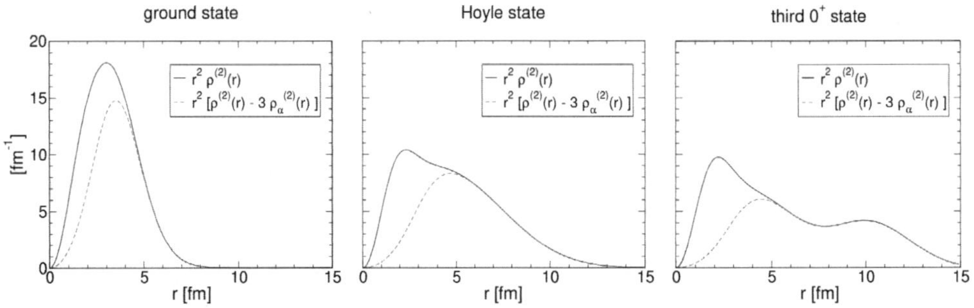

Figure 5. Two-body densities obtained in the cluster model for the ^{12}C ground state, the Hoyle state, and the third 0^+ state. The densities are multiplied by r^2. Solid lines show the full two-body densities, the dashed lines show the results when the contributions to the two-body densities from inside the α-particles is subtracted.

is only smeared out by the finite size of the α-clusters. In microscopic many-body approaches the spin-isospin dependence of the two-body density could be used to validate the assumption of α-clustering. In principle the two-body densities will depend on the interaction used in the many-body approach. However this interaction dependence should only affect the behavior at very short distances. Here we are interested in the long-range correlations.

In the full two-body density also pairs of nucleons from the same α-particle are included. If we assume an α-cluster picture the more interesting observable is the probability to find a pair of nucleons from two different α-particles. In case of the cluster model one can simply subtract the contributions from within the α-particles. These are the dashed curves in Fig. 5. For the ground state the probability then peaks at about 3.5 fm reflecting the Pauli exclusion principle between the α clusters. For the Hoyle state we still find a single peak in the probability distribution at about 5 fm. This means that the α-particles in the Hoyle state are further apart from each other than in the ground state. The probability distribution also extends to much larger distances of more than 10 fm. The fact that there is only a single peak is consistent with the picture of a triangular structure where the distances between all three α-particles is the same. The density distributions of the third 0^+ state however shows a two-peak structure, with one peak at about 5 fm and another smaller peak at about 10 fm. This is what one would expect for a chain structure. The α-particles at the ends of the chain each see the α-particle in the middle of the chain at about 5 fm distance and the one at the other end of the chain at about 10 fm distance.

5.2. Expansion in oscillator basis

In no-core shell model calculations the second 0^+ state is found at much too high energies [17]. To illustrate the challenge for the no-core shell model we analyze the cluster model wave functions in the harmonic oscillator basis. Explicitly expanding our wave functions in the harmonic oscillator basis is not feasible. It is however easy to simply count the number of $N\hbar\Omega$ excitations in the wave function. This occupation probability can be calculated by

$$\mathrm{Occ}(N) = \left\langle \Psi \,\middle|\, \delta\Big(\sum_i (\hat{H}_i^{HO}/\hbar\Omega - 3/2) - N\Big) \,\middle|\, \Psi \right\rangle \qquad (5)$$

as proposed by Suzuki *et al.* [28]. The results are shown in Fig. 6 for the cluster model wave functions of the ground state and the Hoyle state. On the left an oscillator parameter $\hbar\Omega$ of 20 MeV was used. That corresponds roughly to the optimal oscillator parameter for the mean-field solution. In case of the ground state the contributions become very small for N larger

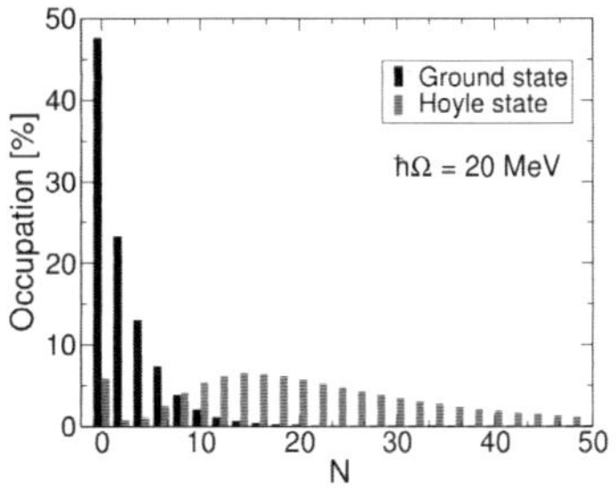

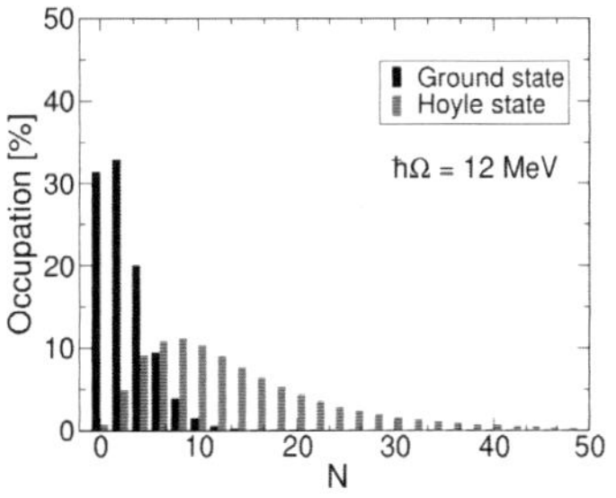

Figure 6. Decomposition of the ^{12}C ground state and the Hoyle state into $N\hbar\Omega$ components for oscillator constants of 20 MeV (left) and 12 MeV (right).

than 8 or 10. It is therefore not surprising that the NCSM calculations for the ground state can be converged. For the Hoyle state however, the distribution extends over a very large range of $N\hbar\Omega$. It is therefore clear that the Hoyle state can not be converged in NCSM calculations with $N_{\max} = 8$ or even 10. The situation looks somewhat better for an oscillator parameter of 12 MeV as shown on the right hand side of Fig. 6. Here the distribution for the Hoyle state peaks at $N = 8$ and decays much more rapidly with N. However, standard NCSM calculations will not be able to reach large enough spaces. Maybe approaches like the importance truncated no-core shell model [29] or the symmetry adapted no-core shell model [30] will allow the description of the Hoyle state within the oscillator basis in the future.

References

[1] Wiringa R B, Stoks V G J and Schiavilla R 1995 *Phys. Rev. C* **51** 38
[2] Feldmeier H, Neff T, Roth R and Schnack J 1998 *Nucl. Phys.* **A632** 61
[3] Neff T and Feldmeier H 2003 *Nucl. Phys.* **A713** 311
[4] Roth R, Neff T and Feldmeier H 2010 *Prog. Part. Nucl. Phys.* **65** 50
[5] Neff T 2011 *Phys. Rev. Lett.* **106** 042502
[6] Adelberger E G *et al.* 1998 *Rev. Mod. Phys.* **70** 1265
[7] Adelberger E G *et al.* 2011 *Rev. Mod. Phys.* **83** 195
[8] Nara Singh B S, Hass M, Nir-El Y and Haquin G 2004 *Phys. Rev. Lett.* **93** 262503
[9] Bemmerer D *et al.* 2006 *Phys. Rev. Lett.* **97** 122502
[10] Confortola F *et al.* 2007 *Phys. Rev. C* **75** 065803
[11] Brown T A D *et al.* 2007 *Phys. Rev. C* **76** 055801
[12] Di Leva A *et al.* 2009 *Phys. Rev. Lett.* **102** 232502
[13] Kim B T, Izumoto T and Nagatani K 1981 *Phys. Rev. C* **23** 33
[14] Langanke K 1986 *Nucl. Phys.* **A457** 351
[15] Kajino T 1986 *Nucl. Phys.* **A460** 559
[16] Nollett K M 2001 *Phys. Rev.* **C63** 054002
[17] Navrátil P, Gueorguiev V G, Vary J P, Ormand W E and Nogga A 2007 *Phys. Rev. Lett.* **99** 042501
[18] Descouvemont P and Baye D 2010 *Rep. Prog. Phys.* **73** 036301
[19] Nörtershäuser W *et al.* 2009 *Phys. Rev. Lett.* **102** 062503
[20] Boykin W R, Baker S D and Hardy D M 1972 *Nucl. Phys.* **A195** 241
[21] Spiger R J and Tombrello T A 1967 *Phys. Rev.* **163** 964
[22] Brune C R, Kavanagh R W and Rolfs C 1994 *Phys. Rev. C* **50** 2205
[23] Chernykh M, Feldmeier H, Neff T, von Neumann-Cosel P and Richter A 2007 *Phys. Rev. Lett.* **98** 032501
[24] Kamimura M 1981 *Nucl. Phys.* **A351** 456
[25] Funaki Y, Tohsaki A, Horiuchi H, Schuck P and Röpke G 2003 *Phys. Rev. C* **67** 051306(R)
[26] Chernykh M, Feldmeier H, Neff T, von Neumann-Cosel P and Richter A 2010 *Phys. Rev. Lett.* **105** 022501
[27] Epelbaum E, Krebs H, Lee D and Meißner U G 2011 *Phys. Rev. Lett.* **106** 192501
[28] Suzuki Y, Arai K, Ogawa Y and Varga K 1996 *Phys. Rev. C* **54** 2073
[29] Roth R and Navrátil P 2007 *Phys. Rev. Lett.* **99** 092501
[30] Dytrych T, Sviratcheva K D, Bahri C, Draayer J P and Vary J P 2007 *Phys. Rev. Lett.* **98** 162503

HITES 2012 IOP Publishing
Journal of Physics: Conference Series **403** (2012) 012029 doi:10.1088/1742-6596/403/1/012029

Status of reaction theory for studying rare isotopes

F M Nunes[1,2] and N J Upadhyay[1]

[1]National Superconducting Cyclotron Laboratory, Michigan State University, East Lansing, MI 48824, USA

[2]Department of Physics and Astronomy, Michigan State University, East Lansing, MI 48824, USA

E-mail: nunes@nscl.msu.edu

Abstract. Reactions are an important tool to study nuclear structure and for extracting reactions relevant for astrophysics. In this paper we focus on deuteron induced reactions which can provide information on neutron shell evolution as well as neutron capture cross sections. We review recent work on the systematic comparison of the continuum discretized coupled channel method, the adiabatic wave approximation and the Faddeev momentum-space approach. We also explore other aspects of the reaction mechanism and discuss in detail difficulties encountered in the calculations.

1. Introduction

Nuclear reactions are one of the most powerful tools to explore the properties of isotopes over the whole nuclear landscape. In addition, from nuclear reaction measurements one can extract capture rates relevant for astrophysics. For these two reasons, it is pivotal to develop the most reliable reaction theory and determine the accuracy with which nuclear reaction predictions can be made.

Rare isotopes have been critical to our understanding of the stability of matter, providing information on the various components of the effective interaction that binds neutrons and protons together. As one explores nuclear properties along isotopic chains, one can establish the isospin dependence of the interaction. As one explores nuclear properties at the limits of stability, one gains insight on the relevance of the continuum.

Rare isotopes also play a very important role in astrophysics. Today it is understood that a good fraction of the heavy elements created in the universe, are produced in the r-process. In this process, a rapid sequence of neutron capture takes us many nucleons away from stability, generating a path through very neutron-rich isotopes. Inputs to astrophysical simulations of the r-process require masses, lifetimes and neutron-capture rates for these exotic nuclei. Since neutron capture on an exotic nucleus is presently not experimentally feasible, alternate methods using nuclear reactions have been proposed. Of primary interest is the use of (d,p) reactions for this purpose.

This is a particularly exciting time for nuclear physics. New rare isotope facilities have either just come online (e.g. RIBF at RIKEN) or are being constructed (e.g. FRIB at MSU). These new facilities, not only have a wider reach into unknown regions, but also have increased rates which will allow for many more, and more detailed, reaction measurements. Since the analysis of the reaction measurements require reaction theory, we underline the importance of having control over the ambiguities in the reaction calculations.

Deuteron induced reactions, being one of the simplest probes, enable inspection of both nuclear structure and nuclear astrophysics. Our work within the TORUS collaboration [1] has focused on benchmarking models for (d,p) reactions and advancing the theory for (d,p) to a new level, with reduced approximations and wider applicability.

Solving the many-body reaction problem A(d,p)B is not feasible whenever the large number of nucleons to be considered is larger than $A \approx 10$. The additional difficulty of the reaction many-body problem, as compared to the bound state problem, arises due to not only the behavior of the scattering wave-function at large distances but also due to the extreme sensitivity of reaction observables to thresholds and the correct asymptotic behavior of bound states in the subsystems. In most cases, it is thus necessary to reduce the many-body problem to a few-body problem, retaining only the relevant degrees of freedom. In particular, for A(d,p)B, this would imply a reduction to a three-body problem $n+p+A$. With this reduction, optical potentials, with their inherent ambiguities, are added to our theory. In a three-body model for A(d,p)B, only nucleon optical potentials are need (apart from the NN interaction). These can be rather well constrained by nucleon elastic scattering and the associated uncertainty can be well estimated. The three-body formulation for A(d,p)B also contains many-body overlap functions which derive from the underlying many-body structure of the nuclei involved, A and B, after integrating unnecessary degrees of freedom. Evidently, (d,p) reactions are used to probe exactly this ingredient.

The three-body scattering problem can be solved exactly within the Faddeev momentum space integral formalism [2, 3] which explicitly includes breakup and transfer channels to all orders. One of the most well established theories for direct nuclear reactions is the Continuum Discretized Coupled Channels (CDCC) method [4, 5]. It includes breakup to all orders by discretizing the projectile continuum into bins. A systematic comparison of CDCC and Faddeev has been completed [6]. Another method, developed specifically for (d,p) transfer reactions, is the so-called adiabatic wave approximation (ADWA) [7]. One can think of ADWA as a simplified CDCC calculation, where the radial behavior of all the continuum components are assumed to be similar to the ground state. It is thus expected to work best if the reaction does not excite higher energies in the projectile, i.e. low energy reactions. Comparison of this method and Faddeev has also been performed in a systematic manner [8]. In this paper we summarize the results of these comparisons, explore other aspects of the reaction mechanism and discuss some challenges of the above mentioned calculations.

2. Comparing methods for (d,p)

The findings in [6] prove that CDCC is able to provide a good approximation to Faddeev for elastic scattering. The explicit coupling to the elastic channel with the neutron-nucleus bound state as done in Faddeev calculations introduces small modifications to the elastic angular distributions mostly at backward angles.

The comparison of CDCC and Faddeev for transfer cross sections is consistent with the comparison of ADWA and Faddeev presented in [8]. We found CDCC and ADWA to be very good approximations of Faddeev for reactions around 10 MeV/u, while the agreement deteriorated with larger beam energies. At larger beam energies observables also become much more dependent on subtle differences in the interactions used in CDCC/ADWA versus Faddeev [6, 8]. For this reason, the shortcomings of either CDCC or ADWA at larger beam energies are tamed by ambiguities in the choice of the energy at which the optical potentials are calculated.

It became clear from the comparisons in [6, 8] that, for projectiles with primarily a loosely bound s-wave ground state, CDCC did not improve the description of transfer when compared

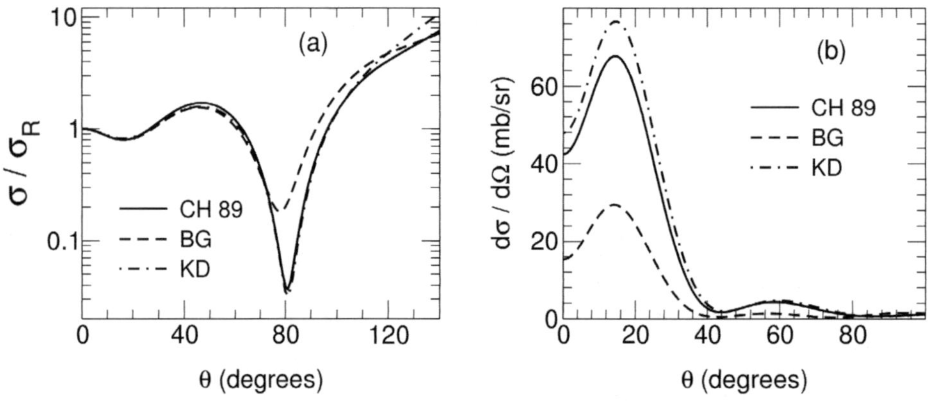

Figure 1. Optical model dependence: (a) elastic angular distribution for p+^{12}C at E_p =6 MeV and (b) transfer angular distribution for ^{12}C(d,p)^{13}C at E_d = 12 MeV. Shown are results using Chapel-Hill89 [11] (solid), Bechetti and Greenlees [13] (dashed) and Konin and Delaroche [12] (dot-dashed).

to the adiabatic model (ADWA). For these cases, ADWA is the preferred tool, given the fact that it is computationally much more efficient than CDCC. It was also found that at larger beam energies, CDCC did not show significant improvement over ADWA. This later point is still rather puzzling since CDCC does have a much better description of the n-p continuum as compared to ADWA.

The situation for breakup is very different. Breakup observables predicted by CDCC are at its best for the higher beam energies explored in [6, 8]. At the low energies, CDCC results are far from the results from Faddeev. We note that to reduce the technical difficulties in the Faddeev approach, the Coulomb interaction was neglected when computing breakup observables (we will come back to this in 3.2). In addition we underscore that exactly the same Hamiltonian for CDCC and Faddeev was used and therefore, when comparing CDCC and Faddeev for breakup, there is no ambiguity in the choice of the interaction.

The CDCC/FAGS comparisons in [6] demonstrate that finding agreement for the elastic for a given target and beam energy does not imply agreement in breakup or transfer. This results from sensitivity of the reaction amplitude to different parts of configuration space. Thus, only by looking at elastic, transfer and breakup simultaneously, the CDCC method can be thoroughly tested.

3. Discussion

We further explore the results obtained in [6], in regards to the dependence of transfer observables on the interactions, dissecting the reaction mechanism for transfer reactions. We also consider the difficulties of calculating breakup observables accurately, namely concerning the treatment of Coulomb and the convergence of the cross sections with increasing model space. The numerical inputs for the CDCC calculations performed in this section and the parameters used for the interactions are the same as those in [6], unless otherwise stated. All calculations were performed using FRESCO [9].

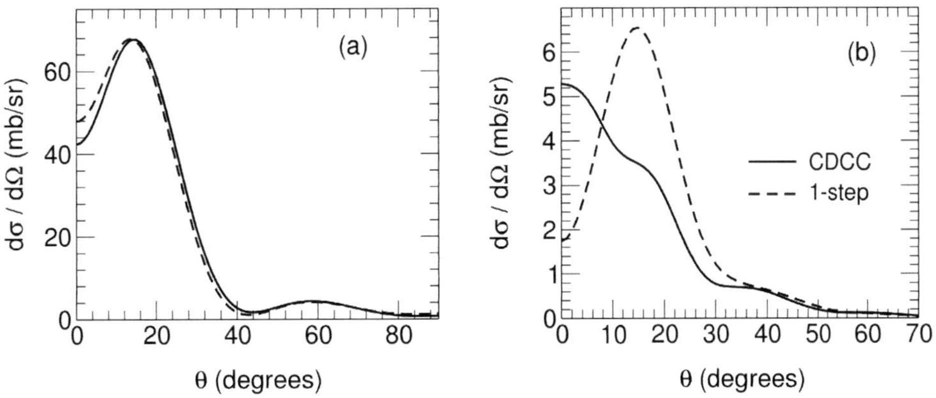

Figure 2. Deuteron continuum effects on transfer cross sections: ^{12}C(d,p)^{13}C at E_d = 12 MeV, (b) ^{48}Ca(d,p)^{49}Ca at E_d = 56 MeV. Comparing full CDCC results (solid) with 1-step approximation (dashed).

3.1. Looking at transfer in more detail

As mentioned in the introduction, the reduction of the many-body problem to a few-body problem brings up the issue of optical potential ambiguities which need to be kept under control. As microscopic approaches improve, it will eventually become possible to extract optical potentials directly from the solutions of the many body problem (e.g. [10]). At present the most reliable way to constrain the optical potential is through elastic scattering. In Fig.1 we show the proton angular distribution for ^{12}C at 6 MeV for three different optical potential parameterizations (fig. a) and the corresponding (d,p) transfer angular distribution as a function of the center of mass angle (fig. b). While CH89 [11] and KD [12] provide similar proton angular distributions for angles smaller than 100 deg (they have been shown to work well for these low energies and low mass targets), larger differences are seen for BG [13]. When the interaction is constrained by elastic scattering, the ambiguity that carries over to transfer is small (the percentage difference at the peak between CH89 and KD in Fig.1(b) is $\approx$ 10%). However, the huge difference in the normalization of the transfer cross section obtained when using BG as compared to CH89/KD emphasizes the well known fact that not constraining the optical potential can introduce large errors in the calculation. It is thus essential that a serious scientific program using (d,p) reactions as a probe for nuclear structure, include systematic measurements of nucleon elastic scattering on the relevant isotopes and at the relevant energies.

When using the CDCC wavefunction to compute transfer amplitudes, it is possible to isolate the effect coming from the direct transfer process and the multi-step process via breakup states in the *np* system. In Fig.2 we compare the importance of breakup as an intermediate step in the transfer process, for (d,p) reactions on ^{12}C at 12 MeV (left panel) and ^{48}Ca at 56 MeV (right panel). While in the first case the effects are minor, for ^{48}Ca(d,p) at 56 MeV, breakup has such a large effect on transfer that it changes completely the shape of the angular distributions. For this reason the traditional DWBA (distorted wave Born approximation) should only be used in cases where breakup has been proven to be insignificant otherwise the extracted information will not be meaningful.

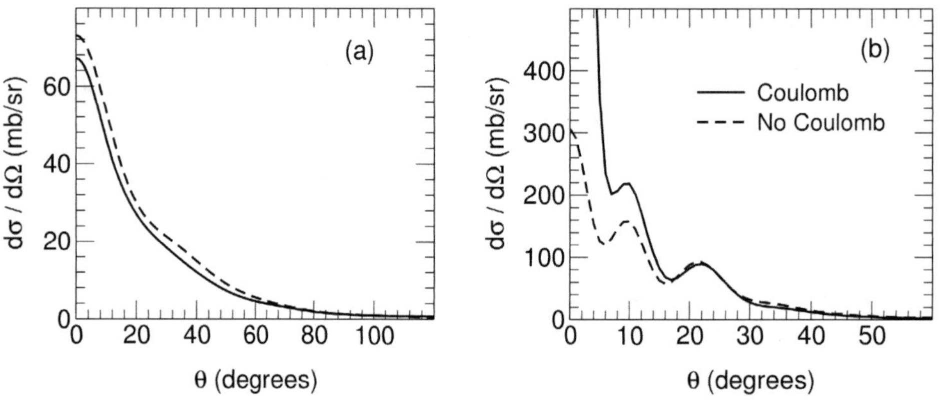

Figure 3. Coulomb effects on the breakup cross sections: ^{12}C(d,pn)^{12}C at $E_{\rm d}$ = 12 MeV, (b) ^{48}Ca(d,pn)^{48}Ca at $E_{\rm d}$ = 56 MeV. Comparing the differential cross sections for the full CDCC results (solid) with those excluding Coulomb interactions (dashed).

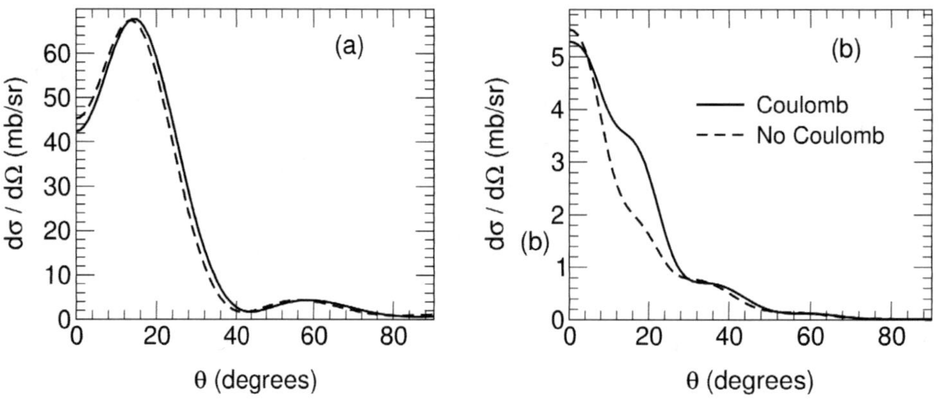

Figure 4. Coulomb effects on the transfer cross sections: ^{12}C(d,p)^{13}C at $E_{\rm d}$ = 12 MeV, (b) ^{48}Ca(d,p)^{49}Ca at $E_{\rm d}$ = 56 MeV. Comparing the differential cross sections for the full CDCC results (solid) with those excluding Coulomb interactions (dashed).

3.2. Looking at breakup in more detail

One important limitation to the Faddeev calculations performed in [8, 6] has to do with the Coulomb interaction. The momentum-space implementation relies on Coulomb screening. While Coulomb screening works well for systems with small charge, as the Coulomb interaction becomes stronger, the screening radius necessary for convergence increases and with it the needed partial waves. Eventually the problem becomes numerically unstable. To avoid this known difficulty, breakup calculations presented in [6] excluded the Coulomb interaction.

The CDCC equations are solved in coordinate space and do not require screening. We thus use them to explore the relevance of the Coulomb interaction in deuteron induced reactions. Fig.3 shows the predictions for the breakup angular distributions as function of the center of mass angle of the outgoing *np* system, for the full CDCC calculation including Coulomb (solid) and for a CDCC calculation where all charges are set to zero (dashed). Expectedly for light

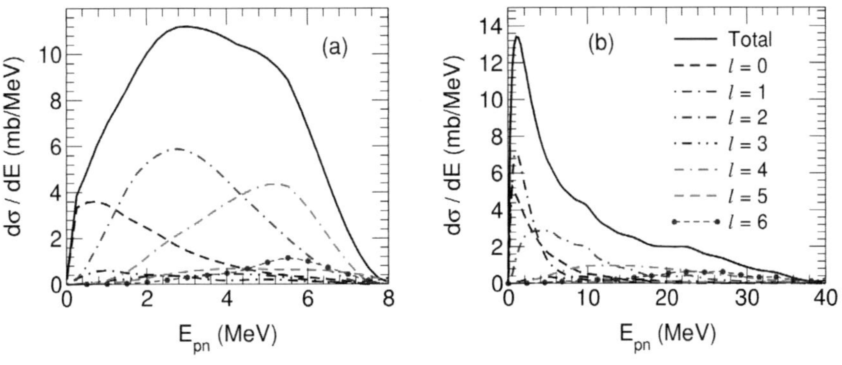

Figure 5. Comparing convergence of the breakup cross sections for (a) ^{12}C(d,pn)^{12}C at $E_d = 12$ MeV and (b) ^{48}Ca(d,pn)^{48}Ca at $E_d = 56$ MeV. Contributions of the different n-p partial waves to the energy distribution, plotted as a function of the n-p relative energy.

systems the Coulomb effects are small. However, for medium mass systems these effects become important, particularly at small angles, which captures the long-range nature of the Coulomb interaction.

In Fig.4 we show the corresponding plot for the transfer angular distributions. Although effects on transfer observables are smaller than for breakup, Coulomb effects increase with charge and these are expected to become even more important for heavy systems. With the techniques of [3], reactions on heavy systems are not accessible. A new technique to handle Coulomb is being proposed within the scope of the TORUS collaboration [14].

CDCC was shown to fail for breakup observables, for reactions performed at low deuteron energies [6]. While CDCC makes a choice of one Jacobi basis set for the expansion of the three-body wave-function, namely that of the deuteron, inspection of the Faddeev calculations demonstrated the importance of the proton and neutron components. Encapsulating these components within the CDCC basis, resulted in very slow convergence. In Fig.5 we show the contribution of each np partial wave (l) to the breakup energy distribution for the reaction on ^{12}C at 12 MeV (fig. a) and for the reaction on ^{48}Ca at 56 MeV (fig. b). For ^{48}Ca, the use of $l_{max} = 6$ was sufficient to obtain convergence, as shown in the comparison with Faddeev [6]. For ^{12}C, the predictions using $l_{max} = 6$ were very far from the results obtained with Faddeev. We should point out that for this particular reaction, calculations with $l_{max} = 8$ become numerically unfeasible with the numerical methods presently used to solve the coupled differential equations in FRESCO.

Moreover, if one estimated an error due to lack of convergence, based on the difference between the CDCC cross section with $l_{max} = 4$ and $l_{max} = 6$, as is customary, one would be mislead to believe in a much smaller error than the real error. Even though Fig.5 appears to suggest that partial waves above $l_{max} = 6$ will not contribute very much individually, the contributions do add up at large np relative energies, and produce a very different shape for the distribution (see [6]. Ultimately, the rate of convergence is too slow.

4. Outlook

In this work we discuss some important developments in reaction theories, based on a three-body formulation, used to analyse deuteron induced reactions. We summarize the results obtained comparing the CDCC method, and its more practical approximation ADWA, with the solution of the Faddeev equations in momentum-space. Presently there is a better understanding of the validity of approximations in ADWA and CDCC when computing breakup and transfer cross sections. The fact that ADWA and CDCC are not good approximations for transfer at intermediate to high energy already calls for an alternative formulation. Most critically, at present no benchmarks exist for heavier systems. Faddeev calculations for (d,p) on heavy nuclei are not possible due to the difficulty of handling the Coulomb interaction. It has thus become clear to the TORUS collaboration that indeed a new formalism [14], more efficient at handling the three-body equations, which provides stable solutions for a wide variety of cases, including a wide range of relevant beam energies (5-50 MeV/u), light and heavy systems, well bound and loosely bound nuclei, is absolutely necessary in the field.

When the input for the reaction theory is well constrained by either data or the underlying microscopic structure, and an accurate solution to the three-body scattering problem is possible for the wide range of interesting cases to be explored, the full scientific potential of the existing and upcoming new generation facilities will be realized.

We thank all members of the TORUS collaboration for useful discussions. The work of N.J.U. was supported by the Department of Energy topical collaboration grant DE-SC0004087. The work of F.M.N. was partially supported by the National Science Foundation grant PHY-0800026 and the Department of Energy through grant DE-FG52-08NA28552.

References

[1] *DOE Topical Collaborations in nuclear theory, Theory Of Reactions of Unstable iSotopes* (www.reactiontheory.org)
[2] Alt E O, Grassberger P, and Sandhas W 1967 *Nucl. Phys.* B **2** 167
[3] Deltuva A 2009 *Phys. Rev.* C **79** 021602(R); 2009 *Phys. Rev.* C **79** 054603
[4] Austern N, Iseri Y, Kamimura M, Kawai M, Rawitsher G, and Yahiro M 1987 *Phys. Rep.* **154** 125
[5] Nunes F M 2011 *Scholarpedia* **6(9)** 10497
[6] Upadhyay N J, Deltuva A and Nunes F M 2012 *Phys. Rev.* C **85** 054621
[7] Johnson R C and Tandy P C 1974 *Nucl. Phys.* A **235** 56
[8] Nunes F M and Deltuva A 2011 *Phys. Rev.* C **84** 034607
[9] Thompson I J 1988 *Comput. Phys. Rep.* **7** 167
[10] Nobre G P A, Dietrich F S, Escher J E ,Thompson I J, Dupuis M, Terasaki J, and Engel J 2010 *Phys. Rev. Lett.* **105** 202502
[11] Varner R L *et al.* 1991 *Phys. Rep.* **201** 57
[12] Koning A J and Delaroche J P 2003 *Nucl. Phys.* A **713** 231
[13] Becchetti, Jr F D and Greenlees G W 1969 *Phys. Rev.* **182** 1190
[14] Mukhamedzhanov AM, Eremenko V and Sattarov A 2012 *Phys. Rev.* C *in press.*

Clustering in nuclei

Yasuyuki Suzuki

Physics Department, Niigata University,Niigata 950-2181, Japan
RIKEN Nishina Center, Wako 351-0198, Japan

E-mail: suzuki@nt.sc.niigata-u.ac.jp

Abstract. I stress a role of clustering in describing the structure of such state that is difficult to understand from the mean field. The examples include the neutron-halo of ^{11}Li, ^{3}H+p and ^{3}He+n clustering in ^{4}He, and a unified description of the ground and first excited 0^+ states of ^{16}O.

1. Introduction

Cluster model has a long history. Especially a microscopic version of the cluster model is known for a long time as a unified theory of both structure and reactions for light nuclei [1]. Quite often a semimicroscopic cluster model taking into account the essence of the Pauli principle has been employed to reproduce those states which are difficult to understand from the shell model. One well-known example is the first excited 0^+ state of ^{16}O. The state together with other low-lying states of ^{16}O are reproduced well by an α+^{12}C cluster model [2] in which the rotational motion of ^{12}C is taken into account. The same model also successfully describes low-energy α+^{12}C scattering [3].

Since it stresses a particular type of nuclear excitations, the cluster model misses many other modes of excitations, e.g. monopole and quadrupole excitations that are well described in a symplectic model [4]. The coupling between the cluster and symplectic models was studied more than 20 years ago [5] using the SU(3) algebra [6], and it was applied to the ^{16}O and ^{20}Ne region [7]. This problem appears to attract some attention even today [8] as experimental observation of the monopole strength reaches higher in the excitation energy.

Some evidence indicating the importance of a cluster concept appears in heavy nuclei as well. The recent observation of the unnatural parity states in ^{212}Po [9] that have enhanced electric dipole transition strength is interpreted in an α+^{208}Pb cluster model [10] in which ^{208}Pb can be excited to the 3^- vibrational state. Note, however, that a microscopic calculation with the cluster model becomes more difficult as the mass number of the nuclear system increases. See, for example, a calculation for the α-decay of the ^{212}Po ground state [11].

I focus on three topics performed from the cluster viewpoint: the neutron halo structure of ^{11}Li [12], ^{3}H(t)+p and ^{3}He(h)+n clustering in the excited states of ^{4}He [13], and the shell-cluster competition in ^{16}O [14]. In each of these I stress different roles of the cluster model. In the first case we want to quantify how much the distortion of ^{9}Li improves the halo nature of ^{11}Li in a multi-cluster approximation. In the second topic I point out the emergence of a cluster structure in ^{4}He based on the results of a four-body calculation for ^{4}He with realistic potentials. In the last case I attempt at describing both the ground and first excited states of ^{16}O in a

^{12}C$+p+p+n+n$ five-body calculation. The basis functions we use are explicitly correlated Gaussians [15] that were proposed more than a half century ago, and further augmented greatly by developing a global vector representation of the angular part responsible for a rotational motion [16, 17, 18, 19]. An optimization of the basis functions is performed by a stochastic variational method [16, 17].

2. Neutron halo structure of ^{11}Li

A basic feature of ^{11}Li can be understood from its Borromean structure of ^{9}Li$+n+n$. It is interesting to know the extent to which the excitation of the ^{9}Li core plays a role in binding the halo neutrons. To examine this problem, we need to break the core. For the moment no *ab initio* calculation is available for ^{11}Li, obviously because ^{11}Li is so fragile that an accurate calculation is beyond a present theory. We used a microscopic multi-cluster approach in which ^{9,10,11}Li are described assuming α, t, and n clusters. That is, ^{9}Li and ^{11}Li are treated as four- and six-body systems of $\alpha+t+n+n$ and $\alpha+t+n+n+n+n$, respectively. The inter-cluster motion is solved as accurately as possible while the intrinsic wave functions of α and t are kept fixed as $0s$ wave functions. In this way we can quantify the distortion of the ^{9}Li core. The nucleon-nucleon interaction employed is the Minnesota potential, and the total wave function is properly antisymmetrized. This model is consistent with an earlier description of ^{7,8,9}Li [20].

The wave function of ^{9}Li is constructed in the $\alpha+t+n+n$ model by carefully optimizing the parameters of the correlated Gaussians. The properties obtained for ^{9}Li are listed in Table 1. They are in excellent agreement with experiment. The basis functions for ^{11}Li are constructed by adding two neutrons to ^{9}Li. They are characterized by Y- and T-type arrangements as well as the orbital and spin angular momenta of the neutrons. Two types of calculations are compared in Table 1. One is a frozen (fr.) model in which the ^{9}Li core is fixed to its ground state configuration. The other one called a full model takes into account the excited configurations, which allows for the distortion of the core. It is seen that a significant gain of the energy is obtained by accounting for the core distortion. The inclusion of the distortion makes the proton radius (r_p) considerably larger (by about 0.3 fm) and the neutron radius slightly smaller. This seems to be understood from a mechanism similar to what we found in ^{6}He [21]: The halo neutron receives more binding by gaining the attraction with the protons inside the ^{9}Li core as the np interaction is more attractive than the nn interaction. For this to happen, the core has to be distorted so as to have more chance that the protons can approach the halo neutrons more frequently. On the other hand the neutron size would shrink because of stronger binding. Though no accurate values on the proton radii were available when we published the results in 2002, our full model result agrees very well with the recently measured radius of ^{11}Li. The binding of ^{11}Li receives approximately equal contributions from p- and s-wave single-particle

Table 1. Properties of ^{9}Li and ^{11}Li. The energies ϵ are given from the two-neutron thresholds. See text for the frozen (fr.) and full models.

Nucleus		ϵ (MeV)	r_p (fm)	r_n (fm)	r_m (fm)	Q ($e$fm^2)	μ (μ_N)	$\sigma_{\rm R}$ (mb)
^{9}Li		-5.91	2.12	2.54	2.41	-3.37	3.40	799±3
	exp.	-6.09	2.11 [22]			-3.06	3.44	796±6
^{11}Li	fr.	-0.12	2.15	3.21	3.15	-3.52	3.21	1054±2
	full	-0.34	2.43	3.09	3.03	-3.71	3.23	1023±2
	exp.	-0.37	2.38 [22]			$\lvert Q\rvert$=3.33 [23]	3.67	1056±14

states. The deviation of the calculated magnetic moment from experiment is probably due to that of t. See Ref. [12] for detail.

3. Clustering due to parity inversion in ^{4}He

^{4}He is a doubly magic nucleus. This system is small enough to make it possible for us to perform an *ab initio* calculation using realistic nucleon-nucleon interactions. Its first excited state with 0^+ is located between the $t+p$ and $h+n$ thresholds. This state is found to have large spectroscopic amplitudes to those decay channels, consistent with the $3N+N$ cluster structure [24]. We find that in this state the spins of the $3N$ and N clusters are coupled to $I=0$, and the orbital angular momentum of the relative motion between them is $\ell=0$ [13].

There are several negative-parity states above the excited 0^+ state. Though they are conventionally considered on the shell-model basis, an inversion doublet picture may be useful to interpret some of their structure. That is, a negative-parity state that has basically the same $3N+N$ configuration may appear near the threshold [25]. Unlike in the case of $\alpha+^{16}$O or $\alpha+^{12}$C, the clusters are, however, not spinless in the present case. Actually when the $3N+N$ clusters are coupled to $I=1$, the odd ℓ value is favored for the relative motion, especially $\ell=1$ is most interesting. The coupling of I and ℓ suggests that some $3N+N$ cluster states with $J^\pi=0^-$, 1^-, 2^- may show up. The calculation using a realistic nucleon-nucleon potential confirms that the lowest $0^-, T=0$ and 2^- states can in fact be well understood from the cluster model rather than the shell model [13]. Figure 1 displays the spectroscopic amplitudes of the three states. As expected, each of three curves shows behavior indicating the $3N+N$ cluster structure. The peaks are centered around 2 fm near the $3N$ surface. The $3N+N$ spectroscopic factors are large, 0.58, 0.52, and 0.53 for the 0^-0, 2^-0, and 2^-1 states.

The negative-parity states of ^{4}He actually have broad widths. Though a bound-state method used in Ref. [13] may be useful to obtain the level spectrum qualitatively, one should in principle treat the continuum more properly. For example, strength functions for electric dipole [26] and spin-dipole operators give us information on the relevant resonance states.

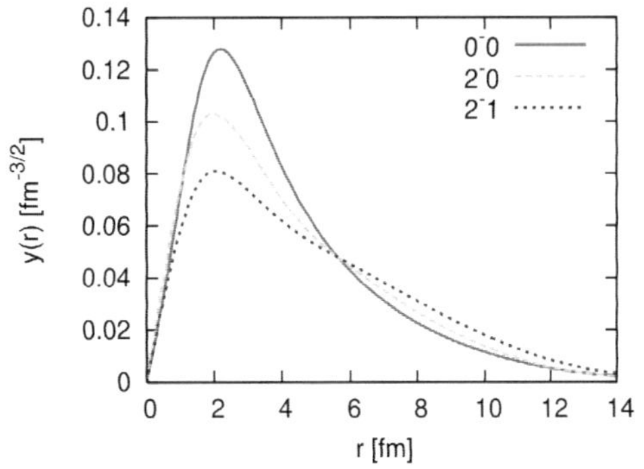

Figure 1. Spectroscopic amplitudes of the three lowest-lying negative-parity states of ^{4}He for the p-wave $h+n$ decay with $I=1$.

4. The 0^+ states of ^{16}O

A description of states that show prominent clustering on the basis of single-particle orbits is extremely hard because it requires in general enormous major-shell excitations. See, e.g.,

Refs. [27, 28]. It is thus obviously very interesting to attempt at describing such states without assuming the clusters *a priori*. The case of ^{4}He discussed in the previous section is one such attempt. Here we focus on the two 0^+ states of ^{16}O. The ground state is again doubly magic, and a typical shell-model state. The first excited state defies a shell-model description. We ask a question of if both states can be described well with a model built on a ^{12}C core plus four valence nucleons $(p+p+n+n)$.

^{12}C is treated by a rotational model. The valence nucleons are constrained not to include those orbits that are occupied by the core. A brief formulation is given in Ref. [14]. The Hamiltonian reads

$$H = T + H_{\text{core}} + \sum_i U_i + \sum_{i<j} V_{ij}, \tag{1}$$

where the potential U_i acts between the core and valence nucleon and V_{ij} is the potential acting among the valence nucleons. Letting ψ_{F} denote the orbits occupied by the core, the total wave function has to satisfy the condition

$$\langle \psi_{\text{F}} | \Psi \rangle = 0 \qquad \text{for all valence nucleons.} \tag{2}$$

As a first attempt we ignore the excitation of ^{12}C, and thus H_{core} is a constant. The potential U_i is determined following the rotational model and its strength is set to reproduce the single-particle energies of ^{13}C. The orbits ψ_{F} are taken as $0s_{1/2}$ and $0p_{3/2}$. The total wave function is assumed to be given as a combination of the correlated Gaussians with the global vector representation for the angular part. Figure 2 displays the energies of the two 0^+ states. The energy convergence is rather slow. This is probably because fulfilling the condition (2) is hard at least in a small basis dimension.

The calculation further continues up to the basis dimension 8500. The ground state is almost converged well, and its energy turns out to be close to experiment. After choosing the basis functions so as to optimize the excited state, the first excited state reaches only slightly above the threshold. The state still contains some forbidden-state components, which pushes up the energy. Estimating this energy loss, we think that the first excited state is obtained just below the α threshold. It appears that we thus obtain the two 0^+ states at the expected positions. An analysis of their structure is in progress.

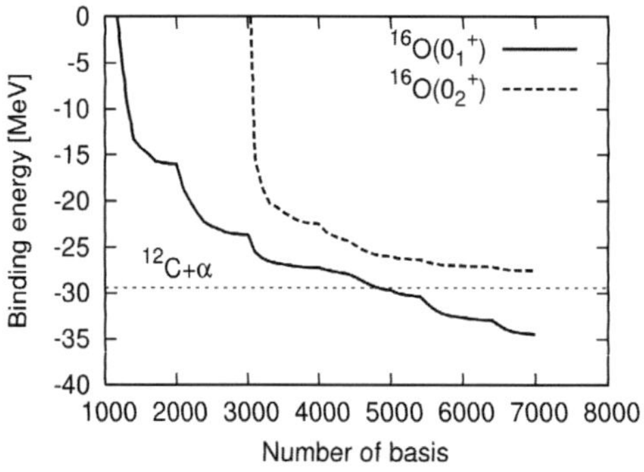

Figure 2. The ground and first excited 0^+ states of ^{16}O as a function of the basis dimension. The dotted line is the calculated α threshold. The Minnesota potential is used. The experimental energies of the two states are 7.16 and 1.11 MeV below the threshold.

5. Summary

In this talk I showed three examples. In a microscopic multi-cluster description of ^{11}Li the distortion of ^{9}Li is vital to gain the binding of the halo neutrons. An *ab initio* calculation using realistic forces confirms that some positive- and negative-parity states of ^{4}He show $3N + N$ cluster structure. Though a unified approach of ^{12}C+$p + p + n + n$ for ^{16}O is quite expensive, we have an indication that the ground and first excited states emerge at the expected energies.

Clustering is an important and useful concept in both structure and reactions. It certainly deviates from a mean field picture. Thus a fully microscopic description of clustering is quite expensive. It is a challenge of next generation to found clustering on innovative microscopic theories. With a progress of research for exotic nuclei, I expect physics of clustering will play a vital role furthermore.

Acknowledgments

It is my great pleasure that I could participate in the symposium in honor of Prof. J. P. Draayer's 70th birthday. My talk at the symposium is due to the collaborations with W. Horiuchi, K. Varga, and R. G. Lovas. I would like to thank all of them. This work is supported in part by the Grants-in-Aid for Scientific Research (C) (Grants No. 21540261 and No. 24540261) of the Japan Society for the Promotion of Science.

References

[1] Wildermuth K and Tang Y C 1977 *A Unified Theory of the Nucleus* (Braunschweig: Vieweg).
[2] Suzuki Y 1976 *Prog. Theor. Phys.* **55** 1751; *ibid.* **56** 111
[3] Suzuki Y, Ando T and Imanishi B 1978 *Nucl. Phys.* **A 295** 365
Suzuki Y and Imanishi B 1981 *Phys. Rev.* C **23** 2414
[4] Rosensteel G and Rowe D J 1977 *Phys. Rev. Lett.* **38** 10
[5] Suzuki Y 1986 *Nucl. Phys.* **A 448** 395
Suzuki Y and Hecht K T 1986 *ibid.* **A 455** 315
[6] Draayer J P and Akiyama Y 1973 *J. Math. Phys.* **14** 1904
Akiyama Y and Draayer J P 1973 *Comp. Phys. Comm.* **5** 405
[7] Suzuki Y 1987 *Nucl. Phys.* **A 470** 119
Suzuki Y and Hara S 1989 *Phys. Rev.* C **39** 658
[8] Yamada T *et al.* 2012 *Phys. Rev.* C **85** 034315
[9] Astier A *et al.* 2010 *Phys. Rev. Lett.* **104** 042701
[10] Suzuki Y and Ohkubo S 2010 *Phys. Rev.* C **82** 041303(R)
[11] Varga K, Lovas R G and Liotta R J 1992 *Phys. Rev. Lett.* **69** 37
Varga K, Lovas R G and Liotta R J 1992 *Nucl. Phys.* **A 550** 421
[12] Varga K, Suzuki Y and Lovas R G 2002 *Phys. Rev.* C **66** 041302(R)
[13] Horiuchi W and Suzuki Y 2008 *Phys. Rev.* C **78** 034305
[14] Horiuchi W and Suzuki Y 2010 *Mod. Phys. Lett.* A **25** 1899
[15] Boys S F 1960 *Proc. R. Soc. London. Ser.* **A 258** 402
Singer K 1960 *ibid.* 412
[16] Varga K and Suzuki Y 1995 *Phys. Rev.* C **52** 2885
[17] Suzuki Y and Varga K 1998 *Stochastic variational approach to quantum-mechanical few-body problems*, Lecture notes in physics Vol. 54 (Berlin: Springer)
[18] Suzuki Y, Horiuchi W, Orabi M and Arai K 2008 *Few-Body Syst* **42** 33
[19] Aoyama S, Arai K, Suzuki Y, Descouvemont P and Baye D 2012 *Few-Body Syst* **52** 97
[20] Varga K, Suzuki Y and Tanihata I 1995 *Phys. Rev.* C **52** 301
[21] Arai K, Suzuki Y and Lovas R G 1999 *Phys. Rev.* C **59** 1432
[22] Nörtershäuer W, Neff T, Sánchez R and Sick I 2011 *Phys. Rev.* C **84** 024307
[23] Neugart R *et al.* 2008 *Phys. Rev. Lett.* **101** 132502
[24] Hiyama E, Gibson B F and Kamimura M 2004 *Phys. Rev.* C **70** 031001(R)
[25] Horiuchi H and Ikeda K 1968 *Prog. Theor. Phys.* **40** 277
[26] Horiuchi W, Suzuki Y and Arai K 2012 *Phys. Rev.* C **85** 054002
[27] Suzuki Y, Arai K, Ogawa Y and Varga K 1996 *Phys. Rev.* C **54** 2073
[28] Neff T and Feldmeier H 2009 *Few-Body Syst* **45** 145

HITES 2012 IOP Publishing
Journal of Physics: Conference Series **403** (2012) 012031 doi:10.1088/1742-6596/403/1/012031

Three puzzles from nuclear astrophysics

W C Haxton

Department of Physics, University of California, Berkeley, and Lawrence Berkeley National Laboratory, MC-7300, Berkeley, CA 94720

E-mail: `haxton@berkeley.edu`

Abstract. I discuss three open problems in astrophysics where nuclear physics can make important contributions: the solar abundance problem, dark matter particle detection, and the origin of the r-process elements.

It is a great pleasure to take part in this celebration of Jerry Draayer's career in physics and to describe three problems in nuclear astrophysics that I find of interest. The descriptions here will be brief, as detailed recent references are available. The topics are 1) the solar abundance problem and its implications for the solar model and future solar neutrino experiments; 2) the elastic scattering of dark matter (DM) particles off nuclei; and 3) the possibility that an early, neutrino-driven r-process (rapid-neutron-capture process) could, in combination with neutron star mergers, account for what we know about the galactic evolution of r-process metals.

1. The solar abundance problem

The Standard Solar Model (SSM) [1, 2] assumes a homogeneous zero-age Sun because of the assumption that the proto-Sun passed through a fully convective Hayashi phase during gas-cloud collapse. The initial composition is divided into hydrogen $X_{\rm ini}$, helium $Y_{\rm ini}$, and metals $Z_{\rm ini}$. The relative abundances of the metals (elements heavier than helium) can be taken from meteoritic data (refractory elements) and from analyses of photospheric absorption lines (volatiles such as C, N, O, Ne, Ar). As $X_{\rm ini}+Y_{\rm ini}+Z_{\rm ini}=1$, the initial solar composition can be fully specified by adjusting $Y_{\rm ini}$, $Z_{\rm ini}$, as well as one other parameter, the mixing length $\alpha_{\rm MLT}$, until the SSM, evolved forward by 4.6 Gyrs, reproduces the present solar radius $R_\odot$, luminosity $L_\odot$, and surface helium abundance Y_S. The metals are important contributors to the opacity and consequently influence a variety of solar properties.

Traditionally the analysis of photospheric absorption lines was done with a 1D model that did not explicitly take into account surface stratification, velocities, and inhomogeneities [3]. More recently 3D parameter-free methods have been introduced, improving the consistency of line analyses [4]. This approach led to a significant reduction in $Z_{\rm ini}$, from ~ 0.0169 to ~ 0.0122, a change that makes the Sun more consistent with similar stars in the local neighborhood. However the agreement between the SSM and various helioseismic properties, such as the sound speed profile and the location of the convective zone boundary, deteriorated with the reduced $Z_{\rm ini}$ (see Fig. 1). The high-energy ^{8}B neutrino flux is also affected, with the high-metallicity GS98-SFII SSM flux being $\sim 22\%$ higher than low-metallicity AGSS09-SFII SSM flux [2].

The solar abundance problem can be summarized as follows: There is a significant discrepancy between a SSM tuned to our best description of the solar surface (a 3D treatment of

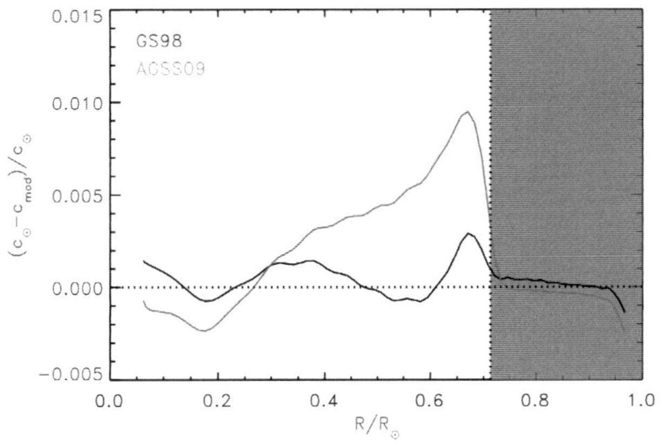

Figure 1. The differences between the sound-speed profile determined from helioseismology (denoted $c_\odot$) and those predicted (denoted c_{mod}) by the high-metallicity GS98-SFII and low-metallicity AGSS09-SFII SSMs. From [2].

photoabsorption lines) and one tuned to reproduce properties of the solar interior (helioseismic observables). The two models have very different Z_{ini}s.

This problem is potentially related to two others illustrated in Fig. 2. The first is the anomalous composition [6] of the gaseous giant planets, Jupiter, Saturn, Uranus, and Neptune. Jupiter and Saturn are enriched in C and N by factors of 4-7, relative to the Sun. The process of planetary formation, which is thought to have occurred late in the evolution of the solar system when only $\sim$ 5% of the nebular gas remained in a thin disk, appears to have scoured approximately 40-90 earth masses ($M_\oplus$) of metal from the disk [5]. The second comes from comparing the Sun's composition with that of solar twins [7], where systematic differences in metal abundances track condensation temperatures, with the solar ratio of volatiles to refractories being 0.05 to 0.10 dex higher than the average twin ratio.

These three problems could have a common origin, planetary disk processes that concentrate ice, dust, and other metal-rich material in the planets. If the depleted gas from which the metals were scrubbed were deposited onto the Sun after the Sun had developed its radiative core (so that the convective envelope at the time resembled the modern one, which contains less than 3% of the Sun's mass), then significant dilution of the convective zone could have occurred. This could produce a Sun with a higher metallicity radiative core and lower metallicity convective surface, invalidating the SSM's assumption of zero-age homogeneity. (Indeed, the excess metal in the planets is similar to the net difference in GS98-SFII and AGSS09-SFII SSM convective-zone metal, about 40 $M_\oplus$.)

There has been a significant amount of recent work focused on this conjecture. The nuclear physics connections include:

(i) Neutrino flux measurements that are sensitive to changes in solar core temperature T_c of 1%, and thus to metallicity differences comparable to those being debated.

(ii) The development of accreting nonstandard solar models (NSSMs), and the demonstration that such models are sharply constrained by solar neutrino and helioseismic measurements.

(iii) The possibility that future solar neutrino flux measurements could provide a direct measurement of the C+N content of the solar core.

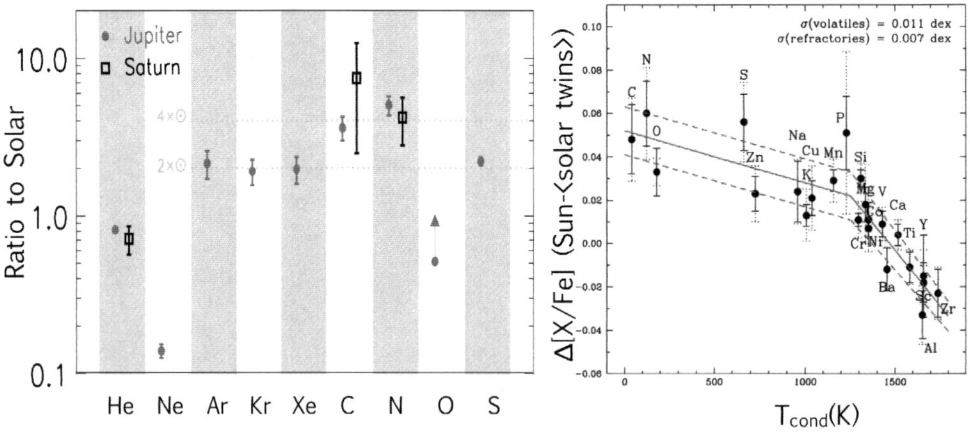

Figure 2. Left panel: Data from the Galileo and Cassini missions showing abundances in the gaseous atmospheres of Jupiter and Saturn, relative to solar. From [5]. Right panel: Deviation between solar metallicities, normalized to iron, from corresponding average quantities for solar twins, displayed as a function of the condensation temperature of the metal. The red line is the best linear fit to the data, with a break in slope allowed at intermediate temperatures, to improve the fit to the low- and high-condensation-temperature data. The red dashed lines form the error envelope. From [7].

In the case of the neutrino fluxes, the most sensitivity to T_c is exhibited by the ^{8}B neutrinos. However, despite current measurements that determine this flux to $\sim 3\%$ and despite SSM model uncertainties of $\sim 14\%$ in this flux that exceed the nominal discrepancy of nearly 22% between the competing models [2], little can be concluded: the AGSS09-SFII and GS98-SFII SSM global fits to the solar neutrino data have identical χ^2s. A metallicity intermediate between the two models would optimize the fit.

The impact of accretion on both neutrino fluxes and helioseismic variables has been evaluated in NSSMs. The work to date [8] varied the time, mass, and composition of the accreted material, but has not allowed the composition to vary in time – a simplification that one could question in view of the right panel of Fig. 2. The results are summarized in Figs. 10 and 11 of [2]. Helioseismic and neutrino observables tightly constrain such NSSMs. The envisioned candidate solution to the solar abundance problem – a low-Z surface consistent with AGSS09 abundances and a high-Z interior consistent with GS98 abundances – can be achieved with metal-free or metal-poor accretion and modest accreted masses $M_{ac} \sim 0.01 M_\odot$. Such models can bring the surface helium abundance into agreement with the data and produce some improvement in the sound-speed figure-of-merit $\langle \delta c/c \rangle$ and in the comparison of model and measured neutrino fluxes. However the lower surface metallicity that accompanies such models forces the convective-zone boundary R_S to move outward, contradicting results from helioseismology.

The third point above – checking interior solar abundances by using solar neutrinos as a probe – is possible because of the CN neutrinos. While the Sun produces 99% of its energy through the pp chain, proton burning also takes place through the CN I cycle, which produces significant fluxes of ^{13}N ($\sim 2.9 \times 10^8$/cm^2/s) and ^{15}O ($\sim 2.2 \times 10^8$/cm^2/s) neutrinos. These fluxes, like the ^{8}B neutrino flux, are very sensitive to T_c, which can be changed by varying any of the approximately 19 SSM input parameters, according to their assigned uncertainties. These uncertainties include the abundances of key metals. But in addition to the response in T_c due to changes in metallicity, the CN cycle has an additional linear dependence on the C+N abundance because these elements catalyze the proton burning.

In [9] a strategy for using the neutrino flux to measure the core C+N content was developed in which a weighted ratio of ^{15}O and ^{8}B fluxes is formed that is nearly independent of T_c, isolating this linear dependence on C+N. The result (including recent improvements in the analysis) is

$$\frac{\phi(^{15}\mathrm{O})}{\phi(^{15}\mathrm{O})^{\mathrm{SSM}}} = \left[\frac{\phi(^{8}\mathrm{B})}{\phi(^{8}\mathrm{B})^{\mathrm{SSM}}}\right]^{0.729} x_{\mathrm{C+N}}\left[1 \pm 0.006(\mathrm{SSM}) \pm 0.027(\mathrm{D}) \pm 0.099(\mathrm{nucl}) \pm 0.032(\theta_{12})\right]. \quad (1)$$

where $x_{\mathrm{C+N}}$ is the number abundance of C+N, normalized to the SSM value – the quantity one wants to constrain. The first factor on the right – effectively our T_c "thermometer" – is now known experimentally to 2%. The uncertainty denoted "SSM" reflects the effects on this relationship when 18 of 19 SSM input parameters are varied. The response of 0.6% reflects the success in constructing a T_c-independent relationship. But there is sensitivity to the 19th parameter, the diffusion coefficient D: Eq. (1) relates a contemporary neutrino flux measurement to the primordial C+N abundance. The composition of the modern Sun's core is altered by the gravitational settling of C+N over 4.6 Gyr, an effect that the SSM predicts is uncertain to 2.7%. The dominant uncertainty in Eq. (1) of 9.9% comes from nuclear physics, particularly the S-factors for ^{7}Be(p,γ) and ^{14}N(p,γ). These S-factor uncertainties can be reduced [10]: with a factor-of-two improvement, nuclear physics would no longer dominate the error analysis. Finally there is a 3.2% uncertainty due to weak interaction parameters.

On the left appears a flux that is currently not measured, but could be determined to a precision of about 7% in the upcoming solar neutrino experiment SNO+ [11], a larger and significantly deeper version of Borexino. The depth – the experiment is under construction in the former SNO cavity at SNOLab – is important, potentially reducing backgrounds due to cosmogenic ^{11}C by a factor of $\sim$ 70, relative to Borexino. When all errors are combined in quadrature, one finds that the solar core abundance of C+N could be determined to $\sim$ 13%, which can be compared to the current debate over differences in $\mathrm{Z_{ini}}$ of $\gtrsim$ 30%.

SNO+ could allow, for the first time in stellar astrophysics, a direct experimental comparison between core and surface compositions, thereby testing the SSM assumption of initial homogeneity. If it were shown that planetary formation affects host-star surface metallicity in a characteristic way – e.g., producing an elevated ratio of volatiles to refractories – this could be important in identifying stars that are more likely to host planets.

2. The nuclear physics of dark matter detection

Several independent astrophysical observations suggest that 85% of gravitating matter is dark, residing beyond the Standard Model. This DM must be long-lived or stable, and must lack strong couplings to itself and to baryons. A leading candidate for the DM is weakly interacting massive particles (WIMPs), slow-moving massive particles (e.g., 10 GeV to 1 TeV) that can elastically scattering off nuclear targets, transferring significant momentum ($\sim$ 100 MeV). Such DM particles can be detected through nuclear recoil in carefully designed detectors mounted in low-background environments, deep underground [12].

Indeed, there is a great deal of experimental activity in DM detection, including several claims of possible discovery, and disagreements among experimentalists about the compatibility of the claims with existing limits. Various natural targets are being used, containing isotopes such as ^{19}F, ^{23}Na, ^{70,72,73,74,76}Ge, ^{127}I, and ^{128,129,130,131,132,134,136}Xe. Thus targets include isotopes sensitive to vector (J$\geq$1/2) and tensor (J$\geq$1) interactions, such as ^{19}F(1/2$^+$) and ^{129}Xe(1/2$^+$); and ^{23}Na(3/2$^+$), ^{73}Ge(9/2$^+$), ^{127}I(5/2$^+$), and ^{131}Xe(3/2$^+$).

DM particles could in principle scatter off any scalar, vector, or tensor static nuclear moment. However analyses almost always treat the nuclear response in the point-nucleus limit, despite momentum transfers comparable to the inverse nuclear size, so that the scattering is

$$\text{spin}-\text{independent (S.I.)} \;\Rightarrow\; \langle g.s.|\sum_{i=1}^{A}(a_0^F + a_1^F \tau_3(i))|g.s.\rangle \quad \text{or}$$

$$\text{spin}-\text{dependent (S.D.)} \;\Rightarrow\; \langle g.s.|\sum_{i=1}^{A}\vec{\sigma}(i)(a_0^{GT} + a_1^{GT}\tau_3(i))|g.s.\rangle. \tag{2}$$

Most often theoretical treatments are "top-down" – calculations motivated by a specific high-energy theory, for which a nuclear response involving a S.I. or S.D. response is then derived.

Recently a different approach was taken in which the most general Galilean-invariant effective DM-nucleon interaction was derived, including operators up to quadratic order in momenta and exchanges of particles of spin-1 or less [13]. That interaction has 11 couplings ($\times$ 2 if isospin is included), in contrast to the S.I. and S.D. cases discussed above:

$$\begin{aligned} L_{ET} &= a_1 1 + a_2 \vec{v}^\perp \cdot \vec{v}^\perp + a_3 \vec{S}_N \cdot (\vec{q} \times \vec{v}^\perp) + a_4 \vec{S}_\chi \cdot \vec{S}_N + i a_5 \vec{S}_\chi \cdot (\vec{q} \times \vec{v}^\perp) + a_6 \vec{S}_\chi \cdot \vec{q}\, \vec{S}_N \cdot \vec{q} \\ &+ a_7 \vec{S}_N \cdot \vec{v}^\perp + a_8 \vec{S}_\chi \cdot \vec{v}^\perp + i a_9 \vec{S}_\chi \cdot (\vec{S}_N \times \vec{q}) + i a_{10} \vec{S}_N \cdot \vec{q} + i a_{11} \vec{S}_\chi \cdot \vec{q}\,. \end{aligned} \tag{3}$$

Here $\vec{S}_N$ and $\vec{S}_\chi$ are the nucleon and DM particle spin, $\vec{q}$ is the three-momentum transfer, and $\vec{v}^\perp$ is the Galilean-invariant velocity operator defined in terms of the ingoing (outgoing) DM particle χ and nuclear target T velocities $\vec{v}_{\chi,\text{in}}$ ($\vec{v}_{\chi,\text{out}}$) and $\vec{v}_{T,\text{in}}$ ($\vec{v}_{T,\text{out}}$), respectively,

$$\vec{v}_T^\perp \equiv \frac{1}{2}\left(\vec{v}_{\chi,\text{in}} + \vec{v}_{\chi,\text{out}} - \vec{v}_{T,\text{in}} - \vec{v}_{T,\text{out}}\right), \tag{4}$$

where the nuclear velocities are defined as averages over the individual nucleon velocities,

$$\vec{v}_{T,\text{in}} = \frac{1}{A}\sum_{j=1}^{A}\vec{v}_{N,\text{in}}(j) \qquad \vec{v}_{T,\text{out}} = \frac{1}{A}\sum_{j=1}^{A}\vec{v}_{N,\text{out}}(j)\,. \tag{5}$$

The coefficients of the eleven effective operators include an isospin dependence that has been suppressed, e.g., $a_1 \equiv a_1^0 + a_1^1\tau_3(i)$, to allow for arbitrary couplings to protons and neutrons.

This starting point provides a vehicle for determining the corresponding most general elastic nuclear response, using the prescription that A-body nuclear currents and charges are the sums over the corresponding one-body nucleon currents and charges. One finds

$$\begin{aligned} H_{ET} &= \sum_{i=1}^{A}\Big[l_0(i)\delta(\vec{x}-\vec{x}_i) + l_0^A(i)\frac{1}{2M}\left(-\frac{1}{i}\overleftarrow{\nabla}_i \cdot \vec{\sigma}(i)\delta(\vec{x}-\vec{x}_i) + \delta(\vec{x}-\vec{x}_i)\vec{\sigma}(i)\cdot\frac{1}{i}\overrightarrow{\nabla}_i\right) \\ &+ \vec{l}_5(i)\cdot\vec{\sigma}(i)\delta(\vec{x}-\vec{x}_i) + \vec{l}_M(i)\cdot\frac{1}{2M}\left(-\frac{1}{i}\overleftarrow{\nabla}_i\delta(\vec{x}-\vec{x}_i) + \delta(\vec{x}-\vec{x}_i)\frac{1}{i}\overrightarrow{\nabla}_i\right) \\ &+ \vec{l}_E(i)\cdot\frac{1}{2M}\left(\overleftarrow{\nabla}_i \times \vec{\sigma}(i)\delta(\vec{x}-\vec{x}_i) + \delta(\vec{x}-\vec{x}_i)\vec{\sigma}(i)\times\overrightarrow{\nabla}_i\right)\Big]\,. \end{aligned} \tag{6}$$

Here each of the associated WIMP amplitudes is a function of the various a_i appearing in Eq. (3) (including the isospin dependence), e.g.,

$$\begin{aligned} l_0(i) &= \left(a_1^0 - i(\vec{q}\times\vec{S}_\chi)\cdot\vec{v}_T^\perp\, a_5^0 + \vec{S}_\chi\cdot\vec{v}_T^\perp\, a_8^0 + i\vec{q}\cdot\vec{S}_\chi\, a_{11}^0\right) \\ &+ \left(a_1^1 - i(\vec{q}\times\vec{S}_\chi)\cdot\vec{v}_T^\perp\, a_5^1 + \vec{S}_\chi\cdot\vec{v}_T^\perp\, a_8^1 + i\vec{q}\cdot\vec{S}_\chi\, a_{11}^1\right)\tau_3(i) \equiv l_0^0 + l_0^1\tau_3(i). \end{aligned} \tag{7}$$

Equation (6) involves five nuclear densities, each of which is familiar from studies of semi-leptonic electroweak interactions,

(i) The isoscalar and isovector vector charge operators $\rho_V^0(\vec{x}) = \sum_{i=1}^A \delta(\vec{x}-\vec{x}_i)$ and $\rho_V^1(\vec{x}) = \sum_{i=1}^A \delta(\vec{x}-\vec{x}_i)\tau_3(i)$ familiar from elastic electron scattering;

(ii) The isoscalar axial-vector charge operator $\rho_A^0(\vec{x}) = \sum_{i=1}^A \frac{1}{2M}\left(-\frac{1}{i}\overleftarrow{\nabla}_i \cdot \vec{\sigma}(i)\delta(\vec{x}-\vec{x}_i) + \delta(\vec{x}-\vec{x}_i)\vec{\sigma}(i)\cdot\frac{1}{i}\overrightarrow{\nabla}_i\right)$ and the corresponding isovector operator that mediates axial-charge $0^+ \leftrightarrow 0^-$ β decay;

(iii) The isoscalar axial-vector spin operator $\vec{j}_A^{\,0}(\vec{x}) = \sum_{i=1}^A \vec{\sigma}(i)\delta(\vec{x}-\vec{x}_i)$ and its isovector counterpart, familiar from inelastic neutrino reactions in the long-wavelength limit (LWL);

(iv) The isoscalar vector velocity current $\vec{j}_{V,v}^{\,0}(\vec{x}) = \sum_{i=1}^A \frac{1}{2M}\left(-\frac{1}{i}\overleftarrow{\nabla}_i\delta(\vec{x}-\vec{x}_i) + \delta(\vec{x}-\vec{x}_i)\frac{1}{i}\overrightarrow{\nabla}_i\right)$ and its isovector counterpart, familiar from electron scattering; and

(v) The isoscalar vector spin-velocity current $\vec{j}_{V,sv}^{\,0}(\vec{x}) = \sum_{i=1}^A \frac{1}{2M}\left(\overleftarrow{\nabla}_i \times \vec{\sigma}(i)\delta(\vec{x}-\vec{x}_i) + \delta(\vec{x}-\vec{x}_i)\vec{\sigma}(i)\times\overrightarrow{\nabla}_i\right)$ and its isovector counterpart.

The vector spin-velocity current appears in Serot's [14] treatment of semi-leptonic electroweak interactions to order $1/M^2$. In that context, because of the properties of this current under time reversal, it is accompanied by a factor of the energy transfer q_0 that vanishes for elastic transitions.

Equation (6) follows from L_{ET}, but could have been written down directly as a nuclear-level effective theory, as it includes all possible local one-body operators of rank ≤ 1 that can be constructed with $\vec{\sigma}$, τ_3, and one gradient.

Nuclear ground states are approximate eigenstates of parity and time-reversal. Thus, ignoring small components of the wave function that break these symmetries, one should retain only multipoles generated from the nuclear densities that transform appropriately under parity and time reversal. The charge operators of Eq. (6) generate even and odd multipoles: because all even and all odd multipoles transform similarly, it is sufficient to consider only the J=0 and J=1 cases. The current operators generate longitudinal, transverse electric, and transverse magnetic multipoles. Again, we can limit our consideration to the lowest contributing even and odd multipoles, as the generalization to other cases is immediate. The multipoles eliminated by symmetry considerations are indicated in Fig. 3.

Fig. 3 also gives the LWLs of the surviving multipoles, which are even vector charge operators (C_0, C_2,...), even vector spin-velocity longitudinal operators (L_0, L_2,....), odd axial spin longitudinal operators (L_1^5, L_3^5, ...), even vector spin-velocity electric operators ($T_2^{\rm el}$, $T_4^{\rm el}$, ...), odd axial spin electric operators ($T_1^{\rm el5}$, $T_3^{\rm el5}$,...), and odd vector spin magnetic operators ($T_1^{\rm mag}$, $T_3^{\rm mag}$,...). Thus there are six independent response functions – not simply the two (S.D., S.I.) conventionally treated.

The $C_{J={\rm even},M}$ operator of Fig. 3 is the standard S.I. response, while $L_{J={\rm odd},M}^5$ and $T_{J={\rm odd},M}^{\rm el5}$ are the longitudinal and transverse components that are usually summed to give the S.D. response. These two contributors to the S.D. response have distinct form factors and couple in distinct ways to WIMP amplitudes, however. The transverse magnetic response $T_{J={\rm odd},M}^{\rm mag}$ is new, generated by the vector velocity current. In the LWL the operator is proportional to the nuclear orbital angular momentum operator $\vec{\ell}(i)$, and thus is explicitly connected with the composite nature of the nucleon (apparent from the accompanying factor of q in the right panel of Fig 3). Thus there are three responses – one proportional to $\vec{\ell}$ and two proportional to $\vec{\sigma}$ in the LWL – that transform under rotations as $\langle J_i|\vec{J}_M|J_i\rangle$. The fifth and sixth responses of Fig. 3 are generated by the spin-velocity current and also reflect nucleon compositeness. They transform as longitudinal and transverse electric projections of the vector spin-velocity current $\sim \delta(\vec{x}-\vec{x}_i)\vec{\sigma}(i)\times\vec{\nabla}$, respectively – just as the two S.D. responses arise as the longitudinal and transverse electric projections of the axial spin current. The first, $L_{J={\rm even},M}$, generates a

	even	odd
vector	C_0	~~C_1~~
axial	~~C_0^5~~	~~C_1^5~~

	even	odd	even	odd	even	odd
axial spin	~~L_0^5~~	L_1^5	~~T_2^{el5}~~	T_1^{el5}	~~T_2^{mag5}~~	~~T_1^{mag5}~~
vector velocity	~~L_0~~	~~L_1~~	~~T_2^{el}~~	~~T_1^{el}~~	~~T_2^{mag}~~	T_1^{mag}
vector spin-velocity	L_0	~~L_1~~	T_2^{el}	~~T_1^{el}~~	~~T_2^{mag}~~	~~T_1^{mag}~~

Response Type	Long-wavelength Limit
$C_{J=\mathrm{even},M}$	$\frac{1}{\sqrt{4\pi}}1(i)$
$L^5_{J=\mathrm{odd},M}$	$\frac{1}{2\sqrt{3\pi}}\sigma_{1M}(i)$
$T^{\mathrm{el5}}_{J=\mathrm{odd},M}$	$\frac{1}{\sqrt{6\pi}}\sigma_{1M}(i)$
$T^{\mathrm{mag}}_{J=\mathrm{odd},M}$	$-\frac{q}{2m_N\sqrt{6\pi}}\ell_{1M}(i)$
$L_{J=\mathrm{even},M}$	$-\frac{q}{3m_N\sqrt{4\pi}}\vec{\sigma}(i)\cdot\vec{\ell}(i)$; $-\frac{q}{m_N\sqrt{30\pi}}[x_i\otimes(\vec{\sigma}(i)\times\frac{1}{i}\vec{\nabla})_1]_{2M}$
$T^{\mathrm{el}}_{J=\mathrm{even},M}$	$-\frac{q}{m_N\sqrt{20\pi}}[x_i\otimes(\vec{\sigma}(i)\times\frac{1}{i}\vec{\nabla})_1]_{2M}$

Figure 3. Left: Parity and time-reversal constraints on multipoles mediating WIMP elastic scattering. Only the lowest even and lowest odd multipoles are shown, as all even and all odd multipoles transform similarly. Multipoles eliminated by parity (time reversal) are indicated by a red (blue) strike. Right: Table of the surviving parity- and time-reversal-allowed responses, and their LWLs.

scalar nuclear response proportional to the spin-orbit interaction $\vec{\sigma}(i)\cdot\vec{\ell}(i)$ as well as a tensor contribution, in the LWL. The scalar response is present for all nuclei (that is, regardless of ground-state spin, like the S.I. charge coupling). But its properties are very different from those of the S.I. operator: $\vec{\sigma}(i)\cdot\vec{\ell}(i)$ produces a coherent isoscalar contribution over closed spin-orbit partner shells, e.g., the closed $1f_{7/2}$ shell for Ge isotopes and the closed $1g_{9/2}$ shell for Xe or I. The sixth response function, $T^{\mathrm{el}}_{J=\mathrm{even},M}$, is the transverse electric projection of the spin-velocity current.

Figure 4 shows response functions calculated in the shell model for natural targets of F, Na, Ge, I, and Xe [13] – these responses could be labeled as S.I., S.D. (transverse axial electric), orbital angular momentum, and spin-orbit, in the LWL. The interactions used for the 2s1d-shell nuclei F and Na, the $2\mathrm{p}1\mathrm{f}_{5/2}1\mathrm{g}_{9/2}$-shell Ge isotopes, and the $3\mathrm{s}2\mathrm{d}1\mathrm{g}_{7/2}1\mathrm{h}_{11/2}$ isotopes of I and Xe were those of Brown and Wildenthal [15], the Madrid-Strasbourg group [16, 17], and Baldridge and Vary [18], respectively. The truncations required for the nuclei beyond the sd-shell are described in [13]: the model spaces for these exploratory calculations were modest, ranging up to $\sim$ 600,000 m-scheme Slater determinants, after reduction by symmetries. Very large differences in experimental sensitivity appear between detector materials, as the interaction and thus the response is varied – the space of possibilities is much larger than in the standard S.I./S.D. description. One might think this is unfortunate – the correlations between different experiments will be harder to anticipate – but I would argue with this view. We know nothing about DM interactions, but we know a lot about nuclear responses. The six nuclear responses provide six nuclear dials that can be used to probe the nature of DM. The experimental DM community might benefit from greater interaction with the nuclear physics community, in formulating strategies for exploiting the available nuclear responses to determine DM particle properties.

3. The search for the r-process site

The search for the site or sites of the r-process, the rapid-neutron-capture process thought to be responsible for synthesizing about half the heavy elements, has been a long and frustrating one. For the past two decades the leading candidate site [19] has been the neutrino-driven wind that blows off the proto-neutron star surface in a core-collapse supernova. As the slightly neutron-rich, high-entropy nucleon gas expands off the star and cools, the protons are locked up

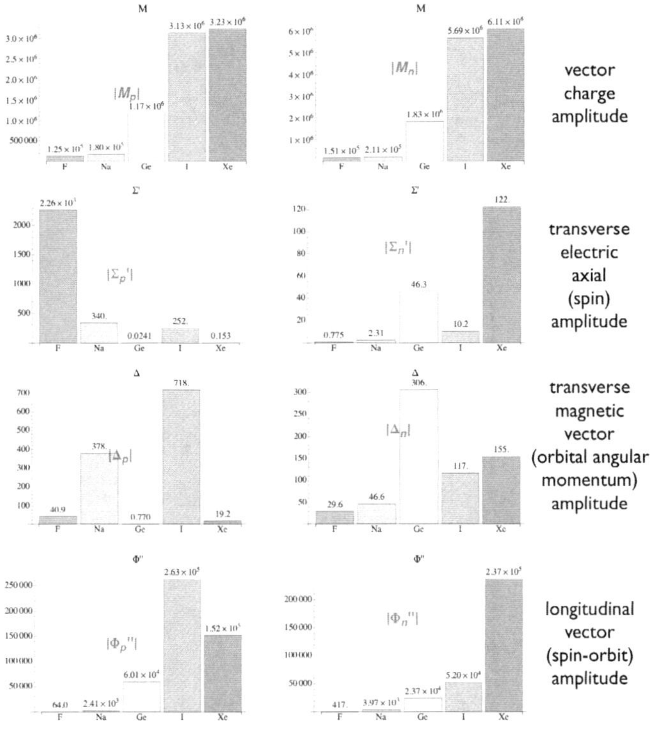

Figure 4. Magnitudes of response function amplitudes are compared for four of the response functions discussed in the text (the S.I. vector response, the S.D. transverse electric axial response, the transverse magnetic vector response, and the longitudinal vector response). The isospin couplings are to protons only (left panels) and neutrons only (right panels). From [13].

in an alpha freeze-out. A small number of seed nuclei can be synthesized through alpha-induced reactions. The remaining free neutrons then capture on these seeds to produce heavy nuclei. If the necessary neutron/seed ratio of $\gtrsim 100$ can be maintained, the synthesis can continue up to the transuranics.

Unfortunately a number of difficulties have arisen in explorations of the neutrino-wind mechanism. It is difficult to keep the number of seeds from growing due to reactions such as $\alpha + \alpha + n \rightarrow {}^9\text{Be} + \gamma$ and $3\alpha \rightarrow {}^{12}\text{C} + \gamma$. It is also difficult to maintain the needed neutron fluence. The wind is needed as the ejection mechanism, but every $\nu_e + n \rightarrow p + e^-$ reaction results in the net loss of two neutrons via $2n + 2p \rightarrow {}^4\text{He} + \gamma$. Consequently, it was recently concluded that the net production in the standard neutrino wind scenario was limited to some N=50 closed-shell nuclei and to some light p-nuclei [20].

This presents quite a puzzle. While there is an alternative and very attractive site that could be responsible for the bulk of the r-process material found in the Milky Way – neutron star mergers – it is unlikely that this mechanism can account for the r-process patterns we see in certain old, metal-poor halo stars (left panel, Fig. 5). Few neutron-star mergers are expected at low metallicity because it takes time for the progenitor stars to evolve, as the right panel of Fig. 5 shows. Furthermore, because metal-poor stars sample the r-process locally at a time

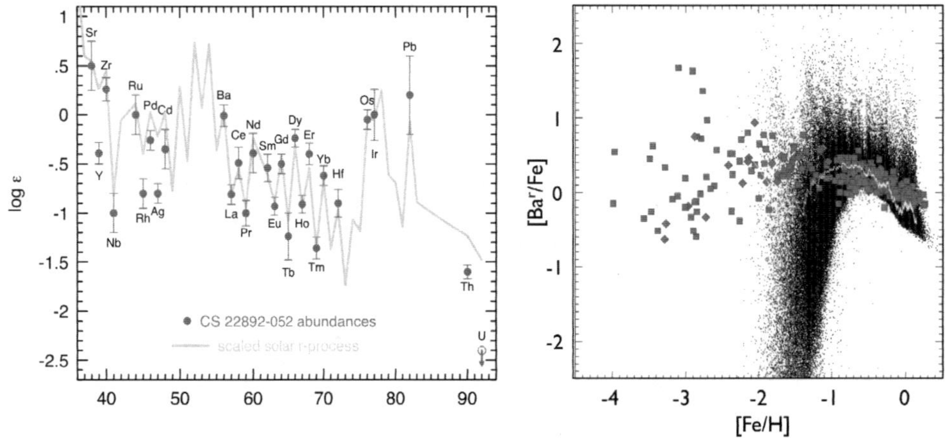

Figure 5. Left: Metal abundances from metal-poor halo star CS 22892-052 compared to the scaled solar distribution. From [21]. Right: The observed r-process fraction of Ba relative to Fe (red squares and diamonds) as a function of [Fe/H], compared to the model calculations (black dots) of the evolution in the galactic rate of neutron star mergers, assuming an average merger rate of 20/Myr and a coalescence timescale of 1 Myr. From [22].

when the galaxy is chemically inhomogeneous, one can estimate the rate of r-process events in the early galaxy from the fluctuations in observed abundances. One finds $\sim 10^{-2}$/y. While the uncertainty in this estimate is large, the rate is typical of core-collapse supernovae and much higher than would be expected for neutron star mergers, even today.

Studies of metal-poor stars have been the main source of new information on the r-process. The metals on the surfaces of such stars are thought to have arisen from a single, or perhaps a few, nearby r-process sites. The correspondence between the abundances in such stars and the scaled r-process distribution of our Sun (left panel, Fig. 5) suggests that the r-process distribution produced in a single event is similar to that averaged over many events.

Because of the difficulty of finding a robust r-process that naturally accounts for both the metal-poor-star data and integrated r-process yields, my collaborators and I recently explored another scenario:

- One r-process component that operates at early times, is associated with supernovae, and depends on the low-metallicity environments found in early massive stars. This early r-process would need to account for the metal-poor-star r-process abundance pattern and the observed local enrichments. It would need to operate until a metallicity of -2.5 to -2.0, the time a more robust r-process would then take over.
- The second process, unconstrained by the metal-poor star data but possibly responsible for the bulk of r-process synthesis, would then need to turn on in a manner consistent with r-process chemical evolution studies. Neutron-star mergers would be the natural candidate for this second r-process.

There is also a candidate mechanism for the first r-process, neutrons produced by neutrino reactions in the He shell of a type II supernova [23]. This mechanism was originally suggested as "the r-process," studied in high-metallicity stars, and found to fail there because the neutron/seed ratio was about a factor of 20 too low [24]. This failure reflected both the relatively

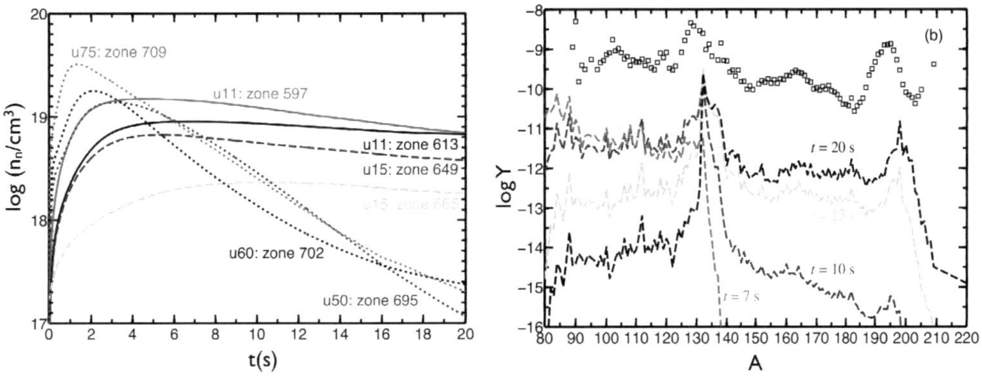

Figure 6. Left: The neutron number density as a function of time generated by neutrino interactions in the He zone of allow-metllicity iron-core supernova. The results are for specific zones in metal-poor progenitors with masses between 11 and 75 $M_\odot$. Right: The resulting number fractions $Y(t)$ at times of 7, 10, 15, and 20 s, compared to the solar r-process pattern (squares). From [25].

weak neutron source and the high abundance of neutron poisons such as ^{14}N in the He shells of the progenitor stars that were studied. The mechanism exploits the fact that neutrons scatter off He without capturing, and thus have an enhanced probability of capture on Fe and other heavy seeds.

We re-examined this r-process in the context of metal-poor stars, using a library of progenitors generated with the Kepler code, and constructing more complete reaction networks. The details are given in [25]. Many of the conclusions of earlier studies were altered: the nuclear physics operated differently (for example, neutral-current production of neutrons was found to be unimportant), and the effects of neutron poisons were found to be less severe. Most important, the study included the effects of neutrino oscillations. With the assumption of an inverted hierarchy and a relatively hard heavy-flavor neutrino spectrum, neutron abundances of $\sim 10^{19}/\mathrm{cm}^3$were produced and maintained for long times in a variety of stars.

The mechanism is unusual: It is a cold r-process, with equilibrium established by neutron capture $\leftrightarrow$ β decay, and with a path farther from the neutron-drip line than in hot r-processes like the ν-driven wind. This has consequences for the abundance pattern and for the time-development of the r-process (see Fig. 6). The time scales are exceptionally long, with the third peak forming only after $\sim$ 20 seconds. Furthermore, synthesis continues after passage of the shock wave, augmented by the compression of the He shell and extending to times $\sim$ 60 seconds (not shown in the figure). The mechanism works well for metallicities of $10^{-4} < Z < 10^{-3}$ in Fe-core supernovae between 11-16 $M_\odot$ and for very massive progenitors of 49-75 $M_\odot$. It naturally accounts for some of the features seen in metal-poor stars, such as a highly variable yield in Eu/Fe. It is also interesting from the neutrino physics viewpoint, turning on (inverted) or off (normal) with the choice of hierarchy. I am not aware of any other example in astrophysics as dramatically dependent on the neutrino hierarchy.

Acknowledgement
I want to thank my key collaborators on the work reported here, including Aldo Serenelli, Liam Fitzpatrick, Ami Katz, Projjwal Banerjee, Alex Heger, and Yong Qian, as well as the organizers

for the opportunity to take part in this celebration of the scientific career of a long-time friend, Jerry Draayer. This work was supported in part by the U.S. Department of Energy under contracts DE-SC00046548 (UC Berkeley) and DE-AC02-98CH10886 (LBNL).

References

[1] Bahcall J N 1989 *Neutrino Astrophysics* (Cambridge: Cambridge University Press)
[2] Haxton W C, Robertson R G H, and Serenelli A M 2013 *Ann. Rev. Astron. Astrophys.* **51** (in press)
[3] Grevesse N and Sauval A J 1998 *Space Sci. Rev.* **85** 161
[4] Asplund M, Grevesse N, Sauval A J, and Scott P 2009 Ann. Rev. Astron. Astrophys. **47** 481
[5] Guillot T 2005 *Ann. Rev. Earth Planetary Sci.* **33** 493
[6] Nordlund A 2009 Solar Twins and Possible Solutions of the Solar and Jupiter Abundance Problems arXiv:0908.3479
[7] Melendez J, Asplund M, Gustafsson B, and Yong D 2009 *Ap. J.* **704** L66
[8] Serenelli A M, Haxton W C, and Peña-Garay C 2011 *Ap. J.* **743** 24
[9] Haxton W C and Serenelli A M 2008 *Ap. J.* **687** 678
[10] Adelberger E G *et al.* 2011 *Rev. Mod. Phys.* **70** 1275
[11] Chen M 2006 *SNO and SNO+* (AIP Conf. Proc. **944**, AIP) p 25
[12] Feng J L 2010 *Ann. Rev. Astron. Astrophys.* **48** 495
[13] Fitzpatrick A L, Haxton W C, Katz E, Lubbers N, and Xu Y, The Effective Field Theory of Dark Matter Direct Detection arXiv:1203.3542
[14] Serot B 1979 *Nucl. Phys.* **A322** 408
[15] Brown B A and Wildenthal B 1988 *Ann. Rev. Nucl. Part. Sci.* **38** 29
[16] Caurier E, Menendez J, Nowacki F, and Poves A (2008) *Phys. Rev. Lett.* **100** 052503
[17] Menendez J, Poves A, Caurier E, and Nowacki F (2009) *Nucl. Phys. A* **818** 139
[18] Baldridge W and Dalton B (1978) *Phys. Rev. C* **18** 539.
[19] Woosley S E and Hoffmann R D 1992 *Ap. J.* **295** 202
[20] Roberts L F, Woosley S E, and Hoffmann R D 1992 *Ap. J.* **722** 1
[21] Sneden C *et al.* 2003 *Ap. J.* **591** 936
[22] Argast D, Samland M, Thielemann F-K, and Qian, Y-Z 2003 *A. & A.* **416** 997
[23] Epstein R, Colgate S, and Haxton W C 1988 *Phys. Rev. Lett.* **61** 2038
[24] Woosley S E, Hartmann D, Hoffman R D, and Haxton W C 1990 *Ap. J.* **356** 272
[25] Banerjee P, Haxton W C, and Qian Y-Z 2011 *Phys. Rev. Lett* **106** 201104

Stardust, supernovae, neutrinos, and the chirality of the amino acids

R N Boyd[1], T Kajino[2,3], and T Onaka[3]
[1]Lawrence Livermore National Laboratory, Livermore, CA 94550, USA
[2]National Astronomical Observatory of Japan, 2-21-1 Mitaka, Tokyo, 181-8588 Japan
[3]University of Tokyo, Department of Astronomy, Graduate School of Science, 7-3-1, Hongo, Bunkyo-ku, Tokyo 113-0033 Japan

E-mail: richard11boyde@comcast.net

Abstract. A mechanism for creating amino acid enantiomerism that involves selection of one chirality by interactions with the neutrinos from a core-collapse supernova is defined. This selection mechanism involves the alignment between the spins of the neutrinos and of the ^{14}N nuclei in the amino acids, or in precursor molecules, which in turn couple to the molecular chirality. Details of the chiral selection, as well as the subsequent chemical evolution and galactic mixing that would ultimately populate the Galaxy with the selected species are discussed. The resulting amino acids could either be the source thereof on Earth, or could have produced the chirality that was ultimately achieved for Earthly amino acids.

1. Introduction

A longstanding puzzle in astrobiology has been the existence of left-handed amino acids in nature and the virtual exclusion of their right-handed forms. This is especially puzzling because most mechanisms suggested for creating this "enantiomerism" would create one chirality in nature locally but would create equal numbers of the other somewhere else. The total dominance of the left-handed forms on Earth is well known, but the left-handed forms appear also to be preferred in the cosmos, based on meteoritic evidence [1-4]. This is especially interesting because the preference for left-handedness apparently persists throughout the cosmos, but both forms were apparently frozen into the meteorites before the right-handed forms could be eliminated. Recent work [5] has confirmed the preference for the left-handed forms.

It is generally accepted that if some mechanism can introduce an imbalance in the populations of the left- and right-handed forms of any amino acid [6], successive synthesis or evolution of the molecules will amplify this enantiomerism to produce ultimately a single form. What is not well understood, though, is the mechanism by which the initial imbalance can be produced, and the means by which it always appears to produce the left-handed chirality for the amino acids.

One suggested mechanism involves processing of a population of amino acids by circularly polarized light [7-9]; this can select one chirality over the other. However, this solution does not easily explain how the physical conditions that would select one form in one place would not select the other in a different location. Another possibility [10] invoked selective processing via the weak interaction; it focused on the β-decay of ^{14}C to produce the selective processing. Another suggestion [11] involving the weak interaction assumed that neutrinos emitted by a core-collapse supernova would selectively destroy the carbon or the hydrogen in the amino acids to produce enantiomerism in the atoms containing the nuclei that were not destroyed. However, it was not possible in either of these studies to show how β-decay or neutrino induced reactions could produce chiral-selective molecular destruction.

In this paper we also invoke the weak interaction to perform selective destruction of one chirality of molecules [12]. The key to this mechanism is the selective processing of the amino acids that exist in the dust grains, or larger meteoroids or comets, that would exist in the vicinity of a core-collapse supernova, but would be specifically related to ^{14}N, a nucleus with spin of 1 that is common to all amino acids. Two features of supernovae are important; the first being the magnetic field that would be established as the star collapsed to a neutron star or a black hole, and the second being the intense flux of neutrinos that would be emitted as the star collapsed. The magnetic fields only have to couple to the molecules strongly enough to produce some orientation of their non-zero spin nuclei. However, the ^{14}N must also couple in some way to the chirality of the molecules; we assume a model to describe this. We also assume that the ^{14}N spin is correlated with the molecular chirality in a way that is consistent through all the amino acids. Then, the neutrinos will preferentially interact with the ^{14}N atoms in one of the chiral forms, and convert the ^{14}N to ^{14}C, destroying that molecule and thus preferentially selecting the other chiral form. These issues are discussed in Section 2.

Although supernovae cannot produce total galactic chirality by themselves, chemical amplification of the chirality-selected, biologically-interesting molecules would drive the dominant form toward total local dominance. Then Galactic mixing, would be able to establish the dominant form throughout the Galaxy. These two mechanisms would make it likely that the Galaxy would be populated everywhere with a preferred chiral form. These issues are described in Section 3. In Section 4 we discuss the constraints that the photons from the supernovae that produce the chiral selection, or from their progenitor stars, might impose on the parameters of the model. Section 5 presents our conclusions.

2. Selective Destruction Mechanism

The key to selective processing in this model is the ^{14}N nucleus. The atomic electrons determine the chirality of the molecule, but the ^{14}N nuclear spin would couple to the electron spin, which would couple to an external magnetic field, thereby aligning the ^{14}N nuclei and providing an orientation thereof. The separation between magnetic substates is μB, where μ is the magnetic moment of ^{14}N, $\sim 6 \times 10^{-7}$ eV/G, and B is the magnetic field generated by the residual electron spin. This would generate a hyperfine interaction that would split the magnetic substates of the nucleus+electron spin. The thermal energy kT for galactic molecular clouds, generally assumed to be at around 10-20K [13], would be about 10^{-3} eV. These two numbers turn out to be comparable, so some orientation of the ^{14}N would be expected to occur from the interaction with the neutron star's or black hole's magnetic field.

The other important feature is that of the chirality of the neutrinos: electron antineutrinos, $\bar{\nu}_e$'s, are right-handed and electron neutrinos, ν_e's, are left-handed [14]. Near one "throat" of the neutron star or black hole the spin of the ^{14}N would be aligned with that of the electron antineutrino, so the total neutrino-nuclear spin would have to be 3/2, but at the opposite throat the two spins would be antialigned, producing a different magnetic substate distribution. Then the reaction

$$\bar{\nu}_e + {}^{14}N \rightarrow e^+ + {}^{14}C$$

would not have the same strength on opposite ends of the neutron star or black hole, as is discussed below, and the resulting destruction of the ^{14}N, and hence of the molecule, would depend on the electronic structure, hence the chirality of the molecule. The above transition is pure "Gamow-Teller" [14] (also for $\nu_e + {}^{14}N \rightarrow e^- + {}^{14}O$, discussed below). This situation is illustrated in figure 1.

The *Q*-value of the reaction for $\bar{\nu}_e + {}^{14}N \rightarrow e^+ + {}^{14}C$ is -1.18 MeV, and for $\nu_e + {}^{14}N \rightarrow e^- + {}^{14}O$ is -5.14 MeV. The high threshold energy for the latter reaction would inhibit it compared to the former, as the energies of the neutrinos emitted in the collapse to a neutron star are of the order of 10 MeV [15-19]. In addition, the cross sections for the neutrino induced reactions increase as the square of the energy above threshold [15], further enhancing the $\bar{\upsilon}_e$ induced reactions. Thus while many electron neutrinos would be capable of converting ^{14}N to ^{14}O, many more electron antineutrinos would be capable of driving ^{14}N to ^{14}C producing a chiral preference for molecules of one chirality by preferentially destroying the molecules with the other chirality [12].

The magnitude of the selection in the extreme situations for $\bar{\upsilon}_e+{}^{14}N \rightarrow e^{+}+{}^{14}C$, i.e., those in which the antineutrino and ^{14}N spins are either aligned or antialigned, can be estimated as follows. If the spins are aligned, the angular momentum of the combined ^{14}N(spin 1) + $\bar{\upsilon}_e$(1/2) system must be 3/2, whereas if they are antialigned it will be ½. The transition produces $^{14}C(0)$. Thus, since the weak interaction can be assumed to be point-like for the present situation, conservation of angular momentum in the aligned case requires some orbital angular momentum transfer. This can be $\ell=1$ for the weak interaction, but this will inhibit the transition compared to that for the antialigned case, in which no orbital angular momentum transfer is required, by roughly an order of magnitude. Since the angular momentum coupling coefficients favor the case in which the spins of the ^{14}N and $\bar{\upsilon}_e$ are antialigned by 2:1 over that in which they are aligned, the antialigned ^{14}N nuclei will suffer selective destruction by roughly a factor of seven (two-thirds of the order of magnitude) over the aligned nuclei.

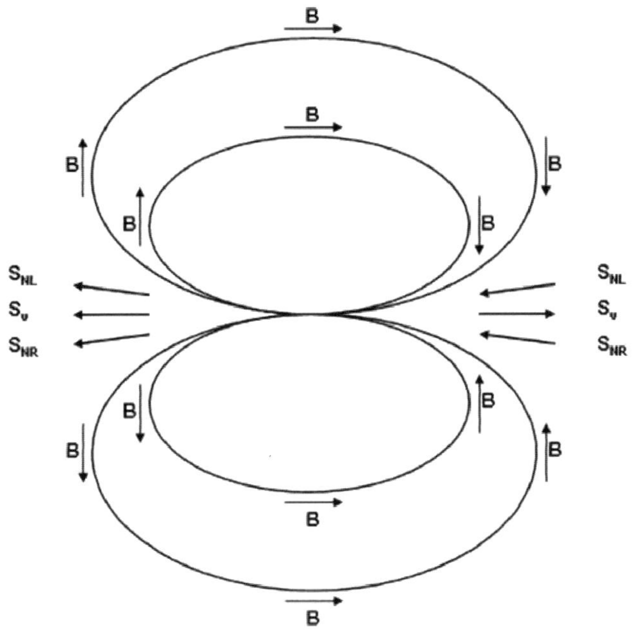

Figure 1. Schematic diagram of the magnetic fields, **B**, surrounding a nascent neutron star and the spins of the ^{14}N nuclei, S_{NL} and S_{NR} (of the left-handed and right-handed molecules, respectively), and of the neutrinos, S_{υ}, that would be emitted.

Although determination of the total effect on the selective destruction of the chiral molecules would require integration over all space, the average effect may not be relevant. If amplification of the enantiomerism produced by any effect progresses most rapidly where the enantiomeric excess is largest, the relevant number will most likely be the maximum enantiomeric excess achieved. We return to this point at the end of section 3. The effect discussed would inevitably produce significant enantiomerism of the chiral molecules and would select the same overall chirality for destruction *for every core-collapse supernova.*

To determine how the ^{14}N spin might couple to the molecular chirality, we have assumed a model that was developed by Buckingham [20, 21] in the context of nuclear magnetic resonance. Paraphrasing from Ref. 20, the magnetic field induces a current density which, in chiral samples, causes a nuclear magnetic moment (odd under time reversal and even parity) to induce a molecular electric dipole moment (even under time reversal and odd parity). The result is an effect that is opposite in sign for left-handed and right-handed enantiomers. The critical feature for this model [20] is the necessity of having a non-zero spin nucleus, i.e., the ^{14}N, for this effect to occur at all.

Fig. 2, shows the result of the Buckingham effect, which skews the populations of the substates, as is indicated, with the skewing going in opposite directions for the LH and RH molecules. Although the magnetic field interacts with the atomic angular momenta, the nuclear spin will tend in the same direction as the atomic angular momentum. Those LH and RH populations will initially be the same at either throat of the neutron star, but a combination of the interactions with the $\bar{\nu}_e$'s and an asymmetry in the neutrino flux at the two throats will select one chirality. The populations of the magnetic substates, as well as those that undergo greater processing, are indicated in fig. 2, with x's indicating the substates that will be preferentially destroyed. As shown, at the magnetic-field-outgoing throat, the +1 substate of the NL molecules will be selected, with the other two substates undergoing relatively greater destruction from the antineutrino interactions. The -1 substate of the NR molecules will be preferentially selected, and it will undergo selective destruction. This will create a preference for the NL molecules over the NR molecules at that throat of the neutron star.

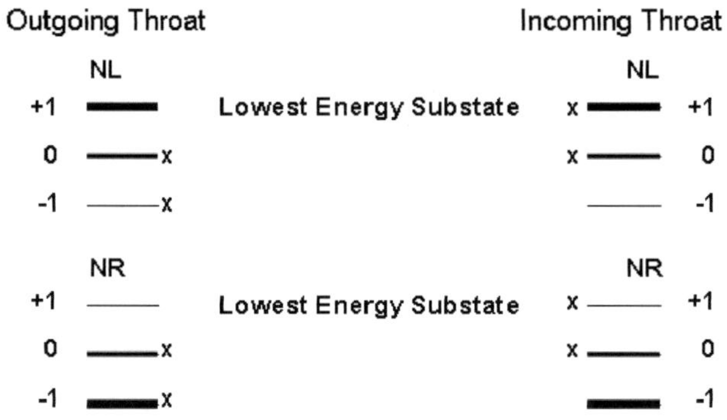

Figure 2. Magnetic substates for ^{14}N nuclei, and their relative populations resulting from the Buckingham effect. Thicker lines indicate larger populations. The "x" indicates a state that will undergo relatively greater processing from the electron antineutrinos. NL refers to left-handed ^{14}N nuclei, and NR to the right-handed ^{14}N nuclei. "Outgoing (Incoming) throat" refers to the throat of the neutron star at which the magnetic field is outgoing (incoming).

At the magnetic-field-incoming throat, the exact opposite selection effect occurs, and if the neutrino fluxes were the same at the two throats, no net chiral selection would occur. However, recent work [22-25] has shown that the extraordinary magnetic field of the neutron star will modify the cross sections for neutrino capture, decreasing them by 20-30 percent near zero degrees on the magnetic-field-outgoing throat and increasing them by a few percent at the opposite throat. This will produce a neutrino flux asymmetry, hence a greater selection of left-handed molecules at the magnetic-field-outgoing throat, and a net overall chiral selection [26]. Note, though, that selection of right-handed molecules would occur at the magnetic-field-incoming throat, and the large spatial separation of the two regions might allow for some isolated grains to contain right-handed molecules, even though the overall selection would be toward left-handed molecules.

3. Spreading Enantiomerism Throughout the Galaxy

Although the supernovae by themselves cannot come close to selecting only one molecular chirality throughout the Galaxy [12], we believe that some level of enantiomerism can be achieved by a combination of chemical evolution and Galactic mixing. As soon as a supernova explodes the enantiomeric material it has produced begins to mix with the racemic material adjacent to its processed volume, driving the racemic material with which it mixes toward partial enantiomerism, possibly on the surfaces of the dust grains [27] or of the meteoroids that also exist in the molecular clouds. This ultimately extends the enantiomeric volume well beyond that which was processed by the supernova neutrinos.

A general discussion of the ability of a collection of molecules exhibiting a small enantiomeric excess to amplify that excess via chemical replication is given in Ref. 27. The replication of molecule X proceeds [28] by autocatalysis and preservation of chirality by $S+T \leftrightarrow X$, which then has its chirality established to become X_L. Replication and amplification then could occur by $S+T+X_L \leftrightarrow 2X_L$. The ability of autocatalysis to produce amplification of enantiomerism has also been demonstrated in the laboratory [12].

Perhaps more specific to cosmic amplification, Ref. 29 developed a model in which chiral replication of complex molecules would occur in the interstellar medium in the warmed (possibly by proximity to a supernova) ice outer shells of grains. Chemical replication would be catalyzed by radicals, for example, H, OH, CO, CH_3, NH, and NH_2, created by the interactions of high energy cosmic rays with preexisting molecules. Subsequent studies might involve radicals that are more complex, and more relevant to amino acid replication. The grains would have to be warmed over the ambient temperature of the molecular clouds so as to enhance the mobility of the heavy radicals.

We believe that either autocatalysis or chemical replication via radicals, or both, could produce the chemical amplification required for the present model, or for that matter, any model that describes production of amino acid chirality in the cosmos, to succeed, and might well be expected to extend the chirality established by the magnetic fields and neutrinos from supernovae.

As the enantiomeric excess of the processed material increases via chemical replication, it will also mix with the rest of the material in the Galaxy, ultimately establishing at least a preference for left-handed amino acids throughout the Galaxy. Details of the processes by which this would occur have been discussed [30] in detail, and include many types of astronomical sources. This mixing would be expected to occur on a much slower timescale than the chemical processing timescale, but would inevitably establish a preferred chirality throughout the Galaxy.

A test of the ubiquity of the left-handed preference will be provided in 2014 when the ROSETTA mission sends a lander onto Comet 67 P/Churyumov-Gerasimenko [31].

4. Potential issues for this model

A concern with this model is with the photon field from the supernova to which the molecules, and the grains on which they are thought to form [27, 32], would be subjected. The solution to this problem lies with the supernovae that result in collapse to a black hole. Recent studies [33, 34] suggest that stars having masses from 8—25 times that of the Sun would be expected to form a neutron star, stars from 25—40 solar masses would form a neutron star, but the fallback would ultimately produce a black hole, and stars having masses more than 40 solar masses would collapse directly to a black hole. Ref. 33 also showed that roughly the same number of neutrinos would be emitted from stars in the latter mass range as for those that collapse to a neutron star, but that a larger fraction of the neutrinos would be electron antineutrinos and that their energies would be higher than in the supernovae from less massive progenitor stars. Both of these effects would enhance the chiral processing. For stars with masses > than 40 solar masses, the result would be a "failed supernova", i.e., the photon flux would be small or nonexistent. The stars in the 25—40 solar mass range would produce some photons, but they would be suppressed compared to the neutron-star case.

Thus the supernovae that go directly to black holes would both produce the neutrinos required to produce molecular enantiomerism and not then destroy the molecules with photons. Because the time required for fallback black hole formation is short, many of these black hole forming supernovae [35] might also be expected to produce sufficiently few photons that they would not destroy the molecules t their neutrinos had processed. Supernovae that produce a neutron star are more frequent than those that produce a black hole [36], but the fraction that does produce a black hole is significant (~20% [33]).]).

Another concern is with the progenitor star. A red giant would encompass the entire region that the supernova could process; these would be the progenitors of the 8-25 solar mass stars. Wolf-Rayet stars, thought to be the progenitors of the supernovae resulting from stars of more than 25 solar masses, are small, but are very hot. However, meteoroids that passed through the Wolf-Rayet clouds would be exposed to the highest temperature regions of the clouds through which they passed, but for short periods of time, and could survive some ablation if they were sufficiently large agglomerations of grains. These would have been formed elsewhere, and could contain molecules formed as they passed through their giant molecular cloud. And there would be a large number of these grains and meteoroids that could produce enantiomerism. They only have to approach within ~one AU of the Wolf-Rayet star to be processed, and this is certainly a large enough volume to include a huge number of grains and meteoroids. Using estimated neutrino-nucleus cross sections [37], we estimate that an object passing within 10^{12} cm of the star would achieve a processing probability of roughly 10^{-9}, and therefore, an enantiomeric excess of 0.5×10^{-9}. If it were large enough to survive if it passed by at 10^{11} cm, it would achieve a maximum enantiomeric excess of roughly 5×10^{-6} percent, or possibly even higher, given the possible enhancements of electron antineutrinos from the massive supernovae.

5. Conclusions

If this model turns out to be correct, the longstanding question of how the organic molecules necessary to create and sustain life on Earth were created will have undergone a strong suggestion that the processes of the cosmos played a role in establishing the molecules of life on Earth. These molecules would appear to have been created in the molecular clouds of the galaxy, with their enantiomerism determined by supernovae, and subsequently either transported to Earth in meteorites or comets, swept up as the Earth passed through molecular clouds, or included in the mixture that formed Earth when the planets were created. Any scenario in which these molecules were created exclusively on Earth in Darwin's "warm little pond," and supported by the experiment of Ref. 38, would find it much more difficult to explain the enantiomerism that is observed on Earth and, apparently, generally in the cosmos.

6. Acknowledgements

This work was supported by the US National Science Foundation grant PHY-9901241, and under the auspices of the Lawrence Livermore National Security, LLC, (LLNS) under Contract No. DE-AC52-07NA27344.

References

[1] Kvenvolden K, Lawless J, Pering K, Peterson E, Flores J, Ponnamperuma C, Kaplan I R and Moore C 1970 *Nature* **228** 923

[2] Bada J L, Cronin J R, Ho, M-S, Kvenvolden, K A, Lawless J G, Miller S L, Oro J and Steinberg S 1983 *Nature* **301** 494

[3] Cronin J R, Pizzarello S, and Cruikshank D P 1998, in *Meteorites and the Early Solar System*, eds J F Kerridge and M S Matthews, University of Arizona Press, (Tucson) p. 819

[4] Cronin J R and Pizzarello S 1997 *Science* **275** 951

[5] Herd C D K, Blinova A, Simkus D N, Wong Y, Tarozo R, Alexander C M O'D, Gyngard F, Nittler L R, Cody G D, Fogel M L, Kebukawa Y, Kilcoyne A L D, Hilts R W, Slater G F, Glavin D P, Dworkin J P, Callahan M P, Elsilla J E, DeGregorio B T and Stroud, R M 2011 *Science* **332** 1304
[6] Glavin D P and Dworkin J P 2009 *Proc Natl Acad Sciences USA* 5487
[7] Bailey J, Chrysostomou A, Hough J H, Gledhill T M, McCall A, Clark S, Ménard F and Tamura M 1988 *Science* **281** 672
[8] Takano Y, Takahashi J, Kaneko T, Marumo K and Kobayashi K 2007 *Earth Planet Sci Lett* **254** 106
[9] deMarcellus P, Meinert C, Nuevo M, Filippi J-J, Danger G, Deboffle D, Nahon L, d'Hendecourt L L S and Meierhenrich U J 2011 *Astrophys J* **727** L1
[10] Mann A K, and Primakoff H 1981 *Origins of Life* **11** 255
[11] Cline D 2005 *Chirality* **17** S234
[12] Boyd R N, Kajino T and Onaka T 2010 *Astrobiology* **10** 561; ibid 2011 *Int J Mol Sci* **12** 3432
[13] Ferriere K M 2001 *Rev Mod Physics* **73** 1031
[14] Boyd R N 2008 *Introduction to Nuclear Astrophysics* (University of Chicago, Chicago) p 136
[15] Keil M Th, Raffelt G G and Janka H-T 2003 *Astrophys J* **590** 971
[16] Esteban-Pretel A, Mirizzi A, Pastor S, Tomas R, Raffelt G G, Serpico P D and Sigl G 2008 *J Phys Conf Series* **120** 052021
[17] Duan H, Fuller G M, Carlson J and Qian Y-Z 2006 *Phys Rev D* **74** 105014-1
[18] Duan H, Fuller G M, Carlson J and Qian Y-Z 2007 *Phys Rev D* **75** 125005-1
[19] Duan H, Fuller G M and Qian Y-Z 2007 *Phys Rev D* **76** 085013-1
[20] Buckingham A D 2004 *Chem Phys Lett* **398** 1
[21] Buckingham A D and Fischer P 2006 *Chem Phys* **324** 111
[22] Horowitz C J and Li G 1998 *Phys Rev Letters* **80** 3694
[23] Arras P and Lai D 1999 Phys Rev D **60** 043001-1
[24] Lai D and Qian Y-Z 1998 *Astrophys J* **505** 844
[25] Maruyama T, Kajino T, Yasutake N, Cheoun M-K and Ryu C-Y 2011 *Phys Rev* D **84** 043001
[26] Famiano M, Boyd R N, Kajino T and Onaka T 2012 manuscript in preparation
[27] Bohlin R C 1975 *Astrophys J* **200** 402
[28] Kondepudi D K and Nelson G W 1985 *Nature* **314** 438
[29] Garrod RT, Weaver S L W and Herbst E 2008 *Astrophys J* **682** 283
[30] Pittard J M 2007 *Diffuse Matter from Star forming regions to active galaxies - A volume honouring John Dyson* ed by Hartquist T W, Pittard J M and Falle S A E G (Springer Dordrecht) 245
[31] Thiemann W H-P and Meierhenrich U 2001 *Orig Life Evol Biosphere* **31** 199
[32] Jenkins E B and Savage B D 1974 *Astrophys J* **187** 243
[33] Lunardini C 2009 *Phys Rev Lett* **102** 231101-1
[34] Heger A, Fryer C L, Woosley S E, Langer N and Hartmann D H 2003 *Astrophys J* **591** 288
[35] Sumiyoshi K, Yamada S and Suzuki H 2007 *Astrophys J* **667** 382
[36] Beacom J, Boyd R N and Mezzacappa A 2001 *Phys. Rev. D* **63** 073011-1
[37] Fuller G M, Haxton W C and McLaughlin G C 1999 *Phys Rev* D **59** 085005-1
[38] Miller S L 1953 *Science* **117** 528

HITES 2012 IOP Publishing
Journal of Physics: Conference Series **403** (2012) 012033 doi:10.1088/1742-6596/403/1/012033

Studying X-ray Burst Nucleosynthesis in the Laboratory

C M Deibel[1,2,3], L Afanasieva[1], M Albers[3], M Alcorta[3], S Almarez-Calderon[3], S Bedoor[4], P F Bertone[3], P Carnelli[5], A A Chen[6], J Chen[3], J A Clark[3], J M Figueira[5], J P Greene[3], C R Hoffman[3], D Irvine[6], C L Jiang[3], B P Kay[3], J Lai[1], H Y Lee[3], J C Lighthall[3,4], S Manwell[6], S T Marley[3,4], C Nair[3], T Palachan-Hazan[3], R C Pardo[3], N Patel[3,7], M Paul[8], B C Rasco[1], K E Rehm[3], A M Rogers[3], D Shetty[4], C Ugalde[3], A Wuosmaa[4], G Zinkann[3]

[1]Department of Physics and Astronomy, Louisiana State University, Baton Rouge, LA 70803 USA
[2]Joint Institute for Nuclear Astrophysics, Michigan State University, East Lansing, MI 48824 USA
[3]Physics Division, Argonne National Laboratory, Argonne, IL 60439 USA
[4]Physics Department Western Michigan University, Kalamazoo, MI 49008 USA
[5]Laboratorio Tandar, Comisión Nacional de Energía Atómica, B1650KNA San Martín, Buenos Aires, Argentina
[6]Department of Physics and Astronomy, McMaster University, Hamilton, ON L8S 4M1, Canada
[7]Department of Physics, Colorado School of Mines, Golden, CO 80401 USA
[8]Racah Institute of Physics, Hebrew University, Jerusalem, Israel

E-mail: deibel@phys.lsu.edu

Abstract. Type I X-ray bursts are the most common explosions in the Galaxy; however, the nucleosynthesis that occurs during the thermonuclear runaway and explosion is poorly understood. In this proceedings we discuss current experimental efforts and techniques that are being used to study X-ray burst nucleosynthesis in the laboratory. Specifically, radioactive ion beam techniques that have recently been developed have allowed the study of some of the most important (α, p) reactions in X-ray bursts for the first time.

1. Introduction

Type I X-ray bursts (XRBs) occur in binary star systems in which hydrogen-rich matter is accreted from a companion, main-sequence star onto the surface of a neutron star at a rate of approximately $10^{-8} - 10^{-10} M_{\odot}/\mathrm{yr}$ (for an in depth review of X-ray binaries see e.g. [1]). During this accretion phase, while temperatures are approximately 0.1 GK, there is a persistent, thermal X-ray emission. However, as matter is accreted, the pressure and density build up on the neutron star's surface until the extremely temperature-dependent triple-α reaction ignites the burst and thermonuclear runaway occurs. Temperatures of $1 - 2$ GK are reached during XRBs, which typically last for approximately $10 - 100$ s while emitting $10^{39} - 10^{40}$ ergs of energy, increasing the luminosity of the system by at least an order of magnitude. After the burst, the accretion phase of the system resumes, and the burst cycle begins again, with typical XRB recurrence times on the order of hours to days. Due to the high density of the neutron star, little if any synthesized material escapes the system and most of the nuclei created during the burst become part of the neutron star crust and seeds for subsequent bursts. As a result, these systems are evolving, with each burst affecting those after it. In addition to the seed nuclei from previous bursts, the initial composition before each burst is also affected by the hot-CNO cycle, which burns hydrogen into helium in between bursts.

While most bursts have regular recurrence times and luminosity peaks, there are many observations of irregular bursts. Certain X-ray bursters are observed to undergo so-called "superbursts", which are thought to be due to carbon burning [2] and are typified by larger peak luminosities (several orders of magnitude greater than their typical XRBs) and longer durations and recurrence times. In addition to superbursts there have been several observations of double-peaked bursts [3, 4, 5], and at least one observation of a triple-peaked burst [6]. Several explanations of these multi-peaked bursts have been proposed including stalling of the nuclear burning front [7], as well as a possible nuclear physics solution [8], discussed below.

After the the triple-α reaction triggers the burst, the nucleosynthesis in an XRB is driven up the proton-rich side of stability by the α, p-process and the rapid proton-capture process (rp-process). The α, p-process consists of a series of (α, p) and (p, γ) reactions. Due to the large Coulomb barrier, the (α, p) reaction strongly depends on the temperature of the burst and this process helps to set the time scale of the rise of the burst luminosity curve through the effects of so-called waiting points. There have been four potential α, p-process waiting points identified: ^{22}Mg, ^{26}Si, ^{30}S, and ^{34}Ar [9]. These (Z-N)/2 = 1 nuclei have low $Q_{p,\gamma}$ values and thus come into a (p, γ)-(γ, p) equilibrium with their Z+1 counterparts and must await β^+ decay, stalling the nucleosynthetic flow. As these nuclei have half-lives on the order of seconds this delay can be significant on XRB time scales. However, if a competing reaction, such as the (α, p) reaction, is fast enough, breakout from the waiting point occurs and nucleosynthesis continues up the chart of nuclides. Therefore, it is important to understand the different processes occurring at a waiting-point nucleus in a self consistent way. Additionally, these waiting points have been suggested as a possible mechanism for the observed double-peaked structure of some XRB light curves, as the stalling of the nucleosynthetic flow may cause a decrease in the energy generated by the burst and thus a dip in the luminosity profile [8].

The rp-process is a series of (p, γ) reactions and β decays that extend the nucleosynthesis up to higher masses. The endpoint of the rp-process in XRBs is thought to be in the SnSbTe region; however, this remains uncertain as on one hand there is evidence for a weak SnSbTe cycle that would move the rp-process path closer to stability [10], while on the other hand some models show leakage out of the SbSnTe cycle to higher mass nuclei [11]. Higher-mass waiting points in the rp-process, which occur at nuclei with long β-decay half-lives and small proton-capture Q values, govern the shape of the decay of the light curve.

In order to understand XRB nucleosynthesis fully, three types of nuclear data are required: well-known masses of the nuclei involved, their β-decay rates, and the rates of reactions that occur in XRBs. Recently, there have been multiple advances in each of these areas (see e.g. [12] for recent mass measurements and [13] for β-decay lifetime measurements). Due to the proton-rich nuclei that are involved in XRB nucleosynthesis, many of the thousands of reactions and processes that occur in these explosive environments are not directly accessible in the laboratory. As a result, theoretical values are used for the nuclear input data in stellar models of XRBs where experimental data does not exist. However, the uncertainty associated with these theoretical values can be very high and those predictions do not always agree with experimental data when they become available.

Some of the unstable nuclei of interest are close enough to stability that they can be accessed using stable beams and targets that are easily produced [14]. However, many of these reactions and nuclei are too far from stability to be studied using stable beams. With the advent of radioactive ion beam (RIB) facilities, some of the processes involving these nuclei can now be studied for the first time. For reaction rates specifically one would ideally like to study them directly [15]; however, even with these new advances in RIB technology, direct measurements are still extremely difficult and often impossible. Therefore, indirect methods with both stable and radioactive ion beams must be used. This proceedings article reviews experimental advances in the determination of (α, p) reaction rates that are important in XRBs, which have just recently been made possible by radioactive ion beam facilities that are now available.

2. Experiment

The (α, p) reactions on all four waiting-point nuclei (^{22}Mg, ^{26}Si, ^{30}S, and ^{34}Ar) have been studied via the time-inverse (p, α) reactions measured in inverse kinematics with RIBs. These RIBs were produced at Argonne National Laboratory using the "in-flight" facility at the Argonne Tandem Linac Accelerator System (ATLAS). Stable beams of ^{24}Mg, ^{28}Si, ^{32}S, and ^{36}Ar were accelerated to energies of approximately 315 - 325 MeV and impinged upon a LN_2-cooled gas target filled with 1.4 atm of D_2 gas. Radioactive ion beams of ^{25}Al, ^{29}P, ^{33}Cl, and ^{37}K with energies of 253, 280, 250, and 275 MeV, respectively, were produced via (d, n) reactions with the stable beams. The radioactive ions of interest were then refocused, rebunched and separated by an analyzing magnet from the stable component of the beam. Typically the low energy tail of the unreacted stable beam is also selected by the analyzing magnet resulting in an unfavorable ratio of approximately 1:1000 radioactive ions/s to stable ions/s. As these two components of the resulting "cocktail" beam are separated in time of flight, a radio-frequency (RF) sweeper is used to eliminate much of the stable beam contamination improving the radioactive to stable beam ratio to better than 1:1. The intensities of the RIBs ranged from $1 - 5 \times 10^4$ ions/s depending on the RIB species.

Changing the energy of the RIB to scan through the energy region of interest for the reaction by changing the energy of the primary, stable beam would be prohibitively time consuming given the difficulty of producing and tuning RIBs. Therefore, in order to change beam energies Au degrader foils of different thicknesses were used. The thicknesses of the Au foils were chosen such that the energy loss of the beam in the target would be approximately the same as the energy loss in the foils (approximately 10 MeV in the case of a 250 MeV ^{33}Cl beam).

Once the RIB has been produced it is delivered to the experimental area where it impinges on a CH_2 target of approximately 650 μg/cm^2. The α particles produced from the (p, α) reactions are detected in an annular double-sided Si detector (DSSD) that is segmented in θ_{lab} (see Fig. 1). The DSSD was placed such that lab angles of $6° - 19°$ were covered for the $p(^{25}\text{Al},^{22}\text{Mg})\alpha$ and $p(^{29}\text{P},^{26}\text{Si})\alpha$ reactions and $8° - 24°$ were covered in the $p(^{33}\text{Cl},^{30}\text{S})\alpha$ and $p(^{37}\text{K},^{34}\text{Ar})\alpha$ reaction studies.

The heavy-ion recoils along with the unreacted beam (from both the RIB and the stable contaminant) were momentum-analyzed by an Enge split-pole spectrograph. The reaction products of interest were separated from other heavy ions by the spectrograph, which was run in gas-filled mode, filled with approximately 15 Torr of N_2 gas in the case of the highest beam energies. The gas-filled mode was used to collapse the charge state distribution of the heavy-ion recoils into a single charge state [16] to maximize detection efficiency at the focal plane, which is crucial in these types of low-statistics experiments, as well as to avoid position overlap of the reaction products of interest and unreacted beam in various charge states. Once the heavy recoils passed through the spectrograph, they were detected at the focal plane by a gas-filled parallel grid avalanche counter (PGAC) and an ionization chamber. These detectors give position and energy loss of the ions, as well as time-of-flight of the ions relative to the RF of the beam, allowing particle identification. By detecting the heavy-ion recoils in coincidence

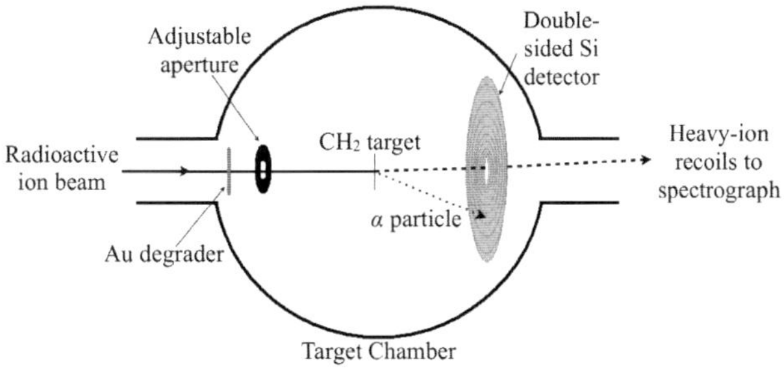

Figure 1. A schematic of the experimental setup.

with the α particles, correcting for efficiencies, and normalizing to the incident beam and target thickness, the cross section of the reaction of interest can be determined.

3. Analysis and Results

The α particles were detected by the DSSD in coincidence with the heavy-ion recoils as discussed above to determine the α-particle yield for the reaction. Background was partially eliminated through particle identification cuts on the time-of-flight and focal plane position (or equivalently magnetic rigidity) of the heavy recoils. Additional background was subtracted by determining the random background rate outside of the coincidence timing cut. Once the α-particle yield was determined it was corrected for various efficiencies and normalized to the incident beam and target areal density.

Unfortunately, as the magnetic rigidity of the high-energy heavy recoils associated with the low-energy α branch is the same as the magnetic rigidity of the unreacted beam, those recoils were blocked from entering the detector to avoid overwhelmingly high count rates and only the high-energy α branch of the reaction could be measured. The efficiency factor for detecting only the high-energy α branch is approximately 0.67 (e.g. for $p(^{33}\mathrm{Cl},^{30}\mathrm{S})\alpha$) as determined via Monte Carlo simulations. The geometrical efficiencies of the DSSD and spectrograph were 0.82 and 0.38, respectively, also determined via Monte Carlo simulations.

Normalization to the incident beam and target was done in several ways. For the $p(^{33}\mathrm{Cl},^{30}\mathrm{S})\alpha$ measurement, the ratio of Rutherford scattering off the carbon component of the CH_2 target at $2.5°$, measured in the spectrograph, to the proton scattering at $8°$, measured in the inner ring of the DSSD, was determined in a separate measurement during the experiment. The proton scattering was then monitored throughout the (p, α) experiment and using the ratio defined above with the calculated Rutherford cross section, normalization to the incident beam and target was achieved. In addition, for all the (p, α) measurements, the beam intensity and composition was checked every few hours by measuring it directly in the spectrograph after attenuation. Both methods of normalization agreed to within error. An additional Si surface barrier detector was used as an added check in both the $p(^{25}\mathrm{Al},^{22}\mathrm{Mg})\alpha$ and $p(^{37}\mathrm{K},^{34}\mathrm{Ar})\alpha$ measurements to monitor the beam intensity and target thickness.

Once the α-particle yield was corrected for efficiencies and normalized, the cross section could then be compared with theoretical calculations, such as those from the NON-SMOKER code [17, 18, 19] (see e.g. Fig. 2 for the case of $p(^{33}\mathrm{Cl},^{30}\mathrm{S})\alpha$ [20]).

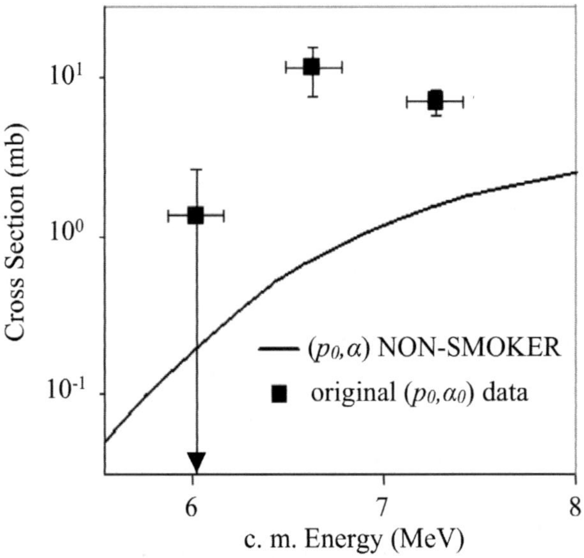

Figure 2. Cross section as a function of c.m. energy for the ^{33}Cl(p_0, α_0)^{30}S data (squares) and the NON-SMOKER calculations [17, 18, 19] for the ^{33}Cl(p_0, α)^{30}S cross section (solid line). The experimental data only include ground state to ground state transitions, while the NON-SMOKER calculations include transitions to excited states. The vertical error bars indicate the uncertainties in the cross sections and the horizontal error bars indicate the energy spread of the beam in the target. Reproduced from [20].

4. Conclusions

It can clearly be seen from Fig. 2 that the experimental data do not agree with theoretical calculations from NON-SMOKER and are up to a factor of four higher than the theoretical cross sections. Indeed the disagreement may be worse than what is shown in Fig. 2 since the measured cross sections are lower limits as they do not contain any possible contribution from transitions to excited states in the final nucleus. Furthermore, for the other reactions measured, preliminary results also show disagreement with theoretical calculations, although not always in the same direction. For example, the measurement of the p(^{37}K,^{34}Ar)α reaction gives cross sections that are significantly lower than NON-SMOKER predictions [21]. Of course, given that the cross section for the forward ^{34}Ar(α, p)^{37}K reaction is only a lower limit, as it does not contain any transitions to excited states in the final ^{37}K nucleus, it is possible that the total cross section for the (α, p) reaction could agree with the theoretical predictions in this case.

This disagreement between experimental measurements and NON-SMOKER calculations is not altogether surprising as these calculations are based on Hauser-Fesbach theory that assumes high level densities, which may not be valid in this mass regime. In this intermediate mass region the level densities for even-even nuclei are expected to be low in the astrophysically relevant energy regime and thus the reaction rate may be dominated by a few isolated resonances. It should be noted that the energies measured for each reaction described here are above this energy region of interest. Given the rapidly decreasing cross section with energy and the low-intensity RIBs available, it was not feasible to measure the cross sections in the region of interest (e.g. $E_{c.m.} = 3.5 - 4.4$ MeV for p(^{33}Cl,^{30}S)α) during these experiments. However, the results of these measurements underscore the need to experimentally determine the rates of the reactions of interest in XRB nucleosynthesis, as the theoretical calculations (e.g. NON-SMOKER) currently used in stellar models do not accurately reproduce measured data.

Although the measurements presented here are above the energy range of interest for XRB nucleosynthesis, the discrepancies between the measured and theoretical cross sections will have implications for stellar models of XRBs if they persist at lower energies. For example, in the case of $^{30}S(\alpha, p)^{33}Cl$, which clearly shows higher cross sections than the theoretical calculations in the energy regime measured, a higher reaction rate would lead to break out of the ^{30}S waiting point. This would imply that the ^{30}S waiting point is not responsible for the double-peaked structure of certain observed luminosity profiles, as the nucleosynthesis flows more readily through ^{30}S. The additional effects of the measured cross sections of the (α, p) reactions on ^{30}S and other waiting-point nuclei on XRB nucleosynthesis will be addressed in a future publication.

In addition to accurate reaction rates, measurements of nuclear masses and β decay rates also continue to be needed and experimental advances on all three of these fronts are progressing as more radioactive ion facilities become available. Stable beam data also continue to provide new insight into XRB nucleosynthesis [14] and are necessary even as RIB capabilities increase. In order to fully understand these astrophysical environments accurate nuclear data is needed as input for stellar models. While theoretical determinations of these parameters fill the gaps where no data exist, until those models can reliably reproduce experimental data, measurements are needed to ensure an accurate description of these stellar events.

Acknowledgments

The authors would like to thank the ATLAS operators and staff for their assistance during these experimental studies. We would also like to acknowledge funding from NSF JINA grant No. PHY0822648 and U.S. DOE contract No. DE-AC02-06CH11357. We would finally like to thank T. Rauscher for helpful discussions regarding NON-SMOKER code calculations.

References

[1] Schatz H and Rehm K E 2006 *Nucl. Phys.* **A777** 601
[2] Cooper R L, Steiner A W and Brown E F 2009 *Astrophys. J.* **702** 660
[3] Sztajno M, van Paradijs J, Lewin W H G, Trümper J, Stollman G, Pietsch W and van der Klis M 1985 *Astrophys. J.* **299** 487
[4] Penninx W, van Paradijs J and Lewin W H G 1987 *Astrophys. J. Lett.* **321** 67
[5] Kuulkers E, Homan J, van der Klis M, Lewin W H G and Méndez M 2002 *Astron. Astrophys.* **382** 947
[6] Zhang G, Mendez M, Altamirano D, Belloni T M and Homan J 2009 *Mon. Not. R. Astron. Soc.* **389** 368
[7] Bhattacharyya S and Strohmayer T E 2006 *Astrophys. J.* **641** L53
[8] Fisker J L, Thielemann F K and Wiescher M 2004 *Astrophys. J. Lett.* **608** 61
[9] Fisker J L, Schatz H and Thielemann F K 2008 *Astrophys. J. SS.* **174** 261
[10] Elomaa V V, Voobjev G K, Kankainen A, Batist L, Eliseev S, Eronen T, Hakala J, Jokinen A, Moore I D, Novikov Y N, Penttilä H, Popov A, Rahaman S, Rissanen J, Saastamoninen A, Schatz H, Seliverstov D M, Weber C and Äyastö J 2009 *Phys. Rev. Lett.* **102** 252501
[11] José J, Moreno F, Parikh A and Iliadis C 2010 *Astrophys. J. Supp. Ser.* **189** 204
[12] Tu X L *et al.* 2011 *Phys. Rev. Lett.* **106** 112501
[13] Nishimura S *et al.* 2011 *Phys. Rev. Lett.* **106** 052502
[14] Setoodehnia K *et al.* 2010 *Phys. Rev. C* **82** 022801(R)
[15] Chen A A *et al.* 2005 *Nucl. Phys.* **A752** 510
[16] Paul M *et al.* 1989 *Nucl. Inst. Meth. Phys. Res.* **A277** 418
[17] Rauscher T and Thielemann F K 2000 *Atomic Data and Nuclear Data Tables* **75** 1
[18] Rauscher T and Thielemann F K 2001 *At. Data Nucl. Data Tables* **79** 47
[19] Non-smoker code, http://nucastro.org/nonsmoker.html
[20] Deibel C M, Rehm K E, Clark J A, Figueira J M, Jiang C L, Kay B P, Lee H Y, Marley S T, Pardo R C, Patel N *et al.* 2009 ATLAS PAC proposal 1279
[21] Deibel C M *et al.* in preparation

The νp-process: critical nuclear physics and astrophysical implications

Carla Fröhlich
Department of Physics, North Carolina State University, Raleigh NC 27695, USA
E-mail: cfrohli@ncsu.edu

Abstract. The neutrino-p-process is thought to occur in the innermost proton-rich layers ejected in core-collapse supernovae. The importance of the νp-process lies in the fact that it may contribute to the abundances of elements above Nickel and possibly the light p-nuclei. The reaction path of the νp-process lies in a region where nuclear masses are partly unknown and all involved reaction rates are based on theoretical predictions. Detailed studies of the νp-process nucleosynthesis and its uncertainty due to the nuclear physics are presented, with a focus on the reaction path at and above ^{56}Ni. The νp-process path is found to be mainly determined by nuclear structure and thus is trajectory independent. Critical nuclear physics input are identified and the impact of uncertainties on the resulting nucleosynthesis is discussed.

1. Introduction

Neutrino-driven winds from core collapse supernovae are an important site for the nucleosynthesis of elements beyond iron. After the explosion, the newly formed, hot proto-neutron star cools emitting neutrinos. The interaction of these neutrinos with the stellar matter results in energy deposition in the outer layers of the proto-neutron star and leads to a supersonic outflow, the neutrino-driven wind. Neutrino-driven winds are typically considered the site for the production of heavy elements through the r-process [1]. However, recent simulations [2, 3, 4, 5] cannot reproduce the extreme conditions required for the r-process (e.g. [6]): The ejecta is proton-rich (electron fraction $Y_e > 0.5$) and the entropy is too low (entropy $s < 100k_B/\text{nuc}$). While these conditions do not allow for the formation of the heaviest elements, lighter elements, e.g. Sr, Y, Zr, can be formed.

In these proton-rich winds, matter is initially hot and fully dissociated. As the matter expands and cools, freeze-out from full nuclear statistical equilibrium (NSE) to quasi-equilibrium (QSE) takes place and nuclei up to ^{56}Ni and even up to ^{64}Ge are formed once the temperature drops below 3 GK. For temperatures between 3 GK and 1.5 GK, sequences of p-capture reactions and β^+-decays dominate. Nuclei with low proton-capture Q-values and long β-decay half-lives act as bottlenecks and and hinder the nucleosynthesis flow towards heavier elements. In the νp-process, the decay of these nuclei is accelerated by (n,p) reactions. These neutrons are produced by anti-neutrino capture reactions on free protons ($\overline{\nu}_e + \text{p} \rightarrow \text{n} + \text{e}^+$). This allows for the synthesis of heavier elements beyond iron through the νp-process [7, 8, 9, 10, 11]

The importance of the νp-process lies in the fact that it is a primary process (i.e. it does not require pre-existing seed nuclei) contributing to the abundances of heavy elements beyond iron. This is in agreement with recent observations of metal-poor stars (e.g., [12]), which exhibit enhanced abundances of elements such as Sr, Y, and Zr. Thus, the νp-process may serve as the Light Element Primordial Process (LEPP) proposed to explain the observed abundances of these elements at the lowest metallicities [13]. In addition, the νp-process may contribute to the light p-nuclei – neutron-deficient stable isotopes above iron which cannot be produced by the s- and r-processes. Recent isotopic data from meteoritic material puts tight constraints on the possible contribution to ^{92}Mo from the proton-rich side [14, 15] but lighter p-nuclei may receive a considerable contribution from the νp-process.

In section 2, we summarize the observational clues for an additional nucleosynthesis process at the lowest metallicities. In section 3, we analyze the details of the νp-process nucleosynthesis both with respect to the nuclear physics inputs and the astrophysical implications. A summary is presented in section 4.

2. Abundance clues

Spectroscopic studies of abundances in metal-poor halo stars can reveal hints about the nucleosynthesis processes in the early Galaxy. These low-mass stars are old and carry in their photosphere the abundance signature of the conditions when they were formed. A subclass of these extremely metal-poor stars with [Fe/H]$\approx$ 3 is strongly enhanced with elements such as Eu ([Eu/Fe]$>$ 0.5). These stars show a heavy element pattern ($56 \leq Z < 83$) which is in extraordinary agreement with the r-process contribution to the solar system abundances [16]. In addition, the abundance pattern in these stars is remarkably robust from star to star. The robustness of the abundance pattern and the remarkable agreement with the solar system r-process contributions imply a universal r-process abundance distribution since these metal-poor stars are thought to be enriched by only one or a few nucleosynthetic events. However, for the lighter heavy elements (elements beyond iron with Z$\leq$ 47) the situation is more complex. The abundances of these elements in the same Eu-enhanced stars do not agree with the solar system r-process contribution, indicating the contribution of an additional nucleosynthesis process (or the r-process is not robust in this region and depends on the conditions). Further evidence for such an additional nucleosynthesis process stems from meteoritic abundances [17, 18]. The larger scatter in [Sr/Ba] in low-metallicity stars (e.g. [19, 20]) is consistent with a second, independent process at low metallicities which produces Sr but no or little Ba. Chemical evolution studies [13] also find a large scatter in Sr, Y, Zr and postulate an additional process, the Lighter Element Primary Process (LEPP) which produces Sr, Y, Zr but little Ba and Eu. In addition, non-correlation of the elements Sr, Y, Zr, Pd, and Ag with Eu can be demonstrated [21, 22].

3. The νp-process

The details of the νp-process nucleosynthesis depend on the nuclear physics inputs (e.g. reaction rates, reaction Q-values, nuclear masses) and on the astrophysical conditions (e.g. the existence and location of the wind termination shock, expansion timescale, and neutrino luminosities and energies). The reaction path of the νp-process partly lies in a region where nuclear masses are unknown [10, 23]. Furthermore, all involved reaction rates are based on theoretical predictions. Since there is not much experimental information (such as level schemes, deformations, decay schemes, scattering and reaction cross sections) available far from stability, the uncertainties in the rates have to be considered in astrophysical studies of the νp-process. A possibility to investigate properties of, and perhaps reactions with, highly unstable nuclei is offered by radioactive ion beam facilities.

The νp-process synthesizes neutron-deficient nuclei through a sequence of rapid (p,γ) reactions, followed by (n,p) reactions or β-decays, where the neutrons are supplied by antineutrino captures on free protons. This is similar to the rp-process in proton-rich thermonuclear burning on the surface of a mass-accreting neutron star [24, 25]. The waiting points in the rp-process (characterized by slow proton-capture reactions and long β-decay lifetimes) are overcome in the νp-process by the (n,p) reactions. In addition, the νp-process occurs at lower proton density $\rho_{\rm p} = \rho Y_{\rm p}$ than the rp-process. Here, ρ denotes the matter density and $Y_{\rm p}$ is the proton abundance. Hence, the νp-process path is slightly closer to stability, initially following the $N = Z$ line and moving towards stability at higher masses. This influences which nuclei are waiting points in the νp-process. We examine how variations in the nuclear input physics influence the νp-process. The focus is on the reaction rates of important reactions. Sensitivities on the nuclear masses will also be discussed. The nucleosynthesis results presented here are based on the explosion of a $15 M_\odot$ star [26], also utilized in [8], which efficiently synthesizes nuclei with $A > 90$. The nuclear reaction network used is the same as in [23]. The reaction rates are taken from a recent REACLIB compilation [27, 28, 29] which contains experimental rates (where available) and theoretical Hauser-Feshbach rates. This is supplemented with weak interaction rates [30, 31, 32] and neutrino and antineutrino capture rates on nucleons and nuclei [33]. The nuclear masses employed for the Hauser-Feshbach rates were taken from [34, 35] and partially replaced by more recent mass measurements of proton-rich nuclei [23, 36].

Here, we focus on reaction rates on nuclei at and above nickel where the reaction rates mostly stem from Hauser-Feshbach predictions. The effect of uncertainties on selected reaction rates on light nuclei has been investigated in Ref. [10]. The (n,p) reactions of special importance in the νp-process are ^{56}Ni(n,p)^{56}Co, ^{64}Ge(n,p)^{64}Ga, and ^{96}Pd(n,p)^{96}Rh [37]. The nucleus ^{56}Ni acts as seed nucleus from which the νp-process starts at temperatures $T \sim 3$GK. The nucleus ^{64}Ge is the first major bottleneck in the νp-process which has to be bridged by (n,p) reactions. Comparing calculations with and without neutrinos, the abundance pattern is very similar up to and including ^{64}Ge. However, without neutrinos the abundance distribution drops precipitously at $A \approx 64$. The νp-process path always proceeds through ^{96}Pd ($N = 50$) which has been predicted to be a "second seed nucleus" for the νp-process [10]. The (n,p) reactions on these three nuclei have been varied individually. In addition, global rate variations of all (n,p) or (p,γ) reactions in a region of nuclei have been performed. For all variations, the rates have been varied by a constant factor, either multiplied or divided by a factor of five. This is appropriate for the global uncertainties of (n,p) rates in this region of the nuclear chart. The regions for the global rate variations are defined by ^{56}Ni, ^{64}Ge, and ^{96}Pd.

For variations of only the ^{56}Ni(n,p)^{56}Co rate we find – in agreement with [10] – that a slower rate results in larger abundances for the nuclei with $A > 96$. A slower rate means that fewer neutrons are captured on ^{56}Ni, and hence, more neutrons are available to be captured on nuclei heavier than ^{56}Ni along the νp-process path [37]. In this case, the νp-process nucleosynthesis can proceed to higher mass numbers (see figure 1, dashed line). Conversely, a faster rate for ^{56}Ni(n,p) leads to the nucleosynthesis ending at lower mass numbers (see figure 1, dotted line) because more neutrons are used on ^{56}Ni and hence are not available anymore elsewhere. A reduced rate for ^{56}Ni(n,p) shifts material from the $60 \lesssim A \lesssim 100$ region to the $A \gtrsim 100$ region. However, it should be noted that an increase in mass number A does not necessarily imply an increase in Z due to the (n,γ) reactions acting at later time on nuclei with large A [37].

As shown in Fig. 2, variations of the ^{64}Ge(n,p)^{64}Ga rate only have much smaller effects on the abundances than for ^{56}Ni(n,p). A reduction of the (n,p) rate on ^{64}Ge affects the abundances in two ways: The abundance of the nucleus ^{64}Ge increases as the flow is inhibited from continuing beyond ^{64}Ge. At the same time, fewer neutrons are used in (n,p) reactions on ^{64}Ge and hence are available for capture on heavier nuclei on the νp-process path. This has the strongest effect on nuclei with $A \gtrsim 100$ as these nuclei have low abundances. Similarly, the relative depletion in the $70 \lesssim A \lesssim 100$ region is small. The increase in abundances for nuclei with $A \gtrsim 100$ is much smaller here than in the case of variations of the ^{56}Ni(n,p)^{56}Co rate. Global variations of all (n,p) rates on and above ^{64}Ge (Fig. 3) exhibit the opposite behavior: A reduction of all (n,p) rates on and above ^{64}Ge reduces the abundances of nuclei with $A \gtrsim 96$, while an increase of all (n,p) rates increases the abundances of nuclei with $A \gtrsim 96$. This behavior is an indication that once the νp-process has reached ^{64}Ge the efficiency of the process is determined by how fast it can proceed, i.e., how much the (n,p) reactions speed up the nucleosynthesis flow compared to β^+ decays [37]. An increase in the abundances of nuclei with $A \gtrsim 96$ is accompanied by a slight decrease in the abundances of nuclei with $64 \lesssim A \lesssim 96$ as this material is shifted to higher mass numbers.

The nucleus ^{96}Pd has been predicted to be a second seed nucleus for the νp-process. However, variations of only the rate of ^{96}Pd(n,p)^{96}Rh have only a small effect on the resulting abundances [37], as seen in Fig. 4. For variations of all (n,p) rates on and above ^{96}Pd the situation is very similar to the ^{64}Ge case: Increased (n,p) rates enable the νp-process to proceed faster and hence lead to increased abundances for $A > 96$. Abundances closely above ^{96}Pd are moved into nuclei with even higher masses when the rates of all (n,p) reactions are increased.

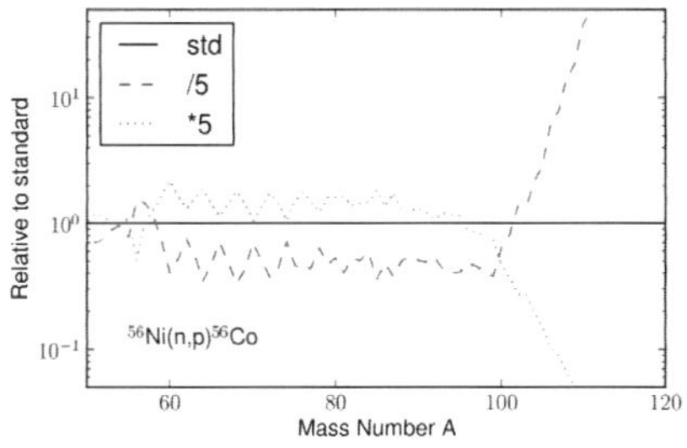

Figure 1. Abundance ratios relative to the standard case from a variation of the ^{56}Ni(n,p)^{56}Co rate.

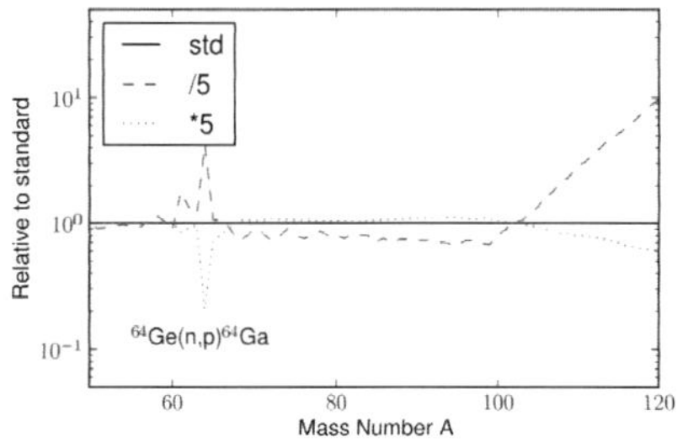

Figure 2. Abundance ratios relative to the standard case from a variation of the ^{64}Ge(n,p)^{64}Ga rate.

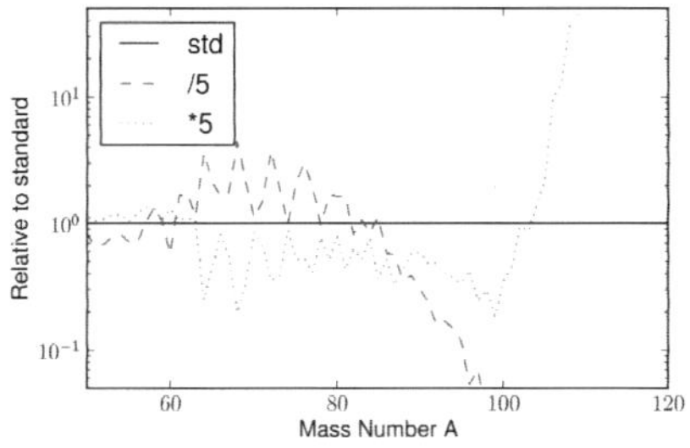

Figure 3. Abundance ratios relative to the standard case from variations of all (n,p) rates above and including ^{64}Ge.

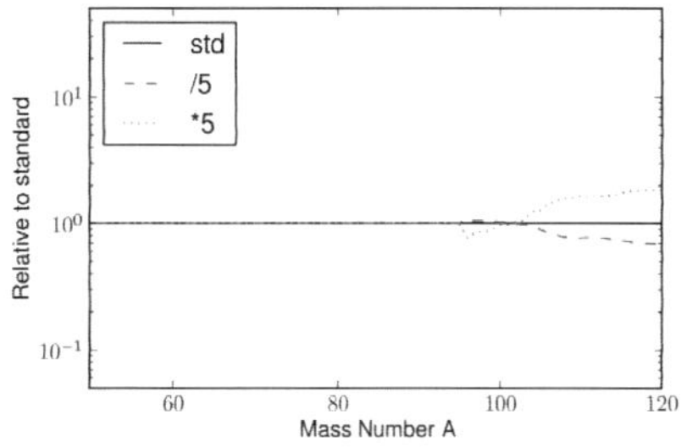

Figure 4. Abundance ratios relative to the standard case from a variation of the ^{96}Pd(n,p)^{96}Rh rate.

Another important nuclear physics input are the nuclear masses. Uncertainties in the nuclear masses affect the proton-capture Q-values and hence determine the ratio between forward capture and reverse photodisintegration rates. This, in turn, determines whether another proton can be captured before an (n,p) reaction (or β-decay) occurs. The nuclear masses used here are taken from AME03 [34, 35], supplemented by the recent mass measurements of [23, 38, 39, 40]. The impact of a few selected or a range of masses on details of the reaction path has been studied in Refs. [10, 41] and in Ref. [23], respectively.

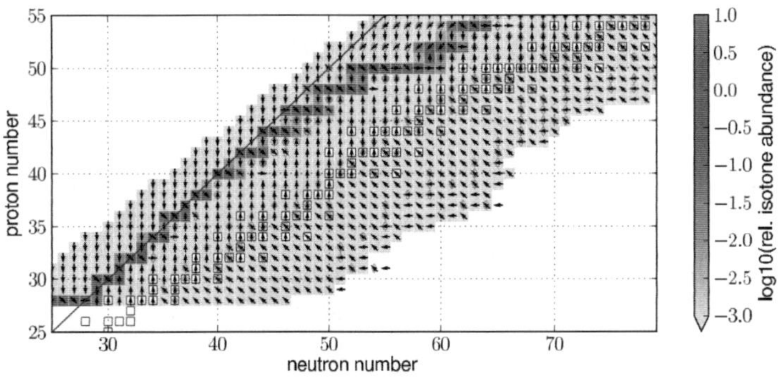

Figure 5. Abundance distribution from (p,γ)-(γ,p) equilibrium in each isotonic chain for typical νp-process conditions. Arrows indicate the dominating net flux per nucleus. The $N = Z$ line is also marked. Circles denote nuclei where uncertainties in the Q-values (masses) impacts the dominating reaction flow. Diamonds mark nuclei where the path direction is uncertain due to uncertainties in the reactions.

The location of the νp-process path is robust and remains unaffected by variations of the astrophysical conditions, such as entropy, expansion timescale, or details of the reverse shock [37, 42, 43, 8, 10, 11]. Variations of reaction rates have at most a small effect on the location of the νp-process path [37, 42, 10]. This is due to the fact the (p,γ)-(γ,p) equilibrium is upheld until late times. Hence, the abundance maximum in each isotonic chain is given by the equilibrium abundance, as seen in figure 5. Figure 6 shows the effective lifetimes for each nucleus. The maximum abundance within a chain is characterized by a low (p,γ) reaction Q-value because the relative speed of forward and reverse rate depends exponentially on the Q-value [37, 29]. The largest flux into the next isotonic chain occurs at these nuclei, which would be waiting points in a pure rp-process [24]. As can be seen in figure 6, these nuclei indeed exhibit longer lifetimes. Nevertheless, if the neutron abundance is sufficiently high they may still be overcome. This implies that a variation of the neutron density or the (n,p) rate on these waiting points will mostly affect how fast nuclei with larger Z can be reached within the timescale of the expansion. If the neutron density is sufficiently high at late times, (n,γ) reactions can become the fastest reactions, even faster than β-decays, and the νp-process path is pushed towards stability at and above the Sn isotopes by (n,γ) reactions [11], providing a strong barrier for the efficient production of elements beyond Sn. The fact that the location of the νp-process path is largely determined by nuclear physics properties make any conclusions on the sensitivity of the final abundances to the nuclear physics inputs robust. A detailed list of the critical nuclear reactions, including implications for experiments, can be found in Ref. [37].

4. Summary

The νp-process nucleosynthesis and its uncertainty due to nuclear physics has been studied. The main trajectory dependence originates from the possible variation of the neutron abundance $Y_{\rm n}$, and thus of the neutron density $\rho_{\rm n} = \rho Y_{\rm n}$, due to different $\nu_{\rm e}$ fluxes experienced during the expansion of the matter. This has a similar impact as a variation of the reaction cross sections of neutron-induced reactions. From detailed reaction variation studies, a different role has been found for ^{56}Ni and the nuclei above the iron group. The nucleus ^{56}Ni - which is the seed nucleus for the νp-process - acts as neutron poison due to its high abundance.

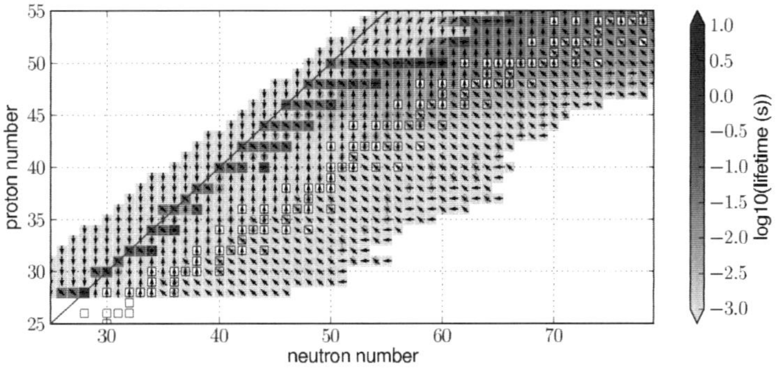

Figure 6. Same as figure 5, but with destruction lifetimes against the dominating reaction.

Neutron capture reactions on nuclei above the iron group govern the flow from one isotonic chain into the next and hence determine how fast nuclei with large Z can be reached. The time evolution of the hydrodynamic quantities determines how long favorable conditions for νp-process nucleosynthesis can be sustained. The location of the νp-process path has been found to mainly depend on the nuclear physics and hence results are independent of the trajectory variations.

Acknowledgments

The author acknowledges fruitful discussions with T. Rauscher, X. Tang, and J.W. Truran. This work is supported in part through the DOE Topical Collaboration "Neutrinos and Nucleosynthesis in Hot and Dense Matter" under contract DE-FG02-10ER41677.

References

[1] Woosley S E , Wilson J R and Mathews G J 1994 *Astrophys. J.* **433** 229–46
[2] Arcones A , Janka H -T and Scheck L 2007 *Astron. & Astrophys.* **467** 1227–48
[3] Hüdepohl L , Müller B , Janka H -T , Marek A and Raffelt G G 2010 *Phys. Rev. Lett.* **104** 251101
[4] Fischer T , Whitehouse S , Mezzacappa A , Thielemann F -K and Liebendörfer M 2010 *Astron. & Astrphys.* **517** 80+
[5] Roberts L F , Woosley S E and Hoffman R D 2010 *Astrophys. J.* **722** 954–67
[6] Hoffman R D , Woosley S E and Qian Y -Z 1997 *Astrophys. J.* **482** 951–62
[7] Fröhlich C , Martínez-Pinedo G , Liebendörfer M , Thielemann F -K , Bravo E , Hix W R , Langanke K and Zinner N T 2006 *Phys. Rev. Lett.* **96** 142502
[8] Pruet J , Hoffman R D , Woosley S E , Janka H -T and Buras R 2006 *Astrophys. J.* **644** 1028–39
[9] Wanajo S 2006 *Astrophys. J.* **647** 1323–40
[10] Wanajo S , Janka H -T and Kubono S 2011 *Astrophys. J.* **729** 46+
[11] Arcones A , Fröhlich C and Martínez-Pinedo G 2012 *Astrophys. J.* **750** 18+
[12] Frebel A et al. 2005 *Nature* **434** 871–73
[13] Travaglio C , Gallino R , Arnone E , Cowan J , Jordan F , and Sneden C 2004 *Astrophys. J.* **601** 864+
[14] Dauphas N , Rauscher T , Marty B and Reisberg L 2003 *Nucl. Phys. A* **719** 287+
[15] Rauscher T 2011 *Proc. of Science* PoS(NIC XI)059
[16] Cowan J , Roederer I U , Sneden C and Lawler J E 2011 *RR Lyrae Stars, Metal-Poor Stars, and the Galaxy* vol 5, ed A McWilliam (Carnegie Observatories Astrophysics Series) p 223
[17] Wasserburg G J , Busso M and Gallino R 1996 *Astrophys. J. Lett.* **466** L109
[18] Qian Y -Z and Wasserburg G J 2000 *Phys. Rep.* **333** 77–108
[19] McWilliam A 1998 *Astronomical J.* **115** 1640–1647
[20] Frebel A and Bromm V 2012 *Preprint* arXiv:1010.1261

[21] Montes F et al. 2007 *Astrophys. J.* **671** 1685–95
[22] Hansen C J , Primas F , Hartman H , Kratz K -L , Wanajo S , Leibundgut B , Farouqi K , Hallmann O , Christlieb N and Nilsson H 2012 *Astron. & Astrophys.* **545** 31+
[23] Weber C et al. 2008 *Phys. Rev. C* **78** 054310
[24] Schatz H et al. 1998 *Phys. Rep.* **294** 167
[25] Schatz H , Aprahamian A , Barnard V , Bildsten L , Cumming A , Ouellette M , Rauscher T , Thielemann F -K and Wiescher M 2001 *Phys. Rev. Lett.* **86** 3471
[26] Janka H -T , Buras R and Rampp M 2003 *Nucl. Phys. A* **718** 269
[27] Rauscher T , code NON-SMOKER^{WEB} v5.0w, 2008; http://nucastro.org/websmoker.html
[28] Cyburt R H et al 2010 *Astrophys. J. Suppl.* **189** 240
[29] Rauscher T 2011 *Int. J. Mod. Phys. E* **20** 1071
[30] Fuller G M , Fowler W A and Newman M J 1982 *Astrophys. J.* **252** 715
[31] Fuller G M , Fowler W A and Newman M J 1982 *Astrophys. J.* **48** 279
[32] Langanke K and Martínez-Pinedo G 2001 *At. Data Nucl. Data Tables* **79** 1
[33] Zinner N T and Langanke K 2004 *private communication*
[34] Wapstra A H , Audi G and Thibault C 2003 *Nucl. Phys. A* **729** 129
[35] Wapstra A H , Audi G and Thibault C 2003 *Nucl. Phys. A* **729** 337
[36] Kankainen A et al 2006 *Eur. Phys. J. A* **29** 271
[37] Fröhlich C , Rauscher T , Tang X and Truran J W *Phys. Rev. C* submitted
[38] Tu X L et al 2011 *Phys. Rev. Lett.* **106** 112501
[39] Haettner E et al 2011 *Phys. Rev. Lett.* **106** 122501
[40] Fallis J et al 2011 *Phys. Rev. C* **84** 045807
[41] Fisker J L , Hoffman R D and Pruet J 2009 *Astrophys. J. Lett.* **690** L135
[42] Fröhlich C and Rauscher T 2012 *AIP Conf. Proc. 11th Int. Conf. Origin of Matter and Evolution of Galaxies* in press
[43] Fröhlich C et al 2006 *Astrophys. J.* **637** 415

Journal of Physics: Conference Series **403** (2012) 012035 doi:10.1088/1742-6596/403/1/012035

Discoveries of Elements 113, 115 and 117

J H Hamilton[1], Yu Ts Oganessian[2] and V K Utyonkov[2] and the collaboration of Lawrence Berkeley National Laboratory, Oak Ridge National Laboratory, and the Research Institute of Atomic Reactors, Russia

[1]Physics Department, Vanderbilt University, Nashville, Tennessee 37235, USA
[2]Flerov Laboratory of Nuclear Reactions, JINR, Dubna, Russia

E-mail: j.h.hamilton@vanderbilt.edu

Abstract. Discovery of two isotopes of the new element 117 in the ^{249}Bk + ^{48}Ca reaction is described. A new ^{243}Am + ^{48}Ca experiment was carried out to firmly establish the discoveries of new elements 115 and 113. A total of thirty one decay chains of ^{288}115 are now observed in the Am reaction. In addition, four new decay chains are assigned to ^{289}115.

Introduction

One of the important frontiers in nuclear research is the discovery of new elements in the Periodic Table and their decay properties. These studies probe nuclear matter under extreme conditions of the Coulomb force at high Z and test nuclear structure predictions of proton and neutron shell closures under such extremes. An Island of Stability is predicted for N = 184 and Z between 114 and 126.

The discoveries of new elements 113, 114, 115, 116 and 118 were reported by bombarding actinide targets up through ^{249}Cf with ^{48}Ca [see ref. 1] [see Fig. 1]. The element 117 was missing because the necessary target, ^{249}Bk, is so short-lived it must be produced in a high flux reactor shortly before the experiment. Beginning in 2005, scientists from the Joint Institute for Nuclear Research in Dubna and Vanderbilt University began discussions with scientists at the High Flux Isotope Reactor, HFIR, to obtain a target of ^{249}Bk. When the possibility of obtaining ^{249}Bk from a reactor irradiation in progress at HFIR became known, a meeting was held at Vanderbilt in 2008 between JINR, ORNL and Vanderbilt scientists to plan a cooperation to obtain and use ^{249}Bk as a target to identify the new element 117. Lawrence Livermore National Laboratory scientists were subsequently invited to join the collaboration. The ^{249}Bk + ^{48}Ca reaction was studied in 2009-10. Five decay chains associated with ^{293}117 were identified along with one longer decay chain assigned to ^{294}117 [2, 3].

Earlier, the reaction ^{243}Am + ^{48}Ca was used to identify the new elements 115 and 113 [4, 5]. In that work, three decay chains were assigned to ^{288}115 and one to ^{287}115. To more definitely establish the ^{288,287}115 decay chains, this experiment was repeated. In the first part of the experiment, 21 new decay chains of ^{288}115 were identified at four different energies to give an excitation function. At the lowest bombarding energy, one new decay chain was observed and assigned to the 2n evaporation channel to ^{289}115 [6]. This new ^{289}115 is an important discovery because it is also populated in the

Journal of Physics: Conference Series **403** (2012) 012035 doi:10.1088/1742-6596/403/1/012035

alpha decay of $^{293}117$ in the Bk reaction to give cross-bombardment verification of these three new elements.

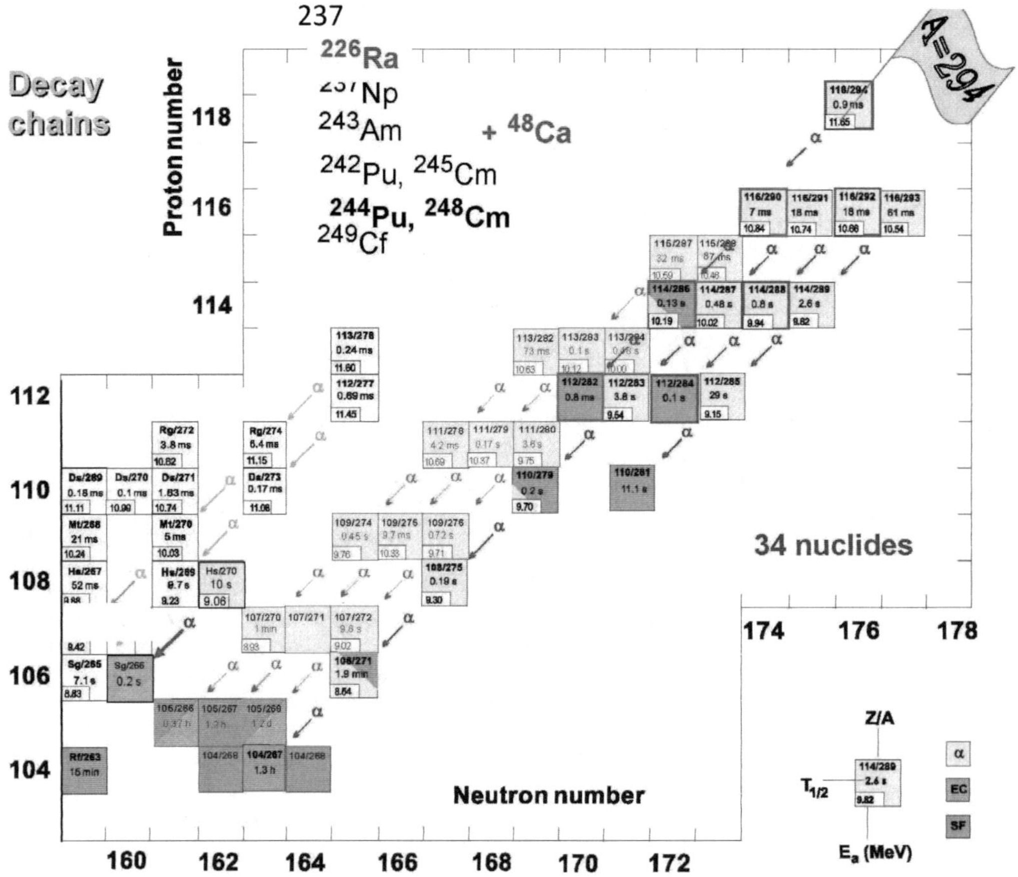

Figure 1. Chart of Nuclides. Shown in yellow are the new elements and isotopes produced in reactions of ^{48}Ca with actinide target. Element 117 is missing.

Experimental Procedures and Results

Details of the procedures and results are found in refs. 2, 3. Here is a brief summary. The ^{249}Bk was produced in the HFIR by irradiating Cm and Am targets for approximately 250 days by a thermal neutron flux of 2.5 x 10^{15} n/cm^2/s. After a three month waiting period to allow the short-lived activities to decay, the ^{249}Bk was chemically separated from the other actinides over a three month period. The procedures were described earlier [2, 3]. The 22 mg of purified ^{249}Bk was shipped to Russia where it was put into a circular target ring at the Research Institute of Atomic Reactors. The ^{249}Bk material thickness was 0.31 mg/cm^2. The wheel was rotated at 1700 rpm to minimize the heating of the target.

The target was bombarded with ^{48}Ca from the U400 cyclotron at the Flerov Laboratory for Nuclear Reactions. The target was irradiated with a ^{48}Ca beam dose of 2.4 x 10^{19} particles over 70 days at an excitation energy of 39 MeV for the reaction. The energy was then lowered to an excitation energy of 35 MeV and the target irradiated with a beam dose of 1.7 x 10^{18} ^{48}Ca particles for 60 days.

The evaporation residues were separated from the ^{48}Ca beam particles, scattered particles and transfer reaction products by the Dubna gas-filled recoil separator. After passing through two time-of-flight detectors to insure the ERs were from the separator, the ERs were implanted into a 4 x 12 cm^2 semiconductor with 12 position sensitive strips surrounded by eight 4 x 4 cm^2 side detectors with no position sensitivity. The uncertainties in the energies of alpha particles detected only in the side detectors was the order of 0.4 MeV. The ERs, alphas and spontaneous fission events all were required to be in the same pixel with all the events in ≤ 1 mm full width and half maximum. The registration efficiency for alpha particles was 87% and for a single fission fragment 100%.

As shown in Fig. 2, when an ER with an energy in the right range followed by an alpha particle in the same pixel (≤ 1 mm) and with the expected energy and half life for the 117 isotope, the beam was cut off and the subsequent alpha decays and SF were observed with very low background.

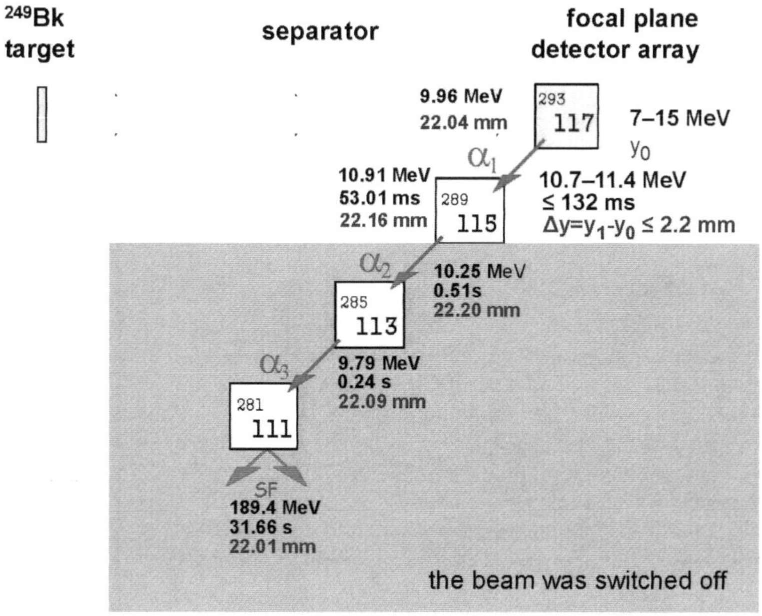

Figure 2. Low background detection where the beam is cut off after the right ER and first alpha decay observed within pre-selected ranges as shown.

At an excitation energy of 39 MeV, five decay chains of $^{293}117$ were observed and at E* = 35 MeV, one long chain of six alphas followed by SF for the decay of $^{294}117$. These results are shown in Fig. 3 [2, 3]. Note the good agreement of the experimental alpha energies and lifetimes shown in black with the theoretical ones [7] shown is blue below the experimental ones. Note for $^{294}117$, since only one event was observed, lifetimes were estimated from the decay times for each alpha decay.
The systematics of the known and our new alpha energies and half lives are shown in Fig. 4. Note all the half lives of our 11 new isotopes are significantly longer than those previously known for elements 111 to 115 because their extra one to two neutrons make these isotopes closer to N = 184. Thus, these data give strong support for the predicted Island of Stability around N = 184. Note $^{294}117$ (N = 177), the most neutron-rich isotope observed, has $T_{1/2}$ ≈ 78 ms while $^{294}118$ (N = 176) with only one less neutron has a half life of only 0.9 ms.

By using the same experimental procedures as in the 117 work [2, 3] and earlier 115 work [4, 5], the ^{243}Am + ^{48}Ca experiment was repeated at five energies, from ^{48}Ca 240 to 253 MeV, over a period of many months in 2011-2012. Beam doses of 3.3 to 11.7 x 10^{18} particles were achieved at each of the five energies. In the first experiment, a total of 21 new decay chains of ^{288}115 and at the lowest energy, one new event assigned to the new isotope ^{289}115 were observed as shown in Fig. 5 [6]. These 21 new decay chains are in full agreement in terms of alpha energies and half lives with the three chains reported earlier [4, 5]. The experiment was continued and seven additional decay chains of ^{288}115, three additional ones of ^{289}115 and one of ^{287}115 [8] were observed. These 31 decay chains of ^{288}115 are the largest number of events providing the most accurate known alpha energies and lifetimes of any superheavy element above Z = 110. The four new ^{289}115 decay chains are extremely important because they are the same chains seen in the decay of ^{289}115 populated in the alpha decay of ^{293}117 produced in the ^{249}Bk + ^{48}Ca reaction. Thus these ^{289}115 data from the Bk and Am reactions provide cross-bombardment verification of the new elements 117, 115 and 113. Since the end of the conference, five additional ^{293}117 and three additional ^{294}117 decay chains have been observed.

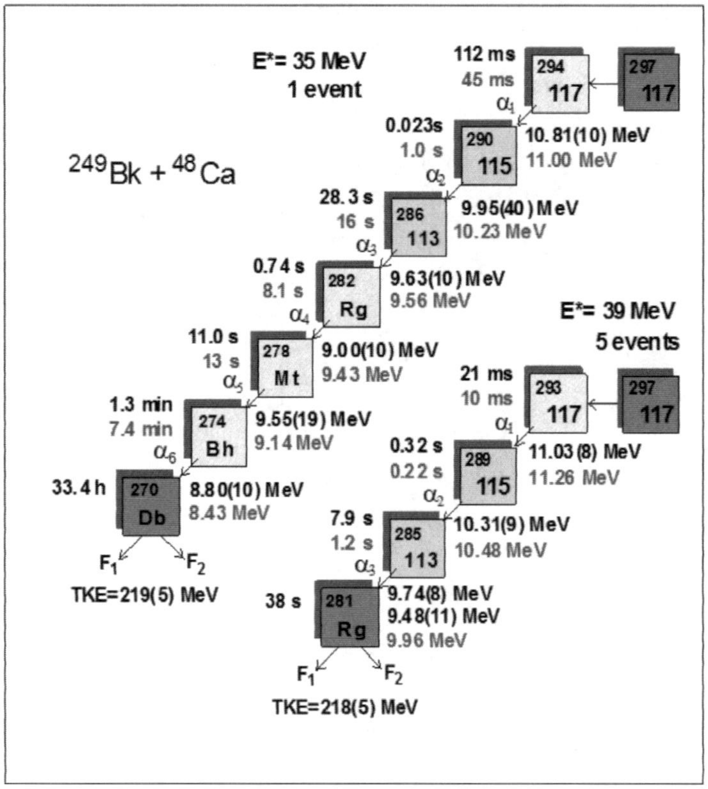

Figure 3. The averages of the five decay chains of ^{293}117 and the one long chain seen in ^{294}117 decay from the ^{249}Bk + ^{48}Ca are shown [2, 3]. Note the good agreement of the macroscopic-microscopic calculated alpha energies (blue) and our experimental ones (black). Figure taken from Ref. [2].

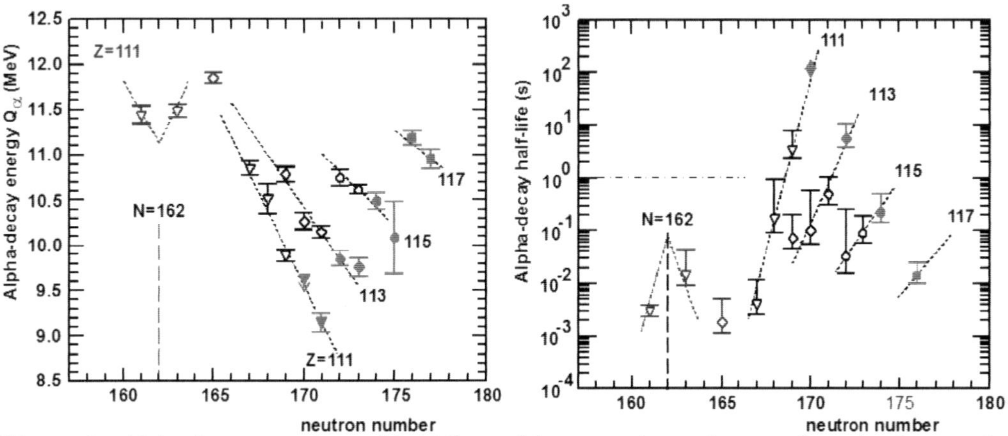

Figure 4. Alpha decay energies and half lives of known and new isotopes from the decays of $^{293,294}117$. The new data are in red. All the half lives of our 11 new isotopes because they are one or two neutrons closer to N = 184 to support its magic character. Figure taken from Ref. [2].

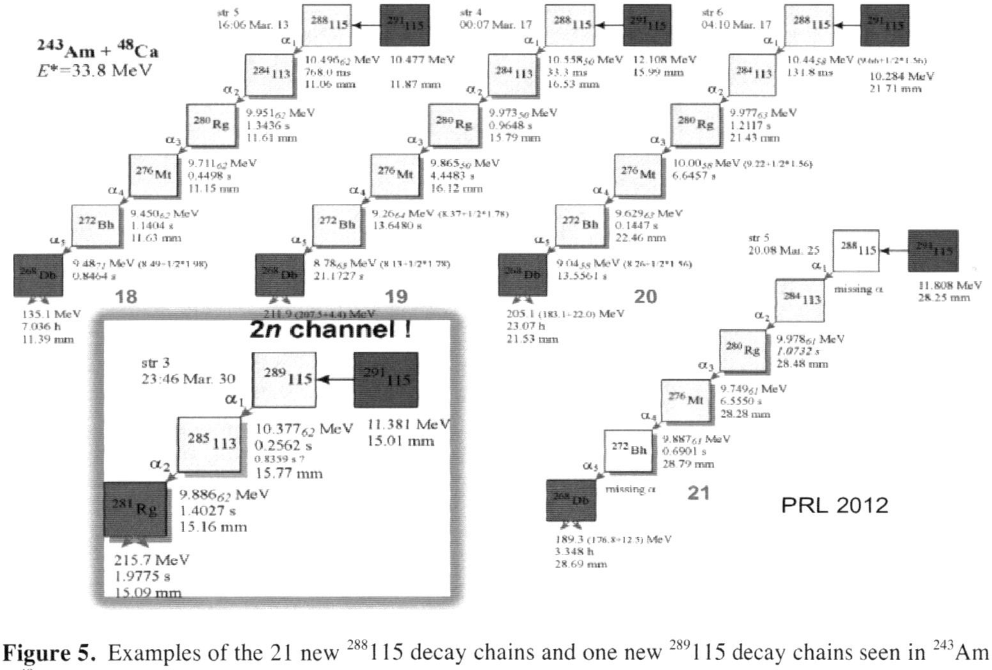

Figure 5. Examples of the 21 new $^{288}115$ decay chains and one new $^{289}115$ decay chains seen in ^{243}Am + ^{48}Ca reaction [6]. Figure adapted from Ref. [6].

Acknowledgement

We are grateful to the JINR Directorate and U400 cyclotron and ion source crews for their continuous support of the experiment. We acknowledge the support of the Russian Foundation for Basic Research Grants No. 11-02-12050 and No. 11-02-12066. Work at Vanderbilt University was supported by the U.S. DOE through Grant No. DE-FG-05-88ER40407. Research at ORNL was supported by the U.S. DOE Office of Nuclear Physics under DOE Contract No. DE-AC05-00OR22725 with UT-Battelle, LLC. Research at Lawrence Livermore National Laboratory was supported by LDRD Program Project. No. 08-ERD-030, under DOE Contract No. DE-AC52-07NA27344 with Lawrence Livermore National Security, LLC. These studies were performed in the framework of the Russian Federation/U.S. Joint Coordinating Committee for Research on Fundamental Properties of Matter.

References

[1] Oganessian Yu Ts, et al. 2007 *J. Phys. G: Nucl. Part. Phys.* **34** R165
[2] Oganessian Yu Ts, et al. 2010 *Phys. Rev. Lett.*. **104** 142502
[3] Oganessian Yu Ts, et al. 2011 *Phys. Rev.* C **83** 054315
[4] Oganessian Yu Ts, et al. 2004 *Phys. Rev.* C **69** 021601(R)
[5] Oganessian Yu Ts, et al. 2005 *Phys. Rev.* C **72** 034611
[6] Oganessian Yu Ts, et al. 2012 *Phys. Rev. Lett.*. **108** 022502
[7] Sobiczewski, A. 2010 *Acta. Phys. Pol.* B **41** 157
[8] Oganessian Yu Ts, et al. to be published

Measurement of $^{17}F + p$ reactions with ANASEN

L E Linhardt[1], L Baby[2], D W Bardayan[3], J C Blackmon[1], H Gardiner[1], E Johnson[2], E Koschiy[2], K T Macon[1], M Matos[1], B C Rasco[1], G Rogachev[2], D Santiago-Gonzalez[2] and I Wiedenhoever[2]
[1] Department of Physics & Astronomy, Louisiana State University, Baton Rouge, LA 70803 USA, [2] Department of Physics, Florida State University, Tallahassee, FL 32306 USA, [3] Physics Division, Oak Ridge National Laboratory, Oak Ridge, TN 37831 USA

Email: llinha2@tigers.lsu.edu

Abstract. Reactions involving radioactive nuclei play an important role in stellar explosions, but those reactions involving short-lived nuclei have only limited experimental information available due to currently limited beam intensities. Several facilities are aiming to provide greater access to these unstable isotopes at higher beam intensities, but more efficient and selective techniques and devices are needed to properly study these important reactions. The Array for Nuclear Astrophysics Studies with Exotic Nuclei (ANASEN), a charged particle detector designed by Louisiana State University (LSU) and Florida State University (FSU), was created for this purpose. ANASEN is used to study the reactions important in the αp- and rp- processes with proton-rich exotic nuclei, providing essentially complete solid angle coverage through an array of 40 silicon-strip detectors backed with CsI scintillators, covering an area of roughly 1300 cm^2. ANASEN also includes an active gas target/detector in a position-sensitive annular gas proportional counter, which allows direct measurement of (α,p) reactions in inverse kinematics. The first in-beam measurements with a partial implementation of ANASEN were performed at the RESOLUT radioactive beam facility of FSU during the summer of 2011. They included stable beam experiments and measurements of the $^{17}F(p,p)^{17}F$ and $^{17}F(p,\alpha)^{14}O$ reactions which are important to understanding the structure of ^{18}Ne and the $^{14}O(\alpha,p)^{17}F$ reaction rate. The performance of ANASEN and initial results from the ^{17}F studies will be presented.

1. Motivation

The most common stellar explosions in the galaxy are novae and x-ray bursts. These events occur in binary star systems, where matter rich in hydrogen and helium accretes from a main-sequence star to its compact partner and ignites in a thermonuclear runaway. In order to understand and correctly model such phenomena, many nuclear reaction rates must be known, especially (p,γ) and (α,p) reactions involving proton-rich nuclei. These critical reactions occur at relatively low energies and are governed by resonant properties near the particle threshold. In the case of (p,γ) reaction rates in novae, typically only a few distinct resonances dominate the reaction rate. However, for (α,p) reactions occurring in x-ray bursts, the temperatures and level densities are generally higher, and resonances also can decay to many excited states. While the statistical model is generally applicable in such cases, the reliability of such models at lower energies is uncertain.

Direct studies of the important reactions are preferred, but indirect studies of the important resonance properties can generally be realized with much less intense beams. Indirect studies can be particularly helpful to identify the most important resonances that can be addressed in the future by direct measurements. The Array for Nuclear Astrophysics Studies with Exotic Nuclei (ANASEN) is a charged-particle detector array designed for direct measurements of (α,p) reaction rates using an active gas target, and for indirect studies of resonance properties through scattering and transfer reactions [1, 2].

2. Array Components

ANASEN uses a position-sensitive proportional counter surrounding the beam axis to enable an active-target mode. Silicon-strip detectors (manufactured by Micron Semiconductor) surround the proportional counter in a barrel configuration with 3 rings of 12 rectangular Super X3 and annular QQQ3 detectors at forward angles form the cap of the barrel. Below in figure 1 is a SolidWorks representation of the inner pieces of ANASEN beside a photo of the assembly.

Figure 1: Solidworks (left) and photo (right) representation of ANASEN.

The Super X3's are double-sided, segmented, 75mm x 40.3mm x 1mm silicon-strip detectors. The back segments are used to determine the energy of particles independent of position to a resolution of typically about 50-70 keV, which is limited in part by the electronics. Comparison of signals from the four front horizontal resistive strips with the full energy measurement provides position resolution of about 1.5mm along the beam axis. The forward angles are covered in quadrants by QQQ3 detectors with an outer radius of 98mm. Each detector has 16 radial front rings, and 16 annular back segments, all of which are non-resistive.

All detectors in the array are backed by trapezoidal-shaped 2-cm thick CsI(TI) crystals (manufactured by SCIONIX), to ensure that the full energy is measured for all particles. There are two different geometries, both types read by two 18mmx18mm pin-diodes, and providing a typical energy resolution of roughly 8%. With the first geometry, one CsI(TI) detector completely backs one Super X3 detector. The second geometry allows for four CsI(TI) detectors to back each QQQ3 detector.

To allow direct measurements of (α,p) reactions, ANASEN includes a gas target capability. When using a gas target, the beam will pass through a thin window and enter a gas-filled recirculating (flow) chamber. In this mode, ANASEN can use helium gas (with a small component of a quenching agent) as a target and detection medium simultaneously. The position-sensitive proportional counter extends along the beam axis for 43 cm. There are five rings of wires. The inner and outer rings are used to prevent electrons produced outside of the detection region from interfering with the signal. The middle ring holds 19 anodes, each separated by a cathode. The remaining two rings are cathodes that help create the independent trapezoidal detection regions of the proportional counter. Figure 2 shows the electric field simulation for the proportional counter configuration of five rings of wires. The 19 carbon-fiber anode wires, 4 kΩ/cm, are equally spaced 3 cm from the beam axis. The position of the traveling particle is determined by charge division method using the signal from both ends of each anode.

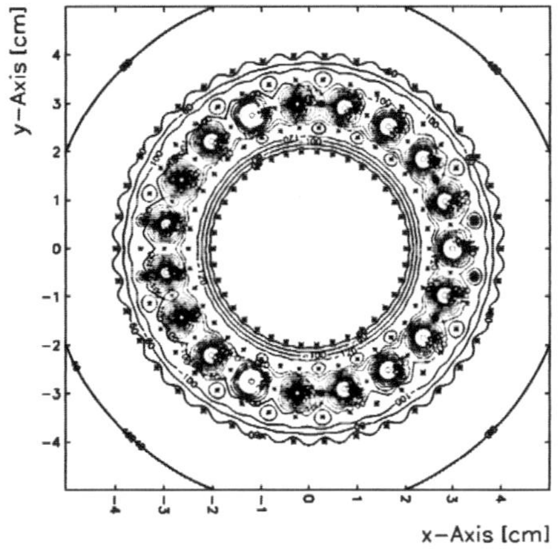

Figure 2: Electric Field Calculation for the Proportional Counter.

With the combination of proportional counter, silicon detectors, and CsI(TI) detectors, there are roughly 800 channels of electronics in the full configuration of ANASEN. To deal with this high channel count in a cost effective manner, ANASEN utilizes Application-Specific Integrated Circuits (ASICs) HINP16C designed by Washington University [3]. Each HINP16C chip handles the pulse-shaping, timing, triggering, and digitization of 16 channels of data. The system as a whole has 2 ASICs HINP16C chips per board, with 16 boards in the system, handling up to 512 channels. External preamplifiers are used to optimize the resolution and dynamic range. We developed a 72-channel preamplifier system for the silicon detectors based on the Indiana University's LASSA preamplifier chip; with a gain of 27 mV/MeV and a 30 ms fall time that matches the shaping of the HINP16C [4].

In addition, a heavy ion recoil detector was developed as an auxiliary detector for experiments measuring inverse mechanics. Recoiling ions at forward angles and with relatively high energies penetrate through a thin Mylar window into an additional chamber attached downstream of ANASEN that forms a gas proportional counter. This counter is fixed at $\Theta_{lab} = 0°$ and can provide coincidence detection of heavy ions with high efficiency for some reactions, though most effectively when a solid target is used that fixes the location of the reaction vertex. The gas in the heavy ion gas ionization chamber can be independently controlled from that in the ANASEN active-target. The chamber itself holds ten pairs of alternating anode and cathode planes oriented perpendicular to the beam axis. We have also developed circuit board anodes that provide independent x and y position of the heavy ion. Each circuit board provides 34 independent signals with 4 mm spacing. Two circuit boards along with eight additional anodes are processed using the 72-channel preamp system and ASICs developed for the silicon array.

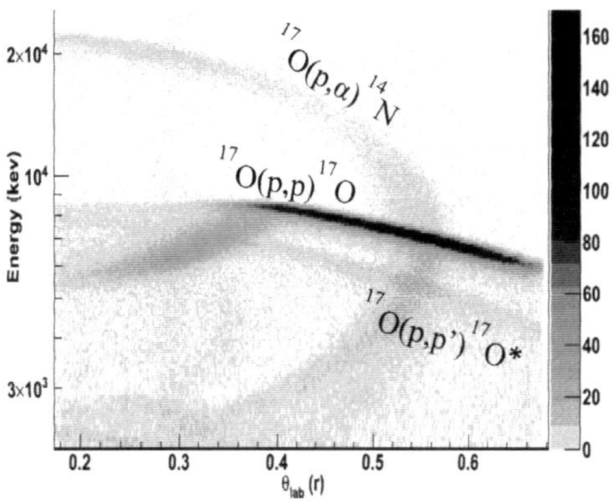

Figure 3: Energy (keV) vs Θ_{lab} (radians) showing the p's and α's detected by the S2 silicon telescope.

3. Commissioning and ^{17}F+p Experiment

ANASEN targets measurements with radioactive ion beams at the ReA3 facility at the National

Superconducting Cyclotron Laboratory [5] and the REOLUT facility at FSU [6]. The initial assembly, testing, commissioning, and first experiments with ANASEN were performed at FSU. We first tested the performance of the electronics with an annular (S2) double double-sided segmented silicon detector telescope and a first implementation of the heavy ion detector by measuring the $^{17}O(p,\alpha)^{14}N$ reaction. Beams of ^{17}O from the tandem accelerator at FSU bombarded solid polyethylene $(CH_2)_n$ targets at four different beam energies between E_{cm}= 1.8-3.0 MeV. The silicon detectors were calibrated with a ^{228}Th source before and after the experiment. All the different reaction channels were clearly separated by kinematics with the silicon strip detectors (shown in Fig. 3), and the better than 5% Z resolution through the heavy ion recoil chamber (shown in Fig. 4). During the experiment, the ^{17}O and ^{14}N was clearly distinguishable and the elastic and inelastic scattering was easily resolved, even with the elastic scattering 10 times stronger than the inelastic.

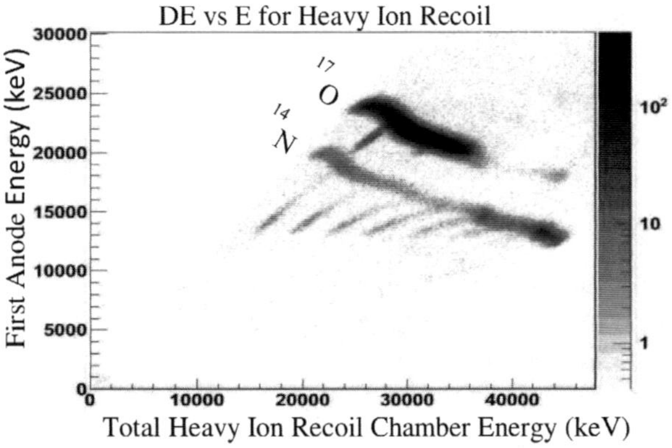

Figure 4: DE vs E of the Heavy Ion Recoil Chamber.

Given the encouraging results from the ^{17}O measurement, we then switched to a ^{17}F beam from RESOLUT to simultaneously study the $^{1}H(^{17}F,p)^{17}F$ and $^{1}H(^{17}F,\alpha)^{14}O$ reactions. This experiment studied the structure of ^{18}Ne and the time inverse of $^{14}O(\alpha,p)^{17}F$ reaction that is the first reaction in the αp process in X-ray bursts. A 55 MeV beam of ^{17}F with an intensity of about 10^5 ion/s bombarded a thick (2 mg/cm^2) polypropylene target. The S2 double-sided segmented silicon-strip detector telescope detected p's and α's at Θ_{lab} = 9.5-28°, covering a relatively large solid angle of interest. The purity of the beam was typically about 70% or better, and the elastic scattering events were easily identified by coincidence between the fluorine in the heavy ion recoil chamber and the protons in the silicon telescope, The center of mass energy for each event was calculated from the measured angle and the energy of the light particle in the silicon telescope, and an excitation function reconstructed. Figure 5 shows the ^{17}F+p elastic scattering yield versus proton center of mass energy (MeV) for Θ_{cm} = 137° with a preliminary R-Matrix fit. Evidence for 2 resonances is seen in the elastic scattering excitation function in the region around E_{cm} ~2.3-2.5 MeV, which corresponds to the most interesting region of excitation energy in ^{18}Ne for the $^{14}O(\alpha,p)^{17}$ reaction. However, additional analysis is required to determine the statistical significance of this observation and to extract resonant properties for the states. We also observed $^{17}F(p,\alpha)^{14}O$ through this experiment, but only a total of roughly 80 events over the entire excitation function. While analysis is still in progress, the statistics are likely not going to be sufficient to draw strong conclusions about possible low energy resonances for this reaction.

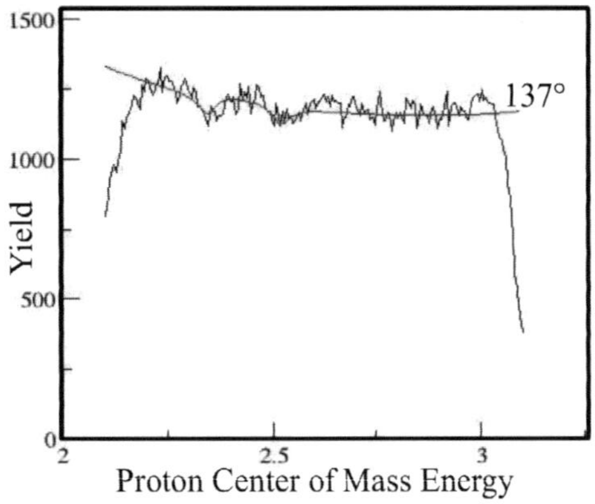

Figure 5: Yield vs E_{cm} for $^{17}F(p,p)^{17}F$ with a preliminary R-matrix fit

4. Summary

The first (α,p) reaction was measured using active target mode of ANASEN in a test experiment with a stable ^{14}N beam on helium gas with a 1% mixture of CO_2. Protons and alpha particles were easily identified by their energy loss in the proportional counter. Figure 6 shows the $^{14}N(\alpha,p)^{17}O_{g.s.}$ excitation function reconstructed from the energy in the silicon array and the trajectory reconstructed from the position in the silicon and in the proportional counter. This excitation function was obtained in 6 hours of beam time and beam intensities of $5x10^5$ ion/s. It is important to note that this complete excitation function was measured simultaneously with a center of mass energy resolution of roughly 70 keV. The next step with ANASEN is to perform the first direct study of $^{14}O(\alpha,p)^{17}F$ directly with a helium gas target.

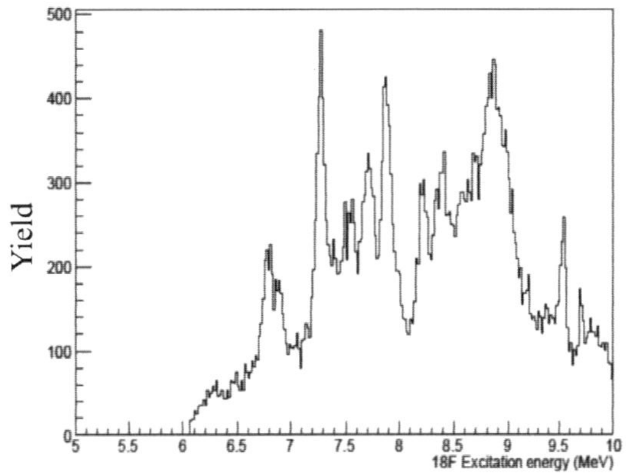

Figure 6: Yield vs ^{18}F excitation energy (MeV) for $^{14}N(\alpha,p)^{17}O_{g.s.}$

Acknowledgements

This work is supported by the National Science Foundation Award #0821308 and by the U.S. Department of Energy Office of Nuclear Physics grant #DE-FG02-96ER40978.

References

[1] M Matos *et al* 2010 *POS (NIC XI)* 226
[2] M Matos *et al* 2012 *14th International Symposium on Capture Gamma-Ray Spectroscopy and Related Topics*, ed *D R* P E Garrett (Singapore: World Scientific)
[3] G L Engel *et al* 2007 *NIM A* **573** 418
[4] B Davin *et al* 2001 *NIM A* **473** 302
[5] http://www.nscl.msu.edu/exp/sr.
[6] http://www.physics.fsu.edu/nuclear/foxlab/nuclear-foxlab.html.

Charged-current weak interaction processes and its impact on proto-neutron star cooling and nucleosynthesis

Gabriel Martínez Pinedo[1,2], **T Fischer**[2,1], **A Lohs**[1], **L Huther**[1]
[1]Institut für Kernphysik, Technische Universität Darmstadt, Schlossgartenstraße 2, 64289 Darmstadt, Germany
[2]GSI Helmholtzzentrum für Schwerioneneforschung, Planckstraße 1, 64291 Darmstadt, Germany

E-mail: gabriel.martinez@physik.tu-darmstadt.de

Abstract. We show that a treatment of charged-current neutrino interactions in hot and dense matter that is consistent with the nuclear equation of state has a strong impact on the spectra of the neutrinos emitted during the deleptonization period of a protoneutron star formed in a core-collapse supernova. We compare results of simulations including and neglecting mean field effects on the neutrino opacities. Their inclusion reduces the luminosities of all neutrino flavors and enhances the spectral differences between electron neutrino and antineutrino. The magnitude of the difference depends on the equation of state and in particular on the symmetry energy at sub-nuclear densities. These modifications reduce the proton-to-nucleon ratio of the neutrino-driven outflow, increasing slightly their entropy. They are expected to have a substantial impact on the nucleosynthesis in neutrino-driven winds, even though they do not result in conditions that favor an r-process. Contrarily to previous findings, our simulations show that the spectra of electron neutrinos remain substantially different from those of other (anti)neutrino flavors during the entire deleptonization phase of the protoneutron star. The obtained luminosity and spectral changes are also expected to have important consequences for neutrino flavor oscillations and neutrino detection on Earth.

1. Introduction

Supernova explosions of massive stars are related to the birth of neutron stars due to the collapse of the stellar core at the end of stellar evolution [1]. The detection of neutrinos from SN1987A [2, 3] confirmed that the $\approx 3 \times 10^{53}$ ergs of gravitational energy gained by the core collapse are emitted as neutrino radiation on time scales of tens of seconds, during which the central protoneutron star (PNS) cools, deleptonizes and contracts. In the delayed neutrino-heating explosion mechanism [4, 1], neutrinos also transport energy from the nascent PNS to the stalled bounce shock. This mechanism remains the most viable scenario to explain supernova explosions as confirmed by recent two-dimensional simulations [5]. Once the explosion sets in, the continuous emission of neutrinos from the PNS drives a low-mass outflow known as neutrino-driven wind [6] that is currently considered the favored site for the productions of elements heavier than iron (e.g. [7]). As neutrinos travel through the stellar mantle, they can suffer flavor oscillations [8], contribute to the nucleosynthesis of several rare isotopes [9, 10] and even drive an r process in the He-shell of metal-poor stars [11] before they are finally detected on Earth.

Accounting for all aspects discussed above requires the knowledge of the spectra of the neutrinos emitted during the cooling phase of the PNS. Due to their low energies $\nu_{\mu,\tau}, \bar{\nu}_{\mu,\tau}$ interact only via neutral-current reactions. Hence, together with the neutron-richness of the PNS surface one expects the following neutrino-energy hierarchy: $\langle E_{\nu_{\mu,\tau}}\rangle > \langle E_{\bar{\nu}_e}\rangle > \langle E_{\nu_e}\rangle$ [12, 13], with $\langle E\rangle$ the average energy of the neutrino spectrum. Early supernova models [14, 15] predicted large energy differences between $\bar{\nu}_e$ and ν_e that resulted in neutron-rich ejecta as required by r-process nucleosynthesis [16]. However, as the treatment of neutrino transport and weak interaction processes improved, the computed energy difference between $\bar{\nu}_e$ and ν_e decreased and the early wind ejecta became proton rich [17, 18]. More recently, it has been possible to perform supernova simulations based on three-flavor Boltzmann neutrino transport for time scales of several tens of seconds [19, 20], covering the whole deleptonization of the PNS. These simulations predict a continuous decrease of the energy difference between neutrinos and antineutrinos of all flavors that became practically indistinguishable after ≈ 10 s. The exact value depends on the progenitor. The proton-richness of the ejecta increases continuously with time and leaves the νp process [21, 22, 23] as the only mechanism for producing elements heavier than iron in neutrino-driven winds.

2. Charged-current neutrino processes at high densities

With the development of three-flavor Boltzmann neutrino transport codes it has been possible to relate the spectra of the emitted neutrinos and the underlying nucleosynthesis to high density neutrino matter interactions and basic properties of the nuclear equation of state. The simulations of ref. [20] have been recently analyzed, showing that the convergence of neutrino and antineutrino spectra at late times is due to the suppression of charged-current processes at high densities [13]. This analysis was based on a set of neutrino opacities that assume a non-interacting gas of nucleons and nuclei. This approximation may be valid during the accretion phase prior to the onset of the supernova explosion when the region from where neutrinos decouple, the neutrinospheres, is located at relatively low densities, $\sim 10^{11}$ g cm^{-3}. However, as the PNS cools the neutrinospheres move to increasingly higher densities where the non-interacting gas approximation breaks down. The nuclear interaction is treated at the mean-field level in equations of state (EoS) commonly used in core-collapse supernova simulations [24, 25]. However, its influence on weak interaction processes is often neglected. These EoS treat protons and neutrons as a gas of quasi-particles that move in a mean-field single particle potential U. Assuming non-relativistic nucleons, which is a good approximation for densities $\rho \leq 5\rho_0$ where $\rho_0 \approx 2.5 \times 10^{14}$ g cm^{-3}, the energy momentum relation closely resembles the non-interacting case [26]:

$$E_i(\boldsymbol{p}_i) = \frac{\boldsymbol{p}_i^2}{2m_i^*} + m_i + U_i, \quad i = n, p, \tag{1}$$

with particle rest-masses m_i. Both the single-particle mean-field potentials and the (Landau) effective masses, m_i^* depend on density, temperature and proton-to-nucleon ratio, Y_e. Importantly, due to the extreme neutron-rich conditions the mean-field potentials for neutron and protons can be very different with their relative difference $U_n - U_p$ directly related to the nuclear symmetry energy [26] (see below).

In order to quantify the impact of the mean field potentials, let us consider (anti)neutrino absorption on neutrons (protons). Assuming zero momentum transfer, i.e. $\boldsymbol{p}_n \approx \boldsymbol{p}_p$ (elastic approximation), the electron(positron) and (anti)neutrino energies are related by:

$$E_{\nu_e} = E_{e^-} - (m_n - m_p) - (U_n - U_p), \tag{2}$$

$$E_{\bar{\nu}_e} = E_{e^+} + (m_n - m_p) + (U_n - U_p). \tag{3}$$

Eqs. (2) and (3) show that the contribution of the mean field potential reduces (increases) the energy of the emitted neutrinos (antineutrinos). The energy difference between neutrinos and antineutrinos is increased by an amount $2(U_n - U_p)$. The opacity, or inverse mean-free path, for (anti)neutrino absorption also changes. In the elastic approximation and assuming extreme relativistic electrons, the opacity for neutrino absorption becomes [27, 26]:

$$\begin{aligned} \chi(E_{\nu_e}) &= \frac{G_F^2 V_{ud}^2}{\pi(\hbar c)^4}(g_V^2 + 3g_A^2) \times \\ & E_e^2[1 - f_e(E_e)]\frac{n_n - n_p}{1 - e^{\beta(\eta_p - U_p - \eta_n + U_n)}}, \end{aligned} \tag{4}$$

with E_{ν_e} and E_e related by equation (2). G_F is the Fermi coupling constant, V_{ud} is the up-down entry of the Cabibbo-Kobayashi-Maskawa matrix, g_V and g_A are vector and axial-vector coupling constants and $n_{p,n}$ the number density of protons or neutrons. f is the Fermi-Dirac distribution function and η is the chemical potential (without rest mass). The emissivity, $j(E_{\nu_e})$, can be obtained from the detailed balance condition $j(E_{\nu_e}) = \exp(-\beta(E_{\nu_e} - \mu_\nu^{\rm eq}))\chi(E_{\nu_e})$, with $\mu_\nu^{\rm eq} = \mu_e - (\mu_n - \mu_p)$ the equilibrium neutrino chemical potential, μ the chemical potential including rest mass and β the inverse temperature. The opacity and emissivity for antineutrino absorption are obtained by exchanging neutron and proton and using equation (3) to relate the positron and antineutrino energies. The quantity $\varphi = \eta - U$ represents the chemical potential for a non-interacting gas of nucleons, that is related to the nucleon number density by the relation:

$$n = 2\int \frac{d^3\boldsymbol{p}}{(2\pi\hbar c)^3}\frac{1}{e^{\beta\left(\frac{p^2}{2m}-\varphi\right)}+1}. \tag{5}$$

Eq. (5) provides a method of determining the mean-field potential, U, when using an EoS that does not provide this quantity, e.g., the EoS of ref. [25]. The difference between proton and neutron chemical potentials is directly related to the nuclear symmetry energy, $S(\rho)$ [28]:

$$\mu_n - \mu_p = 4(1 - 2Y_p)S(\rho) = m_n - m_p + \varphi_n - \varphi_p + U_n - U_p, \tag{6}$$

with $Y_p \ll 1$ the proton fraction. One can observe that the value of $\Delta U = U_n - U_p$, and consequently the energy difference between electron antineutrinos and neutrinos is larger for equations of state with a larger value of the symmetry energy.

Figure 1 shows neutrino and antineutrino opacities and emissivities evaluated at conditions found at the antineutrinosphere for the 18 $M_\odot$ model of ref. [13] at 1 s after bounce. The curves labeled RMF (U_n, U_p) include the contribution of the mean-field potentials $U_n = -7.6$ MeV and $U_p = -14.2$ MeV [25], while the curves labeled $(U_n = 0, U_p = 0)$ assume a non-interacting gas of nucleons, i.e. neglect the contribution of the potentials but still use chemical potentials as given by the EoS. Due to the presence of the mean-field potentials the effective Q-value for electron capture increases with respect to the free case producing neutrinos with substantially lower energy. For the inverse process, neutrino absorption, the opacity is enhanced due to the fact that the produced electron gains an energy $U_n - U_p$ reducing the final-state Pauli blocking of the electron. The situation is completely analogous to (anti)neutrino emission and absorption on heavy neutron-rich nuclei [29]. Using Eq. (4), it can be shown that the opacity for the non-interacting gas, $\chi_{\rm ni}$, is related to the mean-field opacity, $\chi_{\rm mf}$ by $\chi_{\rm mf}(E) = \chi_{\rm ni}(E + U_n - U_p)$. This relationship produces a large enhancement of the neutrino opacity at high densities, $\rho \approx 10^{14}$, where $U_n - U_p \approx 50$ MeV when compared with the non-interacting approximation used in ref. [13]. For antineutrino absorption, positrons are non-degenerate and their spectrum approaches the Boltzmann distribution. Hence, the non-interacting emissivity and mean-field emissivities are related by $j_{\rm mf}(E) = j_{\rm ni}(E - U_n + U_p)$. The mean-field antineutrino opacity is larger at high energies as final-state Pauli blocking of the neutrons becomes less efficient.

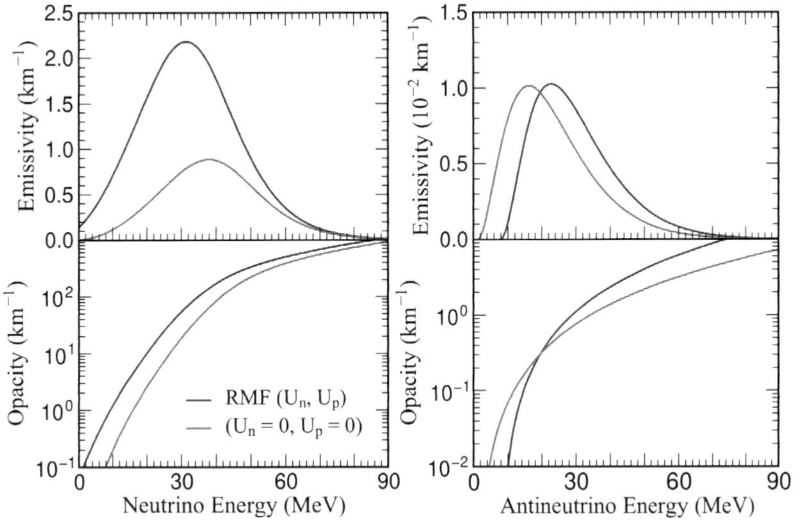

Figure 1. Opacity and emissivity for neutrino (left panels) and antineutrino (right panels), evaluated at conditions $\rho = 2.1 \times 10^{13}$ g cm^{-3}, $T = 7.4$ MeV and $Y_e = 0.035$.

3. Impact in supernova simulations and nucleosynthesis

To explore the impact that the changes in opacities have on the spectra and luminosities of the emitted neutrinos, we have performed core-collapse supernova simulations based on spherically symmetric radiation hydrodynamics with three-flavor Boltzmann neutrino transport for a 15 $M_\odot$ progenitor (see refs [13, 30] for additional details). Because spherically symmetric simulations do not result in explosions for such a massive iron-core progenitor, we enhance the neutrino heating rates in the gain region following the scheme of ref. [20]. It results in the onset of explosion at about 350 ms post bounce. The simulations are evolved from core collapse, through the explosion up to more than 3 seconds after bounce. During core collapse and post-bounce accretion phases, the mean-field potentials, U_n, U_p, are only on the order of several 100 keV in the region of neutrino decoupling, which is located at intermediate densities on the order of 10^{11} g cm^{-3}. Hence, their inclusion does not affect the supernova dynamics prior to the explosion.

After the onset of the explosion the neutrinospheres move to increasingly higher densities reaching values of the order of 10^{13} g cm^{-3}. The left panels of Fig. 2 show the evolution of the luminosity and average neutrino energy for all neutrino flavors. These observables are sampled in a co-moving reference frame at a distance of 1000 km. Using charged-current neutrino opacities that include the mean-field potentials slightly reduces the luminosities for all neutrino flavors. Moreover, as expected from the discussion above, it enhances the differences in luminosities and average energies between electron neutrinos and antineutrinos.

The changes in electron (anti)neutrino spectra and luminosities have important consequences for nucleosynthesis in neutrino-driven winds. The increased difference between average energies of ν_e and $\bar{\nu}_e$ spectra impacts the Y_e of the ejected matter resulting in neutron-rich conditions for the early ejecta and a slight increase in the entropy per nucleon (see fig. 2). In the simulation that neglects the contributions of the mean-field potentials the ejecta are always proton-rich. The decrease in Y_e is not large enough to favor an r-process but results in substantial differences in the production of nuclei around $A \sim 90$ (see fig. 3).

We expect that for times larger than those shown in figure 2 the ejecta becomes proton-rich as the average electron neutrino and antineutrino energy approach each other. However,

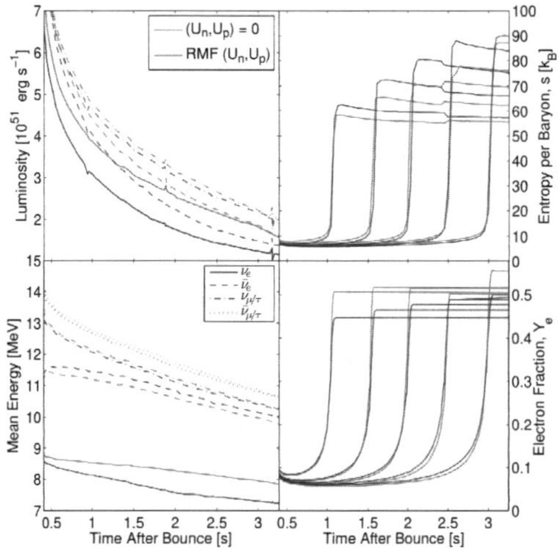

Figure 2. Evolution of the neutrino luminosity (upper left panel) and average energy (lower left panel), proton-to-nucleon ratio, Y_e, (lower right panel) and entropy per nucleon (upper right panel) for several mass elements ejected from the PNS surface. The curves shown in blue use neutrino opacities computed using the mean-field potentials based on the EoS [25] while they are neglected on the red curves. On the left panels solid lines correspond to electron neutrinos, dashed to electron neutrinos, dot-dashed to $\nu_{\mu,\tau}$ and dotted to $\bar{\nu}_{\mu,\tau}$.

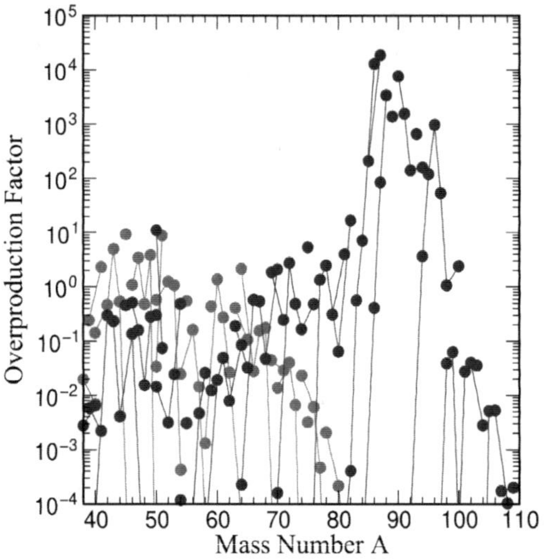

Figure 3. Abundances resulting from nucleosynthesis calculations based on the trajectories shown in figure 2.

contrarily to previous simulations that neglected the contribution of mean-field potentials [20, 19] it will remain a substantial difference between their average energies. This can be related to the relative contribution of charged-current and neutral-current processes to the total neutrino

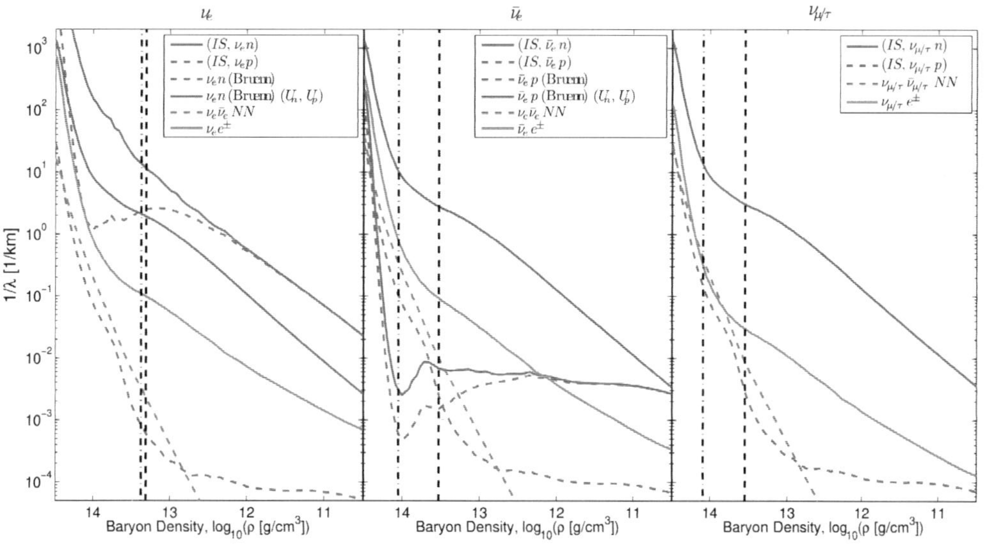

Figure 4. Inverse mean free paths for the individual reactions considered, $(\mathrm{IS}, \nu N)$: isoenergetic neutrino-nucleon scattering, $(\nu_e n,\ \bar{\nu}_e p)$: charged current reactions, $(\nu\bar{\nu}NN)$: N–N–Bremsstrahlung , for ν_e (left panel), $\bar{\nu}_e$ (middle panel) and $\nu_{\mu/\tau}$ (right panel), based on radial profiles and of a 18.0 $\mathrm{M}_\odot$ progenitor. The red lines have been computed using equation (4) including (solid line) and neglecting (dashed line) mean-field potentials. The vertical black dashed and dash-dotted lines mark the position of the transport and energy spheres. See ref. [13] for additional details.

opacity. Fig. 4 shows the position of the neutrinospheres for different neutrino flavors and the opacities as a function of density at 10 s after bounce based on the data from the 18 $\mathrm{M}_\odot$ model of ref. [13]. At this time, charged-current processes (red lines) for $\bar{\nu}_e$ have become almost negligible due to the increase in symmetry energy as the neutrinospheres move to high densities. As a consequence the total opacity for $\bar{\nu}_e$ is dominated by neutral current processes. This explains why the spectra of $\bar{\nu}_e$ and $\nu_{\mu,\tau}$ become very similar at late times as neutral current processes do not differentiate between neutrino flavors. The situation is different for ν_e neutrinos whose opacity is dominated by charged-current neutrino absorption on neutrons. Notice that this is only the case for opacities that consistently consider the mean-field potential (solid red line). If the mean-field potentials are neglected (dashed red lines) neutrino absorption is largely suppressed. The fact that neutral-current processes dominate for $\bar{\nu}_e$, while charged-current processes dominate for ν_e is responsible of maintaining a substantial energy difference between both neutrino species during the whole deleptonization of the protoneutron star.

4. Conclusions

We have shown that a treatment of the charged-current (anti)neutrino opacities, that is consistent with the EoS as suggested by [26], has important consequences for the neutrino-spectra evolution during the PNS cooling phase. The most relevant finding is an increased difference between average energies of ν_e and $\bar{\nu}_e$ that persist during the whole simulation time of 3 seconds after the onset of the explosion. The changes on neutrino spectra are expected to have important consequences for nucleosynthesis, flavor oscillations and neutrino detection on Earth. Our results imply that not only the evolution of the neutrino luminosities [31] but also the spectral differences between ν_e and $\bar{\nu}_e$ are sensitive to the symmetry energy of nuclear matter. Our simulations are based on neutrino opacities computed using the elastic approximation that neglects momentum exchange between nucleons. They need to be extended to consider the full kinematics [26] of the reaction and many-body correlations [32, 33, 34] that are expected to become important at later times than those considered in the present study. Furthermore, it is important to explore the sensitivity of the results to different EoS and in particular to EoS that are consistent with recent constrains on the nuclear symmetry energy [35].

This work has been partly supported by the Deutsche Forschungsgemeinschaft through contract SFB 634, the Helmholtz International Center for FAIR within the framework of the LOEWE program launched by the state of Hesse, the Helmholtz Association through the Nuclear Astrophysics Virtual Institute (VH-VI-417) and the National Science Foundation under project no. PBBSP2-133378.

References

[1] Janka H T, Langanke K, Marek A, Martínez-Pinedo G and Müller B 2007 *Phys. Repts.* **442** 38–74
[2] Hirata K, Kajita T, Koshiba M, Nakahata M and Oyama Y 1987 *Phys. Rev. Lett.* **58** 1490–1493
[3] Koshiba M 1992 *Phys. Repts.* **220** 229–381
[4] Bethe H A and Wilson J R 1985 *Astrophys. J.* **295** 14–23
[5] Mueller B, Janka H T and Marek A 2012 *ArXiv e-prints* (*Preprint* 1202.0815)
[6] Duncan R C, Shapiro S L and Wasserman I 1986 *Astrophys. J.* **309** 141–160
[7] Qian Y Z 2003 *Prog. Part. Nucl. Phys.* **50** 153–199
[8] Duan H, Fuller G M and Qian Y 2010 *Ann. Rev. Nucl. Part. Sci.* **60** 569–594
[9] Woosley S E, Hartmann D H, Hoffman R D and Haxton W C 1990 *Astrophys. J.* **356** 272–301
[10] Heger A, Kolbe E, Haxton W, Langanke K, Martínez-Pinedo G and Woosley S E 2005 *Phys. Lett. B* **606** 258–264
[11] Banerjee P, Haxton W C and Qian Y Z 2011 *Phys. Rev. Lett.* **106** 201104
[12] Keil M T, Raffelt G G and Janka H T 2003 *Astrophys. J.* **590** 971–991
[13] Fischer T, Martínez-Pinedo G, Hempel M and Liebendörfer M 2012 *Phys. Rev. D* **85**(8) 083003
[14] Woosley S E, Wilson J R, Mathews G J, Hoffman R D and Meyer B S 1994 *Astrophys. J.* **433** 229–246
[15] Takahashi K, Witti J and Janka H T 1994 *Astron. & Astrophys.* **286** 857–869
[16] Hoffman R D, Woosley S E and Qian Y Z 1997 *Astrophys. J.* **482** 951–962
[17] Liebendörfer M, Mezzacappa A, Thielemann F K, Bronson Messer O E, Raphael Hix W and Bruenn S W 2001 *Phys. Rev. D* **63** 103004
[18] Buras R, Rampp M, Janka H T and Kifonidis K 2006 *Astron. & Astrophys.* **447** 1049–1092
[19] Hüdepohl L, Müller B, Janka H, Marek A and Raffelt G G 2010 *Phys. Rev. Lett.* **104** 251101
[20] Fischer T, Whitehouse S C, Mezzacappa A, Thielemann F K and Liebendörfer M 2010 *Astron. & Astrophys.* **517** A80
[21] Fröhlich C, Martínez-Pinedo G, Liebendörfer M, Thielemann F K, Bravo E, Hix W R, Langanke K and Zinner N T 2006 *Phys. Rev. Lett.* **96** 142502
[22] Pruet J, Hoffman R D, Woosley S E, Janka H T and Buras R 2006 *Astrophys. J.* **644** 1028–1039
[23] Wanajo S 2006 *Astrophys. J.* **647** 1323–1340
[24] Lattimer J M and Swesty F D 1991 *Nucl. Phys. A* **535** 331–376
[25] Shen H, Toki H, Oyamatsu K and Sumiyoshi K 1998 *Nucl. Phys. A* **637** 435–450
[26] Reddy S, Prakash M and Lattimer J M 1998 *Phys. Rev. D* **58** 013009
[27] Bruenn S W 1985 *Astrophys. J. Suppl.* **58** 771–841
[28] Haensel P, Potekhin A Y and Yakovlev D G 2007 *Neutron Stars 1: Equation of State and Structure* (*Astrophysics and Space Science Library* vol 326) (New York: Springer)
[29] Langanke K and Martínez-Pinedo G 2003 *Rev. Mod. Phys.* **75** 819–862
[30] Martínez-Pinedo G, Fischer T, Lohs A and Huther L 2012 *ArXiv e-prints* (*Preprint* 1205.2793)
[31] Roberts L F, Shen G, Cirigliano V, Pons J A, Reddy S and Woosley S E 2012 *Phys. Rev. Lett.* **108** 061103
[32] Reddy S, Prakash M, Lattimer J M and Pons J A 1999 *Phys. Rev. C* **59** 2888–2918
[33] Burrows A and Sawyer R F 1998 *Phys. Rev. C* **58** 554–571
[34] Burrows A and Sawyer R F 1999 *Phys. Rev. C* **59** 510–514
[35] Lattimer J M and Lim Y 2012 (*Preprint* 1203.4286)

HITES 2012 IOP Publishing
Journal of Physics: Conference Series **403** (2012) 012038 doi:10.1088/1742-6596/403/1/012038

New experimental developments for s- and p-process research

R Reifarth, O Ershova, J Glorius, K Göbel, C Langer, O Meusel, R Plag, S Schmidt, K Sonnabend
Goethe Universität Frankfurt, 60438 Frankfurt, Germany

M Heil
GSI Helmholtzzentrum f. Schwerionenforschung GmbH, 64291 Darmstadt, Germany
E-mail: reifarth@physik.uni-frankfurt.de

Abstract. Almost all of the heavy elements are produced via neutron-induced processes in a multitude of stellar production sites. The remaining minor part is produced via photon- and proton-induced reactions. The predictive power of the underlying stellar models is currently limited because they contain poorly constrained physics components such as convection, rotation or magnetic fields. An important tool to determine such components is the comparison of observed with modeled abundance distributions based on improved nuclear physics input.

The FRANZ facility at the Goethe University Frankfurt, which is currently under construction will provide unprecedented neutron fluxes and proton currents available for nuclear astrophysics. It will be possible to investigate important branchpoint nuclei of the s-process nucleosynthesis path and proton-induced reactions important for p-process modeling.

At the GSI close to Darmstadt radioactive isotopes can be investigated in inverse kinematics. This allows experiments such as proton-induced cross section measurements using a heavy-ion storage ring or measurements of gamma-induced reactions using the Coulomb dissociation method. The future FAIR facility will allow similar experiments on very exotic nuclei, since orders of magnitude higher radioactive ions beams will be possible.

1. Introduction

Almost all of the heavy elements are produced via neutron capture reactions equally shared between s and r process [1, 2]. The remaining minor part is produced via photon- and proton-induced reactions during the p process, see Fig. 1. The predictive power of the underlying stellar models is limited because they contain poorly constrained physics components such as convection, rotation or magnetic fields. The crucial link between the observed abundances [3] and the desired parameters of the stellar interiors are nuclear reaction data. In contrast to the r process [4] the isotopes important for the p process [5] as well as for the s process [6] are stable or not to far off the valley of stability. The nuclear properties of those nuclei are therefore experimentally much easier to access. While indirect measurements, like the investigation of time-reversed reactions, are very often the only choice because of the short half-lives of the investigated isotopes [7, 8], this article is focussing on experimental developments aiming at a better understanding of the s and p processes by direct measurements of important capture reactions.

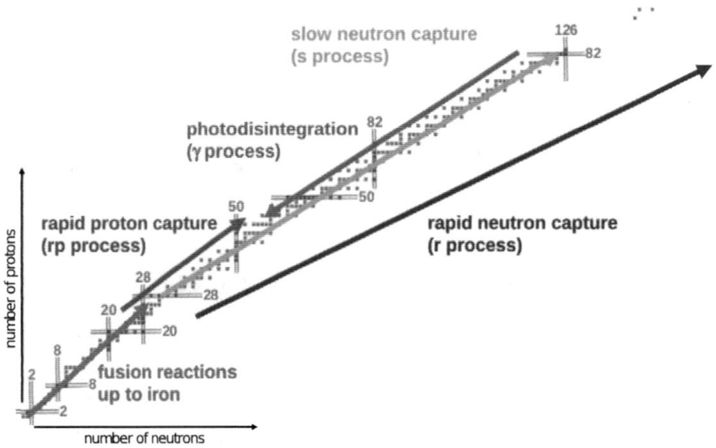

Figure 1. The s- and r-processes start with the iron peak nuclei as seeds. The s-process path follows the nuclear valley of stability until it terminates in the lead-bismuth region. The r-process drives the nuclear matter far to the neutron-rich side of the stability line and upwards until beta-delayed fission and neutron-induced fission occur and recycle the material back to smaller mass numbers. Only a few isotopes on the proton-rich side of the valley of stability get significant contributions from the different models of the p-process.

2. The s process

The modern picture of the main s-process component refers to the He shell burning phase in AGB stars [9]. Nuclei with masses between 90 and 209 are mainly produced during the main component. The highest neutron densities in this model occur during the ^{22}Ne(α,n) phase and are up to 10^{12} cm^{-3} with temperatures around $kT = 30$ keV. The other extreme can be found during the ^{13}C(α,n) phase where neutron densities as low as 10^7 cm^{-3} and temperatures around $kT = 5$ keV are possible. Similarly to the main component, also the weak component referring to different evolutionary stages in massive stars has two phases [10, 11]. Nuclei with masses between 56 and 90 are mainly produced during the weak component. The first phase occurs during the helium core burning with neutron densities down to 10^6 cm^{-3} and temperatures around $kT = 25$ keV. The second phase happens during the carbon shell burning with neutron densities up to 10^{12} cm^{-3} at temperatures around $kT = 90$ keV.

If the rates for neutron capture reactions are comparable to the rate of beta decay of particular nuclei, then the s-process path branches and some fraction of these nuclei are transformed via neutron capture, while another fraction undergoes beta decay. The branching ratio, or relative likelihood, for the different reactions depends on the physical conditions in the interior of the star, like temperature, neutron density, and electron density. Thus, the branching ratios deduced from the isotopic ratios observed in stellar material provide the tools to effectively constrain modern stellar models of the s-process, provided one knows the fundamental rates for neutron capture and beta decay under stellar conditions.

Neutron capture measurements on radioactive isotopes for neutron energies in the keV region represent a stringent challenge for further improvements of experimental techniques. This holds true for the neutron sources, the detection systems and the technology to handle radioactive material. Though the activation method or accelerator mass spectroscopy of the reaction products could be applied in a limited number of cases, experimental facilities like DANCE at LANL (USA) [12], n-TOF at CERN (Switzerland) [13] and the upcoming facility FRANZ in Frankfurt (Germany) [14] are addressing the need for such measurements on the basis of the more universal method of detecting the prompt capture γ-rays, which is required for the application of the neutron time-of-flight (TOF) technique. In particular the FRANZ facility, which will be located close to the new FAIR facility might allow the investigation of radioactive isotopes with half-lives down to tens of days, while present facilities require half-lives of a few hundred days.

In addition to the uncertainties affecting the neutron capture rates [15], weak interaction properties also face severe theoretical problems. Although all the β-decay and EC rates of relevance in the s-process are known under terrestrial conditions, the contribution of thermally populated excited states, as well as atomic effects in the strongly ionized stellar plasma can drastically modify the laboratory values [16]. The calculated β-rates in stellar environments are subject to nuclear uncertainties, which remain difficult to estimate. The uncertainties in the stellar β-decay and EC rates strongly depend on the relevance of the experimentally unknown transitions at a given temperature and density. So far, the stellar decay rates have been estimated in a systematic study [16]. In order to illustrate the effect of the remaining nuclear uncertainties the calculations have been re-iterated modifying the unknown transition rates by an error value of log ft = ± 0.5 [17]. For a typical s-process temperature of $T_8 = 3$ K and electron density of $N_e = 10^{27}$ cm^{-3} the final rate was then altered by a maximum value of 3.16. This maximum variation is obtained for many rates.

Under stellar conditions, ground state and low-lying excited states are in thermal equilibrium. The combined β-decay rates of these states determine, therefore, the β-decay probability of the species. Hence, both rates need to be determined in order to understand the important branchings in the reaction path of the s-process. But only in the case of isomeric excited states with sufficiently long half-lives one may attempt to determine their weak-decay rate experimentally.

3. The p process

The most intensively studied astrophysical site of the p-process is the O-Ne layer of a massive star passed by the shock wave of a type II supernova [18]. Here, the p nuclei are produced in a series of photodisintegration reactions, such as (γ,n), (γ,p), and (γ,α) reactions, from a given seed abundance. This picture is often referred to as γ-process [5]. The reaction network to describe this scenario includes a huge number of reaction rates, which are generally calculated in the framework of the Hauser-Feshbach statistical model. In the last decade, different experimental studies have been carried out to test the reliability of these predictions and their sensitivity to the nuclear physics input parameters, such as γ-ray strength functions and particle-nucleus optical potentials (OP). Radiative capture reactions were observed with in-beam and activation methods to understand the influence of different particle-nucleus OPs and to learn about the inverse reactions being involved in p-process nucleosynthesis. This approach was followed for protons [19, 20, 21], α particles [22, 23, 24] and neutrons [25, 26] at various facilities.

Two major problems in modelling the production of the p nuclei in the γ-process remain unresolved, even though the sometimes huge discrepancies between theoretical prediction and experimental results are taken into account [5]. First, the light and abundant p nuclei (^{92,94}Mo and ^{96,98}Ru) are largely underproduced compared to the solar p-nuclei abundance distribution. This holds true even for the relative overproduction factors of the light isotopes only, see left part of Fig. 2. Second, considering the entire galactic chemical evolution type II supernovae do not contribute sufficiently to the production of the p nuclei [27]. Therefore, additional sites and processes like proton captures in combination with γ-process in type Ia supernovae have to be

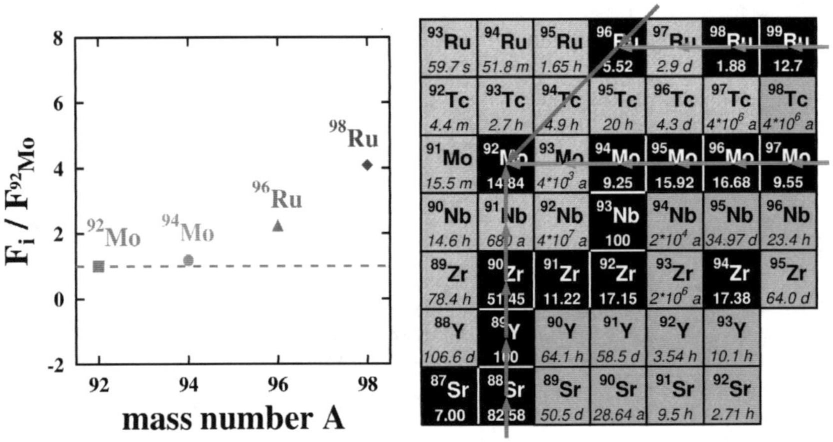

Figure 2. Left: Overproduction factors of the light, most abundant p nuclei during the γ-process normalized to ^{92}Mo [5]. Right: Production mechanisms for the most abundant p nucleus ^{92}Mo. The green and red arrows indicate (γ,n) and (γ,α) reactions. They occur in the γ process. The situation is illustrated exemplarily for a typical temperature of $T_9 = 2.5$ K as predicted by [30]. The blue arrows indicate the production via a chain of radiative proton capture reactions as suggested in [27, 29].

taken into account. This was already proposed in Ref. [28] and, just recently, it was found in two different approaches [27, 29] that it might solve both of the mentioned problems. The conditions during type Ia supernovae as derived in these works are suitable to allow the production of the light p nuclei by a series of radiative proton-capture reactions in addition to the γ-process as illustrated in the right part of Fig. 2. The production via proton-capture reactions (blue arrows) is compared to a selection of photon-induced reactions occurring in the γ-process (red and green arrows). The experimental database in the astrophysical relevant energy range, the Gamow window, is scarce, since the amount of available sample material is usually limited. In addition, the cross sections are small in the Gamow window because of the Coulomb barrier between the charged reaction partners. Therefore, the signature of the reaction exceeds the background only if the intensity of the projectiles is high enough.

4. (n,γ) reactions

As already discussed in section 1, it is desirable to improve the currently available experimental possibilities for neutron capture experiments. Spallation or photo-neutron sources require large accelerators, but a small accelerator as used for the recent ^{60}Fe activation at FZK [31] is best suited for neutron experiments in a university environment. This solution has the additional advantage that the neutron spectrum can be tailored to the specific energy range of interest.

Among the different options for producing neutrons in the keV region, the ^{7}Li(p,n)^{7}Be reaction with a threshold of 1.881 MeV is by far the most prolific. Near the threshold one can also take advantage of the fact that kinematically collimated neutrons can be produced in the energy range up to 100 keV. The current approach at FRANZ is therefore to use the existing experience with this method to produce neutrons by upgrading the proton source as well as high current lithium targets. The well-known setup for ToF measurements at the former Forschungszentrum Karlsruhe had a flight path of about 80 cm and about 10^4 neutrons/s/cm^2

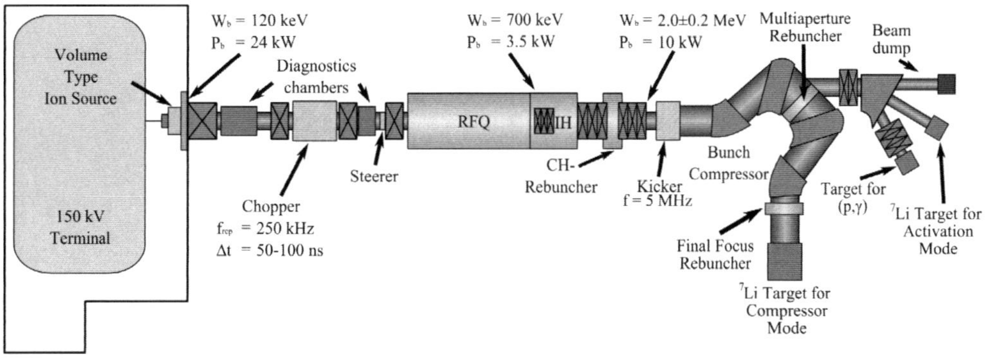

Figure 3. Schematic layout of the Frankfurter Neutron Source FRANZ.

at the sample position with proton currents of $\approx$ 2 μA [32, 33]. During activation measurements (DC) 10^9 neutrons/s at proton currents of $\approx$ 100 μA are typically produced [34].

The Stern-Gerlach-Zentrum recently founded at the Goethe University Frankfurt allows to build and operate larger experiments now in accelerator physics, astrophysics and material science research. It was decided to develop an intense neutron generator within the next years. The proton driver LINAC consists of a high voltage terminal already under construction to provide primary proton beam energies of up to 120 keV. A volume type ion source will deliver a DC beam current of 100-250 mA at a proton fraction of 90%. A low energy beam transport using four solenoids will inject the proton beam into a RFQ while a chopper at the entrance of the RFQ will create pulse lengths in the range of 100 ns at a repetition rate of up to 250 kHz. A drift tube cavity, witch delivers variable end energies between 1.8 and 2.2 MeV will be installed downstream of the RFQ. Finally a bunch compressor of the Mobley type forms a proton pulse length of 1 ns at the Li target. The maximum energies of the neutrons will be adjustable between $\approx$50 keV and $\approx$500 keV by the primary proton beam energy (see Fig. 3).

In a first step average beam currents of 0.1 mA are possible with a repetition rate of 250 kHz. Those conditions are very well suited for time-of-flight measurements in the astrophysically interesting keV-regime, as illustrated in the left part of Fig. 4. The second step would be to focus on improvements of the lithium target technology with the goal to increase the proton beam on target and hence the neutron flux. Improved cooling technologies allow targets with stable lithium layers that can handle up to 2 mA. This implies that without any major changes of the experimental setup compared to FZK (apart from the neutron production) an increase in neutron flux by a factor of 1000 can be achieved.

An interesting branching of the s-process is the region around ^{85}Kr, see right part of Fig. 4. On the one hand, the ^{84}Kr/^{86}Kr ratio can be determined in presolar SiC grains originating from T-AGB stars [35, 36]. The competition between β^--decay and neutron capture at ^{85}Kr is reflected in the observed ^{84}Kr/^{86}Kr ratio. The interpretation of the observed ratio as a measure of stellar parameters is currently hampered by the only very poorly known ^{85}Kr(n,γ) cross section [37, 15]. On the other hand, the decay of the long-lived ^{87}Rb (48 Gyr) reflects the abundance ratio of the otherwise s-only isotopes ^{86,87}Sr. A detailed understanding of the s- and r-contribution to ^{87}Rb would allow the determination of the onset of the s- and r-process in the history of our universe [38]. Also here, the biggest remaining uncertainty is the ^{85}Kr(n,γ) cross section. While some attempts have been made to investigate s-process branchings indirectly investigating the time-reversed reaction [39], this method is very difficult to apply in the case of ^{86}Kr(γ,n) because of the unfavourable decay properties of ^{85}Kr.

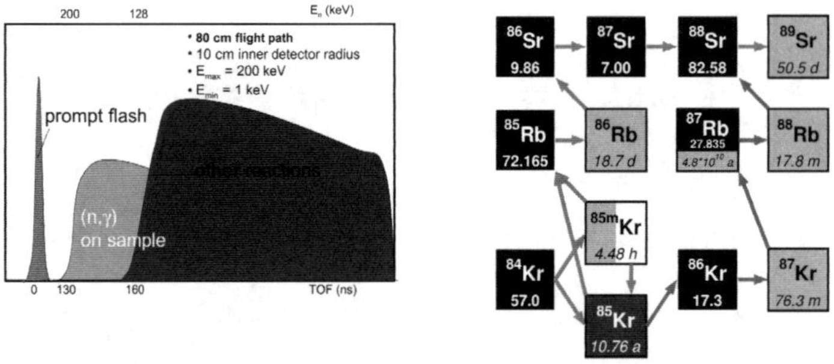

Figure 4. Left: Schematic TOF spectrum of the setup shown in Figure 3 for a maximum neutron energy of 200 keV. The TOF region corresponding to neutron energies between 130 and 200 keV is basically free of beam-related background. Right: Reaction network during s-process nucleosynthesis around the important branch point ^{85}Kr. The green arrows correspond to the situation, where no neutron capture occurs at ^{85}Kr, while the red arrows denote the neutron capture branch.

A direct measurement of the neutron capture cross section of ^{85}Kr using the time-of-flight method is therefore highly desirable. The FRANZ facility will be able to provide sufficient neutron flux for a measurement using a 4π-array to calorimetrically detect the γ-rays following the neutron capture [40]. One of the big challenges of such experiments is the production of the sample. In this particular case, the sample could be produced via α-irradiation of stable, metallic ^{82}Se. Since the ^{82}Se$(\alpha,\gamma)^{86}$Kr channel is suppressed at high energies, the ^{82}Se$(\alpha,n)^{85}$Kr reaction is by far dominating at high energies. The second highest cross section at high energies, ^{82}Se$(\alpha,p)^{85}$Br, also leads to ^{85}Kr via β^--decay with $t_{1/2} \approx 3$ min, thus ^{85}Kr can be produced almost isotopically pure. Krypton stays trapped inside selenium until heated [41]. The temperature during the irradiation has to stay below $\approx 50°$ C. After the transport of the irradiated sample to a sample preparation site, the freshly produced krypton can be released by heating the selenium sample to the melting point. The amount of ^{85}Kr required for an experiment at FRANZ is $\approx 10^{18}$ atoms. This corresponds to a maximum number of 2.5×10^{19} ^{4}He ions on target.

5. (p,γ) reactions

New experiments concerning direct (p,γ) reactions are currently under development at the experimental storage ring at GSI, where reaction measurements in inverse kinematics are possible close to the Gamow window of the p-process [42]. In addition, it seems feasible to facilitate FRANZ for proton-induced reactions at low energies. As already shown in Fig. 3, one beam line is foreseen to use the proton beam directly for proton-induced reactions. In the current layout, the FRANZ facility provides protons in an energy range of 1.8 MeV to 2.2 MeV. Thus, it covers the low-energy part of the Gamow window at typical temperatures for p-process nucleosynthesis. For the measurement of proton-induced reactions the accelerator will be operated in cw mode using the RFQ and IH structure. The time structure of the beam is determined by the intrinsic repetition rate of 175 MHz. This translates into a time of 5.7 ns between two bunches of 1 ns width. It is possible to distinguish between events in coincidence to beam pulses and the breaks in between for a detector system providing time resolutions in the sub-ns range, thus, providing a tool for background suppression.

The proton beam current in cw mode reaches up to 20 mA in the current design of FRANZ. Compared to Van-de-Graaff accelerators the current is enhanced by a factor of 100 to 1000. Therefore, measurements of very small reaction cross sections and/or with very little sample material can be realized. However, the sample has to be designed in a way that the corresponding high areal power densities resulting in high heat loads can be withstood and sputtering of the sample is avoided. It is planned to investigate (p,γ) reaction in the region around the neutron shell closure at $N = 50$, which includes the production chain of ^{92}Mo via proton captures (Fig. 2).

References

[1] Anders E and Grevesse N 1989 *Geochim. Cosmochim. Acta* **53** 197
[2] Thielemann F K, Fröhlich C, Hirschi R et al 2007 *Progress in Particle and Nuclear Physics* **59** 74–93
[3] Lodders K 2003 *Meteoritics and Planetary Science Supplement* **38** 5272
[4] Thielemann F K, Mocelj D, Panov I et al 2007 *International Journal of Modern Physics E* **16** 1149
[5] Arnould M and Goriely S 2003 *Physics Reports* **384** 1–84
[6] Käppeler F, Gallino R, Bisterzo S and Aoki W 2011 *Reviews of Modern Physics* **83** 157–194
[7] Langer C, Lepyoshkina O, Aksyutina Y et al 2010 *PoS* **Nuclei in the Cosmos XI** 224
[8] Ershova O, Adrich P, Alvarez-Pol H et al 2010 *PoS* **Nuclei in the Cosmos XI** 232
[9] Lugaro M, Herwig F, Lattanzio J C, Gallino R and Straniero O 2003 *Ap.J.* **586** 1305
[10] Travaglio C, Gallino R, Arnone E, Cowan J, Jordan F and Sneden C 2004 *Ap. J.* **601** 864
[11] Heil M, Käppeler F, Uberseder E, Gallino R and Pignatari M 2007 *Prog. Nucl. Part. Phys.* **59** 174
[12] Reifarth R, Bredeweg T A, Alpizar-Vicente A, et al 2004 *Nucl. Instr. Meth. A* **531** 528
[13] Abbondanno U, Aerts G, Alvarez-Velarde F et al 2004 *Physical Review Letters* **93** 161103
[14] Reifarth R, Chau L P, Heil M et al 2009 *PASA* **26** 255
[15] Bao Z Y, Beer H, Käppeler F et al 2000 *Atomic Data Nucl. Data Tables* **76** 70
[16] Takahashi K and Yokoi K 1987 *Atomic Data Nucl. Data Tables* **36** 375
[17] Goriely S 1999 *Astron. Astrophys.* **342** 881
[18] Hayakawa T, Iwamoto N, Kajino T, Shizuma T, Umeda H and Nomoto K 2008 *ApJ* **685** 1089–1102
[19] Spyrou A, Lagoyannis A, Demetriou P, Harissopulos S and Becker H W 2008 *Phys. Rev. C* **77** 065801
[20] Güray R T, Özkan N, Yalçın C et al 2009 *Phys. Rev. C* **80** 035804
[21] Kiss G G, Gyürky G, Elekes Z et al 2007 *Phys. Rev. C* **76** 055807
[22] Somorjai E, Fülöp Z, Kiss A Z et al 1998 *A&A.* **333** 1112–1116
[23] Spyrou A, Becker H W, Lagoyannis A, Harissopulos S and Rolfs C 2007 *Phys. Rev. C* **76** 015802
[24] Yalçın C, Güray R T, Özkan N et al 2009 *Phys. Rev. C* **79** 065801
[25] Dillmann I, Domingo-Pardo C, Heil M et al 2010 *Phys. Rev. C* **81** 015801
[26] Marganiec J, Dillmann I, Domingo-Pardo C, Käppeler F and Walter S 2010 *Phys. Rev. C* **82** 035806
[27] Kusakabe M, Iwamoto N and Nomoto K 2011 *ApJ* **726** 25
[28] Howard W, Meyer B and Woosley S 1991 *ApJ* **373** L5
[29] Travaglio C, Röpke F K, Gallino R and Hillebrandt W 2011 *ApJ* **739** 93
[30] Rauscher T 2006 *Phys. Rev. C* **73** 015804 (*Preprint* `arXiv:astro-ph/0510710`)
[31] Uberseder E, Reifarth R, Schumann D et al 2009 *Physical Review Letters* **102** 151101
[32] Wisshak K, Guber K, Käppeler F et al 1990 *Nucl. Instr. Meth. A* **292** 595 - 618
[33] Reifarth R, Heil M, Käppeler F et al 2002 *Phys. Rev. C (Nuclear Physics)* **66** 064603
[34] Reifarth R, Arlandini C, Heil M et al 2003 *Astrophysical Journal* **582** 1251
[35] Pignatari M, Gallino R, Amari S and Davis A M 2006 *Memorie della Societa Astronomica Italiana* **77** 897
[36] Lewis R, Amari S and Anders E 1994 *Geochim. Cosmochim. Acta* **58** 471
[37] Abia C, Busso M, Gallino R, Dominguez I and Straniero Oand Isern J 2001 *Ap. J.* **559** 1117
[38] Beer H and Walter G 1984 *Astrophys. Space Sci.* **100** 243–253
[39] Sonnabend K, Mohr P, Vogt K et al 2003 *Ap. J.* **583** 506
[40] Couture A and Reifarth R 2007 *Atomic Data and Nuclear Data Tables* **93** 807
[41] Suzuki K, Blessing G, Qaim S et al 1982 *The Int. J. of Applied Radiation and Isotopes* **33** 1445
[42] Zhong Q, Aumann T, Bishop S et al 2010 *Journal of Physics: Conference Series* **202** 012011

Configuration mixing in the quark model

Roelof Bijker[1], **Jacopo Ferretti**[1] **and Elena Santopinto**[2]

[1] Instituto de Ciencias Nucleares, Universidad Nacional Autónoma de México, AP 70-543, 04510 Mexico DF, Mexico

[2] I.N.F.N., Sezione di Genova, via Dodecaneso 33, 16164 Genova, Italy

E-mail: bijker@nucleares.unam.mx

Abstract. We discuss the flavor asymmetry and the strangeness content of the proton in an unquenched quark model. It is shown that the inclusion of hadron loops leads automatically to an excess of $\bar{d}$ over $\bar{u}$, but gives very small contributions to the strange magnetic moment and strangeness radius of the proton, both of which are in agreement with experimental data.

1. Introduction

There is strong evidence for the existence of exotic degrees of freedom (other than valence quarks) in hadrons, in particular for the need to include the effects of quark-antiquark pairs (hadron loops). The importance of these quark-antiquark configurations (or higher Fock components in baryon wave functions) is evident from measurements of the $\bar{d}/\bar{u}$ asymmetry in the nucleon sea [1], parity-violating electron scattering experiments [2, 3], the proton spin crisis [4], as well as from analysis of helicity amplitudes [5] and strong couplings of baryon resonances [6, 7].

The aim of this contribution is to review a recently proposed extension of the constituent quark model in which the effects of quark-antiquark pair creation ($u\bar{u}$, $d\bar{d}$ and $s\bar{s}$) are taken into account in an explicit form via a $^3P_0$ coupling mechanism [8, 9]. We discuss applications to the flavor asymmetry [9] and the strangeness of the proton [10]

2. Unquenching the quark model by configuration mixing

In the constituent quark model (CQM) hadrons are described as a system of constituent (or valence) quarks and antiquarks, qqq for baryons and $q\bar{q}$ for mesons. Despite the success of the quark model, there is strong evidence for the existence of exotic degrees of freedom (other than valence quarks) in hadrons from CQM studies of electromagnetic and strong couplings of baryons that are, on average, underpredicted by CQMs [11]. More direct evidence for the importance of quark-antiquark components in the proton comes from measurements of the $\bar{d}/\bar{u}$ asymmetry in the nucleon sea [1, 12], the proton spin crisis [4] and PVES experiments [13].

The unquenched quark model developed in [8, 9] is motivated by earlier studies on extensions of the quark model in which the $q\bar{q}$ pairs are created in the $^3P_0$ state with the quantum numbers of the vacuum [14, 15]. The present approach is based on a three-quark CQM to which the higher Fock components are added by configuration mixing, thereby employing a $^3P_0$ model for the coupling between the qqq and $qqq - q\bar{q}$ configurations. The pair-creation mechanism is inserted at the quark level and the one-loop diagrams are calculated by summing over a complete set of intermediate baryon-meson states. Under these assumptions, the baryon wave function consists of a zeroth order three-quark configuration $\mid A\rangle$ plus a sum over all possible higher Fock components due to the creation of $^3P_0$ quark-antiquark pairs

$$\mid \psi_A\rangle \;=\; \mathcal{N}\left[\mid A\rangle + \sum_{BCl}\int d\vec{k}\,\mid BC\vec{k}\,lJ\rangle\frac{\langle BC\vec{k}\,lJ\mid T^{\dagger}\mid A\rangle}{M_A - E_B - E_C}\right] . \tag{1}$$

Here A denotes the initial baryon, B and C represent the intermediate baryon and meson, and M_A, E_B and E_C are their respective energies, $\vec{k}$ and l the relative radial momentum and orbital angular momentum of B and C, and J is the total angular momentum $\vec{J} = \vec{J}_B + \vec{J}_C + \vec{l}$. The operator $T^{\dagger}$ creates a quark-antiquark pair in the $^3P_0$ state with the quantum numbers of the vacuum: $L = S = 1$ and $J = 0$ [8, 9, 16]. The coupling vertex $\langle BC\vec{k}\,lJ\mid T^{\dagger}\mid A\rangle$ can be derived in explicit form in the harmonic oscillator basis [16, 17].

In order to calculate the effects of quark-antiquark pairs on an observable, one has to evaluate the contribution of all possible intermediate states. By using a combination of group theoretical and computational techniques, the sum over intermediate states is carried out up to saturation and not only for the first few shells as in previous studies [14, 15]. Not only does this have a significant impact on the numerical results, but it is necessary for consistency with the OZI-rule and the success of CQMs in hadron spectroscopy [8, 18, 19].

The unquenching of the quark model is useful only if it preserves the phenomenological successes of the CQM. In applications to mesons it was shown that the inclusion of $q\bar{q}$ pairs does not destroy the good CQM results [18] and preserves the OZI hierarchy [19]. For baryons we studied the effect of the unquenching on the magnetic moments - historically one of the successes of the CQM - and found that, after renormalization, the good results for the magnetic moments are preserved [8].

The results presented in this contribution were obtained using harmonic oscillator wave functions up to four oscillator shells for the intermediate baryons and mesons with a single oscillator parameter for the baryons and another one for the mesons. All parameters were taken from the literature, without attempting to optimize their values in order to improve the agreement with experimental data [8, 9, 10].

3. Flavor asymmetry

The flavor asymmetry of the proton provides an important test for models of nucleon structure since it measures the difference between $\bar{d}$ and $\bar{u}$ quarks, and hence probes directly the existence of higher Fock components $qqq - q\bar{q}$ in the proton wave function.

The flavor asymmetry of the proton $\mathcal{A}(p)$ is related to the Gottfried integral S_G for the difference of the proton and neutron electromagnetic structure functions as

$$S_G \;=\; \int_0^1 \frac{F_2^p(x) - F_2^n(x)}{x}dx \;=\; \frac{1}{3} - \frac{2}{3}\int_0^1 \left[\bar{d}_p(x) - \bar{u}_p(x)\right] dx \;=\; \frac{1}{3}[1 - 2\mathcal{A}(p)] . \tag{2}$$

Under the assumption of a flavor symmetric (or rather flavor independent) sea one obtains the Gottfried sum rule $S_G = 1/3$, whereas any deviation from this value is an indication of the $\bar{d}/\bar{u}$ asymmetry of the nucleon sea. All experiments show evidence that there are more $\bar{d}$ quarks in the proton than there are $\bar{u}$ quarks [1].

In the unquenched quark model, the flavor asymmetry of the proton can be calculated directly from the difference of the number of $\bar{d}$ and $\bar{u}$ sea quarks in the proton, even in the absence of explicit information on the (anti)quark distribution functions. The inclusion of the $q\bar{q}$ pairs leads automatically to an excess of $\bar{d}$ over $\bar{u}$ quarks [9]. In the UCQM, the flavor asymmetry for the proton $\mathcal{A}(p)$ is calculated to be 0.151, which corresponds to a value of the Gottfried integral of 0.232, in remarkable agreement with the experimental data (see Fig. 1).

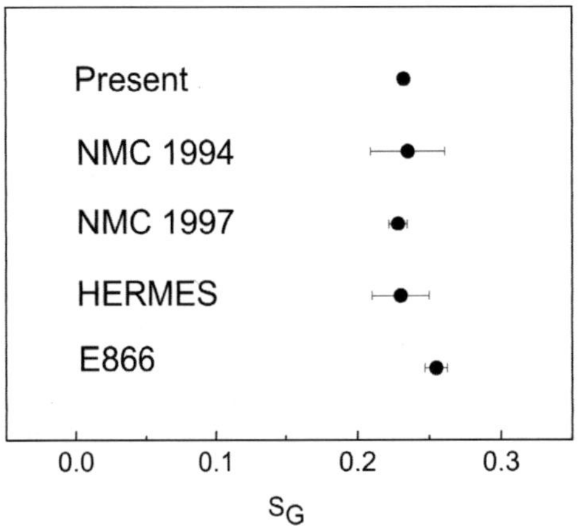

Figure 1. Comparison between calculated value of the Gottfried integral S_G and the experimental data from NMC 1994 [20], NMC 1997 [21], HERMES [22] and E866 [23]. Figure taken from Ref. [9].

The main contribution to the flavor asymmetry of the proton is due to pion loops, especially the $n\pi^+$ intermediate state, thus confirming in an explicit calculation the explanation given in Refs. [24] in the context of the meson-cloud model. In addition, we find that there are important contributions from the $\Delta\pi$ channel and, especially, from the off-diagonal terms $p\pi^0$-$p\eta_8$ and $p\pi^0$-$p\eta_1$ which together are of the order of 15-20 % of that of the $N\pi$ channel. The contribution of the intermediate vector mesons is very small due to a cancelation between the $n\rho^+$ and the $\Delta\rho$ channels and the cross terms $p\rho^0$-$p\omega_8$ and $p\rho^0$-$p\omega_1$ [25].

Since the unquenched quark model is valid not only for the proton, but for all baryons (ground state or resonance), it is straightforward to calculate the flavor asymmetries of the other octet baryons. In Table 1, we show a comparison of some predictions for the flavor asymmetry of the Σ^+ and Ξ^0 hyperons relative to that of the proton. In the unquenched quark model, the flavor asymmetry of the proton is predicted to be of the same order as that of the Σ^+ hyperon and much larger than that of the cascade particle

$$\mathcal{A}(p) \sim \mathcal{A}(\Sigma^+) \gg |\mathcal{A}(\Xi^0)| \ . \tag{3}$$

This behavior is very different from that obtained in the chiral quark model $\mathcal{A}(\Sigma^+) = 2\mathcal{A}(p) = 2\mathcal{A}(\Xi^0)$ [26], the balance model $\mathcal{A}(\Sigma^+) > \mathcal{A}(\Xi^0) > \mathcal{A}(p)$ [27], and the octet model

Table 1. Relative flavor asymmetries of octet baryons.

Model	$\mathcal{A}(\Sigma^+)/\mathcal{A}(p)$	$\mathcal{A}(\Xi^0)/\mathcal{A}(p)$	Ref.
Unquenched CQM	0.833	−0.005	[9]
Chiral QM	2	1	[26]
Balance Model	3.083	2.075	[27]
Octet couplings	0.353	−0.647	[28]

Table 2. Strange magnetic moment of the proton

Reference	μ_s [μ_N]
SAMPLE [43]	$0.01 \pm 0.29 \pm 0.31 \pm 0.07$
Global Analysis [38]	$0.12 \pm 0.55 \pm 0.07$
Global Analysis [39]	0.29 ± 0.21
LQCD [40]	-0.046 ± 0.019
LQCD [41]	$-0.017 \pm 0.025 \pm 0.007$
Two-component [37]	0.315
Unquenched QM [10]	0.0018

$\mathcal{A}(p) > |\mathcal{A}(\Xi^0)| > \mathcal{A}(\Sigma^+)$ [28]. The values for the chiral quark model and the balance model were taken from [29].

In order to distinguish between the predictions of the different models and to obtain a better understanding of the non-perturbative structure of QCD, new experiments are needed to measure the flavor asymmetry of hyperons. In particular, the flavor asymmetry of charged Σ hyperons can obtained from Drell-Yan experiments using charged hyperon beams on the proton [28] or by means of backward $K^{\pm}$ electroproduction [30].

4. Strangeness of the proton

Other interesting observables that directly probe the sea quarks of the nucleon are provided by the strange form factors of the proton. Even though the net strangeness of the nucleon is zero, the nucleon may have a nonzero distribution of strangeness, in the same way as the neutron has a nonvanishing charge distribution while its electric charge is zero.

Experimentally, the strange form factors can be determined by combining the measurement of parity-violating asymmetries of the proton with information on the electromagnetic form factors of the proton and neutron [31, 32].

$$G^s_{E,M}(Q^2) = \left(1 - 4\sin^2\theta_W\right) G^{\gamma,p}_{E,M}(Q^2) - G^{\gamma,n}_{E,M}(Q^2) - G^{Z,p}_{E,M}(Q^2) , \tag{4}$$

where G^s_E and G^s_M denote the strange electric and magnetic form factors, respectively. The first measurements of the parity-violating asymmetries showed surprisingly large and positive values of G^s_M, *e.g.* the SAMPLE collaboration found $G^s_M = 0.61 \pm 0.17 \pm 0.21 \pm 0.19$ at $Q^2 = 0.1$ $(\mathrm{GeV/c})^2$ [33]. However, more recent experiments by the HAPPEX [2, 34, 35], PVA4 [3] and G0 [36] collaborations showed that the contribution of strange quarks to the electric and magnetic form factors, G^s_E and G^s_M, is much smaller and compatible with zero within the experimental errors.

In the majority of theoretical analyses, the strangeness content to the nucleon is discussed in terms of its static properties, the strange magnetic moment and the strangeness charge radius

$$\begin{aligned} \mu_s &= G^s_M(0) , \\ r^2_s &= -6 \left. \frac{dG^s_E(Q^2)}{dQ^2} \right|_{Q^2=0} . \end{aligned} \tag{5}$$

A compilation of theoretical values of μ_s and r^2_s shows that, even though calculations in a variety of hadronic models and effective hadronic theories vary widely both in absolute value and sign, most studies agree on a small negative strangeness radius and a moderate negative strange magnetic moment [32, 37]. On the contrary, the results of a combined fit of the strange electric

Table 3. Strangeness radius of the proton

Reference	r_s^2 [fm^2]
HAPPEX [34]	$0.098 \pm 0.108 \pm 0.026$
HAPPEX [2]	$-0.006 \pm 0.042 \pm 0.021$
HAPPEX [2]	0.012 ± 0.044
PVA4 [3]	$-0.053 \pm 0.040 \pm 0.020$
G0 [36]	$0.015 \pm 0.038 \pm 0.019 \pm 0.018$
Global Analysis [38]	0.014 ± 0.096
Global Analysis [39]	0.019 ± 0.038
LQCD [42]	$-0.007 \pm 0.004 \pm 0.002 \pm 0.021$
LQCD [41]	$-0.0024 \pm 0.0015 \pm 0.0007$
Two-component [37]	0.005
Unquenched QM [10]	0.012

and magnetic form factors measured by SAMPLE, PVA4 and HAPPEX at $Q^2 \sim 0.1$ (GeV/c)^2 yielded the opposite sign for μ_s [38, 39]. Recent lattice calculations give small negative values for both μ_s and r_s^2 [40, 41, 42] (see Tables 2 and 3).

In the unquenched quark model, the strange magnetic moment of the proton is given by the expectation value of the operator

$$\vec{\mu}_s = \frac{1}{e_s} \sum_i \mu_{i,s} \left[2\vec{s}(q_{i,s}) + \vec{\ell}(q_{i,s}) - 2\vec{s}(\bar{q}_{i,s}) - \vec{\ell}(\bar{q}_{i,s}) \right] \; . \tag{6}$$

Here, $\mu_{i,s}$ is the magnetic moment of the strange quark i (not be confused with the strange magnetic moment of the proton μ_s). The strange magnetic moment of the proton is entirely due to the sea quarks. There is a contribution from the quark spins of the $s\bar{s}$ pair 0.0012 μ_N, as well as from its orbital motion 0.0006 μ_N. Both contributions are small and give a total strange magnetic moment $\mu_s = 0.0018\ \mu_N$ [10].

Similarly, the strange radius of the proton is calculated as the expectation value of the operator

$$r_s^2 = \sum_{i=1}^{5} \left(\vec{r}_{i,s} - \vec{R}_{CM} \right)^2 \; , \tag{7}$$

where $\vec{r}_{i,s}$ and $\vec{R}_{CM}$ are the coordinates of the strange quark i and the center of mass, respectively. The strangeness radius of the proton is calculated to be $r_s^2 = 0.012$ fm^2 [10].

In conclusion, the effects of the higher Fock components on the strange magnetic moment and the strange radius of the proton are found to be very small. Tables 2 and 3 show that the results for the unquenched quark model are compatible with the latest experimental data and recent lattice calculations.

5. Summary and conclusions

In this contribution, we presented some recent work on unquenching the quark model by configuration mixing between the valence quark and $qqq - q\bar{q}$ higher Fock configurations. As application, we studied the flavor asymmetry and the strangeness of the proton.

The inclusion of the $q\bar{q}$ pairs leads automatically to an excess of $\bar{d}$ over $\bar{u}$ quarks, in agreement with the observed flavor asymmetry of the proton. The results for the flavor asymmetry of the proton are dominated by the $N\pi$ channel, but with important contributions from the $\Delta\pi$ channel and the off-diagonal $N\pi$-$N\eta$ terms.

On the other hand, the effects of the higher Fock components on the strange magnetic moment and the strangeness radius of the proton are found to be very small, in agreement with the latest experimental data and recent lattice calculations.

Even though different models of hadron structure may show similar results for the properties of the proton, often their predictions for the other octet baryons exhibit large variations. Therefore, in order to be able to distinguish between the predictions of different models of hadron structure and to obtain a better understanding of the non-perturbative structure of QCD new experiments are needed to measure the flavor asymmetry and spin content of other octet baryons.

Acknowledgments

It is a great pleasure to dedicate this contribution to Jerry Draayer on the occasion of his 70th birthday. The results for the unquenched quark model were obtained by a combination of group theoretical and computational techniques, an area to which a large part of this meeting was dedicated in honor of Jerry's seminal contributions.

This work was supported in part by grants from CONACyT (78833) and PAPIIT-DGAPA (IN113711), Mexico, and INFN, Italy.

References

[1] Garvey G T and Peng J C 2001 *Prog. Part. Nucl. Phys.* **47** 203
[2] Acha A *et al* 2007 *Phys. Rev. Lett.* **98** 032301
[3] Baunack S *et al* 2009 *Phys. Rev. Lett.* **102** 151803
[4] Bass S D 2008 *The spin structure of the proton* (Singapore: World Scientific)
[5] Aznauryan I G *et al* 2008 *Phys. Rev. C* **78** 045209
Aznauryan I G *et al* 2009 *Phys. Rev. C* **80** 055203
[6] Capstick S and Roberts W 1994 *Phys. Rev. D* **49** 4570
[7] Bijker R, Iachello F and Leviatan A 1997 *Phys. Rev. D* **55** 2862
Bijker R, Iachello F and Leviatan A 2000 *Ann. Phys. (N.Y.)* **284** 89
[8] Bijker R and Santopinto E 2009 *Phys. Rev. C* **80** 065210
[9] Santopinto E and Bijker R 2010 *Phys. Rev. C* **82** 062202(R)
[10] Bijker R, Ferretti J and Santopinto E 2012 *Phys. Rev. C* **85** 035204
[11] Capstick S and Roberts W 2000 *Prog. Part. Nucl. Phys.* **45** 241
[12] Kumano S 1998 *Phys. Rep.* **303** 183
[13] Armstrong D S and McKeown R D 2012 *Annu. Rev. Nucl. Part. Sci.*, to be published [arXiv:1207.5238].
[14] Zenczykowski P 1986 *Ann. Phys. (N.Y.)* **169** 453
[15] Geiger P and Isgur N 1997 *Phys. Rev. D* **55** 299
[16] Roberts W and Silvestre-Brac B 1992 *Few-Body Systems* **11** 171
[17] Ferretti F 2011 *Ph. D. Thesis* University of Genova, unpublished.
[18] Geiger P and Isgur N 1990 *Phys. Rev. D* **41** 1595
[19] Geiger P and Isgur N 1993 *Phys. Rev. D* **47** 5050
[20] Amaudruz P *et al* 1991 *Phys. Rev. Lett.* **66** 2712
Arneodo M *et al* 1994 *Phys. Rev. D* **50** R1
[21] Arneodo M *et al* 1997 *Nucl. Phys. B* **487** 3
[22] Ackerstaff K *et al* 1998 *Phys. Rev. Lett.* **81** 5519
[23] Towell R S *et al* 2001 *Phys. Rev. D* **64** 052002
[24] Thomas A W 1983 *Phys. Lett. B* **126** 97
Henley E M and Miller G A 1990 *Phys. Lett. B* **251** 453
[25] Bijker R and Santopinto E 2011 *J. Phys.: Conf. Ser.* **322** 012014
[26] Eichten E J, Hinchcliffe I and Quigg C 1992 *Phys. Rev. D* **45** 2269
[27] Zhang Y J, Deng W Z and Ma B Q 2002 *Phys. Rev. D* **65** 114005
[28] Alberg M, Henley E M, Ji X and Thomas A W 1996 *Phys. Lett. B* **389** 367
Alberg M, Falter T and Henley E M 1998 *Nucl. Phys. A* **644** 93
[29] Shao L, Zhang Y J and Ma B Q 2010 *Phys. Lett. B* **686** 136
[30] Osipenko R, private communication.
[31] McKeown R D 1989 *Phys. Lett. B* **219** 140
Beck D H 1989 *Phys. Rev. D* **39** 3248

[32] Beck D H and Holstein B R 2001 *Int. J. Mod. Phys. E* **10** 1
Beck D H and McKeown R D 2001 *Annu. Rev. Nucl. Part. Sci.* **51** 189
[33] Spayde D T *et al.* 2000 *Phys. Rev. Lett.* **84** 1106
[34] Aniol K A *et al.* 2006 *Phys. Rev. Lett.* **96** 022003
[35] Ahmed Z *et al.* 2012 *Phys. Rev. Lett.* **108** 102001
[36] Androic D *et al* 2010 *Phys. Rev. Lett.* **104** 012001
See http://www.npl.uiuc.edu/exp/G0/Backward
[37] Bijker R 2006 *J. Phys. G: Nucl. Part. Phys.* **32** L49
[38] Young R D *et al.* 2006 *Phys. Rev. Lett.* **97** 102002
[39] Liu J *et al.* 2007 *Phys. Rev. C* **76** 025202
[40] Leinweber D B *et al.* 2005 *Phys. Rev. Lett.* **94** 212001
[41] Doi T *et al.* 2009 *Phys. Rev. D* **80** 094503
[42] Leinweber D B *et al.* 2006 *Phys. Rev. Lett.* **97** 022001
[43] Hasty *et al.* 2000 *Science* **290** 2117

Implications of nonzero θ_{13} for the neutrino mass hierarchy

D J Ernst[1,2], B K Cogswell[1,2], H R Burroughs[1], J Escamilla-Roa[1], D C Latimer[3]

[1] Department of Physics and Astronomy, Vanderbilt University, Nashville, Tennessee 37235
[2] Department of Physics, Fisk University, Nashville, Tennessee 37208
[3] Department of Physics, University of Puget Sound, Tacoma, Washington 98416

E-mail: david.j.ernst@vanderbilt.edu

Abstract. The Daya Bay, RENO, and Double Chooz experiments have discovered a large non-zero value for θ_{13}. We present a global analysis that includes these three experiments, Chooz, the Super-K atmospheric data, and the $\nu_\mu \to \nu_e$ T2K and MINOS experiments that are sensitive to the hierarchy and the sign of θ_{13}. We report preliminary results in which we fix the mixing parameters other than θ_{13} to those from a recent global analysis. Given there is no evidence for a non-zero CP violation, we assume $\delta = 0$. T2K and MINOS lie in a region of L/E where there is a hierarchy degeneracy in the limit of $\theta_{13} \to 0$ and no matter interaction. For non-zero θ_{13}, the symmetry is partially broken, but a degeneracy under the simultaneous exchange of both hierarchy and the sign of θ_{13} remains. Matter effects break this symmetry such that the positions of the peaks in the oscillation probabilities maintain the two-fold symmetry, while the magnitude of the oscillations is sensitive to the hierarchy. This renders T2K and NOνA, with different baselines and different matter effects, better able in combination to distinguish the hierarchy and the sign of θ_{13}. The present T2K and MINOS data do not distinguish between hierarchies or the sign of θ_{13}, but the large value of θ_{13} yields effects from atmospheric data that do. We find for normal hierarchy, positive θ_{13}, $\sin^2 2\theta_{13} = 0.090 \pm 0.020$ and is 0.2% probable it is the correct combination; for normal hierarchy, negative θ_{13}, $\sin^2 2\theta_{13} = 0.108 \pm 0.023$ and is 2.2% probable; for inverse hierarchy, positive θ_{13}, $\sin^2 2\theta_{13} = 0.110 \pm 0.022$ and is 7.1% probable; for inverse hierarchy, negative θ_{13}, $\sin^2 2\theta_{13} = 0.113 \pm 0.022$ and is 90.5% probable, results that are inconsistent with two similar analyses.

1. Introduction

Neutrinos undergo the phenomenon of flavor oscillations; a neutrino created in a particular flavor (electron, mu, or tau) will change its flavor, a phenomenon not accounted for in the standard model of the electroweak interaction. A possible explanation of this behavior is neutrino oscillations. The neutrinos are assigned masses. The created neutrino has definite flavor, but these flavor eigenstates are not the physical free particles, the mass states. This leads to a phenomenology that is parameterized by three mixing angles, a CP violating phase, and two mass-squared differences,$\Delta_{ij} =: m_i^2 - m_j^2$.

Experiments have found that neutrino oscillations approximately break into two 2×2 oscillations, termed solar and atmospheric oscillations. The solar oscillations are predominantly governed by the solar data and the KamLAND experiment, and they predominantly determine θ_{12} andΔ_{21}, while the atmospheric oscillation parameters θ_{23} andΔ_{32} are predominantly

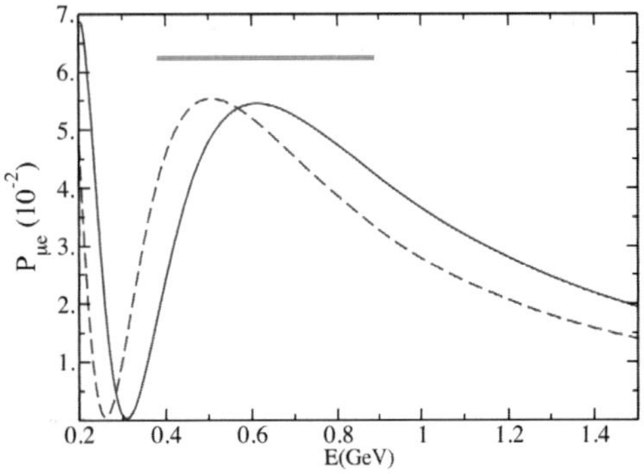

Figure 1. $\mathcal{P}_{\mu e}$ in vacuum versus neutrino energy for the T2K experiment. The solid curve depicts the normal hierarchy, positive θ_{13} as well as the inverse hierarchy, negative θ_{13} results. The dashed curve depicts the normal hierarchy, negative θ_{13} as well as the inverse hierarchy, positive θ_{13} results. The horizontal bar represents the range where the muon interaction rate is greater than 50% of its maximum.

determined by the atmospheric data and the long-baseline muon disappearance data. The last mixing angle θ_{13} and the CP-violating phase δ mix the two 2×2 oscillations.

An outstanding question in neutrino phenomenology is that of the hierarchy, whether the small mass-squared difference lies below (normal hierarchy) or above (inverse hierarchy) the large mass-squared difference. A related question, as we will show, is whether the value of θ_{13} is positive or negative. We utilize the convention [1] that $-\frac{\pi}{2} \leq \theta_{13} < +\frac{\pi}{2}$ and $0 \leq \delta < \pi$. This is the more convenient parameterization for discussing the symmetries of interest here. Recent long-baseline reactor experiments [2, 3, 4] have accurately measured the value of θ_{13} and found it to be large, $\sin^2(2\theta_{13}) \approx 0.1$. Before these measurements, the world's data demonstrated very little preference for either the hierarchy or the sign of θ_{13} [5]. The goal of this work is to examine the effect that the recent knowledge of θ_{13} has on the hierarchy question and on the sign of θ_{13}. This report is preliminary in that we will fix the mixing parameters except θ_{13} at values from a recent analysis [6]. No significant evidence for CP violation has been found in other work [6, 7] utilizing this same data set so we set the CP phase δ to zero to simplify the calculations. In future work, we will include the CP phase and then marginalize it, a more thorough and correct treatment.

2. Symmetries and Degeneracies

The task of extracting neutrino mixing parameters from data is complicated by the existence of symmetries that lead to degeneracies [8], where degeneracies mean different mixing parameters yield the same oscillation probability. The MINOS [9], T2K [10], and the future NOνA [11] experiments are at an L/E where such a symmetry nearly exists. In the limit of $\theta_{13} = 0$ and no matter effects, these experiments would be insensitive to the hierarchy. The symmetry pertinent to these experiments, using our bounds on the mixing angles, is the combined interchange of hierarchy and a change of the sign of θ_{13}, a four-fold degeneracy. In Fig. 1 we depict the vacuum oscillation probability $\mathcal{P}_{\mu e}$ for the T2K experiment for full three-neutrino mixing using the parameters [6]$\Delta_{21} = 7.54 \times 10^{-5}$ eV^2, $\theta_{12} = 0.557$,$\Delta_{31} = 2.42 \times 10^{-3}$ eV^2, $\theta_{13} = 0.1$, and $\theta_{23} = 0.674$.

We see that the four-fold degeneracy is broken by non-zero θ_{13}. However, the solid curve represents both the normal hierarchy, positive θ_{13} as well as the inverted hierarchy, negative θ_{13} results while the dashed curve represents the inverted hierarchy, positive θ_{13} and the normal hierarchy, negative θ_{13} results. Thus non-zero θ_{13} breaks the four fold degeneracy down to two two-fold degeneracies. Hierarchy and linear in θ_{13} effects require interference between the two mass-squared difference terms, and this is the smallest value of L/E where such effects appear. These effects are greatest at the peak of the first oscillation generated by the small solar mass-squared differences. For vacuum oscillations this occurs [12] when $L/E = 1.6 \times 10^4$ km/GeV.

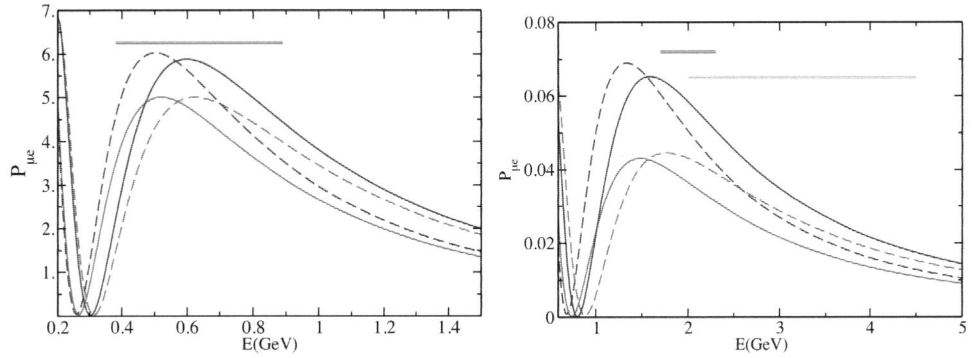

Figure 2. $\mathcal{P}_{\mu e}$ versus neutrino energy for the T2K experiment, left, and the MINOS/NOνA experiments, right. The blue curves depict the normal hierarchy, the red curves the inverse hierarchy. The solid curves depict positive θ_{13}, the dashed curves negative θ_{13}. For the left curve, the horizontal brown bar represents the range where the muon interaction rate is greater than 50% of its maximum, while for the right curve the brown bar represents the same quantity for the NOνA experiment, and the turquoise bar represents the MINOS experiment.

Including matter effects yields $L = 4.0 \times 10^3$ km and $E = 0.23$ GeV as typical ideal values. Extrapolating from the results in Ref. [13] to the now known value of θ_{13} gives a large $\mathcal{P}_{\mu e}$ with a 20% difference between $\mathcal{P}_{\mu e}$ for $\delta = 0$ and $\delta = \pi/2$ for such an experiment.

We present in Fig. 2 $\mathcal{P}_{\mu e}$ versus neutrino energy including the MSW matter effects for MINOS [9], T2K [10] , and NOνA [11]. The effect of including the matter interaction of the neutrino is to alter the magnitude of the oscillations while leaving the position of the peaks nearly unchanged. Thus spectral information that measures the position of the peaks maintains approximately the two-fold degeneracy of the vacuum oscillations. The magnitude of the signal then breaks this symmetry to distinguish the hierarchy while it alone does not distinguish well the sign of θ_{13}.

There is a complementarity between the T2K and NOνA experiments. T2K is situated to

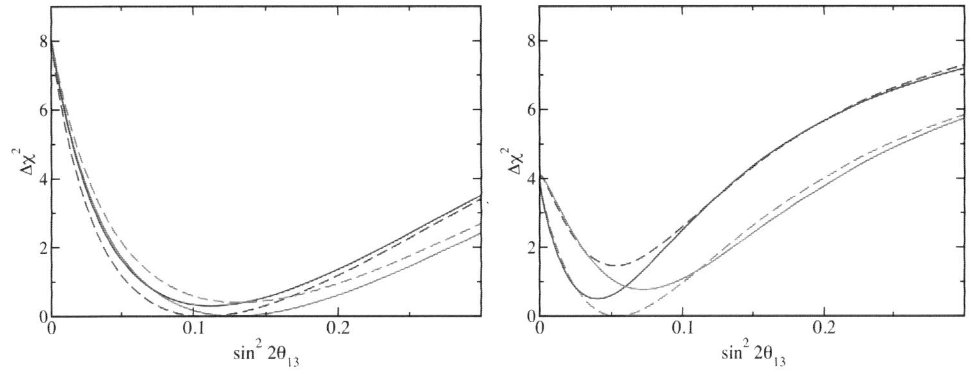

Figure 3. $\Delta\chi^2$ versus $\sin^2 2\theta_{13}$ for T2K, left, and MINOS, right. Solid curves are normal hierarchy, dashed inverse hierarchy. Blue curves are positive θ_{13}, red curves negative θ_{13}.

Table 1. Results for T2K, MINOS, and the combination. For the four cases of hierarchy and the sign of θ_{13}, we present the best fit values of $\sin^2 2\theta_{13}$, its 90% error bars, the maximum confidence level at which $\theta_{13} \neq 0$, and the probability that each case is the correct choice.

experiment	hierarchy	sign θ_{13}	$\sin^2 2\theta_{13}$	CL, $\theta_{13} \neq 0$	% probable
	normal	+	$0.11^{+0.20}_{-0.11}$	99.1%	22.1%
T2K	normal	−	$0.11^{+0.29}_{-0.09}$	99.2%	24.9%
	inverse	+	$0.13^{+0.11}_{-0.10}$	99.4%	28.8%
	inverse	−	$0.13^{+0.12}_{-0.10}$	99.3%	24.3%
	normal	+	$0.040^{+0.190}_{-0.040}$	80.8%	21.5%
MINOS	normal	−	$0.053^{+0.215}_{-0.053}$	75.6%	16.3%
	inverse	+	$0.073^{+0.208}_{-0.073}$	83.1%	28.0%
	inverse	−	$0.054^{+0.194}_{-0.054}$	85.8%	34.3%
	normal	+	$0.064^{+0.103}_{-0.045}$	99.5%	18.4%
T2K+MINOS	normal	−	$0.072^{+0.064}_{0.045}$	99.5%	17.9%
	inverse	+	$0.093^{+0.114}_{-0.059}$	99.7%	33.2%
	inverse	−	$0.080^{+0.118}_{-0.055}$	99.7%	30.5%

determine the peak of the oscillation probability, ideal for breaking the four-fold degeneracy. NOνA has a longer baseline and larger matter effects thus increasing the difference in the magnitude of the oscillations, giving it a greater sensitivity to the hierarchy.

3. Long-Baseline Experiments

Data from T2K and MINOS $\nu_\mu \rightarrow \nu_e$ appearance experiments are available, but both lack significant statistical quality. The$\Delta\ \chi^2$ versus θ_{13} results are presented in Fig. 3. For T2K, the left graph, we see that the quality of the fit (the value of$\Delta\ \chi^2$ at the minimum) is governed by the spectral information, while the position of the minimum is determined by the height of the oscillation probability and is thus correlated with the hierarchy. For MINOS, we see that the longer baseline increases the matter effects. For small θ_{13} the two fold symmetry is present. However, for the now known value of $\sin^2 2\theta_{13} \approx 0.1$, MINOS is sensitive only to the hierarchy. From Fig. 2 we see that this is because MINOS is in a region where the magnitude of the data is important but spectral information is limited as its data do not extend down in energy sufficiently far to be sensitive to the position of the peaks.

In Table 1 we present results from our analysis of the T2K, MINOS and combined experiments. Our analyses agree quite well with that of the experimentalists. For T2K we utilize a new estimate of the earth's matter density [14] of 2.6 gm/cm^3 rather than the 3.2 gm/cm^3 used by the experimentalists. For the present T2K data we see relatively little distinction between the hierarchy or the sign of θ_{13}. There is a statistically insignificant favor (disfavor) for the inverse hierarchy and positive θ_{13} (normal hierarchy and positive θ_{13}).

We see in Fig. 3 that the$\Delta\ \chi^2$ curves are not quadratic. Hence the use of normal statistics is not valid. We calculate the error bars and the probability that $\theta_{13} \neq 0$ using the Bayesian method from [15] with the prior taken as a constant for the independent variable $\sin^2 2\theta_{13}$. Both experiments provide an indication that θ_{13} is not zero, T2K at the 99% CL and MINOS at the 80% CL, the combined at the 99.5% CL. They both show a small dependence on hierarchy and the sign of θ_{13}, with the combined result preferring the inverse hierarchy at 64% probability.

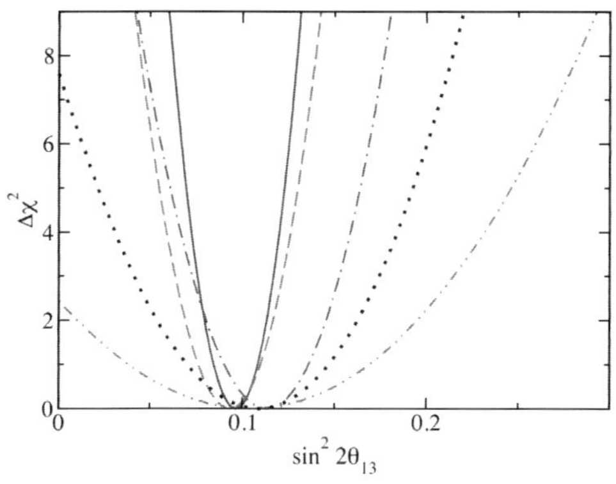

Figure 4. $\Delta\chi^2$ versus $\sin^2 2\theta_{13}$ for the Chooz experiment (dot-dot-dash green curve) with modified flux, the Double Chooz experiment (dotted black curve), the RENO experiment (indigo dot-dash), the Daya Bay experiment (red dashed), and the combination of all four experiments (solid blue).

4. Reactor Experiments

Recent reactor experiments, Double Chooz [2], Daya Bay [3], and RENO [4], have measured the value of θ_{13}. The obtained value of θ_{13} is surprisingly large. Insufficient information is given to perform an *ab initio* analysis of the Daya Bay and RENO experiments. However, we find that if we take the data to be given by the points in Fig. 4 for Daya Bay and Fig. 3 for RENO, utilize the systematic errors given, and take the statistical errors as implied by the number of events in each detector, we get results nearly identical to those given by the experimentalists.

In Fig. 4 we present$\Delta\ \chi^2$ for these reactor experiments, plus results for the original Chooz [16] experiment, in which we use an increase in the flux by 3.3% [17]. This change makes Chooz consistent with the recent experiments. Daya Bay and RENO, by virtue of having identical near and far detectors, provide a definitive non-zero value for θ_{13}. Reactor experiments measure $\mathcal{P}_{ee}$ and have no sensitivity to hierarchy or the sign of θ_{13}. All four of these experiments are consistent with each other, all indicating a value for $\sin^2 2\theta_{13} \approx 0.1$.

In Table 2 we present results for the reactor experiments. The result for the combination is $\sin^2 2\theta_{13} = 0.097^{+0.019}_{-0.020}$. θ_{13} is non-zero at the confidence level of CL= $1.0 - 1. \times 10^{-9}$ %, or 6.2 σ. We find a somewhat smaller significance because we use the Bayesian approach from Ref. [15].

5. Super-K Atmospheric Experiment

The implications of the Super-K atmospheric experiment [18] for θ_{13} have been discussed in Ref. [5]. The features discussed there are visible in the left panel of Fig. 5. There are two

Table 2. Implications of the reactor experiments Chooz, Double Chooz, Daya Bay, RENO and the combination. We present the best fit values of $\sin^2 2\theta_{13}$ and its 90% error bars, the maximum confidence level at which $\theta_{13} = 0$ is excluded.

experiment	$\sin^2 2\theta_{13}$	CL, $\theta_{13} \neq 0$
Chooz	$0.104^{+0.091}_{-0.094}$	94%
Double Chooz	$0.109^{+0.061}_{-0.064}$	99.7%
Daya Bay	$0.093^{+0.026}_{-0.027}$	$1. - 6. \times 10^{-6}$ %
RENO	$0.113^{+0.037}_{-0.038}$	$1. - 1. \times 10^{-4}$ %
Combined	$0.097^{+0.019}_{-0.020}$	$1. - 1. \times 10^{-9}$ %

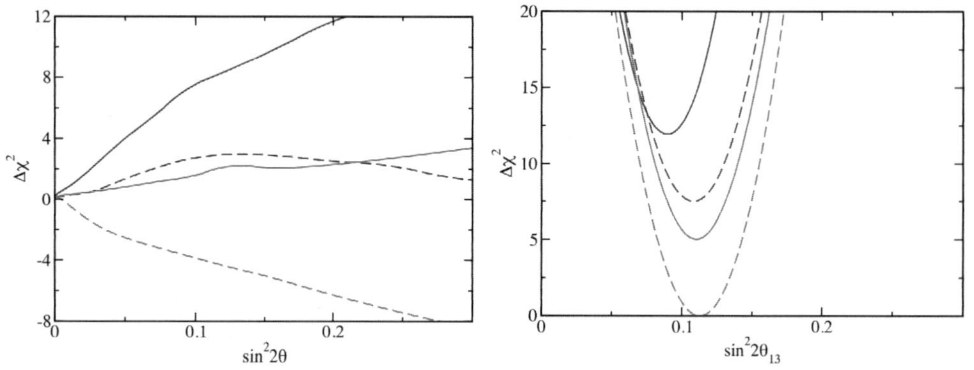

Figure 5. [Left]Δ χ^2 versus $\sin^2 2\theta_{13}$ for the Super-K atmospheric experiment. [Right]Δ χ^2 versus $\sin^2 2\theta_{13}$ for all experiments. The curves are coded as in Fig. 2.

subsets of the atmospheric data that play a role in determining θ_{13} and in discerning the hierarchy. One is data in the 3 to 10 GeV range with neutrinos passing through matter. The MSW interaction produces resonances in this energy region for the normal hierarchy, but not the inverse hierarchy. The data do not indicate the existence of these resonances, and thus, for normal hierarchy, atmospheric data constrain the magnitude of θ_{13} with no such constraint for the inverse hierarchy. The second subset that constrains θ_{13} is data at large L and small E in the range of L/E where there is interference between the oscillations arising from the two mass-squared differences. These interference terms are linear in θ_{13}. Combining the quadratic resonance terms and the linear terms, it was found that for normal hierarchy and positive θ_{13} the two terms are of the same sign and produce a constraint on θ_{13} that sets the bound more stringently than did the Chooz experiment. However, for normal hierarchy and negative θ_{13} the two terms are of opposite sign and of nearly equal magnitude and thus tend to cancel. For the inverse hierarchy there are only the linear in θ_{13} terms, and they favor the negative hierarchy, see the left panel of Fig. 5. We note that these results differ from those presented in Fig. 24 of Ref. [19] realizing that their $\cos\delta = -1$ branch is our negative θ_{13} branch. We note that they keep higher-order terms beyond the subdominant approximation. However, the expansion is divergent precisely in the region where the linear terms become significant and thus can never fully capture the effects of the linear in θ_{13} terms.

Table 3. Results for a global analysis for the four combinations of hierarchy and sign of θ_{13}. The last column gives the percent probability that the choice of hierarchy and θ_{13} are correct.

hierarchy	sign θ_{13}	$\sin^2 2\theta_{13}$	% probable
normal	+	0.090 ± 0.020	0.2%
normal	−	0.108 ± 0.023	2.2%
inverse	+	0.093 ± 0.022	7.1%
inverse	−	0.113 ± 0.022	90.5%

6. Conclusions

We present our results for an analysis that includes all the experiments in the right panel of Fig. 5 where we show$\Delta\ \chi^2$ for the four cases of hierarchy and sign of θ_{13}. The location of the minima are found to be only slightly affected by the choice of hierarchy or sign of θ_{13}. For the now known value of $\sin^2 2\theta_{13} \approx 0.1$, atmospheric data distinguish between hierarchy and the sign of θ_{13} at a not negligible level. The normal hierarchy, positive θ_{13} case is disfavored, while the inverse hierarchy, negative θ_{13} case is favored with the other two cases lying in the middle. This is made quantitative in Table 3. By marginalizing over $\sin^2 2\theta_{13}$ we calculated the relative probability of each of the four cases. The probability that normal hierarchy, positive θ_{13} is correct is 0.2%, a fairly strong indication that it is not correct. The inverse hierarchy, negative θ_{13} case is 90.5% probable that it is correct, a not negligible preference, while the normal hierarchy, negative θ_{13} and the inverse hierarchy, positive θ_{13} cases are 2.2% and 7.1% probable, respectively.

We compare our best fit value $\sin^2 2\theta_{13} = 0.113$ for negative hierarchy, negative θ_{13} to $\sin^2 2\theta_{13} = 0.103$ found for normal hierarchy, positive θ_{13} or negative hierarchy, negative θ_{13} from Ref. [6] and to $\sin^2 2\theta_{13} = 0.094$ found for positive or negative hierarchy and negative θ_{13} for Ref. [7]. There is disagreement on the sign of θ_{13} and the hierarchy question. We find a larger hierarchy difference than do the other two. Since the present T2K and MINOS experiments have little preference for the hierarchy and the sign of θ_{13}, the differences between the analyses arise from the atmospheric analysis.

These differences may be resolved by future Super-K atmospheric data. Although the Super-K data are statistically significant on the whole, the role of θ_{13} in the analysis is a small effect. For example, the bins that indicate the presence or absence of the MSW resonance, are statistically not too significant. Further, T2K and NOνA have not produced data of statistical significance. With the large value for θ_{13} their future data have great promise to help resolve this situation.

Acknowledgments

The work of D.J.E. is supported, in part, by US Department of Energy Grant No. DE-FG02-96ER40975, the work of B.K.C. is supported, in part, by US Department of Education Grant No. P200A090275, and the work of J.E-R. is supported, in part, by CONACyT, Mexico.

References

[1] Latimer D C and Ernst D J 2005 *Phys. Rev.* D **71** 017301
[2] Abe Y *et al.* [Double-Chooz Collaboration] 2012 *Phys. Rev. Lett.* **108** 131801
[3] An F P *et al.* [Daya Bay Collaboration] 2012 *Phys. Rev. Lett.* **108** 171803
[4] Ahn J *et al.* [RENO Collaboration] 2012 *Phys. Rev. Lett.* **108** 191802
[5] Escamilla-Roa J, Latimer D C and Ernst D J 2010 *Phys. Rev.* C 82 028501; Roa J E, Latimer D C and Ernst D J 2009 Phys. Rev. Lett. **103** 061804
[6] Forereo D V, Tortóla M and Valle J W F 2012 arXiv:1205.4018 [hep-ph]
[7] Fogli G, Lisi E, Marrone A, Montanino D and Palazzo A 2012 arXiv:1205.5254 [hep-ph]
[8] Barger V, Marfatia D and Wisnant K 2002 *Phys. Rev.* D **65** 073023; Latimer D C and Ernst D J 2005 *Mod. Phys. Lett.* A **20** 1663
[9] Adamson P *et al.* [MINOS Collaboration] 2012 *Phys. Rev. Lett.* **108** 191801
[10] Abe K *et al.* [T2K Collaboration] 2011 *Phys. Rev. Lett.* **107** 041801
[11] Nowak [NOνA Collaboration] 2012 *AIP Conf. Proc.* **1441** 423
[12] Latimer D C and Ernst D J 2005 *Phys. Rev.* C **71** 062501
[13] Latimer D C, Escamilla J and Ernst D J 2007 *Phys. Rev.* C **76** 05502
[14] Hagiwara K, Okamura N and Sendai K-k 2011 *JHEP* **1109** 082
[15] Hunter H R, Cogswell B K, Escamilla-Roa J, Latimer D C and Ernst D J 2012 *Phys. Rev.* C **85** 068501
[16] Apollonio M *et al.* 2003 *Eur. Phys.* C **27** 331
[17] Mueller Th A *et al.* 2011 *Phys. Rev.* C **83** 054615
[18] Hosaka J *et al.* [Super-K Collaboration] 2006 *Phys. Rev.* D **74** 032002
[19] Fogli G L, Lisi E, Marrone A, Palazzo A and Rotunno A 2006 *Prog. Part. Nucl. Phys.* **57** 742

HITES 2012 IOP Publishing
Journal of Physics: Conference Series **403** (2012) 012041 doi:10.1088/1742-6596/403/1/012041

Nuclear Physics in Box

Thomas Luu
Lawrence Livermore National Laboratory
E-mail: tluu@llnl.gov

Abstract. I provide a short overview of the current status of nuclear physics calculations using lattice Quantum Chromodynamics (LQCD). I demonstrate, at a very high level, how LQCD calculations are performed and how nuclear scattering data are extracted from these calculations, emphasizing the overlap between traditional nuclear many-body theory and LQCD calculations. I look at the $\Omega^-\Omega^-$ system as a concrete example, and in so doing demonstrate the predictive nature of LQCD calculations as applied to nuclear physics.

1. Why Lattice QCD?

It has long been accepted that Quantum Chromodynamics (QCD), the theory governing the behavior of quarks and gluons, is the underlying theory of the nuclear force, and that nuclear physics itself resides in a regime of QCD where the quarks and gluons are strongly interacting. The strongly interacting nature of QCD in this regime is no surprise: the quarks and gluons form bound hadronic states categorized as mesons ($q\bar{q}$) and baryons (qqq). The interaction between baryons and mesons, in turn, has its origins from QCD as well. The composite nature of these hadrons, coupled with the strongly interacting behavior of its quark and gluon constituents, makes tracing the lineage of the nuclear force directly from QCD a daunting task.

Fortunately there has been a slew of empirical data related to nucleons and their reactions that have allowed the interaction between nucleons to be constrained to high precision. With all this data, a natural question arises: Is QCD needed for nuclear physics? The answer is, of course, *yes*. In addition to nucleons and nuclei, the spectrum of QCD includes a diverse range of exotic mesons and baryons with strangeness content. For the former, we will shortly start probing these systems with the upcoming 12-GeV upgrade at the Thomas Jefferson National Accelerator Facility. Having a theoretical understanding of these exotic mesons, coupled with empirical data, will elucidate the nature of confinement. Baryons with non-zero strangeness are difficult to produce terrestrially, yet are known to play significant roles in astrophysical environments. Here QCD offers the means in which to understand the properties and behavior of hyperons and hyper-nuclei without resorting to any model assumptions. Even in the strangeness $S = 0$ sector there are aspects of the nuclear force that are unknown but could be resolved from QCD, such as the three-neutron interaction.

Because of the strongly-interacting nature of QCD, no formal methods are available for calculations of nuclear physics observables. As such, numerical methods must be used to tackle QCD in this regime. To date, lattice QCD (LQCD) offers the only means in which to numerically do this. The promise of LQCD is that, with sufficient high-performance computing (HPC) resources and algorithm development, the properties of few-hadron systems that are poorly known or inaccessible by experiments (like the ones mentioned above), will be calculable. As described in later sections, this promise is beginning to come true.

2. Nuclear physics in a box: A primer on LQCD

Lattice QCD is a formulation of QCD in which space and time are discretized on a lattice and the theory is confined to a finite cubic volume with fields subject to periodic boundary conditions. Strictly speaking, the volume is a hyper-cube, as the time extent is also made finite. In this manner the theory becomes amenable to numerical calculations. A lattice QCD calculation is performed using the Monte Carlo method with Markov Chain importance samplings of the QCD vacuum that are generated with a distribution prescribed by QCD, and physical observables are then measured on these samplings. As the measurements are stochastic in nature, there is an inherent uncertainty that is associated with each calculation. However, with increasing number of measurements there is a corresponding smaller statistical uncertainty in the calculation. With sufficient computational resources, and a careful account of the discretization artifacts imposed by the lattice, calculations that are not currently possible with pencil and paper are numerically feasible. Lattice QCD represents the best opportunity for answering longstanding questions in nuclear physics. But because of the diverse scales associated with LQCD calculations of nuclear observables, LQCD is computationally intensive. Because of this, LQCD calculations have historically been done using larger-than-physical quark masses to alleviate the computational demands.

2.1. Working in imaginary time

A central tenet of LQCD is the calculation of propagators using the path-integral formalism, where all possible paths are summed with corresponding weights depending on the path's particular action. The time-evolution of propagators is dictated by the systems eigen-energies E_j appearing in the argument of an exponent, e^{iE_jt}, and can be written generically

$$\text{propagator} = \sum_j^{\infty} C_j e^{iE_jt} ,$$

where C_j is some c-number and the tower of eigen-energies E_j share in common a particular symmetry (usually represented through quantum numbers). The goal of most LQCD calculations is the extraction of the energies E_j (or at least the lowest few). As the extraction of complex arguments is difficult numerically, LQCD calculations are done in *imaginary* time where $t \to it$. The consequences of working in imaginary time are complex and too numerous to discuss here, but generally speaking LQCD calculations in imaginary time draw strong analogs with statistical mechanics. Propagators are now called correlators, and their time evolution is given by

$$\text{correlator} = \sum_j^{\infty} C_j e^{-E_jt} \quad (1)$$

$$\to C_0 e^{-E_0t} \quad t \gg 1 . \quad (2)$$

The time-evolution is now dictated by a series of decaying exponentials. Given sufficient time, the ground state contribution to the correlator becomes dominant as the higher states have fallen off exponentially faster.

2.2. Interpolators

To impose a specific symmetry on the correlators, lattice QCD practitioners form interpolating operators that have the same quantum numbers as the system in question. These interpolating

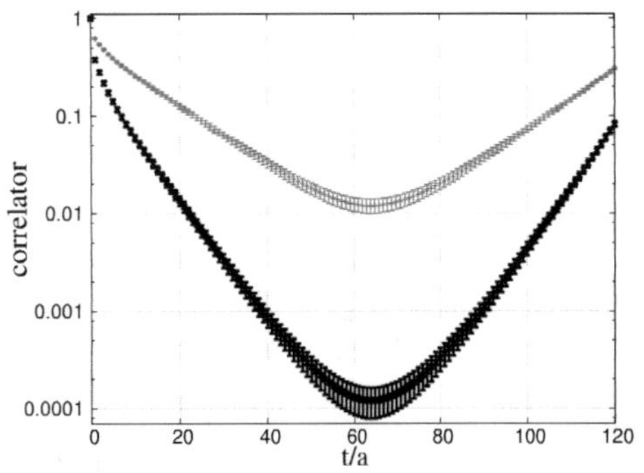

Figure 1. Examples of single pion (top, red) and $I = 2$ two-pion (bottom,black) correlators. The x-axis is plotted in units of the lattice spacing a and is therefore dimensionless. The correlators have been normalized to 1 and $t = 0$. The symmetric nature of these correlators comes from the finite-time extent in the LQCD calculations.

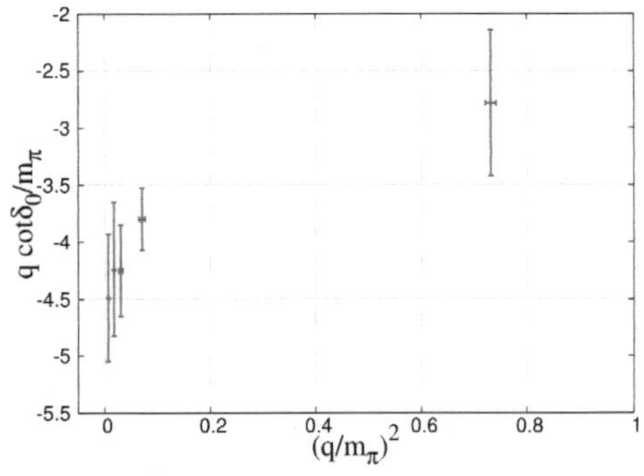

Figure 2. The phase shift for the $I = 2$ two-pion system calculated from LQCD with pion mass $m_\pi \sim 390$ MeV using multiple volumes. Data was taken from ref. [1] and only the zero CM momentum results were used to generate this plot, and statistical and systematic errors were combined in quadrature. Axes are plotted in dimensionless units.

operators represent sinks and sources within the lattice, and the probability of propagating from the source to the sink with these operators is given by the correlator that now has the specific symmetries imposed by the interpolating operators. The energies that dictate the time-evolution of this correlator represent the eigen-energies of a system with these same quantum numbers. For example, to determine the lowest eigen-energy of the π^+ system, interpolating operators representing the up and anti-d quark are formed (e.g. delta-functions at specific lattice sites), and the correlator is calculated using these interpolating operators. The operators themselves do NOT represent a π^+, but should have sufficient overlap with the true (but not known *a priori*) π^+ system[1]. After sufficient time the lowest energy E_0 is extracted (e.g. by fitting to a single exponential), from which the mass of the pion is determined, $m_\pi = E_0$. Figure 1 shows examples of correlators for the single pion and isospin $I = 2$ two-pion systems. Note that as these calculations are fully relativistic, they include the rest mass of the systems. Therefore the the decay of the two-pion system should be nearly double that of the single pion when plotted on a semi-log scale, as is the case in fig. 1.

2.3. Finite-volumes vs. Continuum physics

The finite-volume nature of LQCD calculations, coupled with its imaginary-time formalism, makes it seem at first sight impossible to extract continuum physics observables, such as scattering phase shifts. Lüscher showed from a field theoretic point of view, however, that there is a simple relation between the finite-volume energies of a two-particle system to the same system's continuum scattering phase shift[2] [2, 3],

$$q\cot\delta_0 = \frac{2}{\sqrt{\pi}L}\, \mathcal{Z}_{0,0}\left(1;\tilde{q}^2\right) \quad , \tag{3}$$

where $\mathcal{Z}_{0,0}$ is the generalized zeta function, δ_0 is the s-wave scattering phase shift, L is the length of the cube, $\tilde{q} = qL/2\pi$, and q is related to the energy of the interacting two-particle

[1] The 'perfect' overlap would be one where C_0 is nonzero and all other C_j are zero.
[2] This relation is only valid below the inelastic threshold.

state, $E = \sqrt{q^2 + m_1} + \sqrt{q^2 + m_2}$, which is calculated in LQCD. Equation 3 demonstrates that given the masses m and two-body energies E (ground state and excited states) of a system calculated from LQCD at different volumes L^3, the phase sift δ_0 can be mapped out. Analogous finite-volume formulae exists for higher partial waves [4]. Figure. 2 shows a phase shift example for the $I = 2$ two-pion system calculated from LQCD using multiple volumes and energy states. The data was taken from ref. [1]. Other examples of phase shifts, as well as effective range parameters, calculated from LQCD can be found in refs. [5, 6, 7].

Given the phase shifts, one can then extract effective range parameters through the relation

$$q\cot\delta_0 = -\frac{1}{a_0} + \frac{r_0}{2}q^2 + \dots \,, \tag{4}$$

where a_0 and r_0 are the scattering length and effective range, respectively, and higher order terms in the effective range expansion have been dropped. For the $I = 2$ two-pion system, the following effective range parameters were found [1]:

$$m_\pi a_0 = .23 \pm .019 \tag{5}$$

$$m_\pi r_0 = 12.9 \pm 3.23 \,, \tag{6}$$

where the statistical and systematic errors have been combined in quadrature.

2.4. Recent developments in Nuclear Physics from LQCD

Because of the recent advances in HPC resources, LQCD calculations are constantly pushing the envelope towards calculations using physical quark masses. Recent calculations of single hadron resonances have been performed at the physical point [8], but for multi-hadron systems extrapolations from unphysical masses are made to the physical point. Examples include calculations of certain nucleon-hyperon systems [9], as well as the spectrum of nuclei and hyper-nuclei at the 'SU3' point, where all quark masses are degenerate and set at the strange quark mass [10]. Understanding the interactions of these systems will determine their roles in the nuclear matter equation of state, which ultimately impacts black hole formation, supernova evolution, and neutron star composition. Even without extrapolations, LQCD results at larger-than-physical quark masses provide us with a deeper understanding of the rich nature of QCD in a world where fundamental input parameters are slightly different than what is observed. For example, bound states not observed in nature are prevalent in LQCD calculations where the light quarks (up and down) are larger than physical [11, 12, 13]. Calculations of the nuclear potential have been attempted [14, 15], though these calculations suffer from uncontrolled approximations and are model-dependent.

One of the most promising aspects of LQCD is its power to probe systems and their interactions that are experimentally inaccessible. A prime example is the three-pion interaction, which was found to be slightly repulsive [16]. Studies of the three-nucleon interaction are underway. Also promising is the use of LQCD to calculate nuclear matrix elements. A recent example here involves the first calculation of the neutral current parity violating parameter, $h_{\pi NN}$, from LQCD [17].

3. The $\Omega^-\Omega^-$ system as a specific example

The strangeness $S = -6$ $\Omega^-\Omega^-$ system is an interesting testbed for nuclear physics. The Ω^- particle itself consists of three strange quarks bound with spin=3/2. Typically the Ω^- is used to 'set the scale' in LQCD calculations, in that the calculated Ω^- mass is tuned to its physical value. And because the Ω^- has isospin $I = 0$, it's coupling to the light quarks are suppressed. This, in turn, means that the light-quark dependence of the $\Omega^-\Omega^-$ system is suppressed, and therefore current LQCD calculations of this system should be 'near' physical.

Table 1. Angular momentum J designations and their octahedral decompositions Γ. Parity is also assigned where applicable.

J	Γ
0	A_1^+
1/2	G_1
1	T_1^-
3/2	H
2	$E^+ \oplus T_2^+$
5/2	$H \oplus G_2$
3	$A_2^- \oplus T_1^- \oplus T_2^-$

The $\Omega^-\Omega^-$ system has been looked at before from a quark model perspective, but calculations have been inconclusive, with results suggesting that the system can be weakly repulsive [18] or strongly bound [19]. This discrepancy can be resolved by LQCD. The results presented below draw off of work presented in ref. [20].

3.1. The Ω^- interpolating operator

The interpolating operator for the Ω^- particle is built from linear combinations of the following object:

$$\Omega_{\mu_1\mu_2\mu_3} = \epsilon_{abc} s^a_{\mu_1} s^b_{\mu_2} s^c_{\mu_3} ,$$

where Roman letters refer to color indices, Greek letters refer to spin indices, s are the strange-quark fields, and repeated indices are summed. To build a state with definite spin, appropriate combinations of the spin indices must be made. However, as calculations are done within a cubic volume, standard angular momentum designations no longer apply. Instead of an infinite number of possible spin states (e.g. 1/2, 3/2, 5/2, etc.), within a cubic volume there are only three irreducible representations that, in the continuum limit, connect to odd-half spin: G_1, G_2, and H. These irreps are part of the double cover octahedral point group. Table 1 shows the decomposition of the few lowest lying angular momentum states into the octahedral irreps [21].

The spin-3/2 Ω^- particle therefore transforms under the H basis within a cubic volume, and the linear combination of spinor indices must account for this. Basak et al. have determined these combinations [22]. For example, the interpolating operator with good H symmetry that most closely transforms as $|3/2, 3/2\rangle$ in the continuum is given by Ω_{111} [3]. For a $|3/2, -3/2\rangle$, one would use Ω_{222}.

3.2. Angular momentum coupling

The two-particle $\Omega^-\Omega^-$ system's total angular momentum must be made from coupling the individual spin and angular momentum of its constituent parts. As is done in nuclear physics, this is accomplished through standard angular momentum coupling coefficients (i.e. Clebsch-Gordan coefficients). However, as was the case for the spin mentioned above, standard angular momentum designations are different in a cubic volume and are decomposed into five distinct irreps of the octahedral group: A_1, A_2, T_1, T_2, and E. Table 1 show how these states are decomposed into continuum angular momentum states. The coupling of these irreps utilize analogous Clebsch-Gordan coefficients. The number of Clebsch-Gordan coefficients in this case

[3] Spin indices are in the Pauli-Dirac basis.

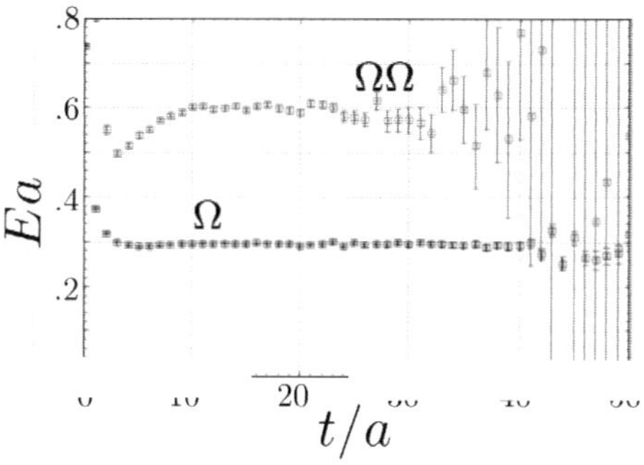

Figure 3. Effective mass plots for the single Ω^- and $\Omega^-\Omega^-$ A_1^+ system. Axes are plotted in units of the lattice spacing and are therefore dimensionless. These results show the extracted lowest energies of the single Ω^- and $\Omega^-\Omega^-$ correlators, as a function of time.

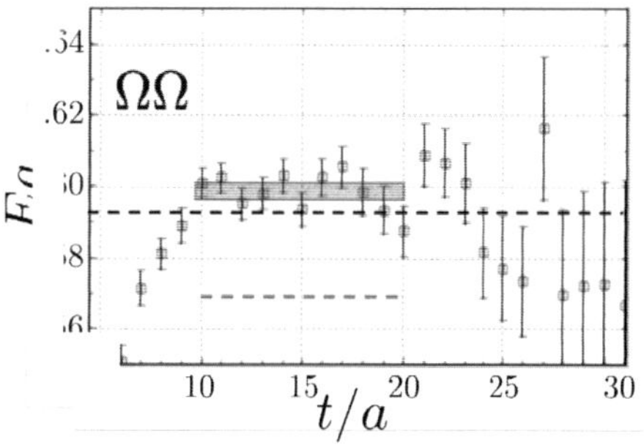

Figure 4. Close-up of the $\Omega^-\Omega^-$ effective mass. Band shows the fit result and range of fit window. The long dashed line corresponds to the two-Ω^- threshold (i.e. twice the Ω^- mass), and the short (red) dashed line represents the results if the system were bound by 100 MeV.

are finite, as there are only a finite number of irreps in the octahedral group. The coupling coefficients used in this work were taken from ref. [22].

The $\Omega^-\Omega^-$ system of interest is a s-wave system, and therefore the total angular momentum J of the system is dictated by the coupling of the particle's individual spins. Antisymmetry dictates that J can be either 0 or 2 in this case, which would correspond to the A_1^+ symmetry or E^+/T_2^+ symmetry, respectively, in a cubic volume. The s-wave component (A_1) of the problem is enforced by projecting the individual Ω^- particles to zero momentum during the LQCD calculation.

3.3. Results

In fig. 3 the effective mass plots are shown for the single-Ω^- and $\Omega^-\Omega^-$ A_1^+ system. These plots essentially show the fitted lowest energy extracted for the corresponding Ω^- and $\Omega^-\Omega^-$ correlators. Not surprisingly, the value for the $\Omega^-\Omega^-$ mass is nearly twice the value of the single Ω^- mass. In fig. 4 a close-up of the $\Omega^-\Omega^-$ effective mass is shown, as well as the values of the two-Ω^- threshold (long black dashed line). The shaded band in the figure provides the fitted energy as well as the time-extent of the fit. As the band falls above the two-Ω^- threshold, we identify this state as a scattering state with slight repulsion. These results do not see a deeply bound 100 MeV state, as this would have fallen near the short (red) dashed line in the figure.

Using eqs. 3 and 4, the scattering length can be extracted for this interacting system and the following prediction is made,

$$a_0 = .15 \pm .22 \text{ fm} , \tag{7}$$

where the statistical and systematic errors have been combined in quadrature. It is important to note that though these calculations have produced only a scattering state, the presence of a deeply bound state is not completely ruled out, particularly one that is so bound as to not be resolved with the current lattice spacings (a = .12 fm) used in these LQCD calculations. Future calculations with finer lattice spacings are needed to definitively rule out the presence of a deeply bound state.

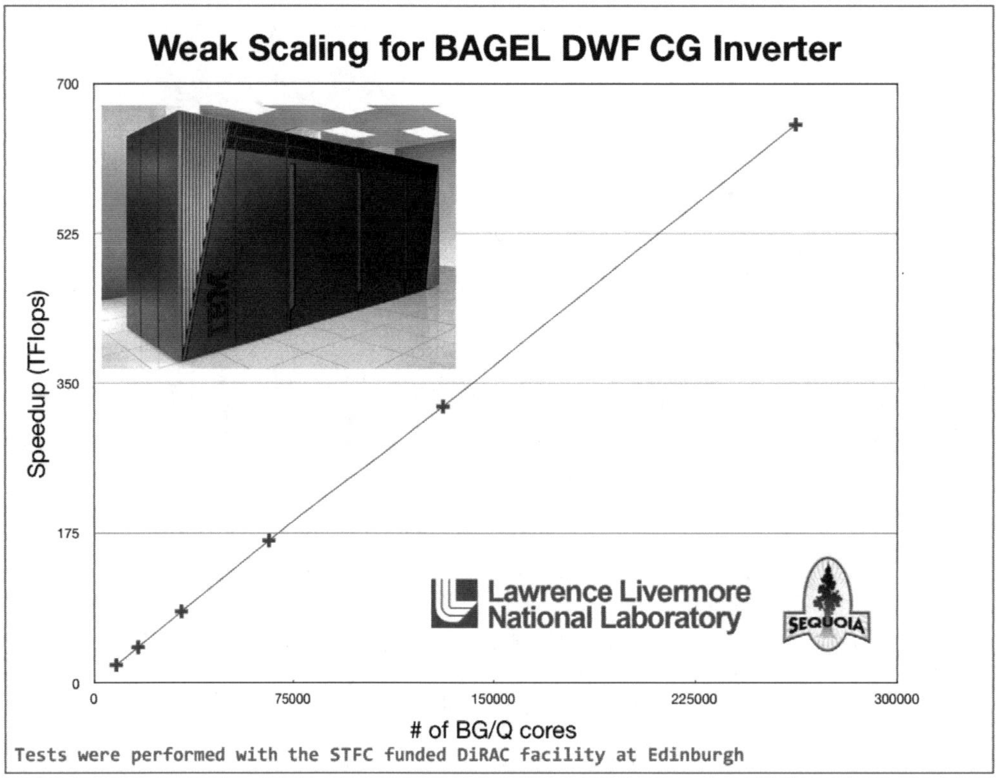

Figure 5. Weak scaling (i.e. lattice size grows with increasing number of cores) of optimized conjugate gradient inverter for Domain Wall Fermion discretization on the Sequoia machine. Near linear behavior suggests optimal scaling.

4. What's on the horizon

Much of the recent progress in LQCD calculations of nuclear observables has been enabled, in part, by the recent gains in HPC. This trend is anticipated to continue, with the end result being the seamless coupling of nuclear physics, LQCD, and HPC. As an example of the promise that HPC will bring to LQCD and nuclear physics, fig. 5 shows the weak scaling behavior of an inversion routine (commonly used within LQCD codes) on LLNL's recently installed 16 Petaflop Sequoia machine. The linear relation between performance and number of cores demonstrates optimal scaling behavior for this code. Because of the large computational power of this machine, LQCD can readily perform calculations at, or near, the physical point. Calculations of the $\Omega^-\Omega^-$ system using Sequoia would reduce current uncertainties by at least an order of magnitude. Anticipated results with petascale (and beyond) resources of nuclear physics observables and their uncertainties have been estimated in ref. [23]

The significance of HPC gains goes beyond just calculating nuclear phenomena at the physical point. It will also allow for seamless integration of nuclear many-body effective theories to couple with LQCD calculations, thereby extending the applicability of LQCD calculations. At the same time, nuclear physics becomes rooted to the fundamental theory of QCD, and answers to longstanding questions about the nuclear force will be addressed. Most importantly, nuclear physics, predominantly data driven historically, becomes a predictive science.

Acknowledgments
The author thanks members of the NPLQCD collaboration, from which much of the work presented here was done by this collaboration. The author also thanks J. Wasem, M. Buchoff, and C. Schroeder for their contributions to this paper related to the parity violating interaction, $\Omega^-\Omega^-$ interaction, and scaling calculations on Sequoia. This work performed under the auspices of the U.S. Department of Energy by Lawrence Livermore National Laboratory under Contract DE-AC52-07NA27344.

References
[1] Beane S *et al.* (NPLQCD Collaboration) 2012 *Phys.Rev.* **D85** 034505 (*Preprint* 1107.5023)
[2] Luscher M 1986 *Commun.Math.Phys.* **105** 153–188
[3] Luscher M 1991 *Nucl.Phys.* **B354** 531–578
[4] Luu T and Savage M J 2011 *Phys.Rev.* **D83** 114508 (*Preprint* 1101.3347)
[5] Dudek J J, Edwards R G, Peardon M J, Richards D G and Thomas C E 2011 *Phys.Rev.* **D83** 071504 (*Preprint* 1011.6352)
[6] Beane S R *et al.* (NPLQCD Collaboration) 2008 *Phys.Rev.* **D77** 094507 (*Preprint* 0709.1169)
[7] Sasaki K, Ishizuka N, Yamazaki T and Oka M (PACS-CS Collaboration) 2010 *Prog.Theor.Phys.Suppl.* **186** 187–192
[8] Fodor Z and Hoelbling C 2012 *Rev.Mod.Phys.* **84** 449 (*Preprint* 1203.4789)
[9] Beane S, Chang E, Cohen S, Detmold W, Lin H W *et al.* 2012 (*Preprint* 1204.3606)
[10] Beane S, Chang E, Cohen S, Detmold W, Lin H *et al.* 2012 (*Preprint* 1206.5219)
[11] Beane S *et al.* (NPLQCD Collaboration) 2012 *Phys.Rev.* **D85** 054511 (*Preprint* 1109.2889)
[12] Beane S *et al.* (NPLQCD Collaboration) 2011 *Phys.Rev.Lett.* **106** 162001 (*Preprint* 1012.3812)
[13] Inoue T *et al.* (HAL QCD Collaboration) 2011 *Phys.Rev.Lett.* **106** 162002 (*Preprint* 1012.5928)
[14] Inoue T *et al.* (HAL QCD Collaboration) 2012 *Nucl.Phys.* **A881** 28–43 (*Preprint* 1112.5926)
[15] Ishii N *et al.* (HAL QCD Collaboration) 2012 *Phys.Lett.* **B712** 437–441 (*Preprint* 1203.3642)
[16] Beane S R, Detmold W, Luu T C, Orginos K, Savage M J *et al.* 2008 *Phys.Rev.Lett.* **100** 082004 (*Preprint* 0710.1827)
[17] Wasem J 2012 *Phys.Rev.* **C85** 022501 (*Preprint* 1108.1151)
[18] Wang F, Ping J l, Wu G h, Teng L j and Goldman J T 1995 *Phys.Rev.* **C51** 3411 (*Preprint* nucl-th/9512014)
[19] Zhang Z, Yu Y, Ching C, Ho T and Lu Z D 2000 *Phys.Rev.* **C61** 065204
[20] Buchoff M I, Luu T C and Wasem J 2012 *Phys.Rev.* **D85** 094511 (*Preprint* 1201.3596)
[21] Mandula J E, Zweig G and Govaerts J 1983 *Nucl.Phys.* **B228** 91
[22] Basak S *et al.* (Lattice Hadron Physics Collaboration (LHPC)) 2005 *Phys.Rev.* **D72** 074501 (*Preprint* hep-lat/0508018)
[23] Young G, Dean D and Savage M 2009 Forefront questions in nuclear science and the role of high performance computing http://science.energy.gov/ascr/news-and-resources/workshops-and-conferences/grand-challenges/

Effects of Quark Spin Flip on the Collins Fragmentation Function in a Toy Model

H H Matevosyan[1], A W Thomas[1] and W Bentz[2]
[1]CSSM and ARC Centre of Excellence for Particle Physics at the Tera-scale, School of Chemistry and Physics, University of Adelaide, Adelaide SA 5005, Australia
[2]Department of Physics, School of Science, Tokai University, Hiratsuka-shi, Kanagawa 259-1292, Japan

Abstract. The recent extension of the NJL-jet model to hadronization of transversely polarized quarks allowed the study of the Collins fragmentation function. Both favored and unfavored Collins fragmentation functions were generated, the latter purely by multiple hadron emissions, with 1/2 moments of opposite sign in the region of the light-cone momentum fraction z accessible in current experiments. Hints of such behavior has been seen in the measurements in several experiments. Also, in the transverse momentum dependent (TMD) hadron emission probabilities, modulations of up to fourth order in sine of the polar angle were observed, while the Collins effect describes just the linear modulations. A crucial part of the extended model was the calculation of the quark spin flip probability after each hadron emission in the jet. Here we study the effects of this probability on the resulting unfavored and favored Collins functions by setting it as a constant and use a toy model for the elementary single hadron emission probabilities. The results of the Monte Carlo simulations showed that preferential quark spin flip in the elementary hadron emission is needed to generate the favored and unfavored Collins functions with opposite sign 1/2 moments. For the TMD hadron emission modulations, we showed that the model quark spin flip probabilities are a partial source of the higher rode modulations, while the other source is the Collins modulation of the remnant quark from the hadron emission recoil.

1. Introduction

The recent interest in the transverse momentum structure of hadrons has prompted renewed efforts both in experimental and theoretical studies. Notably, experiments using semi-inclusive deep inelastic scattering (SIDIS) have advanced our understanding of the underlying partonic structure of the nucleon in momentum space. For example, the naively time-reversal odd Sivers distribution function has been measured to be non-zero [1, 2, 3]. In the factorization regime, the SIDIS cross-sections can be expressed as convolutions of parton distribution functions with elementary parton-probe scattering amplitudes calculable using perturbative QCD and parton fragmentation functions [4, 5]. The complete tree-level expressions at leading order have been presented in Ref [6]. Thus, to access the underling partonic structure of the nucleons, a precise knowledge of the fragmentation functions are needed. The particularly interesting case of the T-odd Collins fragmentation function $H_1^\perp$ has been widely explored both in theoretical models [7, 8, 9, 10] and experiments [11, 12, 1, 3, 13, 14], with the first direct measurements of the Collins fragmentation mechanism being performed by the Belle collaboration using hadron pair production in e^+e^- collisions [11, 12]. Further, the experimental studies from HERMES,

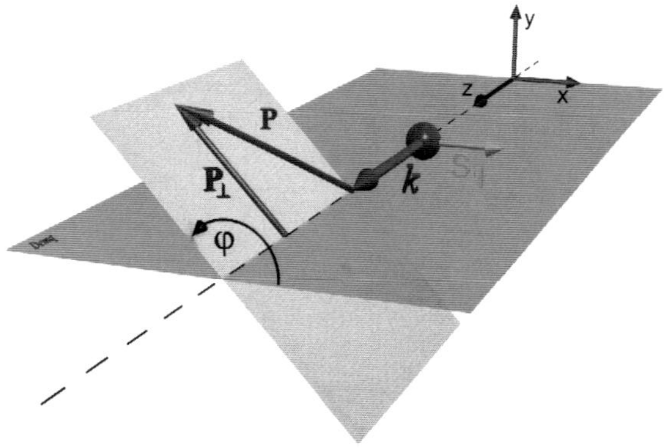

Figure 1. Illustration of the three dimensional kinematics of transversely polarized quark fragmentation. The fragmenting quark's momentum $\boldsymbol{k}$ defines the z-axis with its transverse polarization spin vector $\boldsymbol{S}_q$ along x axis. The emitted hadron has momentum p with the transverse component $\boldsymbol{P}_\perp$ with respect to the z-axis. The polar angle of hadron's momentum P with respect to the zx plane is denoted by φ.

COMPASS and JLab are suggesting that the unfavored Collins functions have a similar size and an opposite sign to that of the favored ones [1, 3, 13, 14].

The theoretical calculation of Collins function for pions and later also kaons within the spectator model of Refs. [7, 8, 9] made predictions for the favored Collins function using the mechanism of interference of one-meson-loop amplitudes with the tree level amplitude. However, these model calculations could not make direct calculations of the unfavored Collins function, as only a single hadron emission was included. Recently, the Nambu–Jona-Lasinio (NJL) jet model of Refs. [15, 16, 17, 18, 19] has been extended to calculate the Collins fragmentation functions, both favored and unfavored, using Monte Carlo (MC) simulations in the quark-jet hadronization mechanism [20]. Here the transverse polarization of the fragmenting quark has been introduced, and the light cone spinors were used to calculate the quark spin flip probabilities in each hadron emission step. In the hadronization of a transversely polarized quark, the Collins effect describes a modulation of the unpolarized hadron fragmentation function with a Collins fragmentation function term that is proportional to the sine of the angle φ of the hadron's transverse momentum and the quark's spin in the $\gamma^* N$ frame, as depicted in Fig. 1. The corresponding expression using the "Trento Convention" [21], with polarized quark q carrying momentum $\boldsymbol{k}$ and spin $\boldsymbol{S_q}$ fragmenting to unpolarized hadron h of mass m_h, carrying the light-cone momentum fraction z and transverse momentum $\boldsymbol{P}_\perp$ with respect to quark's momentum $\boldsymbol{k}$ can be written as

$$D_{h/q^\uparrow}(z, P_\perp^2, \varphi) = D_1^{h/q}(z, P_\perp^2) - H_1^{\perp h/q}(z, P_\perp^2)\frac{P_\perp S_q}{z m_h}\sin(\varphi), \tag{1}$$

where the unpolarized fragmentation function is denoted $D_1^{h/q}(z, P_\perp^2)$ and $H_1^{\perp h/q}(z, P_\perp^2)$ is the Collins function. When integrated over $P_\perp^2$, the polarized fragmentation function can be expressed in terms of the integrated unpolarized fragmentation function $D_1^{h/q}(z)$ and the 1/2 moment of the Collins function $H_{1(h/q)}^{\perp(1/2)}(z)$

$$D_{h/q^\uparrow}(z, \varphi) \equiv \int_0^\infty dP_\perp^2 \; D_{h/q^\uparrow}(z, P_\perp^2, \varphi) = \frac{1}{2\pi}\left[D_1^{h/q}(z) \; - 2H_{1(h/q)}^{\perp(1/2)}(z) S_q \sin(\varphi)\right] dz \, d\varphi, \tag{2}$$

where

$$D_1^{h/q}(z) \equiv \pi \int_0^\infty dP_\perp^2 \; D_1^{h/q}(z, P_\perp^2), \tag{3}$$

$$H_{1(h/q)}^{\perp(1/2)}(z) \equiv \pi \int_0^\infty dP_\perp^2 \frac{P_\perp}{2 z m_h} H_1^{\perp h/q}(z, P_\perp^2). \tag{4}$$

In the Monte Carlo simulations of Ref. [20], we used as input the elementary unpolarized and Collins fragmentation functions for single hadron production, calculated within NJL model, where we used the mechanism of Ref. [8] for the elementary Collins function. The full polarized fragmentation function was then calculated in MC simulations with multiple hadron emission. The resulting polarized fragmentations, when integrated over the hadron transverse momentum squared ($P_\perp^2$), exhibited the Collins modulation both for favored and unfavored hadron production. Here the unfavored Collins function's 1/2 moment is comparable in size but opposite in sign to that for the favored fragmentation. For the transverse momentum dependent (TMD) polarized fragmentation functions, at fixed values for the hadron's light cone momentum fraction z and $P_\perp^2$, exhibited modulation with a fourth order polynomial in $\sin(\varphi)$, even in simulations with only two produced hadrons.

In this work we have two goals, focusing on pion fragmentations from light quarks only. First, we examine the dependence of the sign and the magnitude of the 1/2 moment of unfavored fragmentation function on the quark spin flip probability used in the NJL-jet model MC simulations. Second, we examine the origins of the higher order Collins modulations in the TMD polarized fragmentation functions.

2. NJL-jet Model with transversely polarized quark

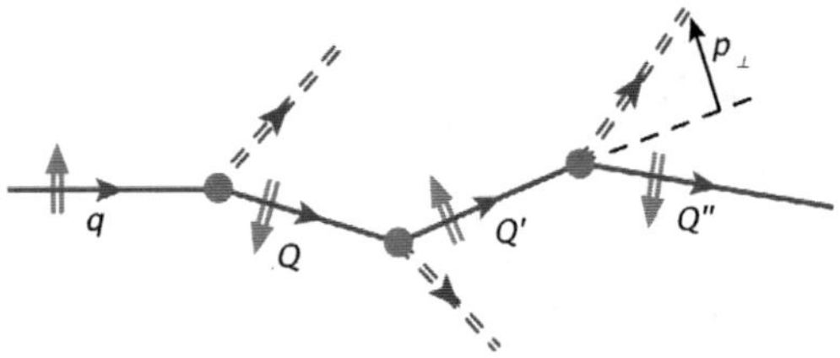

Figure 2. NJL-jet model including transverse momentum and quark polarization transfer. Here the orange double-lined arrows schematically indicate the spin direction of the quark in the decay chain.

Schematically, the NJL-jet model with the transversely polarized quark is depicted in Fig. 2. In MC simulations of Ref. [20], in each hadronization step, we used the elementary polarized fragmentations to sample the momentum of the production hadron, z_1, $p_\perp$ and φ_1, with respect to the fragmenting quark in that step (here φ_1 is defined with respect to the spin of the fragmenting quark). Using the momentum we calculated and recorded the hadron's momentum, z, $P_\perp$ and φ, with respect to the initial fragmenting quark. Further, we also determined the momentum of the remnant quark using momentum conservation, while the probability of the remnant quark's spin flip was sampled using the spin non-flip and flip probabilities $|a_1|^2$ and $|a_{-1}|^2$. These probabilities were calculated using Dirac spinors of the transversely polarized quark, that in turn can be expressed in terms of the Lepage-Brodsky spinors in helicity basis [22, 23]. Explicit these probabilities are proportional to

$$|a_1|^2 \sim l_x^2,\ |a_{-1}|^2 \sim l_y^2 + (M_2 - (1-z)M_1)^2, \tag{5}$$

where M_1 and M_2 are the masses of the fragmenting and the remnant quarks, and $\boldsymbol{l}_\perp \equiv (l_x, l_y) = -\boldsymbol{p}_\perp$ is the transverse momentum of the remnant quark with respect to the momentum direction of the fragmenting quark.

In this work, to explore the role of the spin-flip probability, we perform MC simulations with various fixed values of this probability. This allows us to clearly demonstrate the role of the quark spin flip during the hadronization process on the resulting fragmentation functions. Also, using constant for this probability instead of the one with the dynamical factors of Eq. (5) will help to disentangle the sources of the higher order modulation effects in $\sin(\varphi)$ observed in TMD fragmentations.

For input to the Monte Carlo framework, the elementary probability of a transversely polarized quark q to emit a single hadron h carrying light-cone momentum fraction z, transverse momentum $\boldsymbol{p}_\perp$ (with magnitude $p_\perp$ and polar angle φ with respect to the emitting quark's spin), can be written as

$$d_{h/q^\uparrow}(z, p_\perp^2, \varphi) = d_1^{h/q}(z, p_\perp^2) - \widetilde{H}_1^{\perp h/q}(z, p_\perp^2)\frac{p_\perp S_q}{z m_h}\sin(\varphi), \tag{6}$$

where $d^{h/q}(z, p_\perp^2)$ is the elementary unpolarized fragmentation function and $\widetilde{H}_1^{\perp h/q}(z, p_\perp^2)$ is the elementary Collins fragmentation function. The unpolarized fragmentation function has been calculated within the NJL model in tree level approximation [18]. The corresponding expression for an unpolarized quark emitting a pseudoscalar meson can be written as

$$d_1^{h/q}(z, p_\perp^2) = G_{hQ}(p_\perp^2)\frac{C_q^h}{16\pi^3}\, g_{hqQ}^2\, z\, \frac{p_\perp^2 + [(z-1)M_1 + M_2]^2}{\left[p_\perp^2 + z(z-1)M_1^2 + zM_2^2 + (1-z)m_h^2\right]^2}, \tag{7}$$

where the subscripts on the constituent masses, M_1 and M_2, denote the flavors of the fragmenting and the remnant quarks respectively, that are also indicated by the subscripts q and Q. C_q^h and g_{hqQ} denote the isospin factor and quark-meson coupling constant, that have been determined within the NJL model [16, 18], while $G_{hQ}(p_\perp^2)$ is the dipole regulator for the transverse momentum dependance. The details of the derivations and the model parameters can be found in Ref. [18]

The model elementary Collins function within the NJL model has been also calculated in Ref. [20], using the mechanism of interference between tree level amplitude and amplitude with gauge link coupling for single hadron emission, outlined in Refs. [8, 9]. Here we are interested in exploring the general properties of the quark-jet hadronization mechanism, thus the details of the particular models are unimportant. To further simplify the model and help reduce the required number of the Monte Carlo events that allow for a reliable extraction of the higher order modulations by enhancing the azimuthal modulation signal with respect to the statistical fluctuations, we employ a toy model for the elementary Collins function taking it promotional to the unpolarized one, with a coefficient chosen such that the polarized probability is still positive for all values of the corresponding arguments

$$d_{h/q^\uparrow}^{(toy)}(z, p_\perp^2) = d_1^{h/q}(z, p_\perp^2)(1 - 0.9\sin\varphi), \tag{8}$$

where we assume that $\widetilde{H}_1^{\perp h/q}(z, p_\perp^2)\frac{p_\perp}{m_h z} = 0.9\ d_1^{h/q}(z, p_\perp^2)$ and $S_q = 1$. Within the NJL-jet model, the elementary splitting functions are renormalized such that quark's total probability of emitting a hadron in each step is one: $\sum_h \int dz\ dp_\perp^2/2\ d\varphi\ \hat{d}_{h/q^\uparrow}(z, p_\perp^2, \varphi) = 1$, the sum is over all the hadrons, while a quark of given flavor can emit directly. These renormalized splittings will be used in the next two sections as input to the Monte Carlo simulations of the quark-jet hadronization process.

3. The Quark Spin Flip Probability Effects on 1/2 Moments of the Collins Fragmentation Functions

In this section we explore the properties of the 1/2 moments of the Collins functions, Eq. (4), for pions emitted in the hadronization of an up quark as a function of the quark spin flip probability SF in each hadron emission step. We perform Monte Carlo simulations to extract the polarized quark fragmentation functions $D_{h/q^\uparrow}$ as functions of z, $P_\perp^2$ and φ within the NJL-jet framework, as described in Ref. [20]. In this section, we limit the number of emitted hadrons in each quark decay chain to $N_{Links} = 2$. This allows us to cleanly see the effects of the quark spin flip on the

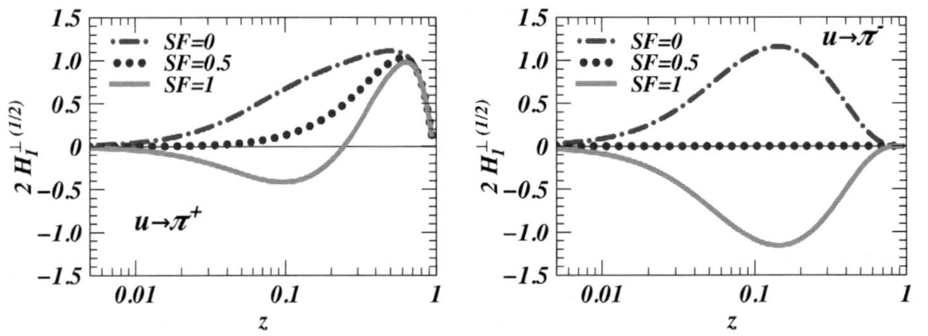

(a) (b)

Figure 3. Fitted values for twice the Collins function 1/2 moment, $2H_1^{\perp(1/2)}$ for $u \to \pi^+$ (a) and $u \to \pi^-$ (b) as a function of z from Monte Carlo simulations with three different values of the remnant quark spin flip probability SF in each elementary emission, where the number of emitted hadrons in each quark decay chain is fixed to $N_{Links} = 2$.

emission of the second hadron in the decay chain, where the unfavored fragmentation functions are first generated.

Next, to extract the integrated unpolarized fragmentation function (3) and the 1/2 moment of the Collins fragmentation functions (4), we use the relation to the $P_\perp^2$ integrated transversely polarized fragmentation written in Eq. (2). Thus, we perform a minimum-χ^2 fit to $D_{h/q^\uparrow}(z, \varphi)$ using a form $F(c_0, c_1) \equiv c_0 + c_1 \sin(\varphi)$ for fixed values of z, identifying the fitted coefficients with the unpolarized and Collins terms in Eq. (2). The resulting fits for the $P_\perp^2$ integrated pion polarized fragmentation functions with the functional form of $F(c_0, c_1)$ perfectly describe the results of the MC simulations, with average χ^2 per number of degrees of freedom (NDF) very close to unity. The unpolarized fragmentations are not affected by the quark spin flip probability SF and have been studied in detail in our previous work [17, 18]. Thus we only present here the results for the Collins functions.

The results from fits for $H_1^{\perp(1/2)}(z)$ for fragmentations of a u quark to π^+ and π^- are depicted in Fig. 3, where each curve depicts the results for a different value of SF. The results for π^+ in the subplot (a) show that, if $SF \leq 0.5$, then $H_1^{\perp(1/2)}$ is always positive, crossing zero and becoming negative at small values of z for $SF > 0.5$. For unfavored π^- fragmentation, the effects of SF are more dramatic: $H_1^{\perp(1/2)}$ is positive (same sign as that for π^+) for $SF < 0.5$ and vanishes for $SF = 0.5$. Thus, in order to generate a non-zero unfavored Collins fragmentation function, the quark spin non-flip and flip must be unequally probable. Finally, for $SF > 0.5$ the $H_1^{\perp(1/2)}$ is negative, at $SF = 1$ becoming exactly the negative of the results for $SF = 0$. This is easy to understand, as the unfavored functions are only generated after the initial hadron emission, so we can consider only two hadron emission for simplicity. In case where $SF = 1$ (flip of the quark's spin after each emission), the angle φ in Eq. (6) that is measured with respect to the spin of the initial quark acquires an additional phase π compared to the simulations with $SF = 0$ (no quark spin flip) since the modulation in each hadron emission step is generated with respect to the azimuthal angle of the currently fragmenting quark, resulting in the Collins term having the opposite sign.

Thus, the representative features of the full model results in Ref. [20], where the 1/2 moments behave similarly to the scenario where $SF > 0.5$, qualitatively do not depend on the details of the particular input unpolarized and Collins fragmentation functions, but rather represent the characteristics of the quark-jet model with preferential spin flip probability ($|a_{-1}|^2 > |a_1|^2$ in Eq. (5)).

4. Higher Order Modulations in TMD Collins Function

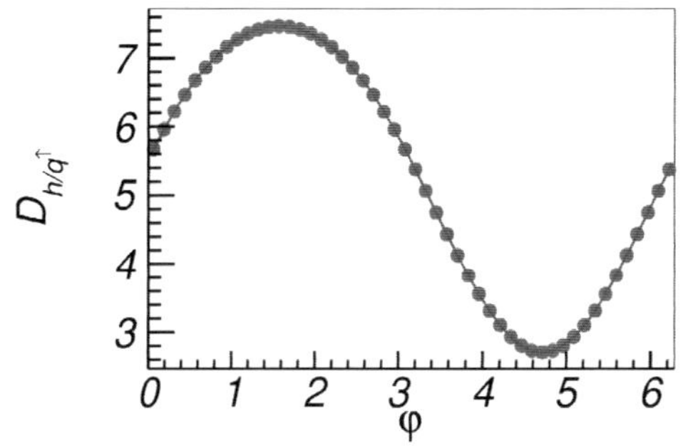

(a)

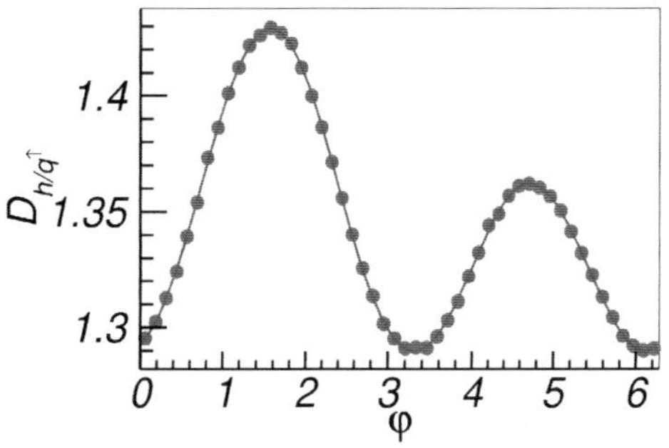

(b)

Figure 4. The histograms (blue dots) for the polarized number density $D_{\pi^+/u^\uparrow}(z, P_\perp^2, \varphi)$ for fixed values of $P_\perp^2 = 0.04$ GeV^2 (a) and $P_\perp^2 = 0.25$ GeV^2 (b) with $z = 0.3$ for both cases, as a function of the azimuthal angle φ, from Monte Carlo simulations using the toy model with a fixed number of hadrons emitted in each decay chain, $N_{Links} = 2$. The minimum-χ^2 fits with a second order polynomial in $\sin(\varphi)$ (red lines) are also depicted.

In the previous section we showed that setting quark spin flip probability $SF = 1$ can be considered as the asymptotic solution for generating the largest value of the unfavored Collins function that mimics the behavior of our model calculations in Ref. [20]. Thus in this section we will concentrate on simulations with $SF = 1$, that will allow us to easily extract the Collins modulations from the unpolarized fragmentation functions.

In this section we consider the full transverse momentum dependent fragmentation functions. As mentioned in Ref. [20], the polarized fragmentation function can no longer be successfully described with a first order polynomial in $\sin(\varphi)$ for all values of z and $P_\perp^2$, even for MC results with just two hadrons emitted, $N_{Links} = 2$. The plots in Fig. 4 depict the MC result $D_{\pi^+/u^\uparrow}(z, P_\perp^2, \varphi)$ (in blue dots) for a fixed value of $z = 0.3$ and $P_\perp^2 = 0.04$ GeV^2 (a), and $P_\perp^2 = 0.25$ GeV^2 (b). The minimum-χ^2 fits with a second order polynomial in $\sin(\varphi)$ are depicted as red lines.

Here we analyze the dependence of the order of these modulations on the number of hadrons produced, N_{Links}. For that, we consider minimum-χ^2 fits of the corresponding polarized fragmentation functions with polynomials in $\sin(\varphi)$ of various orders, for fixed values of z and $P_\perp^2$. To quantify the quality of the fits, we use as a measure the χ^2/NDF, that should be close to 1 for a good fit. Our MC simulations are performed with 100 discretization points for both z and $P_\perp^2$ in the interval $[0, 1)$ and 100 discretization points for φ in the interval $[0, 2\pi)$. Thus, the number of the "data points" to fit, for all distinct values of z and $P_\perp^2$, as well as all the possible pion fragmentations of u quark (to 3 types of pions) are $N_F = 3 \times 10^4$. The histograms in Fig. 5 depict the results for these fits for simulations with $N_{Links} = 2$ (a), $N_{Links} = 3$ (b) and $N_{Links} = 6$ (c). Here the horizontal axis denotes the χ^2/NDF and the vertical axis is the number of the fits that yield this value, while the different lines represent the results for fits with different order polynomials in $\sin(\varphi)$. The results for $N_{Links} = 2$, depicted in Fig. 5(a), show that the linear form of Eq. (1) is inadequate to describe the polarized fragmentations for a significant number of fixed values of z and $P_\perp^2$. The quadratic form, on the other hand, provides an excellent description of the simulations, and it can be confirmed by the histograms that cubic fits don't improve the resulting values of chi^2/NDF. Similarly, for the simulation with $N_{Links} = 3$ of Fig. 5(b), the cubic polynomial yields excellent fits, while the lower order polynomials are inadequate to fully describe the simulations. Finally, for the simulations with $N_{Links} = 6$, depicted in Fig. 5(c), at least a fourth order polynomial is needed for good fits, though there is a marginal improvement with fits using sixth order polynomials.

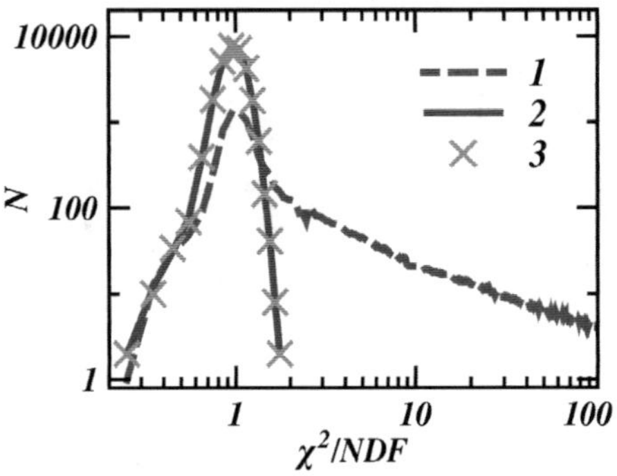

(a)

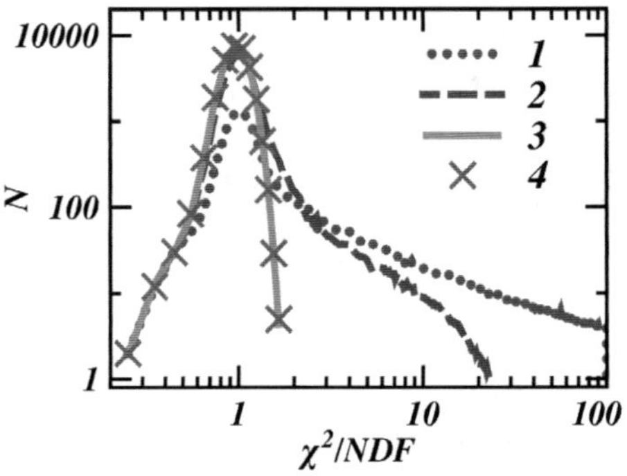

(b)

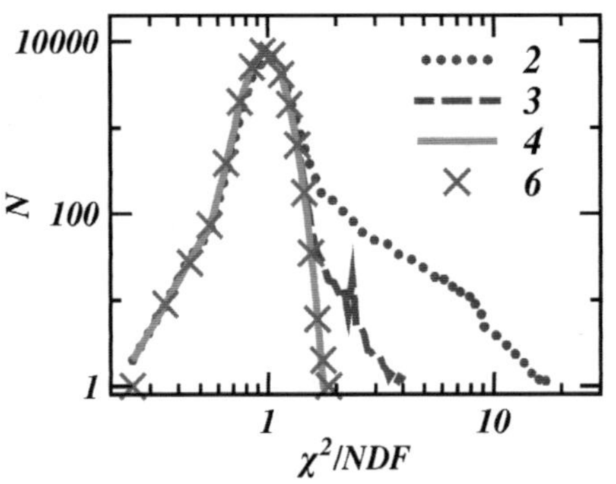

(c)

Figure 5. The histograms of the values N_F of χ^2/NDF for all the hadrons produced by a u quark with polynomials of different orders in $\sin(\varphi)$ for MC simulations for values of N_{Links} of 2 (a), 3 (b) and 6 (c). The numbers in the captions denote the orders of the polynomials used for the corresponding fits and NDF for each fit equals to 100 (the number of fitted points in φ) minus this order plus one.

A simple explanation of these modulations within the quark-jet picture can be easily given. Here, after the initial hadron emission, the hadron emission probability is modulated according to the elementary polarized fragmentation. Thus, the remnant quark, having transverse recoil momentum with respect to the initial quark that is equal to minus that of the emitted hadron, also acquires opposite signed sine modulation in the distribution of its transverse momentum. In the next hadron emission step, the emitted hadron probability is again modulated according to the elementary polarized fragmentation, but now with respect to the currently emitting quark, that itself has a sine modulation in its polar angle distribution. Thus, the produced hadron probability acquires a second order sine modulation in φ with respect to the initial quark, as does the remnant quark in this step, and so on. Thus, in the simulations with N_{Links} produced hadrons, the resulting polarized fragmentation functions should be modulated with a polynomial of order N_{Links}, as we have seen from the results of Figs. 5 (a) and (b), while the simulations with $N_{Links} = 1$ yield the trivial linear modulation. We have to also note from our earlier discussions of the mechanism for this quark recoil traverse momentum induced modulations, that the hadrons emitted in n-th step will only have a emission probability modulation of the order n in $\sin(\varphi)$. Also, we have shown in our previous work, that the remnant quark looses almost all of its light-cone momenta within just a few hadron emission steps [17], so that the hadrons produced with higher and higher orders of modulations will have smaller and smaller average values of z. This explains the fact that the linear form gives reasonably good fits for significant number of slices even in the simulations with large number of N_{Links}. Moreover, the results in Fig. 5(c) for $N_{Links} = 6$ also can be explained using this argument, as our simulations, with only 100 uniform discretization points in z, do not probe in detail the very low z region, where the sixth order polynomial modulations will be significant. Thus the corresponding results can be well described with just fourth order polynomials. Also, the size of the statistical errors and relatively small number of the discretization points in φ would affect the sensitivity to this higher order terms in the fits.

In the full model calculations, where the quark spin flip probabilities were taken to be those of Eq. (5), fourth order modulations of the polarized fragmentation functions appear in simulations with just two produced hadrons in each quark decay chain. The additional two orders of modulation in $\sin(\varphi)$ originate from the quark spin flip probabilities themselves, as those depend on the squares of the components of the recoil quark transverse momentum, that can be easily expressed in terms of $\sin^2(\varphi)$ using the transverse momentum conservation relation of the produced hadron $\vec{p}_\perp$ and the remnant quark $\vec{l}_\perp$ with respect to the fragmenting quark: $\vec{l}_\perp = -\vec{p}_\perp$ and $p_x^2 = p_\perp^2 \cos^2(\varphi)$, $p_y^2 = p_\perp^2 \sin^2(\varphi)$. In practice, we saw that the modulations of order higher than fourth are very hard to extract from our simulations with the present discretization of variables and the number of MC events calculated (of order 10^{11} in current work and 10^{12} in Ref. [20]). This is a purely computational obstacle, that can be easily solved by running the simulation on an order of magnitude more computational nodes than used for this work (only 50 Intel Core-i7 cores ran for about two weeks), although it is hard to justify the need for such a computationally expensive exercise.

5. Conclusions

In this article we examined, using a toy model, the role of the quark spin flip in the quark-jet hadronization picture on the pion Collins functions produced in Monte Carlo simulations. This work was put forward to compliment and elucidate the findings for the 1/2 moment of the Collins function in our previous work in Ref. [20] and also to provide the toy model analyses needed for explanations of the higher order sine modulations in the transverse momentum dependence of the polarized fragmentation functions, first observed in the same work.

First, the dependence of the 1/2 moments of the Collins functions, $H_1^{\perp(1/2)}$, on the quark spin flip probability (SF) for simulations with only two hadron emissions was examined in Section 3. It was shown that $H_1^{\perp(1/2)}$ has a very strong dependence on SF, both for favored and unfavored fragmentation functions. In particular, a non-zero unfavored $H_1^{\perp(1/2)}$ is only possible if $SF \neq 0.5$, with the results for $SF = 0.5 \pm a$, for any $0 < a \leq 0.5$, having equal magnitude and opposite sign. The results with $SF = 1$ have the largest magnitude and opposite sign to the input elementary favored Collins function, similar to the results of Ref. [20]. For the favored fragmentation functions, the results for $SF = 0$ are of the same sign as the elementary favored Collins function for all values of z. As the SF increases from 0 to 0.5, the low-z region gets increasingly suppressed. For $SF > 0.5$, they start positive (with respect to the elementary Collins function), then cross zero and become negative at small values of z, with minimum negative value decreasing as SF increases towards 1. The results for $SF > 0.5$ again resemble those acquired in the full model calculations of Ref. [20], dictating that the qualitative features of $H_1^{\perp(1/2)}$ are generated by the quark spin flip process in the hadronization process and don't depend on the particular elementary Collins functions used as input to the MC simulations.

In Section 4 we examined the higher order modulations of the transverse momentum dependent polarized quark fragmentation functions for $SF = 1$, in order to determine the role of the model spin flip probabilities of Eq. (5), used in Ref. [20], in the mechanism that generates these modulations. We studied the order of the polynomials in $\sin(\varphi)$ needed to describe the polarized fragmentation functions from MC simulations with various numbers of the produced hadrons N_{Links}. We concluded that the order of these modulations is equal to N_{Links} when SF is kept fixed, though the numerical examinations of modulation orders larger than 4 were limited by the variable discretization and statistical fluctuation effects. This part of the modulation was attributed to the modulation of the remnant quark in the decay with the recoil to the hadron emissions. Thus, the modulations that rise by additional two orders with a unit increase in N_{Links} observed in Ref. [20] are attributed to the form of the model SF that depends on the components of transverse momentum of the remnant quarks.

Acknowledgements
This work was supported by the Australian Research Council through Grants No. FL0992247 (AWT), No. CE110001004 (CoEPP), and by the University of Adelaide. H.H.M. would like to thank the "National Center for Theoretical Sciences, Taipei, Taiwan, R.O.C." for their kind hospitality during his visit, when a part of this work has been carried out.

References

[1] Airapetian A *et al.* (HERMES Collaboration) 2005 *Phys.Rev.Lett.* **94** 012002
[2] Avakian H *et al.* (CLAS Collaboration) 2004 *Phys.Rev.* **D69** 112004
[3] Bradamante F (COMPASS Collaboration) 2011 (*Preprint* 1111.0869)
[4] Collins J C, Soper D E and Sterman G F 1985 *Nucl.Phys.* **B261** 104
[5] Collins J C, Soper D E and Sterman G F 1988 *Adv. Ser. Direct. High Energy Phys.* **5** 1–91
[6] Mulders P and Tangerman R 1996 *Nucl.Phys.* **B461** 197–237
[7] Amrath D, Bacchetta A and Metz A 2005 *Phys. Rev.* **D71** 114018
[8] Bacchetta A, Gamberg L P, Goldstein G R and Mukherjee A 2008 *Phys. Lett.* **B659** 234–243
[9] Gamberg L P, Goldstein G R and Oganessyan K A 2003 *Phys.Rev.* **D68** 051501
[10] Artru X 2010 (*Preprint* 1001.1061)
[11] Seidl R *et al.* (Belle Collaboration) 2006 *Phys.Rev.Lett.* **96** 232002
[12] Seidl R *et al.* (Belle Collaboration) 2008 *Phys.Rev.* **D78** 032011
[13] Aghasyan M, Avakian H, Rossi P, De Sanctis E, Hasch D *et al.* 2011 *Phys.Lett.* **B704** 397–402
[14] Aghasyan M and Avakian H (CLAS Collaboration) 2011 *J.Phys.Conf.Ser.* **295** 012063
[15] Ito T, Bentz W, Cloet I C, Thomas A W and Yazaki K 2009 *Phys. Rev.* **D80** 074008
[16] Matevosyan H H, Thomas A W and Bentz W 2011 *Phys.Rev.* **D83** 074003
[17] Matevosyan H H, Thomas A W and Bentz W 2011 *Phys. Rev.* **D83** 114010
[18] Matevosyan H H, Bentz W, Cloet I C and Thomas A W 2012 *Phys.Rev.* **D85** 014021
[19] Casey A, Matevosyan H H and Thomas A W 2012 *Phys. Rev. D* **85**(11) 114049
[20] Matevosyan H H, Thomas A W and Bentz W 2012 *Phys.Rev.* **D86** 034025
[21] Bacchetta A, D'Alesio U, Diehl M and Miller C 2004 *Phys.Rev.* **D70** 117504
[22] Lepage G P and Brodsky S J 1980 *Phys. Rev.* **D22** 2157
[23] Brodsky S J, Pauli H C and Pinsky S S 1998 *Phys.Rept.* **301** 299–486

Lattice Gauge Theory for Nuclear Physics

Kostas Orginos
Department of Physics, College of William and Mary, Williamsburg, VA 23187-8795, USA and Jefferson Laboratory, 12000 Jefferson Avenue, Newport News, VA 23606, USA.
E-mail: kostas@wm.edu

Abstract. Quantum Chromodynamcs (QCD) is now established as the theory of strong interactions. A plethora of hadronic physics phenomena can be explained and described by QCD. From the early days of QCD, it was clear that low energy phenomena require a non-perturbative approach. Lattice QCD is a non-perturbative formulation of QCD that is particularly suited for numerical calculations. Today, supercomputers have achieved performance capable of performing calculations that allow us to understand complex phenomena that arise from QCD. In this talk I will review the most recent results, relevant to nuclear physics. In particular, I will focus on results relevant to the structure and interactions of hadrons. Finally, I will comment on the opportunities opening up as we approach the era of exaflop computing.

1. Introduction

From the subatomic scales to the Universe itself, nuclear physics plays an important role in our understanding of the world around us. This is especially evident in this conference by the breadth of the presentations, which spanned across many different topics, ranging from the spectrum, structure and interaction of hadrons, to nuclear structure, and astrophysics. A common feature of all these topics is the enormous complexity of the problems we attempt to tackle. Often, numerical calculations with the aid of world class supercomputers are necessary to achieve the desired results.

At the core of nuclear physics, however, lies a very simple theory: Quantum Chromodynamics (QCD). The interactions of hadrons originate from this theory resulting the complex emergent phenomena we observe at low energies. Nuclear physics is a direct consequence of QCD and in principle all properties of atomic nuclei can be calculated from QCD. One of the major motivations of studying the interactions of hadrons in lattice QCD is precisely that: Understand how nuclear physics emerges from QCD, as well as compute from first principles some fundamental properties of nuclear matter. However, the large separation of the characteristic scales of QCD ($\sim$ 1 GeV) and the low nuclear scales of a few MeV or even KeV makes such calculations a daunting task. For this reason only recently lattice QCD calculations directly targeting nuclear physics have started to be undertaken.

From the very early times of lattice Quantum Chromodynamics, the importance of calculating properties of interacting hadrons had been recognized. However, the difficulty associated with such calculations on Euclidean space was also quickly recognized and solutions were proposed [1, 2, 3]. Pioneering work focusing on toy problems started also quite early [4] and since then significant work has been done in calculating elastic scattering phase shifts, investigating the existence of bound states and resonances, as well as studying the properties of multi-hadron systems.

In this talk I am reviewing recent results from lattice QCD calculations of properties of interacting hadrons. Emphasis is given to results relevant to nuclear physics.

2. Hadron Interactions in Euclidean Space

Computing the properties of an interacting hadronic system in Euclidean space requires a special treatment. In principle, one should calculate in Euclidean space appropriate correlation functions which after rotation to Minkowski they correspond to the desired scattering amplitude. However, analytic continuation cannot be simply performed using the Euclidean correlation functions on a discrete set of points. For this reason, it is desirable to extract the required information directly from Euclidean correlation functions. The simplest example where this is achieved is the calculation of the the spectrum of energy levels in a finite box, by observing the exponential decay of Euclidean correlation functions. Significant work has been performed nearly from the beginning of Lattice QCD formulation on understanding how properties of interacting hadrons can be calculated directly from Euclidean correlation functions. Pioneering work can be found in [1, 2, 3, 5]. The basic principle behind this work is the fact that all the information about the interactions of hadrons is encoded in the energy spectrum of the hadronic system in a box. Hence a calculation of this spectrum from Euclidean correlation functions can result in the calculation of the desired scattering amplitude. However, this is only possible where the analytic connection of the finite box spectrum to the desired scattering amplitude can be made.

2.1. Scattering and Bound States

The simplest example where the idea described above can be applied is the calculation of the elastic s-wave scattering phase shift of two hadrons. The formalism described here was developed in [1, 3]. In this case the scattering amplitude is described by a single phase (the scattering phase shift) $\delta(k)$.

$$\mathcal{A}(k) = \frac{4\pi}{\mu} \frac{1}{k \cot\delta(k) - ik} \tag{1}$$

where k is the magnitude of the center of mass scattering momentum and μ is the reduced mass of the two body system. The relation of this phase to the energy levels in a finite box can be easily worked out resulting for the case of equal mass particles of mass μ.

$$E_n = 2\sqrt{k_n^2 + \mu^2} \tag{2}$$

with E_n being the energy spectrum of the two body system in a finite box, and k_n the corresponding scattering momentum and,

$$k \cot\delta(k) = \frac{1}{\pi L} S(\frac{kL}{2\pi}), \qquad S(\eta) = \lim_{\Lambda\to\infty} \left[\sum_{\vec{n}}^{|\vec{n}|<\Lambda} \frac{1}{|\vec{n}|^2 - \eta^2} - 4\pi\Lambda \right] \tag{3}$$

where $\vec{n}$ is a 3-vector with integer components. Thus in Euclidean space one calculates the energy levels E_n using normal spectroscopic methods and extracts the scattering phase shift using the above formula. At small enough scattering momentum, where the effective range expansion is valid, the scattering length $\mathtt{a}$ can be identified through

$$k \cot\delta(k) = \frac{1}{\mathtt{a}} + \frac{1}{2}\mathtt{r}^2 k^2 + \cdots \tag{4}$$

with $\mathtt{r}$ being the effective range of the interaction. Expanding at small k one obtains the shift ΔE_0 of the lowest two particle energy level in a box of size L from the non-interacting two particle lowest energy level in the same box:

$$\Delta E_0 = -\frac{4\pi \mathtt{a}}{mL^3}\left[1 + c_1\frac{\mathtt{a}}{L} + c_1\frac{\mathtt{a}^2}{L^2}\right] + \cdots \tag{5}$$

This formula has been extended to more than two bosons in a box in [6, 7].

Generalization of to higher partial waves was also done in [3]. In this case complications arise due to breaking of rotational symmetry causing partial waves to mix. For s-wave scattering (angular momentum $l = 0$) the lowest wave that mixes in is $l = 4$ and its effects can be ignored in most cases. In general, these mixing effects cannot be ignored, and the scattering phase shifts δ_l can be determined as the zeros of a characteristic polynomial defined by

$$\det\left[e^{2i\delta} - \mathbf{U}\right] = 0 \tag{6}$$

with $\mathbf{U}$ an appropriately constructed matrix for each symmetry sector of the cubic group, and δ a diagonal matrix with entries the phase shifts of each partial wave contributing to the cubic symmetry sector at hand. In general, there are infinitely many partial waves contributing to each cubic symmetry sector, however, in practice the matrix is truncated to a small number of partial waves. For details of this formalism see [3] and the more recent [8].

Generalizations of the above formalism to non-center of mass frames can be found in [9, 10, 11, 12]. In addition, the extension to non-cubic boxes can be found at [13]. Finally, extensions to multi-channel scattering have been worked out in [14].

When two particles interact attractively, the possibility of the formation of a bound state arises. In this case the energy splitting ΔE_0 is negative resulting in an imaginary scattering momentum $k = i\kappa$,

$$\Delta E_0 = 2\sqrt{k^2 + \mu^2} - 2\mu \approx \frac{k^2}{2\mu} = -\frac{\kappa^2}{2\mu}\,. \tag{7}$$

It has been shown [15] that in the case of bound states the volume dependence of ΔE_0 is exponential rather than the power law found for scattering states. Recently, in [16] the finite volume energy shift $\delta(\Delta E_0)$ of the bound state energy ΔE_0 for angular momentum $l > 0$ was computed:

$$\delta(\Delta E_0) \sim \alpha(\frac{1}{\kappa L})\frac{e^{-\kappa L}}{\mu L} + \mathcal{O}(e^{-\sqrt{2}\kappa L})\,. \tag{8}$$

where the function $\alpha(x)$ depends on the partial wave as well as the lattice symmetry sector this partial wave is embedded. A notable result of this work is the fact that while s-wave bound states become more deeply bound in finite volume, p-wave bound states are less bound in small volume.

3. Signal to Noise ratio for baryons

Following the formalism described in the previous section the calculation of elastic scattering phase shifts from lattice QCD is reduced to the calculation of the energy spectrum of two bodies in a finite box. This is in principle straight forward to do. Methods for extracting not only the ground state but several low lying energy levels exist [4, 17]. In the case of baryons, however, the extraction of the spectrum of the multi-body system in a box is quite difficult. Here the problem is two fold. First, the energy gaps in the two (or multi) baryon systems are small requiring large Euclidean time to project out the ground state (or a set of few low-lying states). The cost of quark contractions is such that it makes it very difficult to build a sufficiently large basis of

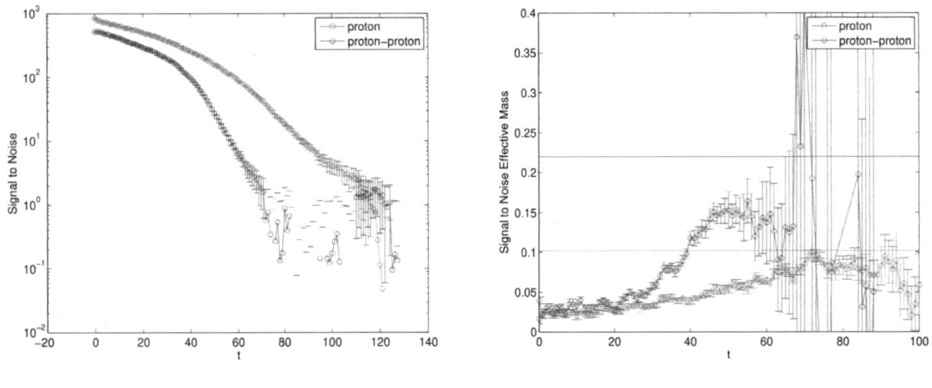

Figure 1. Signal to noise ratio for a single proton and a two proton system. Left, the signal to noise ratio. Right, the logarithmic derivative of the signal to noise ratio.

operators (as is now done for single baryon spectroscopy [18, 19, 20]). Secondly, the statistical noise in the correlators due to Monte Carlo sampling grows exponentially with Euclidean time.

As Lepage showed [21], the signal to noise ratio $\mathcal{R}$ of hadronic correlation functions drops exponentially with Euclidean time with an exponent related to the gap between the mass of the pion and the mass of the the lowest hadronic state contributing to the correlator. In particular, for a system with A nucleons we have,

$$\mathcal{R} \approx e^{-A(m_N - \frac{3}{2}m_\pi)t} \tag{9}$$

where m_N is the nucleon mass and m_π the pion mass and assuming weak interaction between nucleons. Fortunately, as it was shown in [22, 23] this asymptotic bound does not set in at early times leaving some region where reasonable accuracy can be achieved with sufficiently high statistics. In Fig. 1, I show the drop of the signal to noise ratio for the case of a single proton as well as the case of a two proton system in the singlet spin state. In both cases after a region of a mild drop the exponential rapid decay sets in. In the case of the proton, we can follow the signal at sufficiently large time to observe the saturation of the Lepage bound, as it is seen in the right panel of Fig. 1. For the two proton system the correlator becomes too noisy as the Lepage bound is approached, never quite saturating the bound. The data I present here are from NPLQCD collaboration for an ensemble with dynamical anisotropic clover fermions with a 390 MeV pion mass with anisotropy of about 3.5 and volume of $32^3 \times 256$ and a spacial lattice spacing of about 0.125fm.

The issue of the rapid growth of the noise as well as the dense two body spectrum in a box present a major challenge that needs to be overcome, for reaching precision calculations of properties of systems with interacting baryons.

4. Baryon-Baryon Interactions

Pioneering calculations of baryon-baryon scattering lengths have been performed many years ago with the work of Fukugita et al. [24] in the quenched approximation (where quark vacuum polarization effects are ignored). It took ten years for the first calculation for quark vacuum polarization effects to be included. This pioneering work was performed by NPLQCD collaboration [25]. The state of nucleon-nucleon elastic scattering length calculations is summarized in a recent review by Beane *et al* [23]. In addition, HALQCD has been reporting

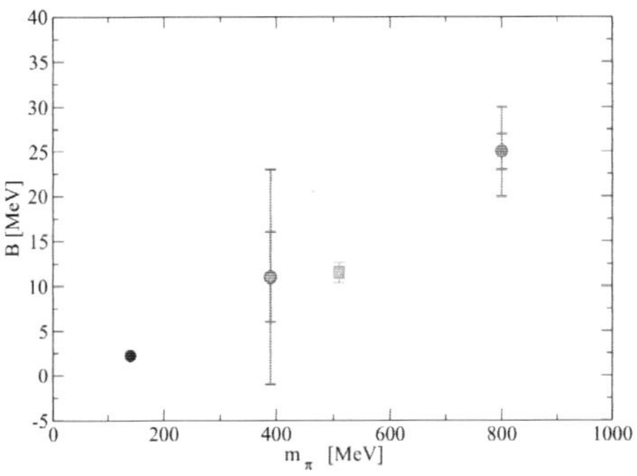

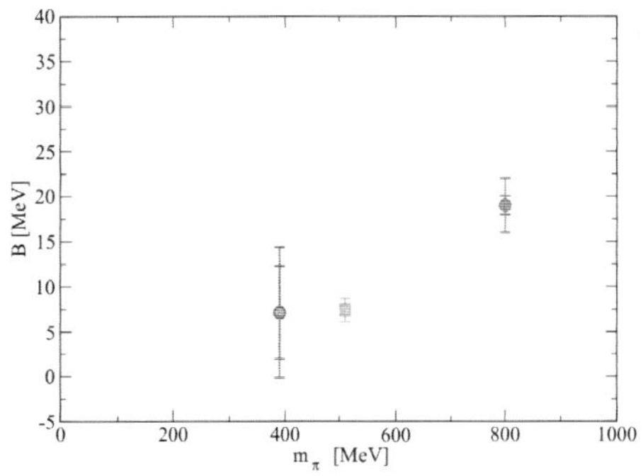

Figure 2. The binding energies of the deuteron (left) and the di-neutron (right) computed by NPLQCD (390 and 800 MeV points) and Yamazaki *et al* (510 MeV point).

results on two-baryon scattering phase shifts in the recent years. Most recently in Ref. [26] results on nucleon-nucleon as well as hyper-nucleon and hyper-hyperon elastic scattering phase shifts were presented.

The issues related to the noise and the complexity of the spectrum of two baryon systems in a box outlined in Sec. 3 are the prime reasons for the sparsity of results in the baryon sector in contrast to the meson sector. As a result, systematic effects due to excited state contamination, heavy quark masses, and lattice spacing effects, are far from being well under control. The fine tuned nature of nuclear physics close to the physical point might be one of the reasons of the clear disagreement between lattice and experimental results. On the other hand, these pioneering results represent the first solid step in the direction of obtaining reliable calculations of these important to Nuclear physics quantities. Clearly, in order to make progress improved methodology, as well as availability of significant computational resources are required.

Recently calculations of the binding energy of the deuteron were performed in the quenched approximation [27]. In the same calculation, a bound di-neutron was observed. The key features of these calculations were the big volumes (up to 12fm), the heavy pion masses (800MeV), and the use of a variational method with 2 operators with wavefunctions designed to have large overlap with the ground state of the system. The combination of these choices makes the noise problem tractable. Furthermore, the same authors preformed similar calculations of the binding energy of the helium nuclei (^3He and ^4He) in quenched approximation [28]. In work performed this year, Yamazaki *et al* [29] also presented results with dynamical quarks (allowing for vacuum polarization effects) for both the deuteron the di-neutron and the helium nuclei (^3He and ^4He). Their result for the ^4He nucleus is in remarkable agreement with experiment, in contrast with the rest of the binding energies which are found to be larger than those in nature. However, these calculations where performed at unphysical heavy quark masses (pions with mass about 500 MeV). Significant further work is required in order to achieve results that are free of systematic errors currently not controlled.

The NPLQCD collaboration has also performed calculations at the flavor SU(3) symmetric point as well as at a point with 390 MeV pion mass and broken SU(3) flavor symmetry. The two calculations, were done with different quark actions (finer grid in the Euclidean time direction). The lighter pion mass point calculations presented in Ref. [30] were performed using an anisotropic quark action that allowed better control of the fitting systematics for extracting binding energies. However, the cost of these calculations was such that statistical errors were substantial. In Ref. [31] the NPLQCD collaboration presented results with very high statistics using an isotropic quark action that allowed better control of the statistical errors. The results with dynamical quarks are summarized in Fig. 2. In this figure it is clear that the lattice calculations by the two different collaborations are in agreement. In addition a general trend of decreasing binding energies with pion mass is observed. This trend is in the right direction for obtaining the experimental result for the deuteron binding energy at the physical pion mass. However, the calculation of the deuteron binding energy at the physical pion mass still remains a daunting task for lattice QCD.

4.1. Strange bound states: H-dibaryon

Two baryon bound states containing strange quarks were also studied in lattice QCD during recent years. Most notable was the work on the h-dibaryon by HALQCD and NPLQCD. The h-dibaryon as a possible six quark state (two of each of the three light flavors in the quark model picture) was proposed by Jaffe in 1977 [32]. This is a strangeness -2, baryon number 2, and angular momentum 0 state. Jaffe's argument was based on the MIT bag model and the perturbative attractive color-spin interactions that form di-quarks in his model. Since then there have been several experimental searches for the h-dibaryon. All this effort is comprehensively reviewed in [33]. Most recently, at RIHC [34] using some input from models, the region between 0 and 95 MeV was excluded for the binding energy of the h-dibaryon. In addition, the possibility of a near threshold resonance is still not excluded by experimental results [35].

Theoretically there have been significant activity attempting to understand the existence of the h-dibaryon within various models. A review of all this work can be found in [36, 33]. Model predictions vary in a wide range, from deeply bound h-dibaryon to a resonance significantly above threshold. In lattice QCD, efforts to address the possibility of existence of the h-dibaryon started early on [37, 38, 39, 40, 41], in the quenched approximation (i.e. ignoring quark vacuum polarization effects).

Recently the first unquenched calculations appeared. NPLQCD collaboration used anisotropic clover improved lattices with two light (up,down) quarks and one strange quark at a single pion mass of 390MeV and a single spacial lattice spacing of approximately 0.125fm. Details of these ensembles can be found in [42]. The calculation was performed on four volumes ranging between 2fm and 4fm and the infinite volume binding energy was calculated by extrapolation using the formalism outlined in Sec. 2. This calculation provides evidence that at 390MeV pion mass the h-dibaryon is bound with binding energy of $B = 13.2 \pm 1.8 \pm 4.0$ MeV where the first error is statistical and the second systematic. In addition to this number, NPLQCD has preliminary results at a lighter pion mass (220MeV). However the precision obtained at this mass is not sufficient to determine the sign of the binding energy. NPLQCD results as well as more details on their calculation can be found in [43, 44].

HALQCD, using very different methodology, also presented results on the h-dibaryon binding energy [45, 46, 26]. Their work is at flavor SU(3) symmetric points and pion masses that range between 469MeV and 1.2 GeV. They also found evidence of a bound h-dibaryon with binding energies ranging from 26MeV up to 37MeV.

The results for the h-dibaryon binding energy discussed here are presented in Fig. 3. As it is evident, the binding energy is decreasing as the pion mass approaches the physical point. In addition, at lighter pion masses the calculations become harder providing less accurate results. For these reasons, although a bound h-dibaryon is observed at heavier pion masses, it is not clear what its fate is at the physical point. Apart from obtaining precision close to the physical point, lattice calculations will have to address and control other remaining systematic errors such as lattice spacing effects, isospin breaking and electromagnetic effects. Given the fact that the h-dibaryon may be a near threshold state (bound or resonance) all these effects have to be carefully addressed before a final result is obtained from lattice QCD. It is encouraging though, that during the last year significant progress has been made. In addition to lattice calculations,

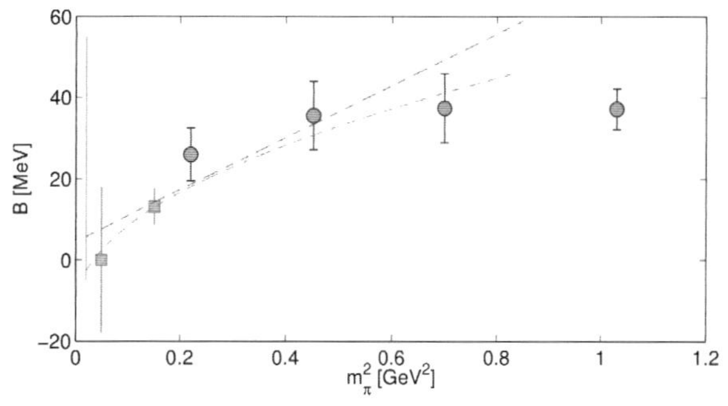

Figure 3. Summary of lattice results for the h-dibaryon binding energy. The squares are results from NPLQCD (Nuclear Physics Lattice QCD) collaboration [43, 44] and the circles are results from HAL QCD (in [45] and in this conference [46]). The solid vertical line marks the physical point. The two dashed lines are plotted to guide the eye (the blue is linear in m_π^2 while the red is linear in m_π).

in the literature recently appeared work that addresses the quark mass dependence of the binding energy of the h-dibaryon [47] using chiral symmetry constraints. In this case, the binding energy also seems to decrease with the pion mass.

5. Hypernuclear physics

The NPLQCD collaboration, using very high statistics and volumes up to 7 fm, recently calculated the spectrum of nuclei and hypernuclei up to atomic number A=4 [31], in lattice QCD. These calculations were performed at the flavor SU(3) symmetric point with pion mass of 800 MeV and a single lattice spacing of approximately $a = 0.145$ fm. The heavy pion masses allowed for good control of the statistical errors. The interpolating fields (quark and hadron level wave-functions) that were used in this work were constructed using an algebraic program. This program constructs both the wave-functions with the relevant quantum numbers, as well as finds the wave-functions with a tractably small number of terms. It is a well known problem that many body wave-functions contain a large number of terms growing exponentially with atomic number. Taking advantage of the underlying quark structure and the Fermi exclusion principle it turns out that particularly simple wave functions can be constructed as shown in [48] [1]. Although these first results look very encouraging, controlling all systematic errors, such as control of excited state contamination, continuum extrapolation, physical up and down quark masses, and large volume extrapolations, will require significant computational resources as well as new clever algorithms to deal with the exponential growth of statistical errors as the physical pion mass is approached.

6. Conclusions

There has been significant progress in the last few years in studies of multi-baryon systems in lattice QCD. The ultimate goal being understanding how nuclear physics emerges from QCD. The challenges in tackling this problem are quite difficult, however, the emergence of new methods, and the expected significant increase in computer cycles in the exaflop era, makes me optimistic that we will continue on this pace of steady progress in the years to come.

Acknowledgments

I would like to thank the organizers for inviting me to take part at this conference in honor of Jerry Draayer's 70th birthday. This work was supported in part by the DOE grants DE-AC05-06OR23177 (JSA) and DE-FG02-04ER41302.

[1] Similar methodology was also presented in [49]

References

[1] Luscher M 1986 *Commun.Math.Phys.* **105** 153–188
[2] Maiani L and Testa M 1990 *Phys.Lett.* **B245** 585–590
[3] Luscher M 1991 *Nucl.Phys.* **B354** 531–578
[4] Luscher M and Wolff U 1990 *Nucl.Phys.* **B339** 222–252
[5] Luscher M 1991 *Nucl.Phys.* **B364** 237–254
[6] Beane S R, Detmold W and Savage M J 2007 *Phys.Rev.* **D76** 074507 (*Preprint* 0707.1670)
[7] Tan S 2008 *Phys. Rev. A* **78**(1) 013636 URL http://link.aps.org/doi/10.1103/PhysRevA.78.013636
[8] Luu T and Savage M J 2011 *Phys.Rev.* **D83** 114508 (*Preprint* 1101.3347)
[9] Gottlieb S A and Rummukainen K 1996 *Nucl.Phys.Proc.Suppl.* **47** 819–822 (*Preprint* hep-lat/9509088)
[10] Christ N H, Kim C and Yamazaki T 2005 *Phys.Rev.* **D72** 114506 (*Preprint* hep-lat/0507009)
[11] Kim C, Sachrajda C and Sharpe S R 2005 *Nucl.Phys.* **B727** 218–243 (*Preprint* hep-lat/0507006)
[12] Feng X, Jansen K and Renner D B (ETM Collaboration) 2010 *PoS* **LAT2010** 104 (*Preprint* 1104.0058)
[13] Miao C, Du X i, Meng G w and Liu C 2004 *Phys.Lett.* **B595** 400–407 (*Preprint* hep-lat/0403028)
[14] He S, Feng X and Liu C 2005 *JHEP* **0507** 011 (*Preprint* hep-lat/0504019)
[15] Luscher M 1986 *Commun.Math.Phys.* **104** 177
[16] Koenig S, Lee D and Hammer H W 2011 *Phys.Rev.Lett.* **107** 112001 (*Preprint* 1103.4468)
[17] Blossier B, Della Morte M, von Hippel G, Mendes T and Sommer R 2009 *JHEP* **0904** 094 (*Preprint* 0902.1265)
[18] Edwards R G, Dudek J J, Richards D G and Wallace S J 2011 *Phys.Rev.* **D84** 074508 (*Preprint* 1104.5152)
[19] Morningstar C, Bell A, Bulava J, Foley J, Juge K *et al.* 2011 *AIP Conf.Proc.* **1388** 34–40 (*Preprint* 1103.2783)
[20] Morningstar C, Bell A, Chen C, Lenkner D, Wong C *et al.* 2010 *PoS* **ICHEP2010** 369 (*Preprint* 1011.6573)
[21] Lepage P G 1989 *TASI'89 ASI 1989* 97–120
[22] Beane S R *et al.* (NPLQCD Collaboration) 2010 *Phys.Rev.* **D81** 054505 (*Preprint* 0912.4243)
[23] Beane S, Detmold W, Orginos K and Savage M 2011 *Prog.Part.Nucl.Phys.* **66** 1–40 (*Preprint* 1004.2935)
[24] Fukugita M, Kuramashi Y, Okawa M, Mino H and Ukawa A 1995 *Phys.Rev.* **D52** 3003–3023 (*Preprint* hep-lat/9501024)
[25] Beane S, Bedaque P, Orginos K and Savage M 2006 *Phys.Rev.Lett.* **97** 012001 (*Preprint* hep-lat/0602010)
[26] Inoue T *et al.* (HAL QCD Collaboration) 2012 *Nucl.Phys.* **A881** 28–43 (*Preprint* 1112.5926)
[27] Yamazaki T, Kuramashi Y and Ukawa A 2011 *Phys.Rev.* **D84** 054506 (*Preprint* 1105.1418)
[28] Yamazaki T, Kuramashi Y, Ukawa A and Collaboration f t P C 2010 *Phys.Rev.* **D81** 111504 (*Preprint* 0912.1383)
[29] Yamazaki T, Ishikawa K i, Kuramashi Y and Ukawa A 2012 (*Preprint* 1207.4277)
[30] Beane S *et al.* (NPLQCD Collaboration) 2012 *Phys.Rev.* **D85** 054511 (*Preprint* 1109.2889)
[31] Beane S, Chang E, Cohen S, Detmold W, Lin H *et al.* 2012 (*Preprint* 1206.5219)
[32] Jaffe R L 1977 *Phys. Rev. Lett.* **38** 195–198
[33] Sakai T, Shimizu K and Yazaki K 2000 *Prog.Theor.Phys.Suppl.* **137** 121–145 (*Preprint* nucl-th/9912063)
[34] Trattner A L 2006 Ph.D. thesis UMI-32-54109
[35] Yoon C J *et al.* 2007 *Phys. Rev.* **C75** 022201
[36] Bashinsky S and Jaffe R 1997 *Nucl.Phys.* **A625** 167–191 (*Preprint* hep-ph/9705407)
[37] Mackenzie P B and Thacker H 1985 *Phys.Rev.Lett.* **55** 2539
[38] Pochinsky A, Negele J W and Scarlet B 1999 *Nucl.Phys.Proc.Suppl.* **73** 255–257 (*Preprint* hep-lat/9809077)
[39] Wetzorke I, Karsch F and Laermann E 2000 *Nucl.Phys.Proc.Suppl.* **83** 218–220 (*Preprint* hep-lat/9909037)
[40] Wetzorke I and Karsch F 2003 *Nucl.Phys.Proc.Suppl.* **119** 278–280 (*Preprint* hep-lat/0208029)
[41] Iwasaki Y, Yoshie T and Tsuboi Y 1988 *Phys.Rev.Lett.* **60** 1371–1374
[42] Lin H W *et al.* (Hadron Spectrum Collaboration) 2009 *Phys.Rev.* **D79** 034502 (*Preprint* 0810.3588)
[43] Beane S *et al.* (NPLQCD Collaboration) 2011 *Phys.Rev.Lett.* **106** 162001 (*Preprint* 1012.3812)
[44] Beane S, Chang E, Detmold W, Joo B, Lin H *et al.* 2011 *Mod.Phys.Lett.* **A26** 2587 (*Preprint* 1103.2821)
[45] Inoue T *et al.* (HAL QCD Collaboration) 2011 *Phys.Rev.Lett.* **106** 162002 (*Preprint* 1012.5928)
[46] Inoue T (for HAL QCD Collaboration) 2011 (*Preprint* 1111.5098)
[47] Shanahan P, Thomas A and Young R 2011 *Phys.Rev.Lett.* **107** 092004 (*Preprint* 1106.2851)
[48] Detmold W and Orginos K 2012 (*Preprint* 1207.1452)
[49] Doi T and Endres M G 2012 (*Preprint* 1205.0585)

The Excited-state Spectrum of QCD through Lattice Gauge Theory Calculations

David Richards
for the Hadron Spectrum Calculation
Jefferson Laboratory, 12000 Jefferson Avenue, Newport News, VA 23606, USA
E-mail: dgr@jlab.org

Abstract. I describe the computation of the spectrum of Quantum Chromodynamics through first principles calculations using lattice gauge theory. I begin outlining briefly the formulation of QCD on a lattice, and then describe some of the challenges in computing the exciting state spectrum of the theory. I then present recent results for the spectrum for mesons containing both the light and heavier charm quarks, and for baryons containing the light quarks. A particular emphasis will be laid on the rôle played by gluons within the spectrum. I conclude by some of the challenges that remain, notably those due to strong decays, and the prospect for future calculations.

1. Introduction

A quantitative description of the spectrum of QCD goes to the core of our understanding of the theory, and is therefore the focus of intense experimental and theoretical effort. The experimental investigation of the spectrum has undergone a resurgence: the observation of new states in the Charmonium system at Belle, BaBar, CDF, D0 and CLEOc with future experiments at BESIII and PANDA, the search for the so-called missing baryon resonances of the quark model at CLAS at JLab@6GeV, and the flagship search for exotic mesons in the GlueX experiment at JLab@12GeV. These experimental initiatives are matched by increasingly vigorous theoretical efforts aimed at obtaining the QCD spectrum from first principles, where lattice QCD enables many quantities in QCD to be *solved* rather than modelled.

The potential of lattice QCD to enhance our understanding of hadronic and nuclear physics was recognized at Jefferson Laboratory when Nathan Isgur established a lattice initiative at the laboratory. Over the last decade, that initiative has become a core activity of the laboratory, comprising the three major components of a successful lattice effort. Firstly, the development of the computational and software infrastructure by both the Theory Center and the High-Performance Computing Group under the auspices of the Department of Energy's SciDAC (*Scientific Discovery through Advanced Computing*) Initiative. Secondly, the development and operation of hardware, based on commodity clusters and latterly heterogeneous architectures, optimized for lattice QCD, and now serving as a national facility. Finally, a rigorous research effort dedicated to addressing the key scientific questions posed by the Jefferson Laboratory program. In this talk, I will focus on the scientific impact of the lattice program, and in particular on our understanding of the excited-state spectrum of both mesons and of baryons.

2. The Excited State Spectrum in Lattice QCD

Lattice gauge calculations solve QCD on a four-dimensional lattice, or grid, of points in Euclidean space. The quarks reside on the points of the grid, whilst the gluons are associated with the links joining those points. Lattice calculations proceed through a Monte Carlo method, in which ensembles of gauge configurations are generated with a probability distribution prescribed by the Euclidean QCD action.

The low-lying spectrum of QCD has long been an important benchmark for lattice QCD calculations, encapsulating both our ability to confront first-principles calculations with well-established observational results, and of our understanding of how QCD changes as the masses of the quarks are varied. A comprehensive picture of resonances requires that we go beyond a knowledge of the ground state mass in each channel, and obtain the masses of the lowest states of a given quantum number. This we can accomplish through the use of the variational method [1, 2]. Rather than measuring a single correlator $C(t)$, we determine a matrix of correlators

$$C_{ij}(t) = \sum_{\vec{x}} \sum_{\vec{y}} \langle O_i(\vec{x}, t) O_j^\dagger(\vec{y}, 0) \rangle,$$

where $\{O_i; i = 1, \ldots, N\}$ are a basis of interpolating operators with given quantum numbers. We then solve the generalized eigenvalue equation

$$C(t)u = \lambda(t, t_0) C(t_0) u$$

to obtain a set of real (ordered) eigenvalues $\lambda_n(t, t_0)$, where $\lambda_0 \geq \lambda_1 \geq \cdots \geq \lambda_{N-1}$. At large Euclidean times, these eigenvalues then delineate between the different masses

$$\lambda_n(t, t_0) \longrightarrow e^{-M_n(t-t_0)} + O(e^{-\Delta M_n(t-t_0)}).$$

where $\Delta M_n = \min\{| M_n - M_i |: i \neq n\}$. The eigenvectors u are orthogonal with metric $C(t_0)$, and a knowledge of the eigenvectors can yield information about the partonic structure of the states.

The (hyper-) cubic lattice does not possess the full rotational symmetry of the continuum. Thus in a lattice calculation, states at rest are classified not according to the spin (J, J_z), but rather according to the irreducible representations (irreps) of the symmetry group of the cube; for states of higher spin, the different continuum degrees of freedom are distributed across several lattice irreps. This reduced symmetry has proved a formidable barrier to the interpretation of lattice calculations by making the assignment of the continuum spins to the energy eigenstates problematic. In our work, the choice of interpolating operators is crucial to the assignment of spins both for mesons and for baryons. For the excited meson spectrum, our starting point is a set of interpolating operators

$$\mathcal{O} = \sum_{\vec{x}} \bar{\psi}(\vec{x}) \Gamma_i \overleftrightarrow{D}_i \overleftrightarrow{D}_j \ldots \psi(\vec{x}) \tag{1}$$

where $\overleftrightarrow{D}$ is a discretized covariant derivative, and $\psi(\bar{\psi})$ a quark (antiquark) field. From this set, we can construct a basis of interpolating operators $\mathcal{O}^{J,M}$ that in the continuum have definite spin,

$$\langle 0 \mid \mathcal{O}^{JM} \mid J', M' \rangle = Z^J \delta_{J,J'} \delta_{M,M'}.$$

It is these operators that are now subduced to the lattice irreducible representations:

$$\mathcal{O}^{[J]}_{\Lambda,\lambda} = \sum_M S^{J,M}_{\Lambda,\lambda} \mathcal{O}^{J,M},$$

where Λ, λ denote the lattice irrep and its row respectively, and $S^{J,M}_{\Lambda,\lambda}$ are calculable coefficients. Whilst the resulting operators $\mathcal{O}^{[J]}_{\Lambda,\lambda}$ interpolate to any state whose spin is subduced to the irrep Λ, we find that in practice the operator interpolates strongly to states of spin J, and that the resulting overlaps Z^J are common across the different subductions; the method is described in detail in ref. [3]. Thus the operators $\mathcal{O}^{[J]}_{\Lambda,\lambda}$ retain a strong memory of their continuum antecedents, and this property enables us to identify the spins of the states for the effectively single-hadron states we have observed in the following calculation.

Our ability to calculate correlation functions efficiently for the large basis of operators that this methodology entails has been greatly advanced through the development of a new method, "distillation" [4], that enables all the elements of the correlation matrix to be computed, with time-sliced sums at both source and sink.

The final ingredient in the spectroscopy effort has been the generation of so-called anisotropic lattices, with finer temporal than spatial discretisation, enabling the fall-off of the principle correlators introduced above to be discerned at short temporal separations. This has been accomplished through the generation of lattices with two flavors of fully dynamical light and a dynamical strange quark, using the clover fermion action. An important milestone was the tuning of the parameters of the action, beginning with the three-flavor theory [5], and the development of a means of specifying the values of the quark masses, in particular those for the light u/d and s quarks, in a way that enables lattice calculations to be extrapolated to the physical values of these masses [6].

3. Excited Meson Spectrum

The new Hall D of the JLab 12GeV upgrade centers on the study of meson states produced in photoproduction reactions in the GlueX experiment. Photoproduction has been proposed, within QCD-motivated models, as a favorable method for the production of exotic mesons, that is those mesons having J^{PC} outside the set allowed to a fermion-antifermion pair. The "hybrid" hypothesis is that these quantum numbers arise from an excited gluonic field, in addition to a quark-antiquark pair.

The availability of the large ensembles of $N_f = 2 + 1$ anisotropic clover lattices, together with the efficient operator-construction admitted by "distillation", has enabled us to compute the isovector meson spectrum for states composed of the light and strange quarks [7, 3], at pion masses down to around 400 MeV. Masses of highly excited states have been calculated, with the quantum numbers reliably delineated, including for the first time in a lattice calculation states of spin 4. Figure 1 summarizes the results on exotic states and compares those results with previous lattice QCD results from Refs. [8, 9, 10, 11, 12, 13]. Further, we see evidence for "hybrid" states, that is those in which the gluons play a vital structural rôle but with non-exotic quantum numbers, as we show below.

3.1. *Isoscalar meson spectrum*

The calculation of the isoscalar meson spectrum, and the study of the hidden-flavor composition of states, is a challenging undertaking, since it requires the computationally demanding task of including quark-annihilation contributions to hadronic correlation functions. The "distillation" method provides a computationally highly efficient way of including these contributions, by enabling the most demanding part of the calculation, that of the so-called perambulators (the parallel transporters of the method) to be computed with using GPUs, with the remaining parts of the calculation performed on CPUs. This feature was exploited to perform a high-precision calculation of the isoscalar spectrum [14], revealing not only the masses but also the hidden flavor-mixing angles for the states.

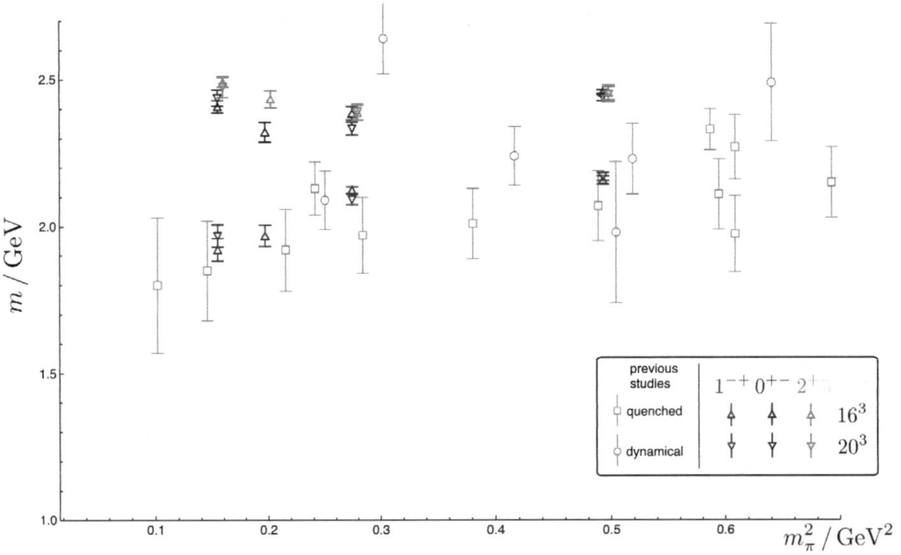

Figure 1. Summary of extracted isovector exotic states at two volumes, and at four quark masses. For comparison 1^{-+} results from Refs. [8, 9, 10, 11, 12, 13] are also plotted.

We employ as our basis of operators

$$\begin{aligned} \mathcal{O}^\ell_A(t) &= \tfrac{1}{\sqrt{2}}\left(\bar{u}\mathbf{\Gamma}^A_t u + \bar{d}\mathbf{\Gamma}^A_t d\right) \\ \mathcal{O}^s_A(t) &= \bar{s}\mathbf{\Gamma}^A_t s, \end{aligned}$$

where u, d, and s are the up, down and strange quark fields, and the $\mathbf{\Gamma}^A_t$ are operators, of the form of eqn. (1), acting in color, Dirac spin, and also in *space* [4], at a time slice, t; in the following we have $SU(2)$ flavor symmetry and therefore use l to denote the light (u, d) quarks. The results for the isoscalar spectrum, together with the isovector results presented above, are summarized in Figure 2. For each value of J^{PC}, we find that the isoscalar energy eigenvalues occur in pairs. For a particular combination of creation and annihilation operators, we are able to diagonalize the resulting two-by-two correlation matrix in flavor space, thereby revealing the hidden flavor-mixing angles, shown as the fraction of black (l) and green (s) in the bars of the figure. For "conventional" mesons, we found near ideal mixing other than in the pseudoscalar and axial states. We also computed, for the first time in a lattice calculation, isoscalar "exotic" states; these appear at a mass scale comparable to their isovector cousins. Though these calculations are as yet incomplete, their implication for the Jefferson Laboratory program is profound; they suggest the presence of numerous states of exotic quantum numbers in the energy range accessible to the GlueX experiment.

3.2. Charmonium Spectroscopy

The emergence of new experimental states in the charmonium system has been an important driver for spectroscopy, and is an especially attractive area for lattice studies where calculations of the spectrum can be performed with high statistical precision. It is thus a particularly fruitful area in which to explore the phenomenology of the spectrum. Figure 3 shows the charmonium spectrum for channels, both exotic and non-exotic, in which we are able to identify "hybrid"

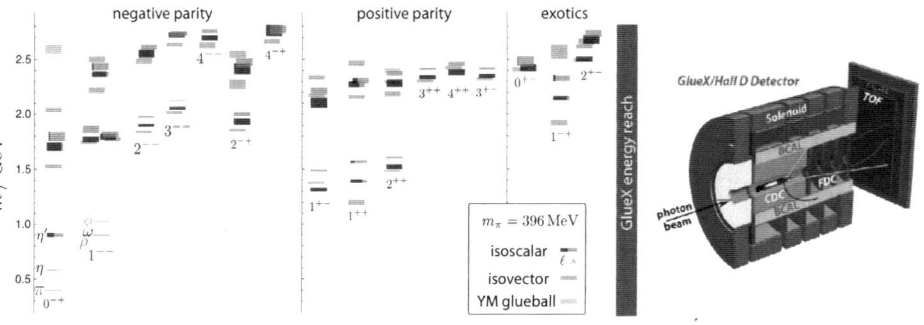

Figure 2. The left-hand panel shows the isoscalar meson spectrum labeled by J^{PC}, where the box height denotes the statistical uncertainty. The light-strange content of each state $(\cos^2\alpha, \sin^2\alpha)$ is given by the fraction of (black, green) and the mixing angle is also shown. Grey boxes indicate the positions of isovector meson states extracted on the same lattice (taken from [3]). Pink boxes indicate the position of quenched glueballs [15]. The right-hand panel shows the GlueX experiment, together with an indication of the expected energy reach.

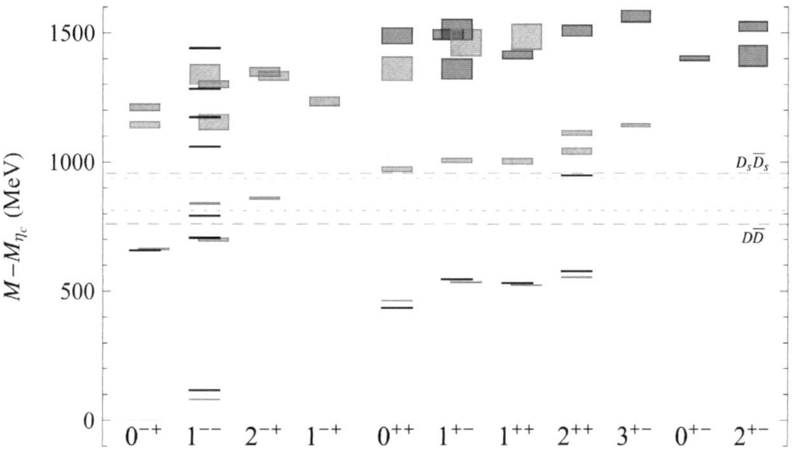

Figure 3. The charmonium spectrum for those channels in which "hybrid" mesons are identified, obtained on a 24^3 spatial volume for a pion mass around 400 MeV. The red and blue boxes are suggested as the members of the lowest-lying and first-excited hybrid multiplets, respectively. The black lines are experimental values, whilst the dashed lines are the lowest non-interacting $D\bar{D}$ and $D_s\bar{D}_s$ levels, obtained on a 16^3 ensemble (short green dashes), and using the experimental masses (long grey dashes).

states [16]. Notably, a lightest and a first-excited hybrid supermultiplet [17] in which the quark-antiquark pairs are coupled to a chromomagnetic gluonic excitation, are identified, indicated by the red and blue bars respectively.

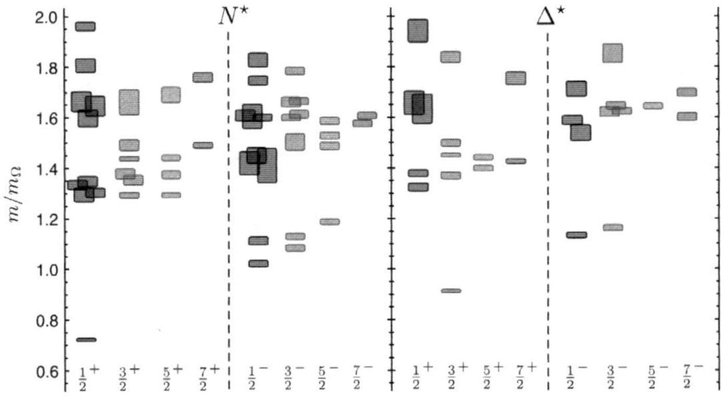

Figure 4. Spin-identified spectrum of Nucleons and Deltas from the lattices at $m_\pi = 396$ MeV, in units of the calculated Ω mass.

4. Excited Baryon Spectrum

The recipe employed to determine the excited meson spectrum can be applied to determine the spectrum of baryons. A basis of continuum interpolating operators is constructed, including up to two derivatives, and which respect classical symmetries, which is then subduced to the appropriate lattice irreps. The overlaps of the interpolating operators with the extracted states is once again key to determining the spin of the states, exploiting the remarkable degree of rotational symmetry observed at the hadronic scale [18]. The nucleon and Δ excited-nucleon spectrum at the lightest of our pion masses is shown in Figure 4, and exhibits a counting of levels consistent with the non-relativistic qqq constituent quark model.

In contrast to mesons, there are no "exotic" quantum numbers for baryons that *demand* a richer structure than that of three quarks with relative orbital angular momentum. As noted in reference [18], the operators associated with an excited gluon field, and which would vanish for a trivial gauge configuration, were not included in the basis. These operators were included in a subsequent reanalysis, shown in Figure 5, for the positive-parity N and Δ spectrum [19], and revealed the presence of "hybrid" baryons dominated by operators with an excited gluon field. Remarkably, the mechanism giving rise to such gluonic excitations appears common to both mesons and baryons, as shown in the right-hand panel of Figure 5, where the excitation energy of both "hybrid" mesons and "hybrid" baryons is shown.

5. Summary

The amalgam of theoretical, algorithmic and computational advances has enabled increasingly sophisticated studies of the spectrum of QCD. These are already having a notable impact, suggesting the important role played by gluonic degrees of freedom in describing the spectrum, the anticipated reach of GlueX to see states of exotic quantum numbers, and the rich spectrum of baryon resonances. The calculations outlined above are incomplete: all but the lightest states are resonances whose study requires the calculation of the momentum-dependent phase shifts. Lattice QCD provides a powerful means of relating infinite-momentum elastic phase information to energy shifts at finite volume [20], and the algorithmic advances discussed here have already been exploited to yield precise calculations of the momentum-dependent phase shifts in non-resonant $I = 2\pi\pi$ scattering [21, 22]. Thus most of the ingredients are in place for calculations of the spectrum that can not only confront but also *predict* the outcomes of future experiment.

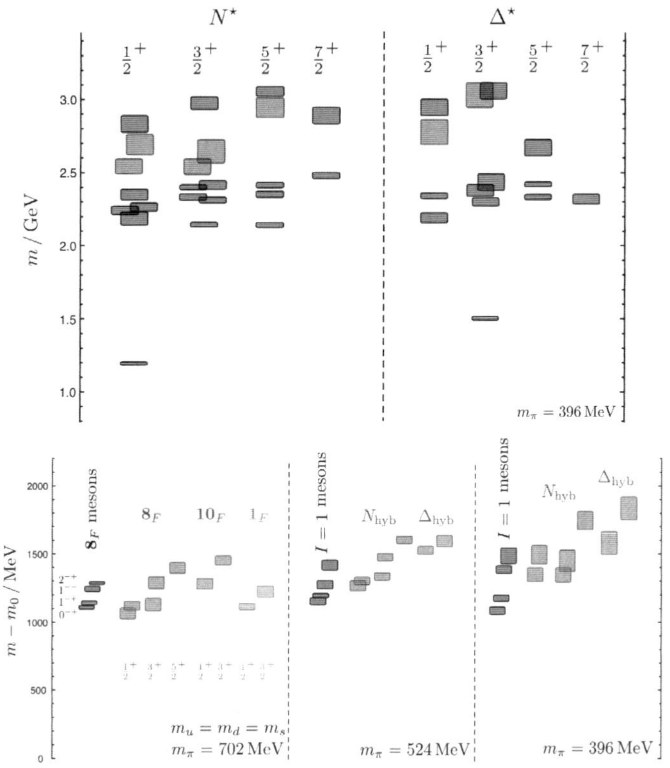

Figure 5. The upper panel shows N and Δ spectrum at $m_\pi = 396$ MeV with the "hybrid" operators included in the basis; blue boxes denote states dominated by such operators and thereby identified as "hybrids". The lower panel shows the excitation spectrum of hybrid mesons and baryons at three values of the light-quark mass; for mesons we show $m - m_\rho$, while for baryons we show $m - m_N$.

Acknowledgements
I am grateful to my colleagues within the *Hadron Spectrum Collaboration*. Chroma [23] and GPU-code from Clark et al. [24, 25] were used to perform this work on clusters at Jefferson Laboratory using time awarded under the USQCD Initiative. Authored by Jefferson Science Associates, LLC under U.S. DOE Contract No. DE-AC05-06OR23177. Gauge configurations were generated using resources awarded from the U.S. Department of Energy INCITE program at the Oak Ridge Leadership Computing Facility at Oak Ridge National Laboratory, the NSF Teragrid at the Texas Advanced Computer Center and the Pittsburgh Supercomputer Center, as well as at Jefferson Lab under the USQCD and LQCD ARRA project.

References
[1] Michael C 1985 *Nucl. Phys.* **B259** 58
[2] Luscher M and Wolff U 1990 *Nucl. Phys.* **B339** 222–252
[3] Dudek J J, Edwards R G, Peardon M J, Richards D G and Thomas C E 2010 *Phys. Rev.* **D82** 034508 (*Preprint* 1004.4930)
[4] Peardon M *et al.* (Hadron Spectrum) 2009 *Phys. Rev.* **D80** 054506 (*Preprint* 0905.2160)

[5] Edwards R G, Joo B and Lin H W 2008 *Phys. Rev.* **D78** 054501 (*Preprint* arXiv:0803.3960)
[6] Lin H W *et al.* (Hadron Spectrum) 2009 *Phys. Rev.* **D79** 034502 (*Preprint* arXiv:0810.3588)
[7] Dudek J J, Edwards R G, Peardon M J, Richards D G and Thomas C E 2009 *Phys. Rev. Lett.* **103** 262001 (*Preprint* 0909.0200)
[8] Bernard C *et al.* 2003 *Phys. Rev.* **D68** 074505 (*Preprint* hep-lat/0301024)
[9] Bernard C W *et al.* (MILC) 1997 *Phys. Rev.* **D56** 7039–7051 (*Preprint* hep-lat/9707008)
[10] Lacock P and Schilling K (TXL) 1999 *Nucl. Phys. Proc. Suppl.* **73** 261–263 (*Preprint* hep-lat/9809022)
[11] Lacock P, Michael C, Boyle P and Rowland P (UKQCD) 1997 *Phys. Lett.* **B401** 308–312 (*Preprint* hep-lat/9611011)
[12] McNeile C and Michael C (UKQCD) 2006 *Phys. Rev.* **D73** 074506 (*Preprint* hep-lat/0603007)
[13] Hedditch J N *et al.* 2005 *Phys. Rev.* **D72** 114507 (*Preprint* hep-lat/0509106)
[14] Dudek J J, Edwards R G, Joo B, Peardon M J, Richards D G *et al.* 2011 *Phys.Rev.* **D83** 111502 (*Preprint* 1102.4299)
[15] Morningstar C J and Peardon M J 1999 *Phys. Rev.* **D60** 034509 (*Preprint* hep-lat/9901004)
[16] Liu L *et al.* (for the Hadron Spectrum Collaboration) 2012 *JHEP* **1207** 126 (*Preprint* 1204.5425)
[17] Dudek J J 2011 *Phys. Rev.* **D84** 074023 (*Preprint* 1106.5515)
[18] Edwards R G, Dudek J J, Richards D G and Wallace S J 2011 *Phys. Rev.* **D84** 074508 (*Preprint* 1104.5152)
[19] Dudek J J and Edwards R G 2012 *Phys.Rev.* **D85** 054016 (*Preprint* 1201.2349)
[20] Luscher M 1991 *Nucl. Phys.* **B354** 531–578
[21] Dudek J J, Edwards R G, Peardon M J, Richards D G and Thomas C E 2011 *Phys. Rev.* **D83** 071504 (*Preprint* 1011.6352)
[22] Dudek J J, Edwards R G and Thomas C E 2012 *Phys. Rev.* **D86** 034031 (*Preprint* 1203.6041)
[23] Edwards R G and Joo B (SciDAC Collaboration) 2005 *Nucl. Phys. B. Proc. Suppl.* **140** 832 (*Preprint* hep-lat/0409003)
[24] Clark M, Babich R, Barros K, Brower R and Rebbi C 2010 *Comput. Phys. Commun.* **181** 1517–1528 (*Preprint* 0911.3191)
[25] Babich R, Clark M A and Joo B 2010 (*Preprint* 1011.0024)

The empirical p-n interactions and atomic masses

Akito Arima[1] and Yu-Min Zhao[2]

[1]Musashi Academy, 1-26-1 Toyotama-kami, Nerima-ku Tokyo 176-8533, Japan
[2]Department of Physics Shanghai Jiao Tong University, Shanghai 200240, P. R. China

Abstract. The systematics of the empirical proton-neutron (p-n) interactions are studied for better understanding of atomic masses. The odd-even staggering of the interaction between the last proton and the last neutron is partly explained by the pairing interaction, further studies on this phenomenon are needed. Two sets of simple and accurate mass formulas are suggested by using the systematics of the empirical p-n interactions.

1. Introduction

I (A. Arima) would like to thank the organizers for inviting me to this Conference in honor of Jerry P. Draayer's 70th birthday. First of all, I wish to congratulate Jerry P. Draayer on his 70th birthday.

K. T. Hecht and A. Adler [1] in Michigan, and K.Shimizu, A. Arima in Tokyo and M. Harvey in Chalk River [2] discovered an interesting degeneracy in nuclear single particle energies in 1969 very independently from each other. This dynamical symmetry is called the pseudo spin symmetry. Furthermore in some areas we found that the pseudo SU(3) symmetry works well. I was very impressed by Jerry Draayer and Ted Hecht, who cooperated to study and develop these two dynamical symmetries [3-5]. Since then, I always have been interested in Jerry's activities.

I would like to talk on the empirical p-n interactions and atomic masses.

There have been great developments theoretically in global mass calculations, for example, the Duflo-Zuker model [6] in 1995, the modified Weizsäcker formula [7] in 2011, etc. [8,9]

These global mass calculations have been applied to 8000～9000 nuclei, and the RMS deviations from experimentally known masses are only 340 keV～580 keV.

Local mass relations, also, have been studied and recognized as very useful to predict nuclear masses. The Garvey-Kelson mass relations [10] are particularly powerful for this purpose. In the study of the local mass relations, it is very essential to study proton-neutron interactions empirically. See the comprehensive reviews Refs. [11,12]. The RMS error (σ) and mean error (ε) of fits given by various mass formulas are shown in table 1.

The empirical p-n interactions $\delta V_{\rm pn}(Z,N)$ are obtained by using the observed masses as follows:

$$\delta V_{\rm pn}(Z,N)=\frac{1}{4}\left[B(Z+1,N+1)-B(Z+1,N-1)-B(Z-1,N+1)+B(Z-1,N-1)\right]$$

Figure 1 shows the empirical p-n interactions taken from Brenner et al [13].

Journal of Physics: Conference Series **403** (2012) 012045 doi:10.1088/1742-6596/403/1/012045

Table 1. The rms errors (σ) and mean errors (ε) of fits given by various mass formulas. σmod and εmod denote errors determined by Moller (Moller and Nix 1988, and 1995) . Taken from Ref. [12].

	1995 data (1768 nuclei)		2001 data (2135 nuclei)		"New" nuclei (382 nuclei)				
	σ	$\bar{\epsilon}$	σ	$\bar{\epsilon}$	σ	$\bar{\epsilon}$	σ_{mod}	$\bar{\epsilon}_{mod}$	R
HFBCS-1 (III.B.4)	0.718	0.102	0.805	0.180	1.115	0.494	1.056	0.460	1.47
HFB-1 (III.B.4)	0.740	0.040	0.822	0.131	1.123	0.510	1.091	0.494	1.47
HFB-2 (III.B.4)			0.674	0.000	0.769	0.377	0.724	0.356	
HFB-2′ (III.B.4)	0.651	−0.039	0.702	0.058	0.857	0.470	0.789	0.437	1.21
FRDM (III.C.3)	0.678	0.023	0.676	0.072	0.655	0.247	0.485	0.202	0.71
TF-FRDM (III.C.5)	0.662	−0.034	0.655	−0.036	0.655	−0.085	0.511	−0.121	0.77
Duflo-Zuker (1995, 1999) (III.D.1)	0.346	−0.010	0.373	0.009	0.479	0.054	0.378	0.028	1.09
Koura *et al.* (2000) (III.D.2)	0.656	0.012	0.682	0.053	0.755	0.200	0.676	0.163	1.03
Nayak-Satpathy (1999) (III.D.3)	0.359	0.000	0.485	0.047	0.837	0.229	0.779	0.208	2.17
Audi-Wapstra (1995) (III.E.1)					0.317	0.053	0.122	−0.002	
Garvey and co-workers (1966, 1969) (III.E.2)	0.277	−0.010			0.717	0.127	0.653	0.096	2.36
Jänecke-Masson (1988) (III.E.2)	0.247	−0.010	0.319	0.010	0.540	0.070	0.451	0.071	1.83
Liran-Zeldes (1976) (III.E.4)	0.534	−0.005	0.586	−0.036	0.722	−0.226	0.554	−0.253	1.04

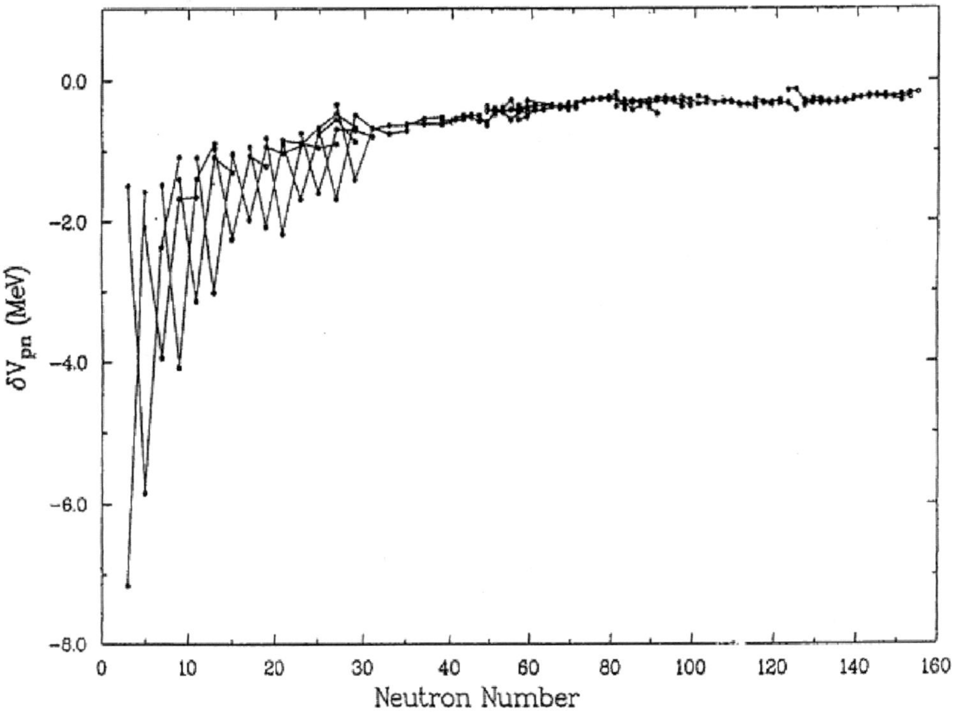

Figure 1. Empirical p-n interactions taken from Ref. [13].

2. Mass predictions by using the empirical p-n interactions

We first generalize the empirical p-n interactions to include the last i protons and j neutrons:

$$\delta V_{ip\text{-}jn}(Z,N) = B(Z,N) - B(Z,N-j) - B(Z-i,N) + B(Z-i,N-j)$$

For $i=1$ and $j=1$, we obtained very similar results to those reported by Basu et al [14] in 1971, as is shown in figure 2.

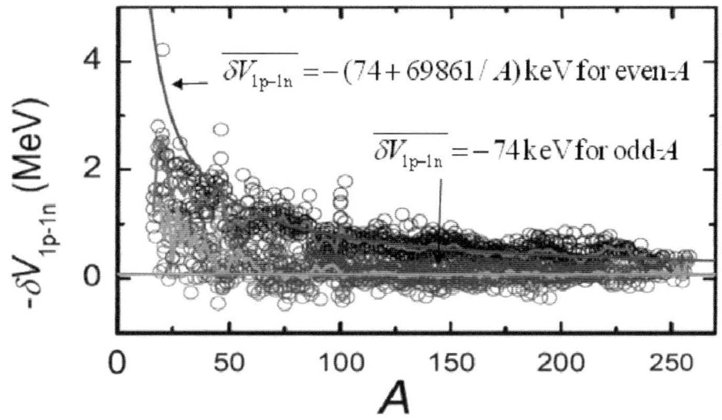

Figure 2. Dependence of $-\delta V_{1p\text{-}1n}$ on nuclear mass number A. Black circles are for even-*A*, and blue circles for odd-*A*.

In order to estimate orders of magnitude of several typical energies, for example the surface energy and the pairing energy, we use the modified Weizsäcker mass formula as follows.

$$E_{\mathrm{LD}}(A,Z) = a_v A + a_s A^{2/3} + E_C + a_{sym} I^2 A + a_{pair} A^{-1/3}\delta_{np} + \Delta_W$$

$$\delta_{np} = \begin{cases} 2-I & \text{for even } Z \text{ and even } N \\ I & \text{for odd } Z \text{ and odd } N \\ 1-I & \text{for odd } Z \text{ and even } N \text{ with } Z<N \\ 1 & \text{for odd } Z \text{ and even } N \text{ with } Z>N \\ 1-I & \text{for even } Z \text{ and odd } N \text{ with } Z>N \\ 1 & \text{for even } Z \text{ and odd } N \text{ with } Z<N \end{cases}$$

$$E_C = a_C \frac{Z^2}{A^{1/3}}\left(1-0.76Z^{-2/3}\right)$$

where $a_{pair} = 5442.3$ keV and $I = \dfrac{|N-Z|}{A}$.

The results for A(nuclear mass number) > 100 are as follows:

- Surface energy ~ 5 keV.
- Coulomb energy ~ 26 keV.
- Symmetry energy ~ 269 keV.
- Pairing energy ~ $\pm$166 keV.

We can confirm that the Garvey-Kelson relations work very well. This means that

$$\delta V_{1p\text{-}1n}(Z,N) \approx \delta V_{1p\text{-}1n}(Z-1,N+1)$$
$$\delta V_{1p\text{-}1n}(Z,N) \approx \delta V_{1p\text{-}1n}(Z+1,N+1)$$

We found the following formulae for averaged $\overline{\delta V_{1p-1n}}(A)$ (see figure 2):

$$\overline{\delta V_{1\mathrm{p}-1\mathrm{n}}}(A)=\begin{cases}-(74+\dfrac{69861}{A})\ \mathrm{keV} & \text{for even } A\\ -74\,\mathrm{keV} & \text{for odd } A\end{cases}$$

These formulae indicate that there is an energy difference between even-*A* and odd-*A* nuclei which amounts about 440 keV for $A>100$.

It is reasonable to try to explain this difference by using the pairing interaction. However, as mentioned above, the pairing energy estimated by the modified Weizsäcker mass formula is about $\pm$ 166 keV. This means that we can expect about 330 keV as the even-odd energy difference. This difference is smaller than 440 keV found in our present study. We do not know how to explain this remaining difference between even-*A* and odd-*A* nuclei.

Casten and his collaborators [15] pointed out that the contribution from the shell effect to δV_{pn} is important. We then assume the following formulae for $\delta V^{\mathrm{cal}}_{1\mathrm{p}\text{-}1\mathrm{n}}$, $\delta V^{\mathrm{cal}}_{1\mathrm{p}\text{-}2\mathrm{n}}$ and $\delta V^{\mathrm{cal}}_{2\mathrm{p}\text{-}1\mathrm{n}}$:

$$-\delta V^{\mathrm{cal}}_{1\mathrm{p}-1\mathrm{n}}(Z,N)=-\overline{\delta V_{1\mathrm{p}-1\mathrm{n}}}(A)+\Delta_{\mathrm{sh}}(Z,N)+\Delta_{\mathrm{C}}(Z,N)+\Delta_{\mathrm{sym}}(Z,N)$$

$$\begin{aligned}-\delta V^{\mathrm{cal}}_{1\mathrm{p}-2\mathrm{n}}(Z,N)=&-\overline{\delta V_{1\mathrm{p}-2\mathrm{n}}}(A)+\Delta_{\mathrm{sh}}(Z,N)+\Delta_{\mathrm{C}}(Z,N)+\Delta_{\mathrm{sym}}(Z,N)\\&+\Delta_{\mathrm{sh}}(Z,N-1)+\Delta_{\mathrm{C}}(Z,N-1)+\Delta_{\mathrm{sym}}(Z,N-1)\end{aligned}$$

$$\begin{aligned}-\delta V^{\mathrm{cal}}_{2\mathrm{p}-1\mathrm{n}}(Z,N)=&-\overline{\delta V_{2\mathrm{p}-1\mathrm{n}}}(A)+\Delta_{\mathrm{sh}}(Z,N)+\Delta_{\mathrm{C}}(Z,N)+\Delta_{\mathrm{sym}}(Z,N)\\&+\Delta_{\mathrm{sh}}(Z-1,N)+\Delta_{\mathrm{C}}(Z-1,N)+\Delta_{\mathrm{sym}}(Z-1,N)\end{aligned}$$

In these formulae,

$$\Delta_{\mathrm{sh}}(Z,N)=a_{\mathrm{sh}}+2b_{\mathrm{sh}}\left|\delta_{\mathrm{p}}\Omega_N(N_{\mathrm{p}}-\Omega_Z)-\delta_{\mathrm{n}}\Omega_Z(N_n-\Omega_N)\right|$$

where δ_{p} (δ_{n}) equals +1 if the valence protons (neutrons) are particle-like, and -1 if they are hole-like, $\Omega_Z=\sum_{j_Z}(j_Z+\frac{1}{2})$ with j_Z representing angular momenta of single-particle levels for valence protons, and $\Omega_N=\sum_{j_N}(j_N+\frac{1}{2})$ with j_N representing angular momenta of single-particle levels for valence neutrons. Furthermore, we used

$$\Delta_{\mathrm{sym}}(Z,N)=a_{\mathrm{sym}}\frac{1}{A\left(2+|IA|\right)^3}+b_{\mathrm{sym}}A^{-1}$$

$$\Delta_{\mathrm{C}}(Z,N)\approx a_{\mathrm{C}}\left(-\frac{4}{9}Z^{4/3}A^{-7/3}-\frac{2}{3}ZA^{-4/3}+\frac{4}{9}Z^2A^{-7/3}+\frac{4}{9}Z^{1/3}A^{-4/3}\right)$$

We fixed the parameters in those expressions by using the χ^2-fitting procedure. Results thus obtained are shown in table 2.

Table 2. Parameters obtained by the χ^2-fitting procedure. Taken from Ref. [20].

Parameter	For Even-A	For Odd-A
a_{sh}	58.95	15.40
b_{sh}	−0.1444	−0.03157
a_C	−34.80	12.00
a_{sym}	12007	22211
b_{sym}	−179.7	−70.42

Looking at the empirical values of those parameters, we are puzzled by large differences between the cases of even-*A* and odd-*A*. This may reflect the fact that the pairing interaction is not enough to explain the difference between $\overline{\delta V_{1p-1n}}$ for even-*A* and $\overline{\delta V_{1p-1n}}$ for odd-*A*.

Our empirical formula works well to predict nuclear binding energies and other nuclear properties such as separation energies and α-decay energies.

We wish to show our results here. But frankly speaking, we are not yet satisfied by our results. We wish to understand for example why the shell effects are different for the cases of even-*A* and odd-*A*.

Our formulas to predict $B(Z,N)$ by using $\delta V^{\text{cal}}_{\text{1p-1n}}$ and those obtained by using $\delta V^{\text{cal}}_{\text{1p-2n}}$ and $\delta V^{\text{cal}}_{\text{2p-1n}}$ are presented as follows.

$$B^{\text{pred}}(Z,N) = B(Z,N-1) + B(Z-1,N) - B(Z-1,N-1) + \delta V^{\text{cal}}_{\text{1p-1n}}(Z,N)$$
$$B^{\text{pred}}(Z,N) = B(Z,N-1) + B(Z+1,N) - B(Z+1,N-1) - \delta V^{\text{cal}}_{\text{1p-1n}}(Z+1,N)$$
$$B^{\text{pred}}(Z,N) = B(Z,N+1) + B(Z+1,N) - B(Z+1,N+1) + \delta V^{\text{cal}}_{\text{1p-1n}}(Z+1,N+1)$$
$$B^{\text{pred}}(Z,N) = B(Z,N+1) + B(Z-1,N) - B(Z-1,N+1) - \delta V^{\text{cal}}_{\text{1p-1n}}(Z,N+1)$$

$$B^{\text{pred}}(Z,N) = B(Z,N-2) + B(Z-1,N) - B(Z-1,N-2) + \delta V^{\text{cal}}_{\text{1p-2n}}(Z,N)$$
$$B^{\text{pred}}(Z,N) = B(Z,N-2) + B(Z+1,N) - B(Z+1,N-2) - \delta V^{\text{cal}}_{\text{1p-2n}}(Z+1,N)$$
$$B^{\text{pred}}(Z,N) = B(Z,N+2) + B(Z+1,N) - B(Z+1,N+2) + \delta V^{\text{cal}}_{\text{1p-2n}}(Z+1,N+2)$$
$$B^{\text{pred}}(Z,N) = B(Z,N+2) + B(Z-1,N) - B(Z-1,N+2) - \delta V^{\text{cal}}_{\text{1p-2n}}(Z,N+2)$$
$$B^{\text{pred}}(Z,N) = B(Z,N-1) + B(Z-2,N) - B(Z-2,N-1) + \delta V^{\text{cal}}_{\text{2p-1n}}(Z,N)$$
$$B^{\text{pred}}(Z,N) = B(Z,N-1) + B(Z+2,N) - B(Z+2,N-1) - \delta V^{\text{cal}}_{\text{2p-1n}}(Z+2,N)$$
$$B^{\text{pred}}(Z,N) = B(Z,N+1) + B(Z+2,N) - B(Z+2,N+1) + \delta V^{\text{cal}}_{\text{2p-1n}}(Z+2,N+1)$$
$$B^{\text{pred}}(Z,N) = B(Z,N+1) + B(Z-2,N) - B(Z-2,N+1) - \delta V^{\text{cal}}_{\text{2p-1n}}(Z,N+1)$$

Results thus obtained by the present study are shown in tables 3 and 4.

Nuclear masses predicted by the present study are compared with new measurements, together with other predictions in figure 3. Alpha-decay energies are also calculated and are shown in table 5.

Table 3. Pred(1) corresponds to predicted results by using $\delta V^{\rm cal}_{\rm 1p\text{-}1n}(Z,N)$ and experimental database of AME2003, pred(2) are those by using $\delta V^{\rm cal}_{\rm 1p\text{-}2n}$ and $\delta V^{\rm cal}_{\rm 1p\text{-}2n}$. M03 are taken from predicted results in the AME2003 table, and $M_{\rm exp}$ corresponds to new experimental data after 2003. # denotes the number of available paths to predict the masses from the experimental database.

Nucleus	$M^{\rm pred(1)}$	$\sharp^{(1)}$	$M^{\rm pred(2)}$	$\sharp^{(2)}$	M_{03}	σ_{03}	$M_{\rm exp}$	$\sigma_{\rm exp}$
^{83}Ge	-60974.79	2	-60972.83	1	-60901	196	-60804	120
^{84}As	-66299.68	1	-66081.99	1	-66082	300	-65869	119
^{85}As	-63394.92	2	-63388.62	1	-63323	196	-63236	120
^{86}As	-59242.89	3	-59203.68	2	-59150	298	-58860	120
^{89}Se	-59333.50	1	-59542.07	1	-59196	298	-58955	120
^{70}Br	-51077.43	1	-51344.19	1	-51426	306	-51425	15
^{122}Ag	-71045.66	1	-71165.57	1	-71231	205	-71106	38
							-71065	120
^{123}Ag	-69781.58	2	-69781.58	1	-69955	205	-69548	30
							-69377	121
^{124}Ag	-66221.55	3	-66311.87	2	-66471	196	-66200	250
^{138}Te	-65903.88	1	-66048.01	1	-65931	205	-65755	122
^{140}I	-64009.51	1	-64077.94	1	-64273	196	-63596	121
^{141}I	-60430.17	2	-60430.17	1	-60519	196	-60301	128
^{143}Xe	-60376.90	1	-60284.83	1	-60445	196	-60253	124
^{213}Tl	1651.70	3	1639.27	2			1763	61
^{221}Po	20069.08	3	20208.92	3			19783	58
^{222}Po			23085.32	3			22476	67
^{224}At			28046.06	3			27706	59
^{236}Ac			51548.90	2	51508	499	51267	68

Table 4. Selected data among our predicted mass excesses (in units of keV). These unknown masses are either important in the context of astrophysics and nuclear structure or to be measured in the near future. $M^{\rm pred(1)}$ and $M^{\rm pred(2)}$ are as explained in the caption to table 3. AME2003 corresponds to the AME2003 database, and AME2011-preview corresponds to the results in tables of Ref. [18], which is a preview of the recent evaluation by Audi et al. Taken from Ref. [20].

Nucleus	$M^{\rm pred(1)}$	$M^{\rm pred(2)}$	AME2003	AME2011-preview
^{84}Mo	-55687 ± 394	-55681 ± 368	-55806 ± 401	-54502 ± 401
^{86}Tc	-50969 ± 184	-51359 ± 247	-53207 ± 298	-51297 ± 298
^{88}Ru	-54277 ± 276	-54314 ± 184	-55647 ± 401	-54399 ± 298
^{90}Rh	-52378 ± 315	-52163 ± 207	-53216 ± 503	-51959 ± 401
^{92}Pd	-54962 ± 295	-54722 ± 166	-55498 ± 503	-55070 ± 503
^{94}Ag	-52719 ± 465	-52629 ± 285	-53300 ± 503	-52412 ± 641
^{129}Cd	-63500 ± 179	-63360 ± 131	-63202 ± 298	-63314 ± 196
^{135}Sn	-61017 ± 279	-61022 ± 215	-60799 ± 401	-60612 ± 401
^{204}Pt	-17955 ± 221	-17995 ± 111		-18062 ± 401

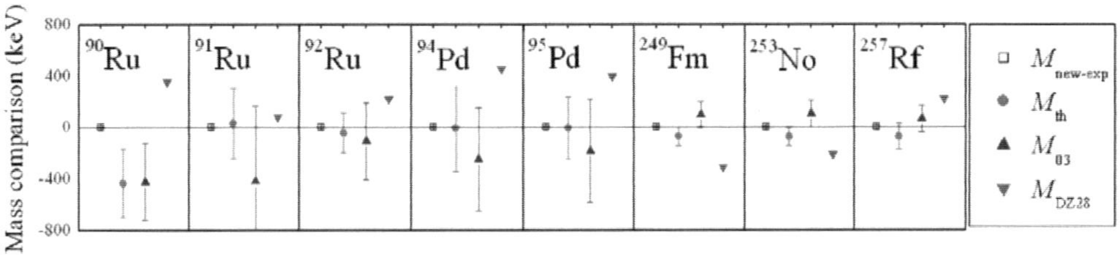

Figure 3. Predicted masses versus new measurements. The solid dots in red are our recent predicted values of masses. M03 refers to the results predicted by AME2003, and DZ corresponds to the results predicted by Duflo and Zuker [6].

Table 5. Predicted alpha-decay energies. Cal-1 are our predicted results, Cal-2 those of Dong et al [19], Cal-3 those of Audi et al [18], and Expt are the new experimental data [18]. Taken from Ref. [21].

Nucl.	Q_α (Cal-1)	Q_α (Cal-2)	Q_α (Cal-3)	Q_α (Expt)
^{248}No	9.022		9.230	
^{249}No	8.916		9.170	
^{250}No		8.801	8.950	
^{251}No	8.696	8.643		8.751
^{252}No	8.567	8.493		8.549
^{253}No	8.406	8.290		8.414
^{254}No	8.238	8.096		8.226
^{255}No	8.405	8.164		8.428
^{256}No	8.590	8.238		8.581
^{257}No	8.440	8.066		8.477
^{251}Lr	9.177		9.534	
^{252}Lr	9.096		9.160	
^{253}Lr	8.892			8.918
^{254}Lr	8.831			8.816

3. Summary

Now I would like to explain why our method works well.

First, the p-n interactions that we exploit include 1p-1n, 1p-2n, and 2p-1n types. These interactions have been much less investigated than the 2p-2n-type proton-neutron interactions. Their systematics have been much less known.

Second, most extrapolation approaches assume that the one- and two-body interactions evolve very slowly and are more or less constant in local regions. In our approach, we go one step forward: we refine the systematics of the p-n interactions by considering shell effects.

Third, if the number of equations to predict the mass of a given nucleus is more than one, our predicted value is given by the average value of all predictions. This reduces the deviations from experimental values. Such an advantage was pointed out by Barea et al [16,17] for the Garvey-Kelson relations. The reason why averaging the predicted results reduces the deviations from experimental values has not yet been well understood.

A brief summary of this report is as follows.

(1) The present method is very easy and practical. The model error is ~150 keV.

(2) The odd-even staggering of $\delta V_{1p\text{-}1n}$ is a good test stone to inspect mass models.

(3) Mass relations between four neighboring nuclei are exploited, and the uncertainties are accumulated more slowly than for the Garvey-Kelson relation.

(4) The odd-even staggering of $\delta V_{1p\text{-}1n}$ puzzles us because the pairing interaction is not enough to explain it.

(5) The empirical values of parameters determined by the χ^2-fitting must be explained from the microscopic point of view. Especially the shell effect must be understood.

(6) The prediction of masses depends on the accuracy of experimental data very much.

We would like to thank Mr. G. J. Fu for his kind help in preparing this paper.

References:

[1] Hecht K T and Adler A 1969 *Nucl. Phys. A* **137** 129
[2] Arima A, Harvey M and Shimizu K 1969 *Phys. Lett. B* **30** 517
[3] Draayer J P, Weeks K L and Hecht K T 1982 *Nucl. Phys. A* **381** 1
[4] Hecht K T, McGrory J B and Draayer J P 1972 *Nucl. Phys. A* **197** 369
[5] Raju R D Ratna, Draayer J P and Hecht K T 1973 *Nucl. Phys. A* **202** 433
[6] Duflo J and Zuker A P 1995 *Phys. Rev. C* **52** R23
[7] Liu M, Wang N and Wu X 2011 *Phys. Rev. C* **84** 014333
[8] Moller P, Myers W D, Swiatecki W J and Treiner J 1988 *At. Data Nucl. Data Tables* **39** 225
[9] Goriely S, Chamel N and Pearson J M 2009 *Phys. Rev. Lett.* **102** 152503
[10] Garvey G T and Kelson I 1966 *Phys. Rev. Lett.* **17** 189
[11] Garvey G T, Gerace W J, Jaffe R L, Talmi I and Kelson I 1969 *Rev. Mod. Phys.* **41** S1
[12] Lunney D, Pearson J M and Thibault C 2003 *Rev. Mod. Phys.* **75** 1021
[13] Brenner D S, Wesselborg C, Casten R F, Warner D D and Zhang J Y 1990 *Phys. Lett. B* **243** 1
[14] Basu M K and Banerjee D 1971 *Phys. Rev. C* **3** 992
[15] Cakirli R B, Brenner D S, Casten R F and Milman E A 2005 *Phys. Rev. Lett.* **94** 092501
[16] Barea J, Frank A, Hirsch J C and Isacker P Van 2005 *Phys. Rev. Lett.* **94** 102501
[17] Barea J, Frank A, Hirsch J G, Isacker P Van, Pittel S and Velazques V 2008 *Phys. Rev. C* **77** 041304
[18] Audi G and Wang M 2011 *private communication*
[19] Dong Tiekuang and Ren Zhongzhou 2005 *Phys. Rev. C* **72** 064331
[20] Fu G J, Lei Y, Jiang H, Zhao Y M, Sun B and Arima A 2011 *Phys. Rev. C* **84** 034311
[21] Jiang H, Fu G J, Sun B, Liu M, Wang N, Wang M, Ma Y G, Lin C J, Zhao Y M, Zhang Y H, Ren Zhongzhou and Arima A 2012 *Phys. Rev. C* **85** 054303

HITES 2012 IOP Publishing
Journal of Physics: Conference Series **403** (2012) 012046 doi:10.1088/1742-6596/403/1/012046

Extension of the periodic system: superheavy, superstrange and antimatter nuclei

Walter Greiner

Frankfurt Institute for Advanced Studies, J. W. Goethe–Universität, D-60438, Frankfurt a.M., Germany

E-mail: greiner@fias.uni-frankfurt.de

Abstract. The extension of the periodic system into various new areas is investigated. Experiments for the synthesis of superheavy elements and the predictions of magic numbers are reviewed. Further on, investigations on hypernuclei and the possible production of antimatter-clusters in heavy-ion collisions are reported. Various versions of the meson field theory serve as effective field theories at the basis of modern nuclear structure and suggest structure in the vacuum which might be important for the production of hyper- and antimatter[1].

1. Introduction

There are fundamental questions in science, like e.g. "How did life emerge?" or "How does our brain work?" and others. However, the most fundamental of those questions is "How did the world originate?". The material world has to exist before life and thinking can develop. Of particular importance are the substances themselves, i. e. the particles the elements are made of (baryons, mesons, quarks, gluons), i. e. elementary matter. The vacuum and its structure is closely related to that. We want to report on these questions, beginning with the discussion of modern issues in nuclear physics.

The elements existing in nature are ordered according to their atomic (chemical) properties in the periodic system, which was developed by Dmitry Mendeleev and Lothar Meyer. The heaviest element of natural origin is uranium. Its nucleus is composed of Z = 92 protons and a certain number of neutrons (N = 128–150). They are called the different uranium isotopes. The transuranium elements reach from neptunium (Z = 93) via californium (Z = 98) and fermium (Z = 100) up to lawrencium (Z = 103). The heavier the elements are, the larger their radii and their number of protons are. Thus, the Coulomb repulsion in their interior increases, and they undergo fission. In other words: the transuranium elements become more unstable as they get bigger. In the late sixties, the dream of the superheavy (SH) elements arose. Theoretical nuclear physicists around S. G. Nilsson (Lund) and from the Frankfurt school [1, 2] predicted that so-called closed proton and neutron shells should counteract the repelling Coulomb forces. Atomic nuclei with these special "magic" proton and neutron numbers and their neighbours could again be rather stable. These magic proton (Z) and neutron (N) numbers were thought to be Z = 114 and N = 184 or 196. Typical predictions of their life-times varied between seconds and many thousand years. Figure 1 summarizes the expectations at the time. One can see the

[1] A similar talk I have given and published in The European Physical Journal D (2012).

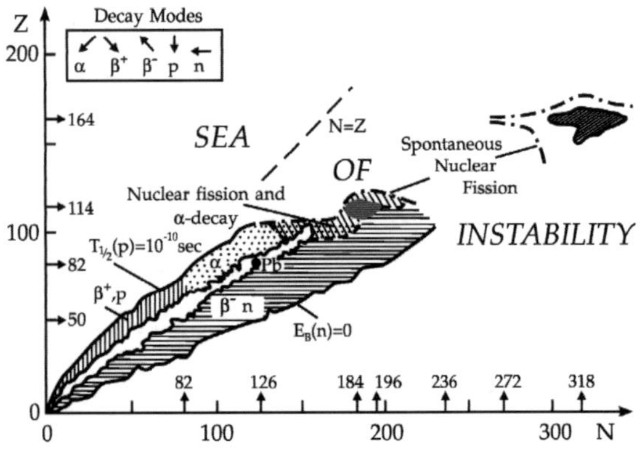

Figure 1. The periodic system of elements as conceived by the Frankfurt school in the late sixties. The islands of superheavy elements (Z = 114, N = 184, 196 and Z = 164, N = 318) are shown as dark hatched areas.

islands of superheavy elements around Z = 114, N = 184 and 196, respectively, and the one around Z = 164, N = 318.

2. Cold Valleys in the Potential

The important question was how to produce these superheavy nuclei. There were many attempts, but only little progress was made. It was not until the middle of the seventies that the Frankfurt school of theoretical physics together with foreign guests (R. K. Gupta (India), A. Sandulescu (Romania)) [3] theoretically understood and substantiated the concept of bombarding of double magic lead nuclei with suitable projectiles, which had been proposed intuitively by the Russian nuclear physicist Y. Oganessian [4]. The two-center shell model, which is essential for the description of fission, fusion and nuclear molecules, was developed in 1969–1972 by W. Greiner and his students U. Mosel [1] and J. Maruhn [5]. It showed that the shell structure of the two final fragments was visible far beyond the barrier, where the nucleus undergoes fusion. The collective potential energy surfaces of heavy nuclei, which were calculated in the framework of the two-center shell model, exhibit pronounced valleys [6].

These valleys provide promising doorways to the fusion of superheavy nuclei for certain projectile-target combinations (Fig. 2). If projectile and target approach each other through those "cold" valleys [3, 7], they get only minimally excited and the barrier, which has to be overcome (fusion barrier) is lowest (as compared to the neighbouring projectile-target combinations). In this way, the correct projectile- and target-combinations for fusion were predicted. Indeed, Gottfried Münzenberg and Sigurd Hofmann and their group at GSI [8] have followed this approach. With the help of the SHIP mass-separator and the position sensitive detectors, which were especially developed by them, they produced the pre-superheavy elements Z = 106, 107, ..., 112, each of them with the theoretically predicted projectile-target combinations, and only with these. Everything else failed. This is an impressive success, which crowned the laborious construction work of many years. The last but one example of this success, was the discovery of element 112 and its long α–decay chain. The Dubna group produced the six isotopes of Z = 113–118 by bombarding ^{244}Pu–^{248}Cf with ^{48}Ca [9]. This is also a cold valley reaction (in this case due to the combination of a spherical and a deformed nucleus), as predicted by Gupta, Sandulescu and Greiner in 1977 [3]. There exist also cold valleys for which both fragments are deformed [7], or have non-axial orientations [10], but these have not been verified experimentally. The cold valleys also play an important role in nuclear fission giving rise to asymmetric and superasymmetric [11, 12] fission and to cluster radioactivity [13].

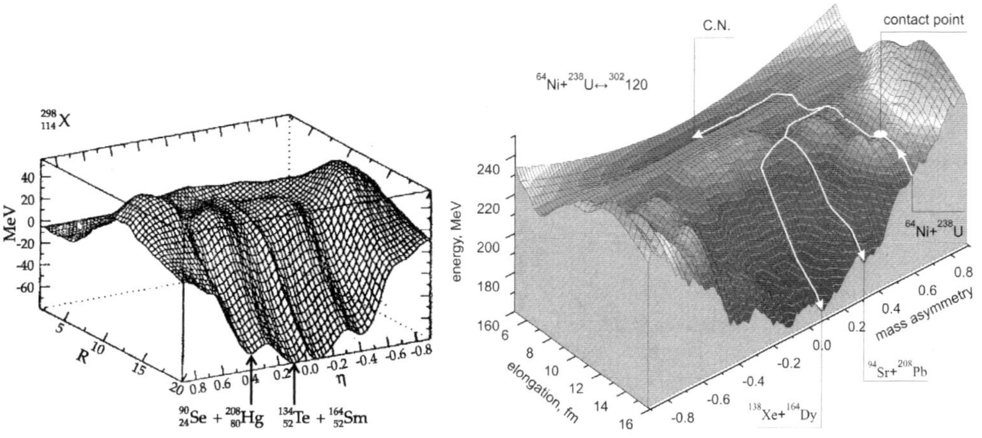

Figure 2. Left: The collective potential energy surface of $^{184}114$ calculated within the two center shell model by J. Maruhn *et al.* shows clearly the cold valleys, which reach up to the barrier and beyond. R denotes the distance between the centers for fragments containing A_1 and A_2 nucleons, while $\eta = (A_1 - A_2)/(A_1 + A_2)$ is the mass asymmetry parameter. Right: Collective potential energy surface of the element $^{302}120$. (C. N. stands for compound nucleus).

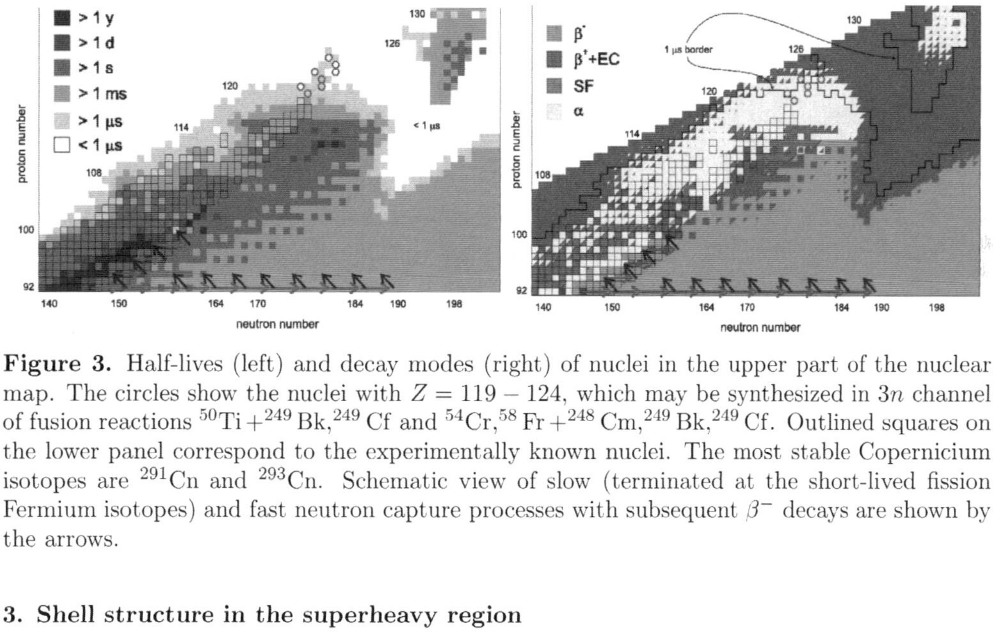

Figure 3. Half-lives (left) and decay modes (right) of nuclei in the upper part of the nuclear map. The circles show the nuclei with $Z = 119 - 124$, which may be synthesized in $3n$ channel of fusion reactions $^{50}\mathrm{Ti} + ^{249}\mathrm{Bk}, ^{249}\mathrm{Cf}$ and $^{54}\mathrm{Cr}, ^{58}\mathrm{Fr} + ^{248}\mathrm{Cm}, ^{249}\mathrm{Bk}, ^{249}\mathrm{Cf}$. Outlined squares on the lower panel correspond to the experimentally known nuclei. The most stable Copernicium isotopes are $^{291}\mathrm{Cn}$ and $^{293}\mathrm{Cn}$. Schematic view of slow (terminated at the short-lived fission Fermium isotopes) and fast neutron capture processes with subsequent β^- decays are shown by the arrows.

3. Shell structure in the superheavy region

Decay properties and stability of heaviest nuclei with $Z \leq 132$ were recently studied within the macro-microscopical approach for nuclear ground state masses and phenomenological relations for the half-lives with respect to α-decay, β-decay and spontaneous fission [14]. It was found (see Fig. 3) that the β-stable isotopes $^{291}\mathrm{Cn}$ and $^{293}\mathrm{Cn}$ with a half-life of about 100 years are the longest-living superheavy nuclei located on the first island of stability. Because of their short half-lives the search in nature of superheavy nuclei may be performed only in cosmic rays. Under

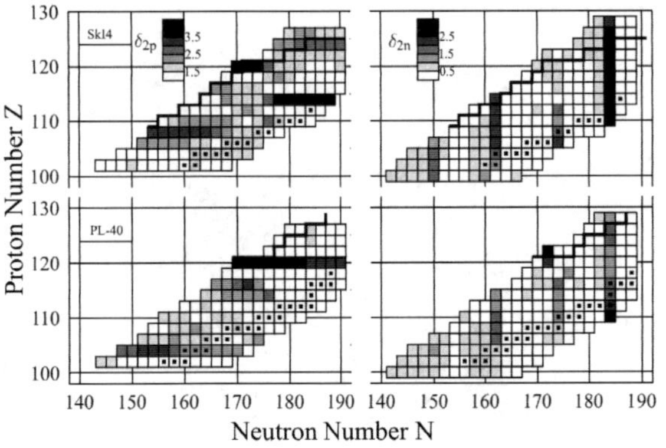

Figure 4. Grey scale plots of proton gaps (left column) and neutron gaps (right column) in the N-Z plane for deformed calculations with the forces SkI4 and PL- 40. Besides the spherical shell closures one can see the deformed shell closures for protons at Z = 104 (PL-40) and Z = 108 (SkI4) and the ones for neutrons at N = 162 for both forces.

terrestrial conditions a measurable amount of superheavies is unlikely to exist. Note, that fusion reactions lead to the proton-rich nuclei. The heaviest synthesized nuclei with $Z = 118$ is situated already quite close to the border of 1 μs half-life. It means that the synthesis and detection of nuclei with $Z > 120$ produced in fusion reactions may be difficult at existing experimental facilities due to their short half-lives (shorter than 1 μs). This prediction should be taken into account while planning new experiments and experimental setups. One may see as well that the nearest neutron-rich isotopes of superheavy elements with $111 \leq Z \leq 115$ to those synthesized recently in Dubna in ^{48}Ca-induced fusion reactions are found to be β^+-decaying. This fact may significantly complicate their experimental identification. However, existence of this area of β^+-decaying nuclei gives us a possibility to reach the center of the island of stability. One of the way to produce ^{291}Cn is the triple β^+ (or EC) decay of ^{291}115 which in turn could be, for example, synthesized after α-decay of ^{295}117 in the reaction $^{48}\text{Ca} +^{249} \text{Bk} \rightarrow^{295} 117 + 2\text{n}$. The proposed method of reaching the island of stability hopefully may be realized in future with the progress in experimental techniques. We found as well the second area of stability of superheavy nuclei (still with shorter half-lives) situated in the region of $Z \sim 124$ and $N \sim 198$. It is separated from the "continent" by the "gulf" of short-living nuclei with half-lives shorted than 1 μs.

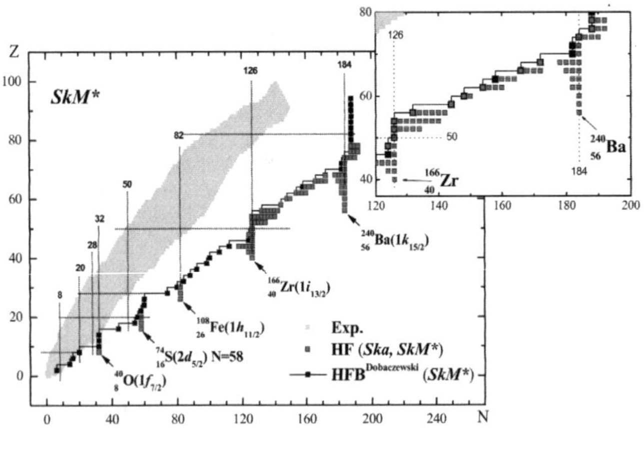

Figure 5. Fragment of the neutron drip line and elements (red squares) that are stable against one neutron emission [18]. One can see the formation of stability peninsulas along neutron magic numbers.

Studies of the shell structure of superheavy elements in the framework of the meson field theory and the Skyrme-Hartree-Fock approach have recently shown that the magic shells in the superheavy region are very isotope dependent [3]. Additionally, there is a strong dependence on the parameter set and the model. Some forces hardly show any shell structure, while other predict the magic numbers Z = 114, 120 and 126. Using the heaviest known even-even nucleus Hassium $^{264}_{156}108$ as a criterium to find the best parameter sets in each model, it turns out that PL-40 and SkI4 produce best its binding energy. However, these two forces make conflicting predictions for the magic number in the superheavy region: SkI4 predicts Z = 114, 120 and PL-40 Z = 120. Most interesting, Z = 120 as magic proton number seems to be as probable as Z = 114. Calculations of deformed systems within the two models [15] reveal again different predictions: Though both parametrizations predict N = 162 as the deformed neutron-shell closure, the deformed proton-shell closures are Z = 108 (SkI4) and Z = 104 (PL-40) (see Fig. 4). Calculations of the potential energy surfaces [15] show single humped barriers; their heights and widths strongly depend on the predicted magic number. Furthermore, recent investigations in a chirally symmetric mean-field theory (see also below) result also in the prediction of these two magic numbers [16, 17]. The corresponding magic neutron numbers are predicted to be N = 172 and to a lesser extend N = 184. Thus, this region provides an open field of research. The charge distribution of the Z = 120, N = 184 nucleus indicates a hollow inside. This may suggest that it might be essentially a fullerene consisting of 60 α–particles and one binding neutron per alpha. The cold valleys in the collective potential energy surface are basic for understanding this exciting area of nuclear physics! It is a master example for understanding the structure of elementary matter, which is so important for other fields, especially astrophysics, but even more so for enriching our "Weltbild", i.e. the status of our understanding of the world around us.

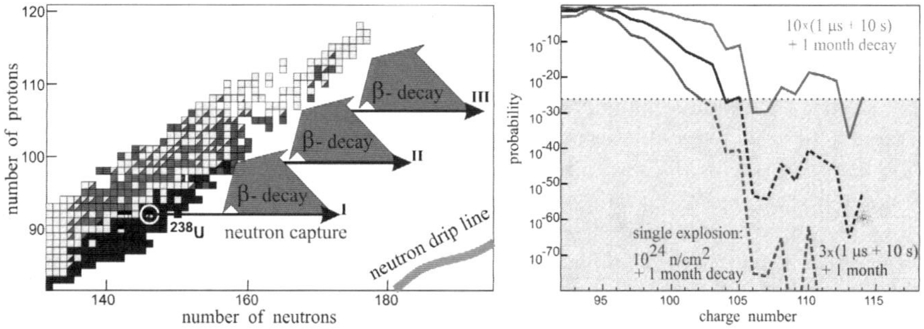

Figure 6. Schematic picture for multiple neutron irradiation of initial ^{238}U material (left) and probability for formation of heavy nuclei (right) in such process (one, three and ten subsequent explosions). Dotted line denotes the level of few atoms.

The investigation of the neutron drip line by extended Hartree-Fock+BCS calculations led to a surprise: extremely neutron rich nuclei along the magic neutron numbers become stable against one-and two-neutron separation [18], see Fig. 5. The standard production of superheavy nuclei by fusing two smaller stable nuclei leads automatically to neutron poor isotopes near the proton drip line (therefore the lifetime of the produced superheavies is so small). Only a few superheavy atoms are produced this way. This leads us directly to the question of how superheavies with larger neutron numbers (and therefore having larger lifetimes: up to thousands of years) can be produced. One also wants to produce such long–living superheavies in macroscopic quantities (milligrams, grams,...) so that they can eventually be used technically. This can be done either by double (or multiple) underground atomic bomb explosions or by pulsed reactors with very high neutron flux ($\simeq 10^{21}$ neutrons/ sec cm^2), see [19, 20] and Figs. 3 and 6.

The idea to take advantage of the shell effects for the production of SH nuclei in the multi-nucleon transfer processes of low-energy heavy ion collisions was proposed in [21]. The shell

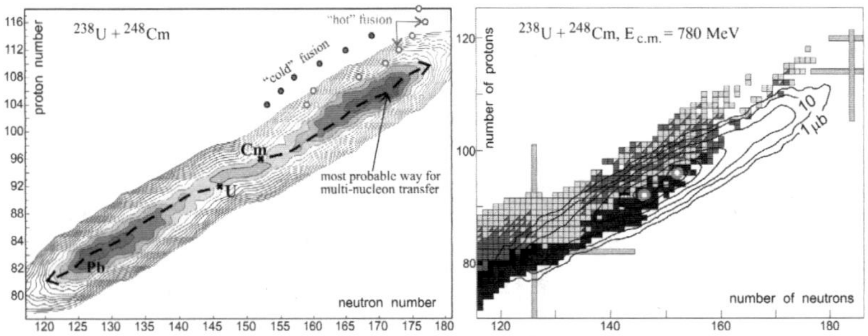

Figure 7. Landscape of potential energy surface (left) and the cross sections (microbarns, logarithmic scale) for production of primary fragments (right) in collision of ^{238}U with ^{248}Cm.

effects are known to play an important role in fusion of heavy ions with actinide targets driving the nuclear system to the quasi-fission channels into the deep lead and tin valleys and, thus, decreasing the fusion probability. On the contrary, in the transfer reactions the same effects may lead to enhanced yield of SH nuclei. It may occur if one of heavy colliding nuclei, say ^{238}U, gives away nucleons approaching to double magic ^{208}Pb nucleus, whereas another one, say ^{248}Cm, accepts these nucleons becoming superheavy in the exit channel, the so called "inverse" (anti-symmetrizing) quasi-fission process. The potential energy surface of the giant nuclear system formed in collision of ^{238}U and ^{248}Cm nuclei is shown in Fig. 7. In low-energy damped collisions of heavy ions just the potential energy surface regulates to a great extent the evolution of the nuclear system driving it to the minimal values of potential energy in the multidimensional space of collective variables. In the course of nucleon exchange the most probable path of the nuclear system formed by ^{238}U and ^{248}Cm lies along the line of stability with formation of SH nuclei which have many more neutrons as compared with those produced in the "cold" and "hot" fusion reactions. Due to fluctuations even more neutron rich isotopes of SH nuclei may be formed in such transfer reactions. The calculated cross sections for formation of primary fragments in low-energy collisions of ^{238}U with ^{248}Cm target are shown in Fig. 7 by the counter lines in logarithmic scale. As can be seen, the superheavy nuclei located very close to the center of the island of stability may be produced in this reaction with rather high cross section of one microbarn. This region of the nuclear map cannot be reached in any fusion reaction with stable projectiles and long-lived targets. Of course, the question arises whether these excited superheavy primary fragments may survive. The calculated cross sections for formation of neutron-rich SH nuclei in low-energy collisions of ^{238}U with ^{248}Cm target are shown in Fig. 8 for final survived fragments. These SH nuclei are located very close to the center of the island of stability and cannot be produced in any fusion reactions with stable projectiles and long-lived targets. These are the shell effects which give us a significant gain as compared to a monotonous exponential decrease of the cross sections with increasing number of transferred nucleons.

We found that the nuclear system consisting of two very heavy nuclei may hold in contact rather long in some cases [22]. During this time the giant nuclear system moves over the multidimensional potential energy surface with almost zero kinetic energy (result of large nuclear viscosity). The reaction time distribution is shown in Fig. 9 for the ^{238}U+^{248}Cm collision. With increase of the energy loss and mass transfer the reaction time becomes longer and its distribution becomes more narrow. The lifetime of a giant composite system more than 10^{-20} s is quite enough to expect positron line structure emerging on top of the dynamical positron spectrum due to spontaneous e^+e^- production from the supercritical electric fields as a fundamental QED process ("decay of the vacuum", Fig. 9) [23]. Formation of the background positrons in these reactions forces one to find some additional trigger for the longest events. For the considered case of ^{238}U+^{248}Cm collision at 800 MeV center-of-mass energy, the detection of the surviving nuclei in the lead region at the laboratory angles of about 25° and at the low-energy border of

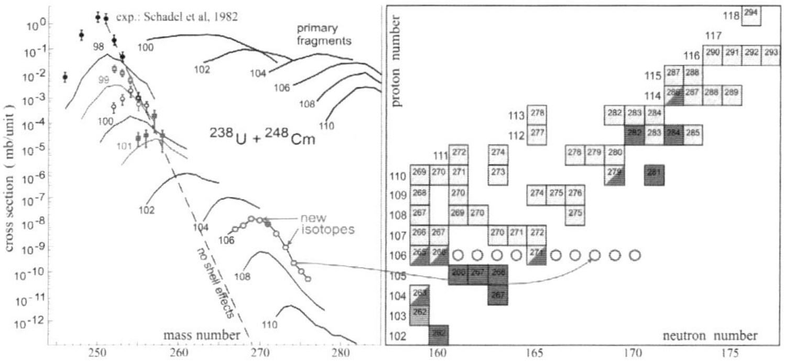

Figure 8. Yield of primary and survived isotopes of SH nuclei produced in collisions of ^{238}U with ^{248}Cm at 800 MeV center-of-mass energy. Dashed line shows the expected locus of transfer reaction cross sections without the shell effects.

their spectrum (around 1000 MeV for Pb) could be a real trigger for the longest reaction time.

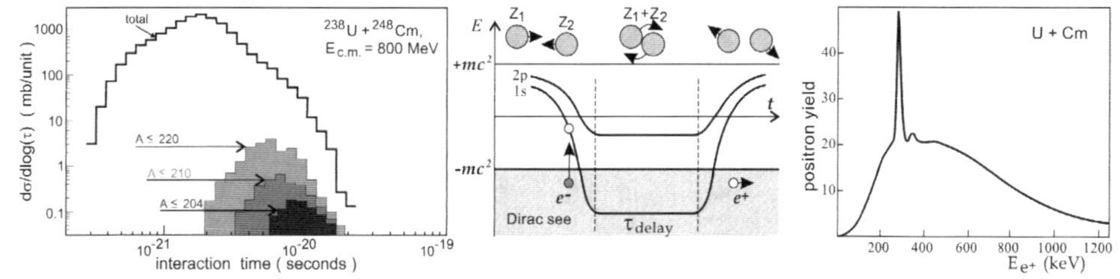

Figure 9. Reaction time distribution (left panel) and spontaneous positron formation (middle panel) in supercritical electric field of long-lived giant quasi-atom formed in collision of U+Cm; the right panel shows the energy distribution of the positron yield.

It was recently found that low-energy collision of actinides may lead to quite an exotic process of three-body clusterization, the so-called true ternary quasifission, leading to formation of two lead-like fragments and some heavy third particle in between [24]. This type of processes is quite possible because the shell effects significantly reduce the potential energy of the three-cluster configurations with two strongly bound lead-like fragments. In Fig. 10 the landscape of the potential energy surface is shown for a three-body clusterization of the nuclear system formed in collision of U+U. It is seen (left panel) that the shell correction at contact configurations makes a very deep minimum for the "lead-calcium-lead" ($A_3 = 50$) clusterization. In the right panel the potential energy is shown as a function of three variables, Z_1, Z_3 (charges of the first and third fragments) and system elongation R (minimized over the neutron numbers) at fixed (equal) deformations of the fragments being in contact. As can be seen, the giant nuclear system, consisting of two touching uranium nuclei, may split into the two-body exit channel with formation of lead-like fragment and complementary superheavy nucleus (the so-called anti-symmetrizing quasifission process which may lead to an enhanced yield of SH nuclei in multi-nucleon transfer reactions [21]). Beside the two-body Pb–No clusterization and the shallow local three-body minimum with formation of light intermediate oxygen-like cluster, the potential energy has the very deep minimum corresponding to the Pb-Ca-Pb–like configuration (or Hg-Cr-Hg) caused by the $N = 126$ and $Z = 82$ nuclear shells. The extreme clustering process of formation of two lead-like doubly magic fragments in collisions of actinide nuclei is a very interesting subject for experimental study. Such measurements, in our opinion, are not too difficult. It is sufficient to detect two coincident lead-like ejectiles (or one lead-like and one

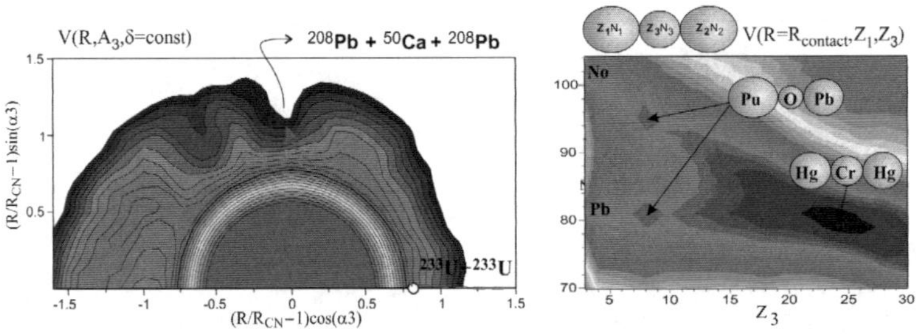

Figure 10. (Left panel) Potential energy (macroscopic plus shell corrections) for ternary quasifission of giant nuclear system formed in $^{233}U+^{233}U$ collision, depending on elongation and mass of third fragment ($\alpha_3 = \pi \cdot A_3/100$, where A_3 is the mass number of the third fragment). (Right panel) Landscape of potential energy of three-body contact configurations of giant nuclear system formed in collision of $^{238}U+^{238}U$.

calcium-like fragments) in U+U collisions to conclude unambiguously about the ternary fission of the giant nuclear system.

4. Extension of the periodic system into the field of hyper- and antimatter

Nuclei that are found in nature consist of nucleons (protons and neutrons) which themselves are made of u (up) and d (down) quarks. However, there also exist s (strange) quarks and even heavier flavours, called charm, bottom, top. The latter has just recently been discovered. Let us stick to the s quarks. They are found in the "strange" relatives of the nucleons, the so-called hyperons ($\Lambda, \Sigma, \Theta, \Omega$). The Λ–particle, e.g., consists of one u, d and s quark, the Θ–particle even of an u and two s quarks, while the Ω (sss) contains strange quarks only.

If such a hyperon is taken up by a nucleus, a hyper-nucleus is created. Hypernuclei with one hyperon have been known for more than 20 years [25]. Carsten Greiner, Jürgen Schaffner and Horst Stöcker [26] theoretically investigated nuclei with many hyperons, hypermatter, and found that the binding energy per baryon of strange matter is in many cases even higher than that of ordinary matter (composed only of u and d quarks). This leads to the idea of extending the periodic system of elements in the direction of strangeness.

One can also ask for the possibility of building atomic nuclei out of antimatter, that means searching e.g. for anti-helium [27], anti-carbon, anti-oxygen. Figure 11 depicts this idea. Due to charge conjugation symmetry, antinuclei should have the same magic numbers and the same spectra as ordinary nuclei. However, as soon as they get in touch with ordinary matter, they annihilate with it and the system explodes. Now the important question arises, how these strange matter and antimatter clusters can be produced. First, one thinks of collisions of heavy nuclei, e.g. lead on lead, at high energies (energy per nucleon $\geq$200 GeV). Calculations with the URQMD-model of the Frankfurt school show that through nuclear shock waves [28] nuclear matter gets compressed to 510 times of its usual value, $\rho_0 \simeq 0.17$ fm^{-3}, and heated up to temperatures of $kT \simeq 200$ MeV. As a consequence, about 10 000 pions, 100 Λ's, 40 Σ's and Θ's and about as many antiprotons and many other particles are created in a single collision. It seems conceivable that it is possible in such a scenario for some Λ's to get captured in a nuclear cluster. This happens indeed rather frequently for one or two Λ–particles; however, more of them get built into nuclei with rapidly decreasing probability only. This is due to the low probability for finding the right conditions for such a capture in the phase space of the particles: the numerous particles travel with all possible momenta (velocities) in all directions.

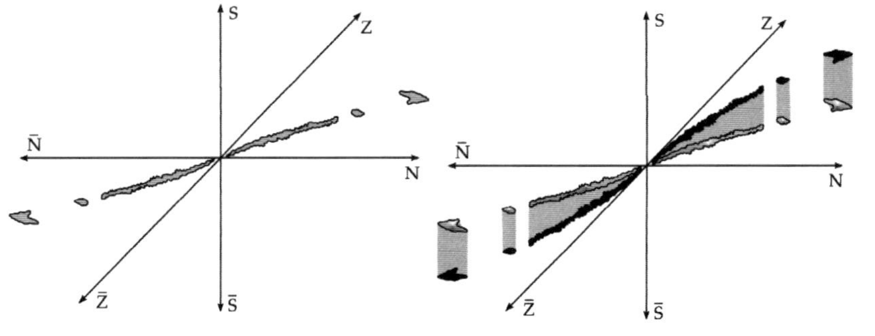

Figure 11. The extension of the periodic system into the sectors of strangeness $S, \overline{S}$ and antimatter $\overline{ZN}$. In the left part of the figure only the stable valley in the usual proton (Z) and neutron (N) plane is plotted, however, extended into the sector of antiprotons and antineutrons. In the right part of the figure it has been indicated, how the stable valley winds out of the Z-N-plane into the strangeness sector. The same can be observed for the antimatter sector.

The chances for hyperons and antibaryons to meet get rapidly worse with increasing number. In order to produce multi–Λ–nuclei and antimatter nuclei, one has to look for a different source.

In the framework of the meson field theory, the energy spectrum of baryons has a peculiar structure, depicted in Fig. 12, left panel. It consists of an upper and a lower continuum, as it is known for electrons (see, e.g. Ref. [29]). Of special interest in the case of the baryon spectrum is the potential well, built of the scalar and the vector potential, which rises from the lower continuum. Naftali Auerbach and collaborators noticed this first [30]. It is known since P. A. M. Dirac (1930) that the negative energy states of the lower continuum have to be occupied by particles (electrons or, in our case, baryons). Otherwise our world would be unstable, because the "ordinary" particles are found in the upper states which can decay through the emission of photons into lower lying states. However, if the "underworld" is occupied, the Pauli-principle will prevent this decay. Holes in the occupied "underworld" (Dirac sea) are antiparticles.

The occupied states of this underworld, including up to 40 000 occupied bound states of the lower potential well, represent the vacuum. The peculiarity of this strongly correlated vacuum structure in the region of atomic nuclei is that — depending on the size of the nucleus — more than 20 000 up to 40 000 (occupied) bound nucleon states contribute to this polarization effect. Obviously, we are dealing here with a highly correlated vacuum. Pronounced shell structure can be recognized [31]. Holes in these states have to be interpreted as bound antinucleons (antiprotons, antineutrons). If the primary nuclear density rises due to compression, the lower well increases while the upper decreases and soon is converted into a repulsive barrier. This compression of nuclear matter can only be carried out in relativistic nucleus-nucleus collision with the help of shock waves, which have been proposed by the Frankfurt school (see W. Scheid et al., Ref. [32]) and which have since then been confirmed extensively (see, e.g. Ref. [33]). These nuclear shock waves are accompanied by heating of the nuclear matter. Indeed, density and temperature are intimately coupled in terms of the hydrodynamic Rankine–Hugoniot equations. Heating as well as the violent dynamics cause the creation of many holes in the very deep (measured from $-M_Bc^2$) vacuum well. These numerous bound holes resemble antimatter clusters which are bound in the medium; their wave functions have large overlap with antimatter clusters. When the primary matter density decreases during the expansion stage of the heavy–ion collision, the potential wells, in particular the lower one, disappear.

The bound antinucleons are then pulled down into the (lower) continuum. In this way

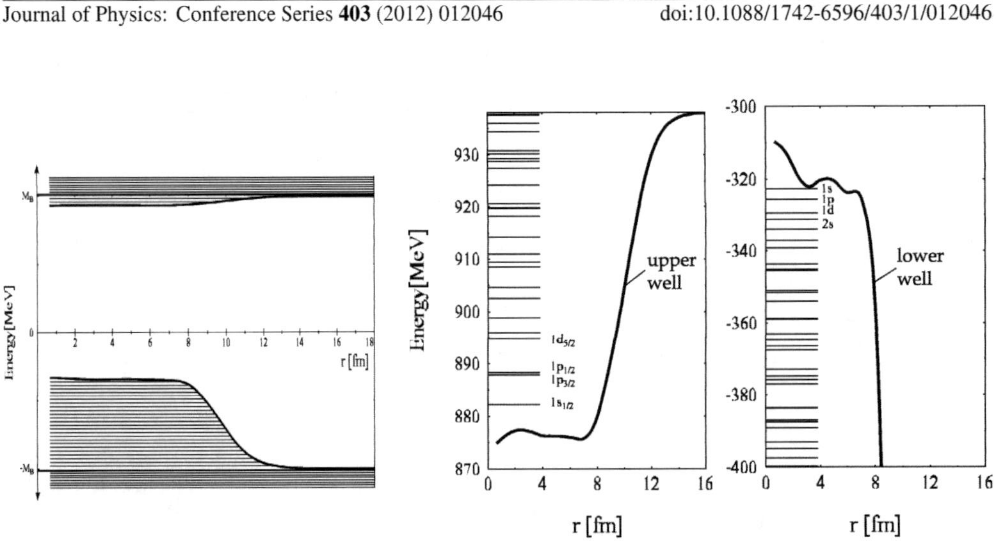

Figure 12. Baryon spectrum in a nucleus (left panel). Below the positive energy continuum exists the potential well of real nucleons. It has a depth of 50-60 MeV and shows the correct shell structure. The shell model of nuclei is realized here. However, from the negative continuum another potential well arises, in which about 40 000 bound particles are found, belonging to the vacuum. A part of the shell structure of the upper well and the lower (vacuum) well is depicted in the two figures on the right.

antimatter clusters may be set free. Of course, a large part of the antimatter will annihilate on ordinary matter present in the course of the expansion. However, it is important that this mechanism for the production of antimatter clusters out of the highly correlated vacuum does not proceed via the phase space. The required coalescence of many particles in phase space suppresses the production of clusters, while it is favoured by the direct production out of the highly correlated vacuum. In a certain sense, the highly correlated vacuum is a kind of cluster vacuum (vacuum with cluster structure). The shell structure of the vacuum levels (see Fig. 12) supports this latter suggestion. Fig. 13 illustrates this idea. Recently the STAR Collaboration at RHIC observed Anti–^{4}He with production rate in excess of coalescent nucleosynthesis production [27].

The mechanism is similar for the production of multi-hyper nuclei $(\Lambda, \Sigma, \Theta, \Omega)$. Meson field theory predicts also for the Λ energy spectrum at finite primary nucleon density the existence of upper and lower wells. The lower well belongs to the vacuum and is fully occupied by Λ's.

Dynamics and temperature then induce transitions ($\Lambda\overline{\Lambda}$ creation) and deposit many Λ's in the upper well. These numerous bound Λ's are sitting close to the primary baryons: in a certain sense a giant multi–Λ hypernucleus has been created. When the system disintegrates (expansion stage) the Λ's distribute over the nucleon clusters (which are most abundant in peripheral collisions). In this way multi–Λ hypernuclei can be formed. Of course this vision has to be worked out and probably refined in many respects. This requires a much more and thorough investigation in the future. It is particularly important to gain more experimental information on the properties of the lower well by (e, p) or (e, p) and also $(\overline{p}_c p_b,\, p_c\overline{p}_b)$ reactions at high energy ($\overline{p}_c$ denotes an incident antiproton from the continuum, p_b is a proton in a bound state; for the reaction products the situation is just the opposite). Also the reaction (p, p' d), (p, p' ^{3}He), (p, p' ^{4}He) and others of similar type need to be investigated in this context. The systematic studies of antiproton scattering on nuclei can contribute to clarify these questions. Various effective theories, e. g. of the Walecka-type on the one side and theories with chiral

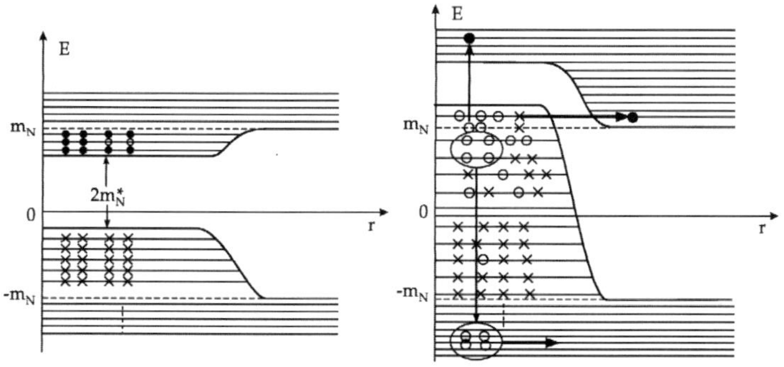

Figure 13. Due to the high temperature and the violent dynamics, many bound holes (antinucleon clusters) are created in the highly correlated vacuum, which can be set free during the expansion stage into the lower continuum (left panel). In this way, antimatter clusters can be produced directly from the vacuum. The horizontal arrow in the right part of the figure denotes the spontaneous creation of baryon-antibaryon pairs, while the antibaryons occupy bound states in the lower potential well. Such a situation, where the lower potential well reaches into the upper continuum, is called supercritical. Four of the bound holes states (bound antinucleons) are encircled to illustrate a "quasi-antihelium" formed. It may be set free (driven into the lower continuum) by the violent nuclear dynamics.

invariance on the other side, have been constructed to describe dense strongly interacting matter [4]. It is important to note that they seem to give different strengths of the potential wells and also different dependence on the baryon density.

According to chirally symmetric meson field theories, the antimatter–cluster production and multi–hypermatter–cluster production out of the highly correlated vacuum takes place at approximately the same heavy-ion energies as compared to the predictions of the Dürr-Teller-Walecka type meson field theories. This in itself is a most interesting, quasi-fundamental question to be clarified. In the future, the question of the nucleonic substructure (form factors, quarks, gluons) and its influence on the highly correlated vacuum structure has to be studied. The nucleons are possibly strongly modified in the correlated vacuum: the Δ resonance correlations are probably important. Is this highly correlated vacuum state, especially during the compression, a preliminary stage to the quark-gluon cluster plasma? To which extent is it similar or perhaps even identical with it?

5. Concluding remarks – outlook

Study of nuclear properties and mechanisms of nuclear reactions in the region of superheavy nuclei remains one of the most challenging problems. In particular, the center of island of stability waits for its discovery, that requires new methods (reactions), which may be used for synthesis of neutron-rich superheavy nuclei. The extension of the periodic system into the sectors of hypermatter (strangeness) and antimatter is of general and astrophysical importance. Indeed, microseconds after the big bang, the new dimensions of the periodic system we have touched upon, certainly have been populated in the course of the baryo- and nucleo-genesis. In the early history of the universe, even higher dimensional extensions (charm, bottom, top) may have played a role, which we did not pursue here. It is an open question, how the depopulation (the decay) of these sectors influences the structure and composition of our world today. Our conception of the world will certainly gain a lot through the clarification of these questions.

References

[1] Mosel U and Greiner W 1968 *Z. Phys.* **217** 256; Mosel U and Greiner W 1968 *Z. Phys.* **222** 261; see also *Memorandum zur Errichtung eines gemeinsamen Ausbildungszentrums für Kernphysik Hessischer Hochschulen*, this was the initiative for building GSI

[2] Nilsson S G, Nix J R, Sobiczewski A, Szymaimageski Z, Wycech S, Gustafson C and Möller P 1968 *Nucl. Phys.* **A 115** 545; Nilsson S G, Thompson S G and Tsang C F 1969 *Phys. Lett.* **B 28** 458

[3] Săndulescu A, Gupta R K, Scheid W and Greiner W 1976 *Phys. Lett.* **B 60** 225; Gupta R K, Săndulescu A and Greiner W 1976 *Phys. Lett.* **B 67** 257

[4] Oganessian Yu Ts, Demin A G, Iljinov A S, Tretyakova S P, Pleve A A, Penionzhkevich Yu E, Ivanov M P and Tretyakov Yu P 1975 *Nucl. Phys.* **A 239** 157

[5] Maruhn J A and Greiner W 1974 *Phys. Rev. Lett.* **32** 548; Maruhn J A 1973, Ph. D. Thesis, J. W. Goethe Universität, Frankfurt an Main

[6] Zagrebaev V and Greiner W 2010 *Giant Nuclear Systems of Molecular Type*, in Clusters in Nuclei - Vol. 1, C. Beck (ed.) *Lecture Notes in Physics* **818** 267

[7] Gupta R K, Münzenberg G and Greiner W 1997 *J. Phys.* **G 23** L13

[8] Hofmann S and Münzenberg G 2000 *Rev. Mod. Phys.* **72** 733

[9] Oganesian Yu Ts, Utyonkov V K and Moody K J 2000 *Sci. Am.* **282**; Oganesian Yu Ts 2007 *J. Phys* **G 34** R165

[10] Mişicu Ş and Greiner W 2002 *Phys. Rev.* **C 66** 044606

[11] Itkis M G, Luk'yanov S M, Okolovich V N, Penionzkevich Yu E, Rusanov A Ya, Salamatin V S, Smirenkin S N and Chubaryan N N 1990 *Yad. Fiz.* **52** 23 [Sov. J. Nucl. Phys. **52** 15]

[12] Gönnenwein F *et al* 2010 *AIP Conference Proceedings* **1224** 338

[13] Poenaru D N and Greiner W 2010 *Cluster Radioactivity* in Clusters in Nuclei 1, C. Beck, ed., vol. **818** of Lecture Notes in Physics, pp. 1–56, Springer

[14] Karpov A V, Zagrebaev V I, Martinez Palenzuela Y, Felipe Ruiz L and Greiner W 2012 *Int. J. Mod. Phys.* **E 21** 1250013

[15] Rutz K, Bender M, Bürvenich T, Schilling T, Reinhard P G, Maruhn J A and Greiner W 1997 *Phys. Rev.* **C 56** 238; Bürvenich T, Rutz K, Bender M, Reinhard P G, Maruhn J A and Greiner W 1998 *Eu. Phys. J.* **A 3** 139; Bender M, Rutz K, Reinhard P G, Maruhn J A and Greiner W 1998 *Phys. Rev.* **C 58** 2126

[16] Serot B D and Walecka J D 1986 *Advances in Nuclear Physics* **16**, Plenum Press, New York; Theis J, Graebner G, Buchwald G, Maruhn J, Greiner W, Stöcker H and Polonyi J 1983 *Phys. Rev.* **D 28** 2286; Klimt S, Lutz M and Weise W 1990 *Phys. Lett.* **B 249** 386; Mishustin I N 1996 *Proc. Int. Conf. Structure of Vacuum and Elementary Matter*, South Africa, Wilderness, World Scientific, Singapore p. 499; Papazoglou P, Zschiesche D, Schramm S, Schaffner–Bielich J, Stöcker H and Greiner W 1999 *Phys. Rev.* **C 59** 411; Mishustin I N, Satarov L M, Stöcker H and Greiner W 2000 *Phys. Rev.* **C 62** 034901

[17] Papazoglou P 1998, Ph. D. Thesis, University of Frankfurt; Beckmann Ch, Papazoglou P, Zschiesche D, Schramm S, Stöcker H and Greiner W 2002 *Phys. Rev.* **C 65** 024301

[18] Gridnev K A, Tarasov V N, Tarasov D V, Gridnev D K, Pilipenko V V and Greiner W 2010 *Int. Jour. Mod. Phys.* **E19** 449; Tarasov V N, Gridnev K A, Gridnev D K, Tarasov D V, Schramm S and Greiner W arXiv:1106.5910v2, submitted to *Int. J. Mod. Phys.*

[19] Botvina A, Mishustin I, Zagrebaev V, and Greiner W 2010 *Int. J. Mod. Phys.* **E19** 2063

[20] Zagrebaev V I, Karpov A V, Mishustin I N and Greiner W 2011 *Phys. Rev.* **C 84** 044617; arXiv:1105.1659

[21] Zagrebaev V I, Oganessian Yu Ts, Itkis M, Greiner W 2006 *Phys. Rev.* **C 73** 031602

[22] Zagrebaev V and Greiner W 2007 *J. Phys. G* **34** 1

[23] Reinhardt J, Müller U, Müller B, Greiner W 1981 *Z. Phys.* **A 303** 173

[24] Zagrebaev V I, Karpov A V and Greiner W 2010 *Phys. Rev.* **C 81** 044608

[25] Povh B 1976 *Rep. Progr. Phys.* **39** 823

[26] Schaffner J, Greiner C and Stöcker H 1992 *Phys. Rev.* **C 46** 322

[27] The STAR Collaboration 2011 *Nature* **473** 353

[28] Stöcker H, Greiner W and Scheid W 1978 Z. Phys. **A 286** 121

[29] Greiner W, Müller B and Rafelski J 1985 *QED of Strong Fields*, Springer Verlag, Heidelberg; Greiner W and Reinhardt J 1998 *Supercritical Fields in Heavy-Ion Physics*, Proc. 15th Advanced ICFA Beam Dynamics Workshop on Quantum Aspects of Beam Physics, World Scientific, Singapore

[30] Auerbach N, Goldhaber A S, Johnson M B, Miller L D and Picklesimer A 1986 *Phys. Lett.* **B 182** 221

[31] Mishustin I N, Satarov L M, Schaffner J, Stöcker H and Greiner W 1993 *J. Phys.* **G 19** 1303; Panda P K, Patra S K, Reinhardt J, Maruhn J, Stöcker H and Greiner W 1997 *Int. J. Mod. Phys.* **E 6** 307

[32] Scheid W, Hoffmann I and Greiner W 1974, in Proc. Symp. Physics with Relativistic Heavy Ions, Lawrence Berkeley Lab, Berkeley

[33] Stöcker H and Greiner W 1986 *Phys. Rep.* **137** 279

Nuclear physics in the coming decade

D J Dean

Oak Ridge National Laboratory, Oak Ridge, TN 37831-6369, USA

E-mail: deandj@ornl.gov

Abstract. The summary talk for this conference is actually a brief status report on the physics of nuclei and what advances we may anticipate during the next decade. We are making progress on multiple fronts to tackle the nuclear quantum many-body problem. Due to the continuing expansion and use of computational science in our field, and due to the new experimental facilities that will soon be available, progress will continue.

1. Introduction

In summarizing this conference in honor of Jerry Draayer's 70th birthday, I tried to perform an extrapolation of the next 10 years (when we celebrate HITES2022) of the physics of nuclei based on where we are today. This kind of discussion is, of course, difficult, as we all know that extrapolation is far more uncertain than interpolation. We heard many excellent presentations at this conference, and one surmises from them that the field is making substantial progress in developing a theoretical understanding of nuclei. I will attempt to briefly indicate some of the progress that is being made and some of the developments that will occur during the next few years.

One goal of our field is to understand nuclei. To that end we develop a comprehensive theory of nuclei; we experimentally determine rare isotope properties and use these as benchmarks to validate our theories and approximations; and we use this validated theory to make predictions where experimental data are not available (e.g., for nucleosynthesis). We have witnessed in the last decade an increasingly sophisticated approach to the nuclear many-body problem. During the next decade, the field will continue to develop effective field theory approaches, computationally scalable many-body techniques, and increasingly accurate energy density functionals for nuclear density functional theory approaches.

During the next decade we should see the completion of second-generation rare isotope experimental facilities such as GSI/FAIR, SPIRAL-II, and HIE-ISOLDE in Europe, FRIB in the US, and an upgraded TRIUMF in Canada, and the continuation of the experimental program at RIBF in Japan and other facilities. Experiments will probe many aspects of neutron-rich nuclei. Furthermore, they should enable a substantial experimental description of nuclear properties along the r-process — the process that creates half of the heavy elements from iron to uranium and is believed to occur during supernova explosions or neutron star mergers.

In developing a comprehensive theory for nuclei, we have increasingly relied on high-performance computing. With calculations performed on these machines, we should be able to constrain theoretical parameters and we should be able to quantify uncertainties of the calculations. Uncertainty quantification, admittedly a buzz word today, is a way of performing numerical analysis to determine the accuracy of a given calculation.

Published under licence by IOP Publishing Ltd

In the following sections, I will expand on each of the areas mentioned in the previous four paragraphs: progress in theory, experiment, computing, and uncertainty quantification.

2. Progress in the theory of nuclei

In her Nobel Lecture, Maria Goeppert-Mayer stated [1] that "[T]he first, basic approach, is to study the elementary particles, their properties, and mutual interaction. Thus one hopes to obtain knowledge of the nuclear forces. If the forces are known, one should, in principle, be able to calculate deductively the properties of individual nuclei." From a different perspective, eloquently expressed by another Nobel Laureate, Philip Anderson [2], the behavior of large and complex aggregates of elementary particles is not easily understood in terms of simple extrapolations. We are making progress in understanding nuclei precisely because of the tension between these two points of view.

The nuclear shell model rests on the idea that for certain nuclei identifiable and relatively large single-particle energy gaps exist between the highest occupied and lowest unoccupied single-particle orbitals in a nucleus. These gaps yield magic numbers in neutrons and protons. Experimental evidence suggests that "shells" as defined in the classical nuclear sense (at particle numbers 2, 8, 20, 50,...) disappear in neutron rich nuclei while others may appear. Part of the reason for this change in structure is the rather simple explanation that the mean field exhibited by very neutron-rich nuclei is not a simple oscillator well. The continuum plays an important role in the very neutron-rich nuclei, and should be incorporated into calculations. In the past few years, progress has been made in this direction across a number of efforts.

The appropriate degrees of freedom that we use in nuclear physics to describe phenomena depend on the energy scale in which we are interested. The theory of quantum chromodynamics (QCD) describes quarks and gluons that make up protons and neutrons (nucleons), and the strong force that we measure between nucleons is the residual of QCD. Significant progress has been made through the applications of effective field theory [3]. Effective field theories describe the nuclear interactions through diagrammatic expansions of a Lagrangian that possesses the symmetries of QCD and also describes the interactions between nucleons and pions. Interestingly, from this approach one naturally obtains three-body forces. One fits parameters of the expansion such that nucleon-nucleon phase-shift scattering data and deuteron properties are obtained [4], and one adjusts the three-body force parameters to obtain the binding of the triton and ^{4}He [5]. Thus far, the two-body forces have been fit at the next-to-next-to-next-to-leading-order (N^3LO), while the three-body forces are fit at the next-to-next-to-leading-order [5].

We have come a long way from the initial nucleon-nucleon interactions of the 1950's. The situation changed dramatically for the better in the mid-1990s when new nucleon-nucleon interactions were accurately fit to scattering data [6, 7]. Today's nucleon-nucleon interactions all fit scattering data with $\chi^2/\text{dof}\approx 1$. Green's Function Monte Carlo (GFMC) calculations [8] utilizing two- and three-body forces paved the way for a modern look at the nuclear many-body problem and have been extremely successful into the p-shell [9]. The three-nucleon force can have a significant effect on nuclear properties. For example, consider calculations of the lifetime of ^{14}C [10] which is anonymously long compared to its neighbors. The no-core shell model calculations utilizing only two-body forces failed to reproduce the lifetime, but one could show that inclusion of the three-body force indeed reproduced the long lifetime. This interesting result shows that more than just a few low-lying states may be affected by the three-nucleon interactions.

Work with coupled-cluster theory also demonstrates three-nucleon force and continuum effects on shell closures in Ca nuclei. Calculations show that the closure at ^{48}Ca cannot be reproduced with the two-body N^3LO interaction [11, 12]. However, inclusion of a three-body force solves this problem. These coupled-cluster calculations utilize an effective three-body potential that

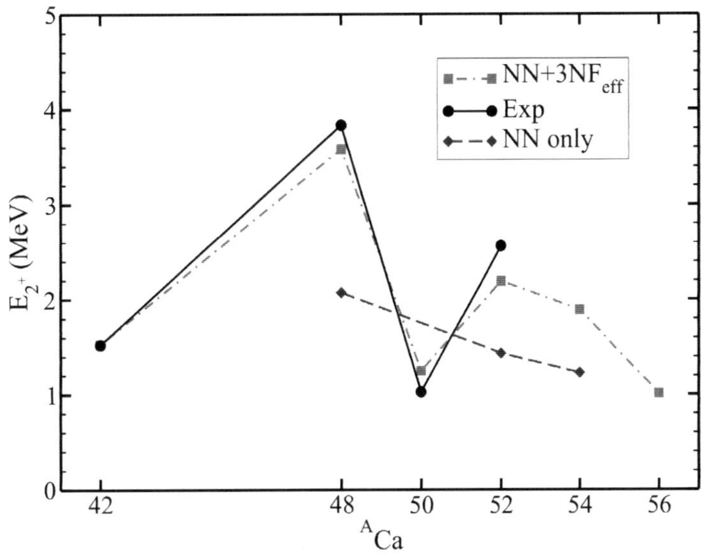

Figure 1. Coupled-cluster results from Ref. [11] for first excited states in Ca isotopes. Results dramatically indicate the effect of a three-body force on the shell closure at ^{48}Ca.

results from integrating one nucleon in the leading-order chiral three-nucleon force over the Fermi sphere with Fermi momentum k_F in symmetric nuclear matter. While not an exact treatment of the three-nucleon potential in coupled-cluster theory (see e.g. [13] for an example of a full coupled-cluster implementation with a three-body interaction), the calculations in Ca isotopes show that the three-body force strongly affects the shell closure at ^{48}Ca, as illustrated in Fig. 1. The coupled-cluster results also point to a neutron drip-line that may reside around ^{60}Ca where continuum effects will be important.

In most cases, the basis used in no-core shell model or coupled-cluster calculations is spherical. The basis lends some simplicity but also requires large numbers of basis states to properly describe highly deformed configurations. Draayer and his colleagues [14] are beginning to pursue the marriage of symmetry-adapted single-particle states and nucleon-nucleon interactions in light nuclei. This nascent approach shows promise to add to our computational tool box for solutions of the nuclear many-body problem.

While some heavier nuclei may be treated within the framework of the no-core shell model or coupled-cluster theory, the full range of nuclei, including well-deformed heavy nuclei, can more naturally be investigated utilizing nuclear density functional theory. Here, the degrees of freedom are quasiparticles and the framework is that of Hartree-Fock-Bogoliubov. Significant progress was made through the Universal Nuclear Energy Density Functional (UNEDF) collaboration in defining the next generation of energy density functionals. The UNEDF-I functional predicts a nuclear landscape with 6900±500 nuclei [15].

3. Rare isotope experiments

Space will not allow for me to adequately cover the experimental landscape today. I will however mention two extremes of measurements in the Sn isotopes as illustrative of the recent past and likely things to come, and a third example from super-heavy nuclei.

The first example is from work performed at the Holifield Radioactive Ion Beam Facility

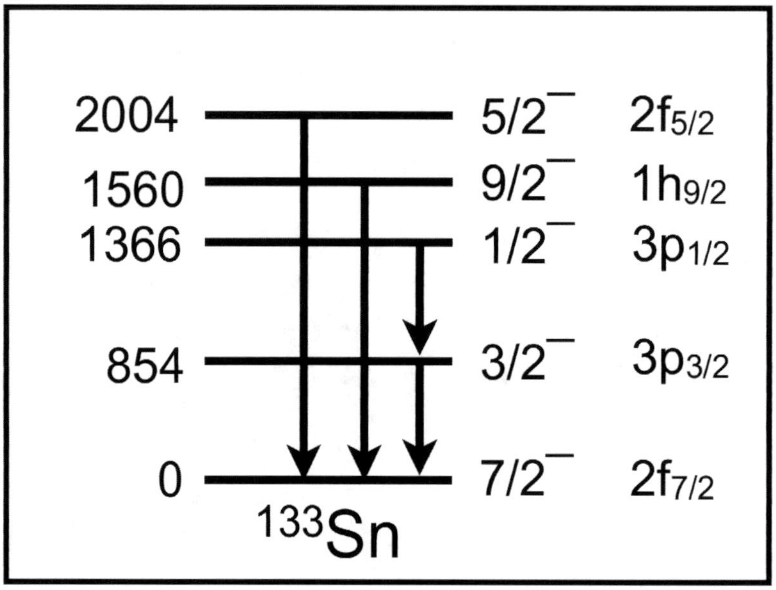

Figure 2. Tentative single-particle states assignments measured in ^{133}Sn. These states were excited through ^{132}Sn(d,p)^{133}Sn reactions at HRIBF. Data are from [16].

(HRIBF) at Oak Ridge [16]. It was long a goal of the facility to produce beams of ^{132}Sn impinging on various targets. Some of the last experiments at the facility included discoveries of the single particle energies of states in ^{133}Sn. These states serve as benchmarks for various theories and models of nuclei. Experiments showed very clearly that ^{132}Sn is indeed doubly-magic. Furthermore, ^{132}Sn is one of the few heavy nuclei experimentally accessible and also close to the r-process path. Fig. 2 represents preliminary level scheme and spin assignments extracted from several ^{132}Sn(d,p)^{133}Sn reactions [16]. Data taken during March, 2012 also indicates evidence for an excited $i_{13/2}$ state in ^{133}Sn [17].

A second example concerns ^{100}Sn. Experiments performed at GSI utilized a ^{124}Xe beam directed at a beryllium target to identify ^{100}Sn and its decay to ^{100}In via a strong Gamow-Teller transition [18]. The measurement of a B(GT) strength of $9.1^{+2.6}_{-3.0}$ to the single 1^+ state in ^{100}In that resides below the Q-window represents years of effort to determine this transition. Future experiments will require more intensity in order to obtain more experimental precision. This experiment was originally motivated by a desire to study more carefully the effects of the nuclear medium on the ratio of the axial-to-vector coupling constants. This rate also feeds into the Sn-Sb-Te cycle which is likely the end-point of the rapid proton capture process.

We are beginning to see confirmation of long-standing predictions (for a review of theoretical efforts see [19]) related to the island of stability in the super-heavy region of elements. Recently, element 117 was discovered at Dubna through the bombardment of a ^{249}Bk target (produced at ORNL's High Flux Isotope Reactor) by a ^{48}Ca beam [20]. Theory predicts a super-heavy island of stability around $N = 184$ and $Z = 118$. Indeed, for the two isotopes known, the half-life of ^{293}Uus is about 14 ms while the half life of ^{294}Uus is about 78 ms (although both numbers have large experimental error bars). Taken as a whole, the super-heavy region from element 105-117 display a trend of increasing stability with larger neutron number. This may be an experimental indication that we have reached the beaches of the island of stability.

4. Trends in Computing

We live in a time of continuing growth in computational capability. Approximately every two years computing power doubles [21]. Twenty years ago, only a few people in our field knew how to extract some physics from the high-performance computers of that day. The fastest of these machines computed at 500 gigaflops (see http://www.top500.org for the first Top500 list in June, 1993). Today, access to 1-10 petaflop (PF) computers is routine.

The hardware changes can be summarized as going from a single vector processor (in the early to late 1980s) to parallel processing in the 1990s and early 2000s. The first wave of change introduced us to parallel computing and message passing protocols such as MPI. The high end computers continued to become larger, with more processors, but also the clock speeds increased regularly, until about the mid-2000s. At that time, multi-core technology began to take over. Today's largest machines utilize multicore processors with graphical processor units (GPUs) attached to each core. Programming languages and algorithms have to continuously evolve in order to keep up with these kinds of hardware changes. This means that the underlying algorithms we utilize to solve our physics problems will also need to evolve. Development of algorithms that can utilize flops efficiently while minimizing data movement will be essential for scalability on these hybrid architecture machines.

The growth in high-end computing power tremendously benefitted the nuclear theory community through a U.S. Department of Energy program called Scientific Discovery through Advanced Computing (SciDAC). The first round of SciDAC included supernovae efforts with some nuclear physics components. The second round of SciDAC funding came in 2006 and one of the big winners in that round was the UNEDF collaboration. UNEDF made significant contributions to the science of nuclei through computational advances [22]. UNEDF was also extremely successful through engaging and training post-docs and students. The third round of SciDAC grants has just been awarded, and the UNEDF follow-on, the Nuclear Computational Low Energy Initiative (NUCLEI), has already begun work. One of the main goals of NUCLEI will involve implementation of uncertainty quantification into various codes and algorithms of the collaboration. The collaboration will also place an emphasis on generating scalable codes for hybrid architectures.

5. Trends in uncertainty quantification

Uncertainty quantification (UQ) is an old and new area of research across the sciences. In fact, UQ has much in common with statistics and numerical analysis. Today, whole journals are devoted to this topic, and our field is just beginning to apply the techniques and ideas, although I would argue that numerical analysis, a major part of UQ, has been with us for a long time.

The basic idea of UQ is to quantify error coming from the following sources: the accuracy of a mathematical model in describing the underlying physics, and the impacts of those underlying model uncertainties on calculations; the error inherent in the numerical algorithm employed to solve a given model; and, the sensitivity of a given numerical solution of the model to the underlying physical parameters. Thus, different sources of error include model approximations, numerical approximations, and the sensivity of the outputs to physical parameter inputs. A fourth part of a calculation's error budget may be statistical if the problem is amenable to some form of Monte Carlo simulation. I will use coupled-cluster theory as an example of the first three of these errors.

Coupled-cluster theory starts from an exponential ansatz for the wave function, $|\Psi\rangle = \exp(T)|\Phi\rangle$ where $T = T_1 + T_2 + \cdots + T_n$ is a n-particle-n-hole (np-nh) excitation operator relative to a Slater determinant state $|\Phi\rangle$. One particular approximation truncates the series at

T_2 (called coupled-clusters in singles and doubles, or CCSD). This turns out to be an extremely powerful approximation, and gives correlation energies converged to the level of 5-7% [23, 24]. Adding T_3 yields converged results at the level of 0.1-1%, but at significant computational cost. Thus the first source of error, model approximations, arises from the level of truncation utilized within the coupled-cluster theory.

Coupled-cluster theory results rest on the solution of a set of nonlinear, coupled algebraic equations. Inherent in such solutions is the necessity of being close to the solution (thus starting from a good mean-field or Hartree-Fock reference Slater determinant is important). A good method for accelerated convergence becomes necessary, and one must choose a cutoff for convergence. This represents an algorithmic, or numerical, source of error In the context of coupled-cluster theory, this type of error is usually small. Another source of numerical error is the choice of single-particle basis states. Convergence requires that one use very large model spaces.

The third source of error involves the physical parameters of the model. For nuclear many-body problems, some of these involve the the physical choice of Hamiltonian and parameters therein. The nucleon-nucleon and three-nucleon potentials are not observables, and thus their choice is not unique. Thus, it becomes important to perform a sensitivity analysis of the parameters within the Hamiltonian. How well constrained are the parameters? This is most evident today in investigations of three-body forces, where fine tuning is often employed to obtain optimal fits. The choice of oscillator strength ($\hbar\omega$) may also affect results unless one proves numerically $\hbar\omega$ independence of solutions.

Each of these sources of errors should be accounted for in calculations. In the future, one will see a lot of work in nuclear physics on UQ implementations embedded within computational approaches that describe nuclei.

6. Predictions for Jerry's 80th birthday

Ten years is not a long time. Some things should happen before 2022. For example, all of the students who attended this meeting should have found gainful employment during the next 10 years. FRIB construction should be either completed or nearing completion. FAIR should be on line. Results from other nuclear physics experiments will have been reported. At the conference, I ended the summary talk by making a few predictions.

- One can hardly fail at predicting that the Higgs will have been discovered. At the time of HITES 2012 a 2σ bump at the right energy had made the news. In September 2012 the ATLAS collaboration at the LHC reported [25] a 5.9 sigma effect at an energy of 125$\pm$0.4 (stat)$\pm$0.4 (sys) GeV, impressively confirming the standard model. A modified prediction would be that the various branching ratios for Higgs decay will be accurately determined. We may find that indeed at these energy scales the standard model is extremely solid.
- We are going to figure out dark matter. Either it is neutralinos (requiring physics beyond the standard model) or something else. Thus far, dark matter searches are coming up empty handed, and significantly constraining minimal supersymmetric extensions to the standard model.
- The 12 GeV upgrade at JLab will have enabled the discovery of the gluonic excitations. These states are considered QCD exotica. They will either be observed with the upgrade, or there is something seriously wrong with our understanding of QCD.
- The r-process will be better understood, but the astrophysical site will continue to be less well understood. Better data closer to the r-process path will enable better modeling of the process.
- There will be an exascale computer somewhere in the world. We are only a factor of 50 away. Someone will find a solution to the energy costs of moving bits around on these larger

machines. Hopefully, this will happen in the U.S., but international competition may yield a different result.

- The largest $0\nu\beta\beta$ decay experiments will have just been built, but finding a signal may not have occurred. A lot will be learned from current generation of experiments and technology developments.
- Jerry and his students and post-docs will continue to make significant progress in applying the symmetry-adapted model spaces to nuclear many-body problems.

Some of these predictions will come true, and we are looking forward to the next birthday party for Jerry to once again measure our progress toward the goal of understanding nuclei.

Acknowledgments

Oak Ridge National Laboratory is managed for the U.S. Department of Energy by UT-Battelle, LLC under contract number DE-AC05-00OR22725.

References

[1] Goeppert-Mayer M 1963 *The shell model* (Nobel Lectures: Physics, IV (Amsterdam: Elsevier)) p 20
[2] Anderson P W 1972 More is different *Science* **177** 393
[3] Weinberg S 1990 Nuclear Forces from chiral Lagrangians *Phys. Lett. B* **251** 288
[4] Machleidt R and Entem D R 2011 Chiral effective field theory and nuclear forces *Phys. Repts.* **533** 1
[5] Gazit D, Quaglioni S and Navratil P 2009 Three nucleon low-energy constants from the consistency of interactions and currents in chiral effective field theory *Phys. Rev. Lett.* **103** 102502
[6] Wiringa R B, Stoks V G J, and Schiavilla R 1995 Accurate nucleon-nucleon potential with charge-independence breaking *Phys. Rev. C* **51** 38
[7] Machleidt R 2001 High-precision, charge-dependent Bonn nucleon-nucleon potential *Phys. Rev. C* **63** 024001
[8] Pudliner B S, Pandharipande V R, Carlson J, and Wiringa R B 1995 Quantum Monte Carlo calculations of $A \leq 6$ nuclei *Phys. Rev. Lett.* **74** 4396
[9] McCutchan E A *et al* 2009 Precise electromagnetic tests of ab initio calculations of light nuclei: states in ^{10}Be *Phys. Rev. Lett.* **103** 192501
[10] Maris P, Vary J P, Navratil P, Ormand W E, Nam H, and Dean D J 2011 Origin of the anomalous long lifetime of ^{14}C *Phys. Rev. Lett.* **106** 202502
[11] Hagen G, Hjorth-Jensen M, Jansen G R, Machleidt R, and Papenbrock T 2012 Evolution of shell structure in neutron-rich calcium isotopes *Phys. Rev. Lett.* **109** 032502
[12] Holt J W, Kaiser N, and Weise W 2010 Density-dependent effective nucleon-nucleon interaction from chiral three-nucleon forces *Phys. Rev. C* **81** 024002
[13] Hagen G, Dean D J, Hjorth-Jensen M, Papenbrock T, and Schwenk A 2007 *Phys. Rev. C* **76** 034302
[14] Draayer J P, Dytrych T, Launey K D, and Langr D 2012 Symmetry-adapted no-core shell model applications for light nuclei with QCD-inspired interactions *Progress in Particle and Nuclear Physics* **67** 516
[15] Erler, J Birge N, Kortelainen M, Nazarewicz N, Olsen E, Perhac A M, and Stoitsov M 2012 The limits of the nuclear landscape *Nature* **486** 509
[16] Jones K L *et al* 2010 The magic nature of ^{132}Sn explored through the single-particle states of ^{133}Sn *Nature* **465** 454
[17] Beene J R *et al* 2013 in preparation
[18] Hinke C B *et al* 2012 Superallowed Gamow-Teller decay of the doubly magic nucleus ^{100}Sn *Nature* **486** 341
[19] Cwiok S, Heenen P H, and Nazarewicz W 2005 Shape coexistence and triaxiality in the superheavy nuclei *Nature* **433** 705
[20] Oganessian Yu Ts *et al* 2010 Synthesis of a new element with atomic number $Z = 117$ *Phys. Rev. Lett.* **104** 142502
[21] Moore G E 1965 Cramming more components onto integrated circuits *Electronics* **38** 4
[22] Bertsch G F, Dean D J, and Nazarewicz W 2007 Computing atomic nuclei *SciDAC Review* **6** 42
[23] Hagen G, Papenbrock T, Dean D J, and Hjorth-Jensen M 2010 Ab initio coupled-cluster approach to nuclear structure with modern nucleon-nucleon interactions *Phys. Rev. C* **82** 034330
[24] Bartlett R J and Musial M 2007 Coupled-cluster theory in quantum chemistry *Rev. Mod. Phys.* **79** 291
[25] ATLAS Collaboration 2012 Observation of a new particle in the search for the Standard Model Higgs boson with the ATLAS detector at the LHC *Phys. Lett. B* **716** 1